Evolution of Crop Plants

Evolution of Crop Plants

Edited by
N. W. Simmonds ScD AICTA FRSE FIBiol

Scottish Plant Breeding Station
Pentlandfield Roslin Midlothian
Scotland

Longman
London and New York

Longman Group Limited
London

Associated companies, branches
and representatives throughout the world

Published in the United States of America
by Longman Inc., New York

First published 1976

Library of Congress Cataloging in Publication Data

Main entry under title:

Evolution of crop plants.

   Includes bibliographies.
   1. Plants, Cultivated – History. I. Simmonds,
   Norman Willison, 1922–
SB71.E88   630.9   75–32563
ISBN 0-582-46678-4

Set in 10 on 11 point Erhardt
and printed in Great Britain
by J. W. Arrowsmith Ltd., Bristol

Editor's introduction

The need for a book such as this became apparent to me some fifteen years ago when I was trying to read up the evolution of crops that might offer illuminating parallels with bananas and potatoes. It quickly became clear that the only compendious work was De Candolle's great book (1886) and that was out of date and pre-cytogenetic. It also became clear that, for a great many crops (and most of those I was interested in), there wasn't even an authoritative essay and, when there was, it often did not answer, or even ask, the interesting questions.

Since that time, the situation has improved, in the sense that there are now several monographs of major crops and a few valuable compendia of essays such as Sir Joseph Hutchinson's *Essays on crop plant evolution* (Cambridge, 1965). But minor crops (and even some major ones) are still very poorly served. Furthermore the literature is all too often very unbalanced: cytogeneticists, archaeologists, historians, geographers, taxonomists, agriculturists, horticulturists and plant breeders all have something to contribute to understanding but they rarely seem to understand (or even read) each other and the study that really collates all the evidence is rare indeed. One result is the almost universal tendency to think of crop plant evolution as something that happened in the past and stopped some time ago. In fact, crop evolution – genetic change in crop populations – is probably at least as rapid now as ever it has been and, in some crops, much more rapid; plant breeders are applied evolutionists (even if they have rarely noticed the fact). One might call this the 'continuum view' of crop evolution.

I concluded therefore that there was need for a work having the following characteristics. It should be comprehensive, covering all the major crops and providing at least an introduction to many minor ones. It should be authoritative, which meant that it had to be multi-authored. It should be as brief as the subject-matter allowed, so that individuals as well as libraries would buy it. And it should allow expression of editorial prejudices in favour both of the continuum view of crop evolution and of the value of diagrams as an aid to concise presentation.

The choice of crops to be treated and the assignment of specific lengths (in the range 2,000–6,000 words) was inevitably a matter of largely arbitrary editorial decision. I tried to balance considerations of agricultural importance and depth of evolutionary understanding and hope that, in the outcome, both authors and readers will feel that a reasonable balance has been achieved.

I decided that, in the interests of coherence and brevity, a standard format would be essential. Experiments with a rather elaborate system of headings and sub-headings convinced me that crop histories were too diverse to be treated thus, so I settled upon a simple but logical sequence of six headings and left authors to adapt it to the needs of their crops. Bibliographies are, inevitably, selective and are intended to provide key- and source-references; I asked authors to refer, as far as reasonably possible, to recent comprehensive reviews for this purpose.

The arrangement of contents presented certain problems. Conventional taxonomic systems offered possibilities but there are several to choose from. Arrangements by economic use had to be discarded because of frequent multiple uses: were Brassicas to be treated as vegetables, fodders or oilseeds? In the outcome, an arbitrary alphabetical arrangement (by families and by genera within families) seemed the most practical and has been adopted; this has the joint merits of ready reference and of placing botanically related crops in proximity. Taxonomists will search in vain for authorities for Latin names and may, I fear, be critical of the omission: it was quite deliberate. Their inclusion would have added nothing to understanding of crop evolution but would have added something to length and much to labour. Taxonomy is one of the foundations of evolutionary understanding but taxonomic process and nomenclatural dispute are irrelevant.

I wondered whether to write a general introductory essay on the subject of crop evolution but decided against doing so. In this field, there is, I think, already perhaps too much generalization from too few examples. This book therefore is, in a sense, an attempt to redress the balance; it concentrates on the particular and, in doing so, reveals, I think, how insecure our knowledge often is, how much more work is needed and how often, even now, the right questions have not yet been asked.

An editor of a book such as this has no light task. Mine has been lightened by many people. My best thanks are due to: authors, for their helpful response to editorial importunities about lengths, deadlines, contents and diagrams; many colleagues, both in Britain and abroad, for advice on possible authors and on points of detail; Messrs Longmans for their efficient, courteous and helpful guidance on technical matters; the Agricultural Research Council, the Department of Agriculture and Fisheries for Scotland and the Scottish Society for Research in Plant Breeding for their approval of the project as one which would take up some public resources; my Secretary, Miss I. M. Hayes, for her outstandingly efficient management of a formidable volume of correspondence; and my wife for her tolerance of a long stretch of week-ends committed to The Book.

We, that is, the editor and authors collectively, hope the work will prove useful. Certainly, it attempts to do something not attempted since De Candolle. Readers must judge of our success and it goes without saying that we shall be pleased to have comments and criticisms for incorporation in any possible revision. We should like to think that the work, lying as it does between the scholarly and the practical, may help in the understanding of past, present and future of our crops. Their future, in a world already hungry and becoming hungrier, is a matter of vital human importance; if we shall have established the essential continuity, linked the scholarly and the practical and shown that past, present and future illuminate each other, we shall be well content.

N. W. Simmonds
Edinburgh
November 1974

Front Board illustrations

Illustrations from the Herbals in the Herbarium Library, Royal Botanical Gardens, Kew

Bananas – Gerard, Herball 1636
Potatoes – Parkinson, Paradisus Terestris 1629
Maize – Mattioli, Herbarz 1562

Back Board illustrations

Bananas – The Jamaica Banana Board
Potatoes – Potato Marketing Board
Maize – Pioneer Hi-Bred International, Inc.

Contents

Editor's introduction
List of Authors

Contents

List of authors

Name
Address   **Chapter(s)**

O. Banga   20, 85
Diedenweg 6, Wageningen, The Netherlands.
Formerly: Institute of Horticultural Plant Breeding,
Wageningen, The Netherlands.

J. Barrau   57
Muséum National d'Histoire Naturelle,
Laboratoire d'Ethnobotanique, 57 Rue Cuvier,
75005 Paris, France.

W. P. Bemis   22
University of Arizona, Tucson, Arizona, USA.

B. O. Bergh   41
Department of Plant Sciences, University of
California, Riverside, Cal. 92502, USA.

D. A. Bond   51
Plant Breeding Institute, Maris Lane, Trumpington,
Cambridge, CB2 2LQ, England.

M. Borrill   38
Welsh Plant Breeding Station, Plas Gogerddan,
Nr. Aberystwyth, SY23 3EB, Wales.

J. W. Cameron   76
Department of Plant Sciences, University of
California, Riverside, Cal. 92502, USA.

C. G. Campbell   69
Canada Department of Agriculture, Research
Station, P.O. Box 3001, Morden, Manitoba, Canada.

G. K. G. Campbell   9
Speen, Aylesbury, HP17 0SL, England.
Formerly: Plant Breeding Institute, Cambridge,
England.

Name
Address   **Chapter(s)**

T. T. Chang   31
International Rice Research Institute, P.O. Box 933,
Manila, Philippines.

B. Choudhury   80
Division of Vegetable Crops and Floriculture,
Indian Agricultural Research Institute, New Delhi
110012, India.

R. B. Contant   12
Department of Applied Plant Sciences, University
of Nairobi, P.O. Box 30197, Nairobi, Kenya.

F. W. Cope   83
Department of Biological Sciences, University of
the West Indies, St Augustine, Trinidad, West
Indies.

D. G. Coursey   23
Tropical Products Institute, 56–62 Gray's Inn Road,
London, WC1X 8LU, England.

D. R. Davies   49
John Innes Institute, Colney Lane, Norwich, NOR
70F, England.

H. Doggett   34
ICRISAT, 1-11-256 Begumpet, Hyderabad 500-016,
A.P., India.

A. Durrant   54
Department of Agricultural Botany, University of
College of Wales, Penglais, Aberystwyth SY23 3DD,
Wales.

C. L. M. van Eijnatten   82
Department of Crop Science, University of Nairobi,
P.O. Box 30197, Nairobi, Kenya.

Alice M. Evans   48, 50
Department of Applied Biology, University of
Cambridge, Cambridge, England.

G. M. Evans   33
Department of Agricultural Botany, University
College of Wales, Penglais, Aberystwyth SY23 3DD,
Wales.

R. Faulkner   87
Forestry Commission, Northern Research Station,
Roslin, Midlothian, Scotland.

Name	
Address	Chapter(s)

M. Feldman 36

Department of Plant Genetics, Weizmann Institute of Science, Rehovot, Israel.

F. P. Ferwerda 75

75 Bennekomse weg, Wageningen, The Netherlands. *Formerly*: Institute of Plant Breeding, Wageningen, The Netherlands.

D. U. Gerstel 79

Department of Crop Science, North Carolina State University, Raleigh, N.C. 27607, USA.

M. M. Goodman 37

Department of Statistics, North Carolina State University, Raleigh, N.C. 27607, USA.

M. P. Gregory 42

Department of Crop Science, North Carolina State University, Raleigh, N.C. 27607, USA.

W. C. Gregory 42

Department of Crop Science, North Carolina State University, Raleigh, N.C. 27607, USA.

M. W. Hardas 55

Indian Agricultural Research Institute, New Delhi 110012, India.

J. J. Hardon 65

Plant Breeding Station, Van der Have B.V., Rilland, The Netherlands.

J. R. Harlan 30, 39

Department of Agronomy, University of Illinois, Urbana, Ill. 61801, USA.

A. M. van Harten 74

Department of Plant Breeding, Agricultural University, 166 Lawickse Allee, Wageningen, The Netherlands.

C. B. Heiser 13, 77

Department of Plant Sciences, Indiana University, Bloomington, Indiana 47401, USA.

J. S. Hemingway 19

Reckitt & Colman Products Ltd., Carrow, Norwich, NOR 75A, England.

J. H. W. Holden 28

Scottish Plant Breeding Station, Pentlandfield, Roslin, Midlothian, EH25 9RF, Scotland.

H. W. Howard 21

Plant Breeding Institute, Maris Lane, Trumpington, Cambridge, CB2 2LQ, England.

T. Hymowitz 45

Department of Agronomy, University of Illinois, Urbana, Ill. 61801, USA.

D. L. Jennings 26, 73

Scottish Horticultural Research Institute, Invergowrie, Dundee, DD2 5DA, Scotland.

J. K. Jones 70

Department of Agricultural Botany, University of Reading, Whiteknights, Reading, RG6 2AS, England.

A. B. Joshi 55

Indian Agricultural Research Institute, New Delhi 110012, India.

Elizabeth Keep 40

East Malling Research Station, Maidstone, Kent ME19 6BJ, England.

P. F. Knowles 11

Department of Agronomy and Range Science, University of California, Davis, California 95616, USA.

E. N. Larter 35

Department of Plant Science, University of Manitoba, Winnipeg, Manitoba R3T 2N2, Canada.

K. Lesins 47

Department of Genetics, University of Alberta, Edmonton, T6G 2E9, Canada.

G. D. McCollum 53

US Department of Agriculture, Agricultural Research Service, Beltsville, Maryland 20705, USA.

I. H. McNaughton 16, 18

Scottish Plant Breeding Station, Pentlandfield, Roslin, Midlothian EH25 9RF, Scotland.

| Name | |
Address	Chapter(s)

N. M. Nayar   67

Central Plantation Crops Research Institute, Vittal-574 243, Karnataka, India.

R. A. Neve   60

Hop Research Department, Wye College (University of London), Ashford, Kent, TN25 5AH, England.

H. P. Olmo   86

Department of Viticulture & Enology, University of California, Davis, California 95616, USA.

J. H. M. Oudejans   66

Da Costalaan 9, Baarn, The Netherlands. *Formerly*: CIBA-GEIGY Ltd., Basle, Switzerland.

L. L. Phillips   56

Department of Crop Science, North Carolina State University, Raleigh, NC 27607, USA.

Barbara Pickersgill   6

Department of Agricultural Botany, University of Reading, Whiteknights, Reading, RG6 2AS, England.

D. L. Plucknett   3

College of Tropical Agriculture, University of Hawaii, Honolulu, Hawaii, USA.

J. W. Purseglove   29

East Malling Research Station, Maidstone, Kent ME19 6BJ, England.

S. Ramanujam   46

Division of Genetics, Indian Agricultural Research Institute, New Delhi 12, India.

C. M. Rick   78

Department of Vegetable Crops, University of California, Davis, California 95616, USA.

W. V. Royes   43

Ministry of Agriculture, Hope Gardens, Kingston 6, Jamaica, West Indies.

E. J. Ryder   14

US Department of Agriculture, Agricultural Research Service, P.O. Box 5098, Salinas, California 93901, USA.

J. D. Sauer   2

Department of Geography, University of California, Los Angeles, California 90024, USA.

N. W. Simmonds   10, 32, 58, 61, 63, 81

Scottish Plant Breeding Station, Pentlandfield, Roslin, Midlothian EH25 9RF, Scotland.

D. Singh   27

UP Institute of Agricultural Sciences, Kanpur UP, India.

D. P. Singh   84

Jute Agricultural Research Institute, Barrackpore, West Bengal, India.

L. B. Singh   3

Horticultural Research Institute, Saharanpur UP, India.

P. M. Smith   88

Department of Botany, University of Edinburgh, King's Buildings, Mayfield Road, Edinburgh EH9 3JH, Scotland.

R. K. Soost   76

Department of Plant Sciences, University of California, Riverside, California 92502, USA.

W. M. Steele   52

Department of Agricultural Botany, University of Reading, Whiteknights, Reading RG6 2AS, England. *Formerly*: Department of Botany, University of Nairobi, Kenya.

W. B. Storey   8, 59

Department of Plant Sciences, University of California, Riverside, California 92502, USA.

K. F. Thompson   17

Plant Breeding Institute, Maris Lane, Trumpington, Cambridge CB2 2LQ, England.

T. Visser   7

Institute for Horticultural Plant Breeding, Mansholtlaan 15, Wageningen, The Netherlands.

R. Watkins   71, 72

East Malling Research Station, Maidstone, Kent ME19 6BJ, England.

List of authors

Sisal and relatives

Agave (Agavaceae-Agaveae)

J. F. Wienk
A. van S'chendellaan 41 Ede The
Netherlands; *formerly* Centre for
Agricultural Research Paramaribo Surinam

1 Introduction

Several species of *Agave* are cultivated for their leaf fibres which provide over 90 per cent of the hard fibres of commerce. The most important one is *A. sisalana*, sisal, followed by *A. fourcroydes* or henequen; *A. cantala*, yielding maguey or cantala, and *A. letonae*, Salvador henequen, are grown to a limited extent. These four species are usually referred to as long-fibre agaves as against the brush-fibre yielding *A. lecheguilla* and *A. funkiana*. *A. amaniensis*, blue sisal, and *A. angustifolia*, dwarf sisal, though of no commercial importance themselves, are valuable as parents in breeding long-fibre agaves.

The cultivated agaves are xerophytic, tropical monocarpic perennials with large, stiff, fleshy, persistent leaves arranged in basal rosettes. They are propagated vegetatively by means of suckers or bulbils, the latter arising on the massive inflorescences after the flowers have fallen; most cultivated species seldom fruit. Harvesting the fibre-containing leaves of sisal, henequen, cantala and Salvador henequen is begun when the lowest leaves that start withering have attained a certain minimum size; only the lower leaves are cut. Cutting is then carried out annually until the plants flower. The fibre is extracted mechanically by decortication but cantala leaves are mostly retted. The brush fibres are produced by scraping the immature leaves of the central bud which is cut when the plants are six years old; the plants will continue to produce central buds, which may be cut twice a year, for another six years. World production of agave fibres in 1972 was estimated at about 760 kt of which 600 kt

Fig. 1.1 Distribution of sisal and its relatives, *Agave* spp.

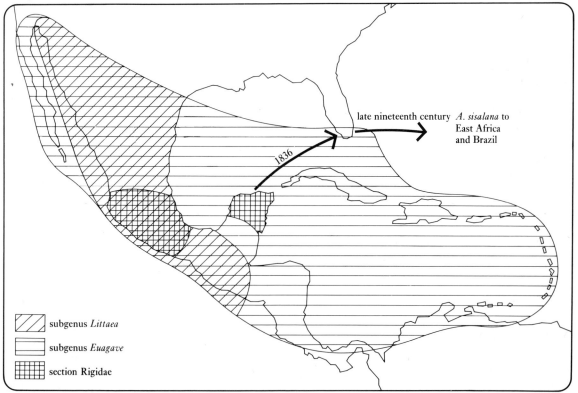

late nineteenth century *A. sisalana* to East Africa and Brazil

1836

subgenus *Littaea*

subgenus *Euagave*

section Rigidae

was sisal. The brush fibres are of little importance.

The agaves are tropical by origin and there are very few commercial plantations outside the tropics. The major sisal producers are Brazil, Tanzania, Mozambique, Angola, Kenya, Madagascar and Haïti. Henequen is grown only in some Central American and Caribbean countries, with Mexico and Cuba accounting for over 95 per cent of total production. Cantala is grown mainly in the Philippines, Salvador henequen in El Salvador and *A. lecheguilla* and *A. funkiana* in Mexico. The fibres of sisal, henequen, cantala and Salvador henequen form the raw material for cordage of which agricultural twines are the most important.

For a general review of economic botany see Purseglove (1972).

2 Cytotaxonomic background

The fibre-bearing agaves belong to a small group of a large and complex genus. The long-fibre types are classified in subgen. Euagave sect. Rigidae, the brush-fibre kinds in subgen. Littaea. The greatest variability in the genus exists in Central Mexico and the widest distribution is found among the members of Euagave. The section Rigidae is more or less confined between latitudes of 15° and 25°N (Fig. 1.1).

The basic chromosome number of the genus is $x = 30$. A polyploid series, complicated by aneuploidy, occurs and somatic chromosome numbers have been found to range between $2n = 58$ and $2n = 180$. The wild and cultivated forms of the section Rigidae include diploids, triploids, tetraploids and pentaploids. This cytological complexity, coupled with vegetative propagation, may well account for the large number of morphologically more or less distinct taxa. The chromosome numbers of the most important species cultivated for fibre or used for breeding long-fibre hybrids are as follows:

Name	$2n$
A. sisalana	$5x = ca. 138–149$
A. fourcroydes	$5x = ca. 140$
A. cantala	$3x = 90$
A. amaniensis	$2x = 60$
A. angustifolia	$2x = 60$

Chromosome numbers of *A. letonae*, *A. lecheguilla* and *A. funkiana* are unknown.

3 Early history

The cultivated agaves originated from wild ancestors in Central America and Mexico but their precise botanical origins are unknown. *A. fourcroydes* and *A.*

letonae were used in pre-Columbian times and the former was extensively cultivated on the Yucatan peninsula of Mexico by the Maya Indians.

Nothing is known about the primary ancestors of the cultivated polyploid species, *A. sisalana*, *A. fourcroydes* and *A. cantala* and the nature of their ploidy (whether auto- or allo-) is still obscure. Moreover, the concept of species in *Agave* leaves much to be desired. Traditional taxonomy has erected many specific names on poorly understood leaf variation (Gentry, 1972). Therefore the confusion is bound to be great and many of the listed species might in fact be synonyms.

Though the Aztec codices illustrate numerous basic and exotic uses of the plants, agaves are not known as fossils. On the other hand the arid and semi-arid conditions of the agaves' natural habitat and their monocarpic habit are likely to have slowed down evolution considerably. The sexual generation time is long and of uncertain outcome; seedling survival is possible only during favourable rainy periods. If the monocarpic parent, with its one burst of flowers and seeds, does not leave progeny, only the suckers have another chance to leave sexual offspring. Sexual generations in such cases can be two, three or more times longer than the monocarpic lifecycle might indicate. Thus gene assortment and recombination may be infrequent; some agave clusters encountered in Central Mexico are perhaps hundreds of years old and still without obvious seeded progeny. However, fragmented distributions may have enhanced the distinctness of such colonies. In other crops, mutation and reassortment of genes in isolation have resulted in new genotypes which, in time, have become genetically distinct from former contemporaries, have lost genetic compatibility and have evolved eventually as distinct species. But whether this applies to the agaves is very much a matter of speculation.

Self- and cross-pollination may occur. The heavy, sticky pollen is shed before the stigma becomes receptive but flowering progresses acropetally and weeks can elapse before the uppermost flowers of the massive inflorescence have opened, so that all stages from the closed bud to the receptive stigma can be encountered. The nectar exuded in the flower tube during anthesis attracts numerous insects (particularly wasps and bees, which are probably the commonest pollinators); bats may also be pollinators. Pollen may also fall by gravity onto the exposed stigmata of lower flowers.

4 Recent history

A. sisalana was taken to Florida from Yucatan in 1836 and it was from this source that many countries cultivating the species obtained their original material. In 1893, sisal bulbils were sent via Hamburg to Tanga in (then) German East Africa, now Tanzania. This introduction was the foundation of the East African sisal industry. The plant was introduced into Brazil (presently the largest producer) at the end of the nineteenth century (Fig. 2.1).

A. fourcroydes has been introduced into many tropical countries but it has never been grown very successfully outside Yucatan.

A. cantala was taken in the early years of the Spanish settlement to the Philippines and later to Indonesia. A wild form is found on the west coast of Mexico but this plant is smaller than the cultivated one now grown in the Far East which must have arisen as the result of human selection.

A. letonae is not known outside El Salvador; *A. funkiana* is found in a very restricted area in the Jaumave valley in Mexico; and *A. lecheguilla* occurs wild in Mexico and in Texas where it is not used commercially.

A. amaniensis was found growing in secondary vegetation at the East African Agricultural Research Station, Amani, Tanzania after the First World War. Its origin is not known but it may have been introduced during German times. It was found to be an undescribed species.

A. angustifolia is found in many tropical countries where it is planted as an ornamental. It has become naturalized in India.

Attempts to improve the long-fibre agaves through breeding have been made in various countries. Such work was initiated in Algeria, Brazil, Indonesia, the Philippines, Puerto Rico, Kenya and Tanzania but, with the exception of that in East Africa, it has not led to useful results and appears not to have been continued. A breeding programme began in Tanzania in 1929; in Kenya, work started fairly recently, so that it is still too early for results.

In East Africa the objects of breeding were a more rapidly growing, long-fibre agave with a higher leaf-number potential than sisal; in most other respects the improved agave should resemble the sisal plant. These include: (*a*) smooth (non-spiny) leaf margins; (*b*) long heavy and rigid leaves of good configuration; (*c*) mean fibre yield per leaf not less than that of sisal; (*d*) adaptability and resistance to pests and diseases; and (*e*) fibre quality comparable with that of sisal (Lock, 1969).

The polyploid species *A. sisalana*, *A. fourcroydes* and *A. cantala*, with a narrowly clonal genetic base, offer little scope for breeding. Fertility is very low and their sexual offspring, if any, invariably have spiny leaf margins; moreover, variation as to growth rate and leaf-number potential is too little to permit selection of more productive plant types.

Though the East African work showed that various interspecific crosses were successful it soon became evident that hybrids between the diploids, *A. amaniensis* and *A. angustifolia*, showed most promise. The results of reciprocal crosses proved that the high rate of leaf production and the high leaf-number potential of *A. angustifolia* can be combined with the long non-spiny leaves of *A. amaniensis*. Most of the F_1 hybrids are fertile, can be selfed, intercrossed or backcrossed with other species and fertility is not lost after further breeding. The first hybrid seedlings were planted in 1936. To improve leaf length the longer lived ones were backcrossed to *A. amaniensis* and some were selfed or backcrossed to *A. angustifolia*. Most selections from among the second-generation hybrids were from selfings or backcrosses to *A. amaniensis*. The results so far show that it is not difficult to obtain high-yielding hybrids with rigid leaves and smooth leaf margins. The greatest difficulty lies in the size and the shape of the leaves; they are often too light or too short, corrugated or otherwise unacceptable. Improvement of leaf characteristics was approached by backcrossing selected second-generation hybrids to *A. amaniensis* and by intercrossing. Although improvement appeared to be possible, further backcrossing has meant a lower rate of leaf unfurling and reduction of leaf-number potential.

The most outstanding clone selected so far, one which meets most of the selection criteria, is hybrid no. 11648, a product of the backcross (*A. amaniensis* × *A. angustifolia*) × *A. amaniensis* (Wienk, 1970).

5 Prospects

The long-fibre agave hybrids selected so far are not yet ideal and more work is needed to correct their shortcomings. A serious defect is susceptibility to *Phytophthora* rot, a disease not known in sisal, henequen or cantala. Both *A. amaniensis* and *A. angustifolia* are susceptible and so are their progeny. Though some variation in susceptibility is present among their hybrids, highly resistant clones are unlikely to be obtained without introducing resistance from other species. *Agave decipiens*, a tetraploid with $2n = 120$, has so far appeared to be completely resistant; it is sexually fertile but its leaves are very short and spiny and of such configuration that at least two generations will be required to obtain an acceptable leaf shape. Chromosome doubling may be

3

2

required at some stage. The resistance might also be introduced to *A. sisalana*.

Whether Tanzania's breeding programme will be continued much longer must depend upon the economic future of the agave fibres. Because of inroads made by synthetics, the future has recently looked very bleak. Sisal production in Tanzania, once the world's largest single producer, declined in the period 1965–72 from 218 to 157 kt/yr. Recently, prices have risen considerably but it remains to be seen whether the rise will provide enough incentive to invest in a long-term programme such as agave breeding; in the past, such fluctuations have always been of short duration, causing production to increase and prices to drop again.

If breeding is carried on, it seems likely that the present limited circle of aneupentaploid *sisalana* clones will be replaced by complex 'interspecific' hybrids, perhaps diploid, perhaps polyploid in constitution.

6 References

Gentry, H. S. (1972). *The Agave family in Sonora. Agric. Handb.* **399**, U.S.D.A., Washington.

Lock, G. W. (1969). *Sisal. Twenty-five years' sisal research.* London (2nd Edn).

Purseglove, J. W. (1972). *Tropical crops. Monocotyledons 1.* London, 7–29.

Wienk, J. F. (1970). The long fibre agaves and their improvement through breeding in East Africa. In C. L. A. Leakey (ed.), *Crop improvement in East Africa.* Commonw. Agric. Bur., Farnham Royal, England, 209–30.

Grain amaranths

Amaranthus spp. (Amaranthaceae)

J. D. Sauer
Department of Geography
University of California
Los Angeles USA

1 Introduction

The three grain Amaranth species are robust annual herbs that were domesticated prehistorically in the highlands of tropical and subtropical America. The tiny seeds are popped or parched and milled for flour or gruel. In taste, nutritional value and yield, the grain compares favourably with maize and other true cereals. However, the crop has declined to a vanishing relic in its homeland; far more Amaranth grain is now produced in Asia (especially in India) than in the Americas. For general reviews of economic botany, see Singh (1961) and Sauer (1967).

2 Cytotaxonomic background

The cultivated species and their probable native regions are:

1   *A. hypochondriacus* (= *A. frumentaceus*, *A. leucocarpus*, etc.) of north-western and central Mexico.

2   *A. cruentus* (= *A. paniculatus*, etc.) of southern Mexico and Central America.

3   *A. caudatus* of the Andes. In the Argentine Andes, the typical form is grown together with a conspicuous mutant that produces club-shaped inflorescence branches with determinate growth, a trait unknown in wild Amaranths. This mutant has commonly been given specific rank (as *A. edulis*) but may better be treated as *A. caudatus* ssp. *mantegazzianus* (Hanelt, 1968).

The wild species that appear most closely related to the above are, respectively:

1   *A. powellii*, a pioneer of canyons, desert washes and other open habitats in the western Cordillera of the Americas. An aberrant form with indehiscent utricles has sometimes been given specific rank as *A. bouchonii*.

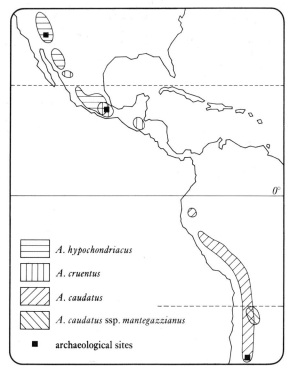

A. hypochondriacus

A. cruentus

A. caudatus

A. caudatus ssp. mantegazzianus

■ archaeological sites

Fig. 2.1 Distribution of the grain amaranths, *Amaranthus* spp.

2   *A. hybridus* (= *A. chlorostachys*, *A. patulus*, etc.), a riverbank pioneer of moister regions of eastern North America and the mild highlands of Central America.

3   *A. quitensis*, a riverbank pioneer of highland and subtropical South America.

All six species are diploids, with $2n = 2x = 32$ consistently reported except that counts of both 32 and 34 are reported for *A. cruentus* and *A. powellii*. Pal and Khoshoo (1974) suggest derivation of *A. cruentus* from *A. powellii* on the basis of their counts of $2n = 34$ for both, but comparative morphology of the species does not support this. Counts of both 32 and 34 are also reported in other sections of the genus, varying between closely related species and within certain species without apparent taxonomic meaning (Grant, 1959; Khoshoo and Pal, 1970).

Many interspecific hybrids, both spontaneous and artificial, have been reported among the grain species and their wild relatives (Murray, 1940*a*; Grant, 1959). Some experimental hybrids show heterosis and nearly normal meiosis but are partially sterile; others have abnormal growth and are totally sterile, including *A. caudatus* × *A. hypochondriacus* (Khoshoo and Pal,

1970; Pal and Khoshoo, 1972, 1974). Introgression between the last two species has not been reported where they are cultivated together.

No spontaneous polyploids are known among the grain Amaranths but colchicine-induced autotetraploids and amphiploids have been bred from some of them. Seed weight in the tetraploids is about double that of the diploids, suggesting agro-economic potential (Murray, 1940*b*; Pal and Khoshoo, 1968).

Amaranths are characteristically wind pollinated but the grain species with colourful inflorescences are occasionally visited by bees (Khoshoo and Pal, 1970). The grain species and their close relatives are monoecious and self-fertile. Arrangement and sequence of anthesis of the unisexual flowers favour a combination of self- and cross-pollination. Each of the many cymes of the inflorescence is initiated by a single staminate flower followed by an indefinite number of pistillate flowers, often over a hundred. Stigmas of the earliest pistillate flowers are receptive before the staminate flower opens; most of the later pistillate flowers develop after the staminate flower has abscissed. However, cymes of different ages are present on each indeterminate inflorescence and pollen transfer among them probably makes selfing more common than crossing.

3 Early history

Wild amaranth seeds were commonly gathered by many prehistoric American Indian peoples. The wild seeds are as nutritious and as large as those of the cultivated species. Archaeological proof of domestication comes with the appearance of pale white seeds, contrasting starkly with the dark brown wild type; the mutation producing this change has never been recorded historically. Associated with the change in colour are improved popping quality and flavour. A small proportion of dark seeds is generally present in the grain crops. Where selection for pale seed colour is relaxed, as when the plants are grown as ornamentals, the dark seeds become predominant.

The earliest record of the pale-seeded crop is from Tehuacan, Puebla, Mexico, where *A. cruentus* appeared about 4000 B.C. and was joined by *A. hypochondriacus* about A.D. 500 (Sauer, 1969). By the fourteenth century A.D., pale-seeded *A. hypochondriacus* was also cultivated by Arizona cliff-dwellers (Bohrer, 1962). The earliest record of *A. caudatus* is from 2,000-year-old tombs in north-western Argentina, where its pale seeds were found mixed with those of *Chenopodium quinoa* and with dark seeds of weed amaranths and chenopods (Hunziker and Planchuelo, 1971).

The three crop species may have been indepen-

5

dently domesticated but there is an alternative possibility, namely that there was a single primary domestication of *A. cruentus* from *A. hybridus*, with the other two domesticates evolving secondarily by repeated crossing of *A. cruentus* with weedy *A. powellii* and *A. quitensis* as the crop spread into their respective territories. Pal and Khoshoo (1974) discount the role of hybridization in evolution of the grain amaranths because of low fertility in their experimental hybrids; however, few of their hybrids were totally sterile and the particular combinations involving *A. cruentus* that are pertinent here were not actually tried.

Evolution of all three domesticates has involved increased size of the whole plant and particularly of the inflorescence, resulting in greatly increased seed yield. All three domesticates also display the effects of selection for striking anthocyanin pigmentation of leaves, stems and inflorescences. Presumably, the intense red colour had ceremonial meaning. At the time of the Spanish Conquest, grain amaranths were important in rituals of the Aztecs and other Mexican peoples. Judging by later ethnographic evidence, ceremonial use of red Amaranths extended from the Pueblo region of the southwestern United States to the Andes and was more widespread than use as a grain crop. The ceremonial dye Amaranths are generally extremely deep red forms of *A. cruentus* with dark seeds; in the Andes some may be *A. cruentus* × *A. quitensis* hybrids.

4 Recent history

In Spanish America after the sixteenth century, grain Amaranth cultivation was regarded as a symbol of paganism and repressed; thus the crop nearly disappeared from history. However, by the early nineteenth century, grain Amaranths had appeared as a staple food crop in the Nilgiri Hills of south India and in the Himalaya; they have since been noted over an increasingly wide area of India as well as across the interior of China to Manchuria and eastern Siberia. Pale-seeded *A. hypochondriacus* constitutes the bulk of the Asiatic crop; dark-seeded *A. hypochondriacus* and pale-seeded *A. caudatus* are minor components. In the 1940s, cultivation of *A. hypochondriacus* was begun in East Africa to supply grain to the local Indian population. The wide latitudinal spread of these species in the Old World presumably required evolutionary changes, because their flowering is controlled by photoperiod.

Amaranthus cruentus has not become established as a grain crop in the Old World. However, dark-seeded, deep red forms of this species have been widely planted in tropical Africa and Asia for over a century as ornamentals, dye plants, fetishes and potherbs.

All three of the domesticated species may have been introduced to the Old World via Europe; they have been grown in European gardens as ornamentals and curiosities for at least 250 years. Only dark-seeded forms of *A. hypochondriacus* were known to have been present in Europe until Hanelt (1968) found a pale-seeded specimen in a sixteenth century German herbarium collection.

5 Prospects

At least some of the grain Amaranths possess the unusual and highly efficient 4-carbon photosynthetic pathway (Tregunna and Downton, 1967; El-Sharkawy, Loomis and Williams, 1968), which may account for the excellent yields claimed. However, at present India seems to offer the only signs of significant expansion of the crop and sustained scientific breeding.

6 References

Bohrer, V. L. (1962). Ethnobotanical materials from Tonto National Monument. In L. R. Caywood (ed.), *Archeological studies at Tonto National Monument, Arizona,* Globe, Arizona, 75–114.

El-Sharkawy, M. A., Loomis, R. S. and **Williams, W. A.** (1968). Photosynthetic and respiratory exchanges of carbon dioxide by leaves of the grain amaranth. *J. Appl. Ecol.*, **5**, 243–51.

Grant, W. F. (1959). Cytogenetic studies in *Amaranthus*. III. Chromosome numbers and phylogenetic aspects. *Canad. J. Genet. Cytol.*, **1**, 313–28.

Hanelt, P. (1968). Beiträge zur Kulturpflanzenflora. I. Bemerkungen zur Systematik und Anbaugeschichte einiger *Amaranthus*-Arten. *Kulturpflanze*, **16**, 127–49.

Hunziker, A. T. and **Planchuelo, A. M.** (1971). Sobre un nuevo hallazgo de *Amaranthus caudatus* en tumbas indígenas de Argentina. *Kurtziana*, **6**, 63–7.

Khoshoo, T. N. and **Pal, M.** (1970). Cytogenetic patterns in *Amaranthus. Chromosomes Today*, **3**, 259–67.

Murray, M. J. (1940*a*). The genetics of sex determination in the family *Amaranthaceae. Genetics*, **25**, 409–31.

Murray, M. J. (1940*b*). Colchicine induced tetraploids in dioecious and monoecious species of the *Amaranthaceae. J. Hered.*, **31**, 477–85.

Pal, M. and **Khoshoo, T. N.** (1968). Cytogenetics of the raw autotetraploid *Amaranthus edulis. Tech. Comm. Nat. Bot. Gdns., Lucknow*, 25–36.

Pal, M. and **Khoshoo, T. N.** (1972). Evolution and improvement of cultivated amaranths. V. Inviability, weakness, and sterility in hybrids. *J. Hered.*, **63**, 78–82.

Pal, M. and **Khoshoo, T. N.** (1974). Grain amaranths. In J. B. Hutchinson (ed.), *Evolutionary studies in world crops: Diversity and change in the Indian subcontinent,* Cambridge University Press, 129–37.

Sauer, J. D. (1967). The grain amaranths and their relatives: A revised taxonomic and geographic survey. *Ann. Mo. Bot. Gdn.*, **54**, 103–37.

Sauer, J. D. (1969). Identity of archaeologic grain amaranths from the valley of Tehuacan, Puebla, Mexico. *Amer. Antiqu.*, **34**, 80–1.

Singh, H. (1961). Grain amaranths, buckwheat and chenopods. *Ind. Coun. Agric. Res., Cereal Crop Ser.*, **1**, 1–46.

Tregunna, E. B. and Downton, J. (1967). Carbon dioxide compensation in members of the *Amaranthaceae* and some related families. *Canad. J. Bot.*, **45**, 2385–7.

Mango

Mangifera indica (Anacardiaceae)

L. B. Singh
Horticultural Research Institute
Saharanpur U.P. India

1 Introduction

The Mango is one of the most important tropical fruits and, in the opinion of many, one of the most delicious fruits in the world. It is cultivated throughout the tropics but is most abundantly produced in India, where it occupies about 0·9 Mha. The mango is a tree which sometimes attains great size (up to 30–40 m tall). It is adapted to seasonal tropical climates at low altitude; a dry season is essential for fruit setting. Purseglove (1968) has reviewed the economic botany of the crop.

2 Cytotaxonomic background

Estimates of the number of species of *Mangifera* vary widely. Mukherjee (1972), in an authoritative review, recognized 41, distributed in two sections. All are southeast Asian and they range from northern India to the Philippines and New Guinea, with a marked centre of diversity in Indochina–Malaysia–Indonesia (Fig. 3.1). All are forest trees, often large, with fibrous-resinous fruits.

Wild *M. indica* occurs in north-eastern India. Botanically it is related to *M. sylvatica*, *M. caloneura*, *M. zeylanica* and *M. pentandra*. These species, and all other *Mangifera* species and cultivars so far examined, have $2n = 2x = 40$ (Mukherjee, 1950; Roy and Visweswariya, 1951). *Mangifera indica* and *M. sylvatica* are karyotypically related. Studies of secondary meiotic associations have suggested secondary polyploidy based on $x = 8$. If so, the polyploidy is presumably ancient for no basic number as low as 8 is known in the Anacardiaceae (where $x = 12, 14, 15, 16$). A single instance of recent tetraploidy ($2n = 4x = 80$) has been reported in a polyembryonic seedling of an Indian cultivar (Roy and Visweswariya, 1951).

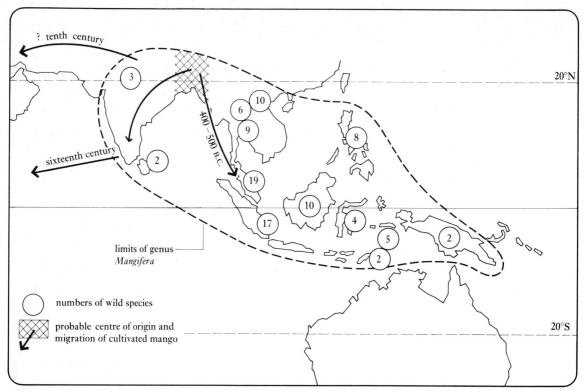

Fig. 3.1 Distribution of wild *Mangifera* spp. and origin of cultivated mango, *M. indica*.

3 Early history

All the evidence (Mukherjee, 1972) points to the local origin of the mango from wild *M. indica*, without the intervention of other wild species. The fruits of several other species are eaten locally but none is of importance. Wild mangoes in general have fruits which are unacceptably fibrous-resinous and may even be poisonous. Human selection must have been for succulence, low fibre, small stone and low resin content. When the mango was domesticated is strictly unknown but dates of 4000 B.P. have been mentioned. The crop spread out of India eastwards in the first instance, perhaps as early as 400–500 B.C. in the hands of Buddhist monks but did not become generally diffused through the islands and in to the Pacific until very much later. Westwards, Mukherjee (1972) believes that it moved as late as the sixteenth century, in the hands of Portuguese travellers, from Goa, by way of East Africa to West Africa and the New World. He noted, however, an alternative possibility, namely that it spread by way of the Sabaean Lane as early as the tenth century.

Mangoes were first grown in the West Indies in the eighteenth century and reached Florida in 1833.

The best mango fruits come from selected clones but it is well to recall that the great majority of trees are unselected seedlings, often more or less wild. Since the crop is outbred, there is usually great variation between seedlings. However, some polyembryony occurs and seemingly true breeding (in effect, apomictic) cultivars are known in southern India. In general, good cultivars must be maintained vegetatively and one form or another of budding is usual. The practice reached India in the sixteenth century, introduced by Europeans, and has since been adapted for the world-wide dissemination of chosen clones by means of budwood.

Outbreeding is encouraged by dicliny and entomophily. Male and female flowers are separated within the inflorescence and insect pollination is essential. In general, mangoes must be self-compatible (as shown by setting of isolated trees) but Sharma and Singh (1970) found evidence of at least some incompatibility.

In general, then, the mango population consists of vast numbers of sexual seedlings, a few polyembryonic lines and many selected clones. India alone is estimated to have 1,000 clones established: elsewhere, far smaller numbers are the rule.

4 Recent history

Breeding was started in India early in this century and the work has been reviewed by Mukherjee *et al.* (1968) and by Singh (1969). On the whole, results have not been encouraging. Hand-crossing is remarkably unrewarding, success rates of the order of three fruits per 1,000 pollinations having been recorded; the life cycle is long, though it can be shortened by well-chosen grafting procedures; polyembryony presents problems; and none of the seedlings produced have proved to be outstanding.

The basic objectives of breeding have been regular bearing (there is normally a strong biennial tendency) and good fruit quality.

As in many other tree fruits, some useful somatic mutants have been picked up; for example, of the cv. Haden in Florida (Young and Ledin, 1954). Induced mutation studies have not been encouraging.

5 Prospects

Should the current interest in growing the crop for overseas export continue, there will undoubtedly be a need for export varieties as well as for improved cultivars for local consumption. Past non-success notwithstanding, it is impossible to believe that no improvement is achievable in a crop in which abundant variability is available, especially if new selection criteria are presented to the breeder. Probably, replacement of the laborious and ineffective hand-pollination by some mass-pollination technique will be essential if adequate populations are to be raised. One such is the 'cage method' whereby parents are enclosed together with house flies as pollinators. It would seem important, in this context, to acquire a better understanding of incompatibility in the crop.

The breeding potential of wild species is unknown but might be guessed to be small unless specific disease resistances were sought. On the other hand, they might have use as sources of dwarfing stocks, as suggested long ago by Furtado (1921). More likely, perhaps, is the possibility that dwarfness (or at least some decrease in stature) might be achieved by mutation, natural or induced, in otherwise desirable clones. Indeed, some clones with trees markedly smaller than usual have long been established.

6 References

Furtado, C. X. (1921). Influence of stock on scion. *Agric. J. India*, 16, 226.

Mukherjee, S. K. (1950). The origin of mango. *Indian J. Genet.*, 2, 49.

Mukherjee, S. K. (1972). Origin of mango (*Mangifera indica*). *Econ. Bot.*, 26, 260–6.

Mukherjee, S. K., Singh, R. N., Majumder, P. K. and Sharma, D. K. (1968). Present position regarding breeding of mango in India. *Euphytica*, 17, 462–7.

Purseglove, J. W. (1968). *Tropical crops. Dicotyledons. 1.* London, 24–32.

Roy, B. and Visweswariya, S. S. (1951). *Cytogenetics of mango and banana.* Maharashtra Association for the Cultivation of Science, Poona.

Sharma, D. K. and Singh, R. N. (1970). Investigation on self incompatibility in *Mangifera indica. Symposium on mango and mango culture.* I.S.H.S., The Hague, The Netherlands.

Singh, L. B. (1969). *Mango (Mangifera indica).* In Ferwerda, F. P. and Wit. F. (eds), *Outlines of perennial crop breeding in the tropics.* Wageningen, 309–27.

Singh, L. B. (1960). *The mango.* London.

Young, T. W. and Ledin, R. B. (1954). Mango breeding. *Proc. Fla. hort. Soc.*, 67, 241.

4

Edible aroids

Alocasia, Colocasia, Cyrtosperma,
Xanthosoma (Araceae)

D. L. Plucknett
College of Tropical Agriculture
University of Hawaii
Honolulu Hawaii USA

1 Introduction

Aroids are important food crops for native people in many parts of the tropics. They are grown for their edible corms or cormels; in some cultivars the leaves and petioles are eaten as spinach or salad. Because they rarely enter into commerce except in local markets, world production is difficult to determine but was recently estimated at about 4 Mt. Most of this production undoubtedly resulted from *Colocasia* and *Xanthosoma* which are used most widely, notably in the Pacific and Caribbean islands and in West Africa. *Alocasia* and *Cyrtosperma* are important mostly as subsistence crops in some Pacific islands, *Cyrtosperma* being the main starch crop of coral atolls. *Colocasia* (some cultivars) and *Cyrtosperma* have special importance as crops for difficult lands, for they can produce large yields under flooded or swampy conditions where, in some cases, even rice could not be grown. *Alocasia, Colocasia* (some cultivars) and *Xanthosoma* are mainly grown as upland rainfed crops.

Except for limited shipments to satisfy migrant food needs, aroids seldom enter into international trade. Few processed products are made from them, although *Colocasia* flour has been marketed successfully in Hawaii and India. Their food value is similar to that of most root crops. Most aroids have calcium oxalate raphides in their corms and leaves; these irritate the mouth and throat. Some varieties are less acrid than others. For reviews of these crops see: Barrau (1957), Coursey (1968), Massal and Barrau (1955), Plucknett (1970), Plucknett *et al.* (1970) and Purseglove (1972).

2 Cytotaxonomic background

Araceae is a large family, with about 100 genera and perhaps 1,500 species. Edible aroid genera are grouped in two sub-families: Lasiodeae (*Cyrtosperma* and *Amorphophallus*) and Colocasioideae (*Alocasia, Colocasia,* and *Xanthosoma*).

Only a few aroids can be considered as economic plants, although some species are harvested for subsistence or emergency food. As is common with vegetatively propagated ancient crops, the taxonomic position of many species is questioned. Major species are as follows:

Species	Vernacular names	2n	Author
A. of Asian origin			
1 *Alocasia indica* (considered by some to be synonymous with *A. macrorrhiza*)		$2x = 28$	Marchant (1971)
2 *Alocasia macrorrhiza*	Giant taro, ta'amu, ape	$2x = 26$ $2x = 28$	Marchant (1971)
3 *Colocasia esculenta* (some authors refer to *C. antiquorum* but Hill, 1939 considered *C. esculenta* to be polymorphic)	Taro, eddo, old cocoyam, keladi, arvi, dasheen (W. Indies)	$2x = 28$ $3x = 42$	Yen and Wheeler (1968); (Plucknett *et al.* (1970); Marchant (1971)
4 *C. esculenta* var. *globulifera*	Dasheen (Pacific and Asia)	$3x = 42$	Plucknett *et al.* (1970)
5 *Cyrtosperma chamissonis* (sometimes called *C. edule* or *C. merkusii*)	Giant swamp taro	?	
B. of South American origin			
1 *Xanthosoma atrovirens*	Tanier, yautia, cocoyam	$2x = 26$	Marchant (1971)
2 *Xanthosoma sagittifolium*	Tanier, yautia, cocoyam	$2x = 26$	Marchant (1971)
3 *Xanthosoma violaceum*	Tanier, yautia, cocoyam	$2x = 26$	Marchant (1971)

Great confusion exists concerning the taxonomy of edible *Xanthosoma*. Other edible species often listed (Haudricourt, 1941, cited by Coursey, 1968) include *X. brasiliense, X. belophyllum, X. caracu, X. jacquini* and *X. mafaffa*, considered by Dalziel (Purseglove, 1972) to be the 'new' cocoyam of West Africa. It may be that, with the exception of *X. brasiliense*, which is grown for its edible leaves, edible *Xanthosoma* should be considered as one polymorphic species (namely *X. sagittifolium*) since most species separations are based on vegetative characteristics.

Despite confusion concerning the status or validity of the many species, it appears that there are probably four major ones: *Alocasia macrorrhiza, Colocasia esculenta, Cyrtosperma chamissonis* and *Xanthosoma sagittifolium*.

3 Early history (Fig. 4.1)

Aroids are ancient crops which owe their development to vegetative mutation or chance seed setting and selection by prehistoric man. All varieties known today were selected and propagated by subsistence farmers of Africa, Asia, Oceania and Latin America, and modern science has played virtually no role in developing or improving cultivars.

The aroids probably originated in swamps or in high rainfall forest areas. Some anthropologists believe that *Colocasia* may have been the first irrigated crop and that ancient 'rice' terraces of Asia were originally constructed for *Colocasia*. Undoubtedly, wild aroids were gathered for food by ancient man who observed that non- or little-acrid types existed and that cast-off plant parts such as small corms, corm fragments and tops grew readily in waste heaps, in moist camp sites and along streams. From this point, it is easy to imagine the next faltering step, planting and cultivating the plants. When did this occur? It is impossible to say for sure, but an estimate of 4,000 to 7,000 years ago seems reasonable.

Alocasia, *Colocasia* and *Cyrtosperma* originated in the Old World. *Alocasia* was probably native to Ceylon (Sri Lanka) or India whence it spread eastwards to Oceania to become a crop in parts of Micronesia, Polynesia and Melanesia.

Colocasia originated in Indo-Malaya (Plucknett *et al.*, 1970). Ethnobotanical evidence favours India as the place of origin (Spier, 1951). It spread eastwards into Southeast Asia, eastern Asia (including China and Japan) and the Pacific Islands. It was recorded as an important food crop in China about 100 B.C. (Plucknett *et al.*, 1970). It moved westwards with megalithic people to Arabia, the eastern Mediterranean and Egypt where it has been known for over 2,500 years. From there it was spread, perhaps more than 2,000 years ago, along the coast to East Africa and across the continent to West Africa by voyagers, explorers and traders. *Colocasia* is known as 'old cocoyam' in West Africa, indicating that it was present there before the introduction of 'new' cocoyam, *Xanthosoma sagittifolium*. *Colocasia* was carried to the Caribbean and tropical America on slave ships as a source of food.

Cyrtosperma originated in southeast Asia, perhaps in Indonesia. It spread throughout the Malayan Archipelago and eastwards to the Pacific, to become a minor crop in Melanesia and parts of Polynesia, but a major food in Micronesia. Its ability to grow in slightly brackish swamps makes it important for coral atolls.

Xanthosoma, a tropical American plant, was developed by Amerindian people of the continent and the Caribbean Islands. During the slave-trading era, *Xanthosoma* was taken to Africa where it became known as 'new' cocoyam. Today in West Africa, *Xanthosoma* is more important than *Colocasia*, ranking third behind cassava and the yams. During the middle and late nineteenth and early twentieth centuries, *Xanthosoma* was spread throughout the Pacific and into Asia where it achieved minor crop status, often at the expense of other root crops, because of its resistance to disease and pests (Massal and Barrau, 1955).

4 Recent history

Cultivars in use are the products of indigenous agriculturists and the role of plant breeding has been almost nil. It has been suggested that attempts be made to hybridize *Xanthosoma* and *Colocasia* in order to transfer disease resistance to *Colocasia*, which has experienced some decline because of diseases such as

Fig. 4.1 Evolutionary geography of the cultivated aroids, *Xanthosoma*, *Colocasia*, *Alocasia* and *Cyrtosperma*.

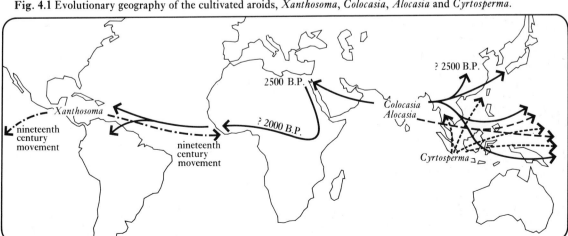

Phytophthora colocasiae and certain viruses.

Colocasia breeding has been attempted in Japan and Hawaii (Plucknett *et al.*, 1970) where seedlings from natural and cross-pollination studies were grown and observed. No new varieties resulted from this work. There are no known plant breeding efforts in the other genera.

Studies of flowering and chance seed-set in *Colocasia* and *Xanthosoma* could be fruitful in providing a basis for plant breeding, for these crops appear to be worthy candidates for improvement. Large differences exist between cultivars in: corm colour; corm quality (especially acridity levels due to raphides); disease susceptibility; ability to grow under varying moisture regimes, including flooding; soil or water salinity levels; and in edibility of corms or cormels or both.

5 Prospects

Aroids are often considered to be static or declining crops. This may be true for *Alocasia* and *Cyrtosperma* and, perhaps, *Colocasia* but not for *Xanthosoma*, which is increasing in some Pacific Islands, West Africa and elsewhere.

The future for the aroids depends largely on four factors: disease and pest resistance of cultivars; food yield and quality; their role in more efficient land use because of an ability to grow in marginal lands; and their importance as emergency, reserve or famine crops.

Of the four, *Colocasia* and *Xanthosoma* appear to be most suitable for wider use. Some *Colocasia* cultivars are resistant to flooding and salinity and are not easily destroyed by storms. The starch is easily digestible and the corm proteins are non-allergenic; it can therefore be made into speciality foods for allergic infants, hospital patients and persons with gastric disorders. Corms, cormels, petioles and leaves may be edible. Disadvantages, of some cultivars, include: acridity of corms and leaves; susceptibility to certain fungal and virus diseases; and a relatively long crop life of 9–15 months. It is a subsistence crop under shifting cultivation and a commercial crop under irrigated or rainfed conditions.

Xanthosoma is more drought tolerant and somewhat more shade tolerant than *Colocasia*. It is a successful intercrop or shade crop for cocoa, rubber, coconut and banana as well as an important crop in shifting cultivation. Its main advantages are remarkable resistance to pests and diseases and high yields under difficult conditions. *Xanthosoma* and *Colocasia* are complementary, not competitive, crops because they are adapted to different agricultural regimes.

Alocasia has several disadvantages namely: some cultivars are very acrid; corms are often coarse; crop life is quite long (2–4 years); and it grows in upland areas suited to upland *Colocasia* or *Xanthosoma*. However, the crop is exceptionally hardy, requires little care and yields are high.

Cyrtosperma fills a special niche in the array of food plants. Its strengths are: extraordinary hardiness; low maintenance requirement; ability to grow in stagnant, often brackish, swamps; and potential as an emergency or famine food. Long crop duration, as much as four years or more, is a disadvantage.

Concerted efforts are needed to collect, preserve and study aroid cultivars around the world. Many clones are being lost each year. Collections are limited and under-financed. There may be more than 1,000 cultivars of *Colocasia* alone; for the other genera it is impossible to estimate how many cultivars exist. Those surviving cultivars represent a legacy of uncounted centuries of activities of subsistence farmers who saw special virtues in variant forms and preserved them; perhaps some of these could be used to help solve our current food problems.

6 References

Barrau, J. (1957). Les Aracées á tubercules alimentaires des Iles du Pacifique Sud. *J. Agric. trop. Bot. appl.*, **4**, 34–51.

Coursey, D. G. (1968). The edible aroids. *World Crops*, **20**, 25–30.

Hill, A. F. (1939). The nomenclature of the taro and its varieties. *Bot. Mus. Leaflet Harv. Univ.*, **7**, 113–18.

Marchant, C. J. (1971). Chromosome variation in Araceae. II. Richardieae to Colocasieae. *Kew Bull.*, **25**, 47–56.

Massal, E. and **Barrau, J.** (1955). Pacific subsistence crops – taros. *South Pacif. Comm. quart. Bull.*, **5**(2), 17–21.

Plucknett, D. L. (1970). The status and future of the major edible aroids: *Alocasia, Amorphophallus, Colocasia, Cyrtosperma,* and *Xanthosperma. Proc. 2nd internl Symp. trop. Root Tuber Crops, Hawaii,* **1**, 127–35.

Plucknett, D. L., de la Pena, R. S. and **Obrero, F.** (1970). Taro (*Colocasia esculenta*). *Field Crop Abstr.*, **23**, 413–26.

Purseglove, J. W. (1972). *Tropical Crops. Monocotyledons* 1. London and New York, 58–74.

Spier, R. F. G. (1951). Some notes on the origin of taro. *Southwestern J. Anthropol.*, **7**, 69–76.

Yen, D. E. and **Wheeler, J. M.** (1968). Introduction of taro into the Pacific: the indications of the chromosome numbers. *Ethnology*, **7**, 260–7.

Note: This chapter is published as Journal series 1799 of the Hawaii Agricultural Experiment Station.

Kapok

Ceiba pentandra (Bombacaceae)

A. C. Zeven
Institute of Plant Breeding (I.v.P.)
Agricultural University Wageningen
The Netherlands

1 Introduction

Before 1940 the kapok tree was an important commercial crop grown for its fibre. In recent years, these fibres have been replaced by synthetics but the kapok tree is still cultivated on a small scale for local use. World trade was about 300 tons in 1965, the main exports coming from Thailand (50%) and Cambodia (30%).

The cells of the inner epidermis of the epicarp form the fibres which are about 1–2 cm long. The air-filled lumen is broad and the wall rather thin. The fibre is therefore fragile which, together with the smoothness of the outer surface, makes spinning impossible. It is, however, very light so has been extensively used in life-jackets and life-belts, in insulated clothing, for upholstery and in sound insulation for aircraft.

For general reviews see Baker (1965) and Zeven (1969).

2 Cytotaxonomic background

Ceiba contains nine species of which eight are tropical American and one, kapok, extends through the tropics of both Old and New Worlds. Kapok itself, *Ceiba pentandra*, is variable, which may be related to its high and variable chromosome numbers ($2n = 72-88$; $x = ?$) (Heyn, 1938; Tjio 1948; Baker, 1965; Zeven, 1969).

The species can be subdivided thus:

1   var. *caribaea* occurs wild and semi-wild in America (S. Mexico, the Caribbean islands, C. America, tropical S. America) and in Africa (from Senegal to C. Africa).
    This variety consists of two types:

    1.1   Forest type: having a tall, unbranched trunk with a high crown and often big buttresses. Cv. Togo is a spineless form with indehiscent fruits and white kapok. Cv. Reuzenrandoe (giant kapok) of Java also resembles this type.

    1.2   Savannah type: having a short, unbranched trunk and a very broadly spreading crown. It may derive from plagiotropic cuttings either obtained from the forest type or from the savannah type itself (Zeven, 1969).

2   var. *indica*, the cultivated type of Southeast Asia and elsewhere. The tree is less robust than var. *caribaea*. It has special forms of branching: pagoda and lanang, both having indehiscent fruits and white kapok.

3 Early history

Toxopeus (1943, 1948, 1950) drew attention to the bicontinental natural distribution of kapok in contrast to its strictly American relatives. Bakhuizen van den Brink (1933) and Chevalier (1949) favoured an American origin. They argued that, in pre-Columbian times, fruits or seeds could have been transported by sea currents to Africa. This view is generally accepted and agrees with the polyploid condition of kapok in comparison with American species, in which lower ploidy levels occur.

Kapok probably reached southeast Asia by way of India from Africa before A.D. 500. It was depicted in Indonesia before A.D. 850 (Steinmann, 1934; Toxopeus, 1941).

Pollination is carried out by bats and bees (Toxopeus, 1936; Van der Pijl, 1956) which visit single or small groups of trees (review in Zeven, 1969). This may result in autogamy, geitonogamy and (occasionally) allogamy. Geitonogamy and allogamy are promoted by protandry (Toxopeus, 1950) but cleistogamy results in autogamy (Jaeger, 1954). In large plantations, bats find it difficult to enter the crowns of the trees where bees are relatively limited in number. There, autogamy is common.

4 Recent history

The recent history is one of decline, owing to the development of substitute synthetic fibres, though the crop retains some local importance.

Breeding work in southeast Asia was disrupted by the war and not resumed. That work (review in Zeven, 1969) indicated that there was considerable potential for improvement by line breeding (implying tolerance of considerable inbreeding) or by isolation of heterotic clones propagated from orthotropic cuttings. Breeding objectives included: short stature, precocity, strong branches, spinelessness, various fruit characters (including indehiscence) and fibre characters (whiteness, buoyancy and resilience).

5 Prospects

The prospect can only be for further decline. If breeding were resumed, it would be directed towards local requirements.

6 References

Baker, H. G. (1965). The evolution of the cultivated kapok tree: a probable West African product. *Research Ser. Inst. Intnl Studies, Univ. Calif.*, **9**, 185–216.

Bakhuizen van den Brink, H. C. (1933). [The Indonesian flora and its first American intruders]. *Natuurk. Tijdschr. Ned. Indië*, **93**, 20–55.

Chevalier, A. (1949). Nouvelles observation sur les arbres à kapok de l'ouest africain. *Rev. int. Bot. appl. trop.*, **29**, 377–85.

Heyn, A. N. J. (1938). [Cytological researches on some tropical cultivated crops and their importance for plant breeding]. *Landbouw*, **12**, 11–92.

Jaeger, P. (1954). Note sur l'anatomie florale, l'anthocinétique et les modes de pollinisation du fromager (*Ceiba pentandra*). *Bull. Inst. franç. Afric. Noire, Dakar*, **16**, 370–8.

Pijl, L. van der (1956). Remarks on pollination by bats in the genera *Freycinetia*, *Duabanga* and *Haplophregma* and on chiropterophily in general. *Acta Bot. Neerl.*, **5**, 135–44.

Steinmann, A. (1934). [The oldest pictures of the kapok tree in Java]. *Trop. Natuur.*, **23**, 110–13.

Tjio, J. H. (1948). Note on nucleolar conditions in *Ceiba pentandra*. *Hereditas*, **34**, 204–8.

Toxopeus, H. J. (1936). [On factors determining yield in kapok]. *Bergcultures*, **10**, 166–75.

Toxopeus, H. J. (1941). [Origin, variation and breeding of some of our cultivated crops]. *Natuurw. Tijdschr. Ned. Indië*, **101**, 19–31.

Toxopeus, H. J. (1943). The variability and origin of *Ceiba pentandra*, the kapok tree. *Contrib. Alg. Proefstat. Landbouw*, **13**, 3–20.

Toxopeus, H. J. (1948). On the origin of the kapok tree, *Ceiba pentandra*. *Ned. Alg. Proefstat. Landbouw*, **56**, pp. 19.

Toxopeus, H. J. (1950). Kapok in C. J. J. van Hall and C. van de Koppel (eds), [*Agriculture in the Indonesian archipelago*. III. *Industrial crops*], The Hague, 53–102.

Zeven, A. C. (1969). Kapok tree, *Ceiba pentandra*. In F. P. Ferwerda and F. Wit (eds), *Outlines of perennial crop breeding in the tropics*, Wageningen, 269–87.

Pineapple

Ananas comosus (Bromeliaceae)

Barbara Pickersgill
University of Reading
Reading England

1 Introduction

The pineapple is a native of the New World but is now cultivated in frost-free areas in the tropics and subtropics of both hemispheres. Annual world production exceeds 3·5 Mt and about two-thirds of this is consumed in the areas in which it is produced. There is a small export trade in the fresh fruit but most commercially grown pineapples are canned or made into juice. Pineapple leaf fibres are used for cloth and cordage in the Philippines and Taiwan and have been used experimentally to produce paper. Pineapple stems and fruits are a possible commercial source of a protease, bromelain, though papain (from *Carica papaya*) remains the only plant protease actually produced commercially.

The pineapple is a perennial which may live for up to 50 years. It produces fruit in four years when grown from seed, but cultivated pineapples are propagated vegetatively and commercial growers take two crops in three years. The edible fleshy portion of the fruit consists of the swollen carpels, the lower portions of the calyx segments and bracts and the inflorescence axis. Modern commercial cultivars are seedless. The plant appears suited to xerophytic conditions. The lower epidermis of the leaf bears trichomes which can absorb water from the atmosphere and may reduce transpiration, while the leaf itself contains a specialized water-storage tissue. The overlapping leaf bases act as reservoirs which impound water and humus, utilized by adventitious roots which grow up between the leaf bases. The soil root system consists of a few long roots, which penetrate to depths of over 1 m and draw water from the lower levels of the soil, and a mass of short roots concentrated round the stem base which utilize water overflowing from the leaf base reservoirs. Pineapples can be grown under an annual rainfall of only 500 mm but they will also tolerate much wetter conditions, up to 5,500 mm.

2 Cytotaxonomic background

Ananas and the closely related monotypic genus *Pseudananas* are distinguished from other genera of the Bromeliaceae by their syncarpous fruit which, in *Ananas*, bears a terminal crown of reduced leaves; *Pseudananas* has no such crown. Both genera share the basic chromosome number $x = 25$ common throughout the Bromeliaceae but *Ananas* is typically diploid ($2n = 2x = 50$) while *Pseudananas* is tetraploid ($2n = 4x = 100$). Both are scantily represented in herbaria since they are native to botanically underexplored parts of lowland South America and their diagnostic inflorescences and fruits constitute singularly intractable material for herbarium specimens. Among the wild species of *Ananas*, *A. bracteatus*, *A. fritzmuelleri* and the recently described *A. parguazensis* seem closely related. Both *A. bracteatus* and *A. fritzmuelleri* produce fleshy, edible (though seedy) fruits,

and *A. bracteatus* has in fact been cultivated in the Paraná river area (Collins, 1960). Fruits of *A. ananassoides* and *A. erectifolius* become nearly dry at maturity, with little flesh. *Ananas erectifolius*, which has long, almost spineless leaves, has, however, been considered a potential fibre crop. The precise ranges of the wild species are not yet known (available records are summarized in Fig. 6.1) and it is possible that other species will be discovered in the future.

Where the species overlap geographically, they are isolated by different ecological preferences (Collins, 1960). There seem to be few, if any, other barriers to hybridization. *Ananas comosus* can be crossed with *A. bracteatus*, *A. ananassoides* and *A. erectifolius*, and the F_1 hybrids are completely fertile. When *A. comosus* was crossed with *Pseudananas*, the F_1 plants were mostly tetraploid, with only a few of the expected triploids, and again fully fertile (Collins, 1960).

Fig. 6.1 Distributions of pineapples, *Ananas*, their wild relatives and early cultivations.

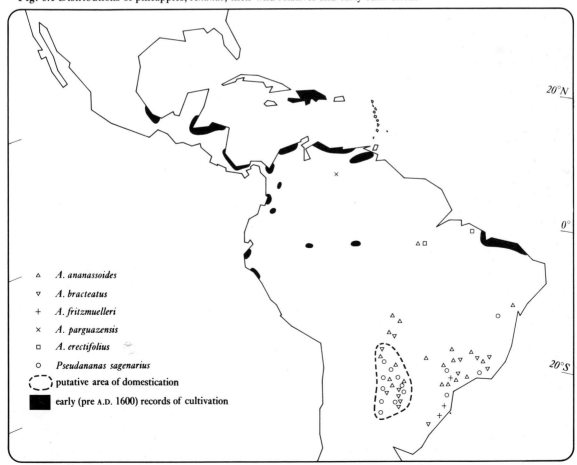

△   *A. ananassoides*

▽   *A. bracteatus*

+   *A. fritzmuelleri*

×   *A. parguazensis*

▫   *A. erectifolius*

○   *Pseudananas sagenarius*

⌐¯⌐ putative area of domestication

■ early (pre A.D. 1600) records of cultivation

3 Early history

Ananas comosus is often stated to be unknown in the wild, though pineapples from abandoned plantings may survive for many years. This poses the problem, common in many crop plants, of distinguishing escapes derived from the crop from primitively wild forms which may be ancestral to the crop. 'Wild' pineapples have been reported from southern central Brazil and were said to differ from 'wild' pineapples in other parts of Brazil and Trinidad but it was not clear whether they were recent escapes from local cultivars or links between an unknown wild *comosus* and modern cultivars (Collins, 1960). 'Wild' pineapples have also been reported from Venezuela.

Collins (1960) considered that the pineapple originated in the Paraná–Paraguay basin, in the area in which *A. bracteatus*, *A. ananassoides* and *Pseudananas* occur wild. He suggested that the pineapple was domesticated by the Tupi-Guarani Indians (who have also been credited with domesticating the peanut) and then spread north and west. By the time Europeans discovered the New World, pineapples were widespread in South America east of the Andes and had reached the West Indies, Mexico and Central America (Collins, 1960; Patiño, 1963; see also Fig. 6.1).

Unfortunately, critical evidence to test Collins's view is lacking. *Ananas comosus* is the only self-incompatible species known in the genus; *Pseudananas* and those wild species of *Ananas* that have been investigated are at least partially self-fertile. Brewbaker and Gorrez (1967) have shown that the pineapple possesses a gametophytic *S*-allele type of incompatibility although, in pineapple, inducing polyploidy does not break down the incompatibility system. It seems unlikely that self-incompatible *A. comosus* could be derived from self-compatible *A. bracteatus* or *A. ananassoides*. It is more probable that 'the ancestor of *comosus* is *comosus*' and, in this context, a biosystematic and cytogenetic investigation of 'wild' and indigenous cultivated *A. comosus* of South America would be of considerable interest. The distribution of wild species not in the direct ancestry of a cultigen may be quite unrelated to the centre of domestication of the crop; thus, tomato, the chili pepper (*Capsicum annuum*), mango and coconut were all domesticated outside the areas of greatest concentration of wild species in their respective genera. Hence Collins's main argument for a Paraná–Paraguay centre for the pineapple is unconvincing.

Sauer (1959) included the pineapple in the complex of vegetatively propagated crops which he considered characteristic of lowland South America, e.g. manioc, sweet potato and other root crops; Sauer also included the peanut which, although not propagated vegetatively, produces its crop underground. He suggested that domestication of these crops began in the tropical forests before the start of seed crop agriculture in Mexico. Although the pineapple can tolerate high rainfall, its xerophytic adaptations hardly suggest that it is a native of the tropical forests. Furthermore, although manioc, sweet potatoes and peanuts have been recorded archaeologically from coastal Peru from about 1000 B.C. onwards, pineapples are absent, apart from a somewhat doubtful painting on a much later wooden vessel. The only archaeological record for the pineapple consists of minute fragments, claimed to represent pineapple seeds and bracts, in two coprolites from caves in the Tehuacán valley of Mexico dated about 200 B.C.–A.D. 700. The pineapple is almost certainly a lowland South American domesticate but it seems to have been a rather late addition to the cultivated plants of the region, possibly because it occurred outside the main centre (or centres) of domestication.

The self-incompatibility of *A. comosus*, compared to the self-compatibility of its edible relative, *A. bracteatus*, may be one reason for its greater success as a cultivated plant. Self-compatible mutants of *A. comosus* produce fruits containing up to 3,000 very hard seeds. Self-incompatible pineapples, in which compatible cross-pollination has been prevented, produce seedless fruits parthenocarpically and these are, obviously, infinitely preferable for the consumer. Pineapples are pollinated by hummingbirds, possibly also by honey bees and pineapple beetles. Natural crossing between clones will produce seedy fruits. However, vegetative propagation, using the leafy crowns of the fruit or slips or suckers from the parent plant, may lead to all plants in one garden having the same genotype and cross-pollination then will produce only seedless parthenocarpic fruit. The Indians of the American tropics were growing seedless pineapples, which they propagated from crowns, by the time of the European conquests. Obviously, a key evolutionary question remains unanswered: are truly wild pineapples parthenocarpic or did parthenocarpy evolve during domestication in response to human selection for seedlessness (as in the bananas)?

The Amerindians do not seem to have been particularly concerned to select against spiny leaves in the pineapple. 'Spineless' forms, in which spines are restricted to the top few inches of the leaf, are dominant to wild-type spiny (Collins, 1960). Spineless types reportedly predominate amongst endemic forms in the upper Amazon basin (Patiño, 1963) but other Indian varieties are spiny. The Indians had also developed varieties which differed in flesh colour, acidity and flavour but there seems to be no particular centre of diversity for the crop.

The pineapple has been supposed to support theories of pre-Columbian trans-oceanic contacts. Heyerdahl claimed that pineapples were present in the Marquesas before Europeans discovered Polynesia. Merrill (1954) discounted this and stated that the first record of the pineapple in Polynesia was 1769, when Captain Cook sowed pineapple seeds in Tahiti. Pineapples have also been identified on wall carvings in ancient Assyria, on pottery models in Egyptian tombs and on murals at Pompeii, but the reproductions that have been published are not convincing and the pre-Columbian presence of the crop in the Old World remains non-proven.

After European discovery of the New World, pineapples spread rapidly throughout the tropics. They were used to provision ships, and were probably a useful anti-scorbutic. The crowns can withstand considerable desiccation and yet provide viable planting material and this, too, must have favoured early establishment in the Old World. They spread by two routes: across the Pacific with the Spanish to the Philippines and with the Portuguese from Brazil to Madagascar and India.

4 Recent history

By far the most important cultivar of pineapple today is Cayenne (also known as Kew or Sarawak): of the ten main pineapple-producing areas of the world, only six use cultivars other than or additional to Cayenne, and the proportion of Cayenne grown is increasing as rapidly as planting material can be made available (Collins, 1960). Collins has traced the early history of this cultivar back to five plants sent by Perrottet from French Guiana to France sometime before 1840, when European interest in 'pineries' and pineapples was running high. The plants were multiplied vegetatively, spread to England and thence to Australia, Florida, Jamaica and Hawaii. Pineapple agriculture is therefore like Far East rubber and New World coffee in having a very narrow genetic base, although mutations have built up during numerous asexual generations and Cayenne now consists of a number of heterozygous clones. Because Perrottet used the name 'mai-pouri' for his pineapple, Collins (1960) suggested that it came originally from the Maipure Indians of the upper Orinoco. However, Py and Tisseau (1965) pointed out that the word 'mai-pouri' also means tapir and may simply have been used to designate a particularly large type of pineapple.

Most of the improvement of the pineapple to date has taken place by simple selection of mutant clones within cultivars and by hybridization between cultivars followed by selection of the highly heterozygous progeny. Most breeding programmes start with Cayenne and are carried out to improve flesh colour, vitamin C content, uniformity of ripening, disease resistance, etc. One of the most serious diseases of pineapple is mealy bug wilt. Cayenne is highly susceptible but some other cultivars and the wild species are resistant.

One cultivar of pineapple, Cabezona, is a natural triploid. Tetraploid clones have been produced artificially, but had smaller fruits than the diploids and lower sugar content. Both triploid and tetraploid pineapples remain self-incompatible, hence seedless.

5 Prospects

The wild species of *Ananas* and *Pseudananas* are now being used in breeding programmes in Hawaii. *Ananas ananassoides* contributes a distinctive flavour, disease resistance and ability to grow in cool, moist conditions. *Ananas bracteatus* is also a source of disease resistance. Pineapple breeding is, however, a slow process. An F_1 plant takes four years to produce fruit and, in the case of an interspecific cross, four backcross generations will probably be necessary to produce a plant suitable for commercial use; thus a new cultivar may be developed 25 years after the initial cross and it then has to be multiplied and distributed. Advances in methods of vegetative propagation enable about 25 daughter plants to be obtained from a single parent but, with a planting density of 40,000 plants/ha (as in Hawaii), a change to a new variety can come about only gradually. This makes the pineapple potentially particularly vulnerable to pest and disease epidemics.

The use of the wild species in breeding programmes seems likely to continue, at least on an experimental basis. It would also seem worthwhile to collect and conserve pineapple clones from the less accessible parts of South America, particularly the upper Amazon and upper Orinoco, where unusual types may have developed. Such collections could be used as necessary to broaden the genetic base of the commercial cultivars and would be essential to fill in some of the gaps in our knowledge of the evolution of the crop.

6 References

Brewbaker, J. L. and Gorrez, D. D. (1967). Genetics of self-incompatibility in the monocot genera, *Ananas* (pineapple) and *Gasteria*. *Amer. J. Bot.*, **54**, 611–16.

Collins, J. L. (1960). *The Pineapple*. London.

Merrill, E. D. (1954). The botany of Cook's voyages. *Chron. Bot.*, 14, 161–384.

Patiño, V. M. (1963). *Plantas cultivadas y animales domesticos en America equinoccial.* I, *Frutales*. Cali, Colombia.

Py, C. and Tisseau, M. A. (1965). *L'Ananas*. Paris.

Sauer, C. O. (1959). Age and area of American cultivated plants. *Actas 33 Congr. Internl Americanistas*, 1, 215–29.

Tea

Camellia sinensis (Camelliaceae)

T. Visser
Institute for Horticultural Plant Breeding
Wageningen The Netherlands

1 Introduction

The tea plant in its natural condition is an evergreen shrub or small tree. When cultivated it is a bush kept low by regular plucking and occasional pruning. The young top shoots – 'two leaves and a bud' – normally form the material from which the tea is manufactured.

Southeast Asia, in particular China and Japan, is the original centre of tea cultivation. Commercial tea plantations were established in India, Sri Lanka (Ceylon) and Indonesia during the nineteenth century and, subsequently, in many other parts of the world. In 1972 the total area was about $1 \cdot 4$ M ha producing some $1 \cdot 4$ Mt of made tea. The crop is now grown under widely varying soil and climatic conditions. A relatively high soil acidity ($pH < 5$) is a specific requirement. Geographically, the crop is widely distributed through the tropics at higher elevations and in the subtropics from 43°N (Caucasus) to 28°S (Argentina).

2 Cytotaxonomic background

The genus *Camellia* includes some 82 species (Sealey, 1958) of which tea (*Camellia sinensis*, formerly *Thea sinensis*) is the most important economically. The plant is diploid ($2n = 2x = 30$) and a few triploids and tetraploids have been found or induced.

For long, only two distinct taxa were recognized. The small-leaved Chinese shrub, *C. sinensis* var. *sinensis* and the large-leaved, tree-like Assam tea plant, *C. sinensis* var. *assamica*, which was discovered in Assam during the 1830s. More recently, extensive studies at the Tocklai Tea Research Institute have shown that more than two types should be distinguished. According to Wight (1962), the China and Assam types deserve specific rank (as *C. sinensis* and *C. assamica*). A third form referred to as the Cambodia type by Kingdon-Ward (1950) or as the Southern form by Roberts *et al.* (1958), has been treated as *C. assamica* ssp. *lasiocalyx*. A fourth form, Wilson's *Camellia*, was specifically ranked as *C. irrawadiensis* by Barua (1965). Yet another kind, called *C. taliensis* by Sealy (1948) and Forest *Camellia* at Tocklai, is a hybrid, *assamica × irrawadiensis* (Roberts *et al.*, 1958).

While *C. taliensis* will probably make acceptable tea, *irrawadiensis* itself makes a product visually like tea but spurious as an infusion. Hybrids between the latter and *assamica* do make tea, even if not of the best Assam quality. It is likely that certain valuable characters of the Darjeeling tea are derived from *irrawadiensis*. *Camellia assamica* subsp. *lasiocalyx* is not generally cultivated but it is of value for tea breeding. Several outstanding clones with a strong *lasiocalyx* element have been found to produce excellent teas in Sri Lanka though the parents from which these clones originated lacked typical Assam flavour. With a view to increasing cold hardiness, recent attempts to cross cultivated tea with *C. japonica* have met with some success.

3 Early history

Camellia has its principal area of distribution in the highlands of southeast Asia, ranging from Nepal northeastwards to Formosa and Japan. *Camellia sinensis* occupies a fan-shaped area between the Brahmaputra and the Yangtze. The three important types of tea (China, Assam and Cambodia) have distributions which bound this area but with overlap between them. They presumably followed three principal routes of dispersal from a centre near the sources of the Irrawady. The indications are that this was a secondary centre, the crop having originated further north, perhaps in the lower Tibetan mountains (Wellensiek, 1937) or even as far north as central Asia (Kingdon-Ward, 1950). It is possible that China and Assam teas had separate origins but Wellensiek (1937) supposed that the small-leaved China tea was the older type from which the large-leaved Assam forms originated, with southern Assam as a secondary centre of origin.

These theories are necessarily rather speculative. The earliest historical records show only that tea was grown in south-eastern China more than 2,000 years ago but the crop is likely to be very much older than this. Whether truly wild teas still exist is not known; probably they do not, though remnants of abandoned village cultivations in Assam and Indochina often give the appearance of being wild. If wild tea does exist, it is likely to have been extensively introgressed by cultivated forms.

4 Recent history

Tea agriculture developed from peasant cultivation to a plantation industry during the nineteenth century. This took place first in India and was made possible by the end of the tea monopoly of the East India Company in 1835. Seed of the so-called wild tea of Assam was mainly used, except in the Darjeeling area where a China type was planted. In Sri Lanka an extensive coffee culture, destroyed by *Hemileia vastatrix* during the 1870s, was rapidly replaced by tea, partly of the Assam–China hybrid and partly of the China type. In Java, commercial tea estates were established from 1878 onwards with imported Assam seed. In the same year planting started in Malawi, this being the first outside Asia. Since then, the crop has been planted in many other countries such as Kenya, Congo, Mozambique, Russia, Turkey and Argentina.

Breeding methodology is dominated jointly by the highly outbred nature of the crop and its potential for vegetative propagation. The first plantations were established with any seed available. Most of the present acreage originates from seed from three sources: (*a*) from selected bushes in tea-producing fields; (*b*) from unselected, unplucked bushes grown for the purpose, usually in producing fields; (*c*) from mass-selected parent bushes in separate seed gardens. Method (*a*) was probably disgenic and (*b*) neutral as to improvement. Method (*c*) gave rise to many 'jat' names which distinguished sources rather than kinds of tea and such selection as there was was probably ineffective.

During the 1930s the tea institutes of the traditional producers initiated breeding programmes, led in Java by Wellensiek (preceded by Cohen Stuart during the 1920s), in Assam by Wight and in Sri Lanka by Tubbs. The first aim of these programmes was systematic selection of mother-bushes on the basis of established criteria (Wellensiek, 1938). Visual appraisal of bush and shoot traits was followed by screening the selections for yield and quality.

The relative importance assigned to the development of clones for actual production and for use as parents of seed-propagated populations varied between programmes. Clonal selection began in Sri Lanka and Java by 1940 but it is only since the late war that any substantial areas of newly bred materials, whether clones or seedlings, have been established in the field. In addition to the institute programmes, estates in both India and Sri Lanka engaged in clonal selection. Other countries such as Russia, Kenya and Japan follow similar methods with varying relative importance attached to clonal and seed propagation.

An initial preference in several countries for seedlings as against clones was partly due to custom, partly to prejudice. Planting seedlings was an established procedure, while efficient methods of vegetative propagation still had to be developed. Now that large-scale multiplication by cuttings has become a routine estate practice, clones have been widely adopted in commercial plantings. So far, the best seedlings obtained in India and Russia, after some 30 years of breeding, appear to yield about 50 per cent more than unselected seedlings. The best clones in Sri Lanka produce, however, twice as much as seedlings from selected clones.

Clone selection is, indeed, the logical *first* step towards improvement. Given the millions of bushes available and the wide variation between them, the finding of outstanding mother-bushes is a matter of applying the proper selection techniques. After propagation and testing, selected clones can be released straight away for experimental estate planting. This is also the proper stage at which to use the best clones as parents. However, the same heterogeneity and heterozygosity which makes clonal selection feasible is also the reason that one cannot hope to make comparable progress in one sexual generation, perhaps not even in several. And clones also have the advantage of speed (Visser, 1969).

5 Prospects

There is now a general emphasis on the selection and planting of clones but, as yet, only a relatively small proportion of the total seedling acreage has been replaced. The consequential long-term trend must be towards diminished genetic variability.

In countries in which the crop originated from limited amounts of imported seed of similar 'jat', variability has possibly been more restricted from the outset than elsewhere. Nevertheless, the great acreage of seedling tea still in existence constitutes a large source of variability. This is however no longer the sole basis for clonal selection. Theory and practice agree in suggesting that the selection rate among seedlings from highly selected parents is much higher than from random seedling populations. In the latter, one selection per 20–40 thousand bushes has been mentioned whereas, in the former, about one selection per thousand seedlings (according to an Indian estimate) may be expected. Future cycles of clonal and sexual selection are therefore likely to involve progressive narrowing of the genetic base.

More detailed information may be forthcoming in the next decade on the inheritance of yield and quality components from the programmes of the several tea institutes. These studies will no doubt have a bearing on selection procedures. Traditional mass selection is based on assumed relations between, for example, the vigour of seedlings in the nursery and their subsequent yield or between bush and shoot traits and yield of mature bushes. The evidence, however, is that the correlation between the performance of nursery seedlings or mature bushes and the performance of their vegetative offspring is weak (Visser, 1969). This is not unexpected but it does not necessarily imply that the methods used are as ineffective as Green (1971) in Kenya presumes. In fact, his trials, when taken together, rather confirm the feasibility of mass-selection (Visser, 1975).

In the long term, as shown above, there is likely to be a substantial loss of genetic variability which will need to be countered by deliberate conservation measures. At present, breeding has hardly touched the variability available and there remain yet unexploited possibilities in the use of polyploids and interspecific hybrids.

6 References

Barua, P. K. (1965). Classification of the tea plant species hybrids. *Two and a Bud, Tocklai Exp. Sta.*, **12**, 13–27.

Green, M. J. (1971). An evaluation of some criteria in selecting large-yielding tea clones. *J. agric. Sci. Camb.*, **76**, 143–56.

Kingdon-Ward, F. (1950). Does wild tea exist? *Nature, Lond.*, **165**, 297–8.

Roberts, E. A. H., Wight, W. and Wood, D. J. (1958). Paper chromatography as an aid to the taxonomy of Thea Camellias. *New Phytol.*, **57**, 211–25.

Sealy, J. (1948). *Camellia taliensis. Curtis bot. Mag.*, **164**, 9684.

Sealy, J. (1958). *A revision of the genus Camellia.* Royal Horticultural Society, London.

Visser, T. (1969). Tea. In F. P. Ferwerda and F. Wit (eds), *Outlines of perennial crop breeding in the tropics*. Wageningen, 459–93.

Visser, T. (1975). Notes on the selection of tea clones. (In the press).

Wellensiek, S. J. (1937). [Travel impressions on tea selection and tea culture]. *Arch. Thee Cult.*, **11**, 201–59. (Dutch, English summary.)

Wellensiek, S. J. (1938). [Researches on quantitative tea selection IV. Mother tree selection]. *Arch. Thee Cult.*, **12**, 63–101. (Dutch, English summary.)

Wight, W. (1962). Tea classification revised. *Curr. Sci.*, **31**, 289–99.

Papaya

Carica papaya (Caricaceae)

W. B. Storey
University of California
Riverside California USA

1 Introduction

The papaya is a small, sparingly branching, soft wooded tree native in tropical America. It is cultivated throughout the tropics both for its edible, melon-like fruit and for its milky latex which contains the proteolytic enzyme, papain. Almost universally, the fruit is eaten fresh. Probably the State of Hawaii, USA, is the only place in the world where it is produced in great quantity for export as well as for local consumption. In 1973, Hawaii exported 7·9 kt of fruit, mostly to the western United States and Japan. The latex, which is used for tenderizing meat and clearing beer and in exfoliative cytology for detecting stomach and intestinal cancers, is produced mainly by several nations in tropical Africa and by Sri Lanka (Ceylon) (Becker, 1958).

2 Cytotaxonomic background

The papaya (*Carica papaya*) is a member of the Caricaceae, a small dicotyledonous family consisting of four genera and 31 species. The genera and the number of included species are: *Carica*, 22; *Jacaratia*, 6; *Jarilla*, 1; *Cylicomorpha*, 2. The first three are indigenous in tropical America, the last in equatorial Africa (Badillo, 1971). To date, chromosome numbers have been counted in only eight species of *Carica*, none in the other three genera. All species counted have $2n = 2x = 18$.

Species having edible fruits are found only in *Carica*. Besides *C. papaya*, they are *C. chilensis, C. goudotiana, C. monoica*, and *C. pubescens*. The fruit is eaten cooked as a vegetable or candied by cooking in sugar syrup to make *dulce* rather than eaten raw. In Peru, the leaves of *C. monoica* are cooked and eaten for greens.

Two putative natural interspecific hybrids are known: *C. monoica* × *pubescens* and *C.* × *heilbornii* of which pistillate trees only are known. Four pheno-typic variants of *C.* × *heilbornii* are known, all of which bear large, seedless, parthenocarpic, edible fruits. They are cultivated by Andean people who maintain them clonally by cuttings. Badillo (1971) classified them as nm. *heilbornii*, nm. *pentagona*, nm. *chrysopetala* and nm. *fructifragrans*. Horovitz and Jiminez (1967) reported producing hybrid plants virtually identical to those of *C.* × *heilbornii* by making reciprocal crosses between *C. pubescens* and *C. stipulata*. Cytological evidence does not support the sometimes stated supposition that *C. papaya* arose as an interspecific hybrid of unknown parentage.

Except for three species of *Carica*, all members of Caricaceae are dioecious. The three exceptions are *C. monoica, C. pubescens* and *C. papaya*. They have sexually ambivalent forms which go through 'sex-reversals' in response to climatic and/or photoperiodic changes during the year. *Carica monoica* is strictly monoecious but at certain times of the year may lack pistillate flowers. Trees of *C. pubescens* exist in three basic sex forms: pistillate, staminate and andro-monoecious. The pistillate and staminate trees are unresponsive to seasonal climatic changes. The andro-monoecious trees are sexually ambivalent, producing staminate, perfect and pistillate flowers in varying proportions at different times of the year.

Carica papaya exists in the same three basic sex forms as *C. pubescens*. The pistillate tree is stable. Staminate and andromonoecious trees may be: (1) phenotypically stable; or (2) phenotypically ambivalent, going through seasonal sex-reversals, during which they produce varying proportions of staminate, perfect and pistillate flowers.

Storey (1958) classified papaya trees in 31 heritable phenotypes on the bases of peduncle length and rami-fication and of seasonal sexual responses. Fifteen of these are variations among staminate trees, and fifteen are variations among andromonoecious trees. The remaining phenotype is the pistillate tree.

Matings among the basic sex forms yield segregat-ing progenies as shown below in which **P** = pistillate, **S** = staminate and **A** = andromonoecious.

Mating	P	S	A
P × **S**	1	1	–
P × **A**	1	–	1
S × **S**	1	2	–
A × **A**	1	–	2
A × **S**	1	1	1
S × **A**	1	1	1

Clearly, staminate and andromonoecious plants   21

are heterozygotes and pistillate ones are recessive homozygotes. A zygotic lethal factor eliminates dominant homozygotes. Although the main features are clear enough, there are three different hypotheses on the genetics of sex determination in the species which are summarized below.

Hofmeyr's (1967) hypothesis involves genic balance. The symbols M_1 and M_2 represent inert or inactivated regions of slightly different lengths on sex chromosomes which are inert or from which vital genes are missing. This accounts for the zygotic lethality of M_1M_1, M_2M_2 and M_1M_2 genotypes. The homologous region, m, is normal. The viable genotypes are: M_1m, staminate; M_2m, andromonoecious; $m\,m$, pistillate.

The greater concentration of genes for femaleness is on the sex chromosomes, the greater concentration for maleness on the autosomes. Thus, the genotype $m\,m$ is pistillate and its homozygosity confers phenotypic stability. Since M_1 is the longer of the inert regions, it is expressed phenotypically as staminate because of the greater influence of the autosomal factors. The shorter M_2 region is less influenced by autosomal genes so the M_2m genotype is expressed phenotypically as andromonoecious. The heterozygosity of M_1m and M_2m renders them susceptible to alteration in phenotypic expression by external influences.

The hypothesis of Horovitz and Jiminez (1967) proposes a reversionary process for sex determination and expression. Its basic assumption is that dioecism is the primitive state in Caricaceae and sex determination is of the classical XX-XY type with heterogametic male and YY lethal to the zygote. At some point in time, a sexually ambivalent form occurred in the genus from which the three present-day exceptions to dioecism (*C. monoica*, *C. pubescens* and *C. papaya*) arose. In *C. papaya*, the ambisexual mechanism built up on the Y chromosomes, giving rise to a modified homologue, the Y_2 chromosome which, in the heterogametic genotype XY_2, is expressed as the sexually ambivalent andromonoecious form. This occurred without any alteration of the X chromosome, which explains the stability of the pistillate form. This hypothesis holds, therefore, that andromonoecism and polygamy followed the evolution of an XX-XY system and are of fairly recent origin.

Storey (1953, 1967, 1969a) hypothesized progressive evolution of dioecism in the family. An unknown ancestor is assumed to have had perfect flowers which evolved as follows. The hypothetical flower must have been much like the perfect, perigynous flowers of *C. pubescens* and *C. papaya*, having a superior, 5-carpelled pistil, a gamopetalous corolla terminating in a 5-parted limb and 10 (2×5) stamens inserted in the throat of the corolla tube. This form is known as *elongata* in *C. papaya*. The staminate flower was derived in the classical way, i.e. by phylogenetic elimination of the functional pistil. The pistillate flower was derived in a succession of extraordinary changes which included: carpellody and connation of the upper cycle of stamens accompanied by abortion of the original carpels to constitute the pistil of a new 5-stamened (*pentandria*) flower; carpellody and connation of the stamens of the *pentandria* flower accompanied by abortion of the pistil to become the pistil of a new pistillate flower; almost complete freeing of the petals; and change of insertion of the ovary from superior to partially inferior (pleurogyny). The pistillate flower, therefore, was derived as a morphological anomaly in that its pistil consists of a whorl of stamens (the lower set) transmuted into carpels at the expense of the original *elongata* carpels and the upper set of stamens.

Sexual differentiation in the form of dioecism followed the derivation of unisexual flowers. Since dioecism seems to be the evolutionary norm in Caricaceae, it is possible that ambisexual forms owe their continued existence to human selection.

In Storey's hypothesis, herein revised, the symbol (SA) represents the sum of the factors involved in transmuting the ancestral androecium into the present-day gynoecium; (sa) represents normal androecium development; (SG) represents the factor or factors responsible for suppression of the gynoecium in the staminate flower; (sg) permits the (SA) factors to function ontogenetically in developing the replacement gynoecium. The symbol l represents the recessive sex-linked zygotic lethal factor that enforces heterozygosity on the staminate and andromonoecious plants; and C represents the factor that prevents crossing-over between the sex-determining factors and the lethal factor, accounting for the non-existence of pistillate plants carrying the l factor. As in the other two hypotheses, heterozygosity permits ambisexuality. The sex-determining genotypes are expressed as follows:

staminate and
andromonoecious: $(sa)\ l\ C\ (SG)/(SA)++(sg)$
pistillate: $(SA)++(sg)/(SA)++(sg)$

For convenience, the sex homologues may be represented as: M^h, andromonoecious or hermaphrodite; M^s, androecious, staminate or male; m, gynoecious, pistillate or female (see Fig. 8.1).

Length and ramification of the inflorescences are secondary sex characters. The alleles which give them expression are on the sex chromosomes but linkage with the sex-determining factors is not absolute. In effect, all staminate and andromonoecious forms are identical. Therefore, the classification of trees as the one form or the other depends largely upon the nature of the inflorescence. In this regard, Storey (1958), and Hamilton and Izuno (1967) agree.

3 Early history (Fig. 9.1)

Botanical evidence indicates that the papaya originated in the lowlands of eastern Central America. There seems to be no record of its history prior to discovery of the New World. However, it must have been cultivated for a very long time by early civilizations for it was known from Mexico to Panama and occurred in a wide range of diverse types. The earliest record (1526) is by G. H. de Oviedo in his book on *The general and natural history of the Indies* (1535). It spread rapidly along early tropical trade routes and, by 1800, was being grown throughout the tropical world, even in some of the then most recently discovered remote islands of the South Pacific.

Evidence from studies of floral anatomy suggests that the prototype of *C. papaya* was perfect-flowered (Storey 1967) but probably had small fruit (50–100 g) like those of the dioecious form of *C. papaya* now naturalized throughout the Caribbean region and known variously by the synonymous names, *C. cubensis, C. jamaicensis, C. portorricensis* and others. Present-day dioecism evolved from the prototype with

the derivation of unisexual flowers (Storey 1967). In view of the evolutionary trend towards dioecism in the family, it seems probable that the present-day forms of *C. papaya* owe their existence to continuous selection and planting by man.

4 Recent history

Papaya production in Hawaii is based on the cultivar Solo, with pyriform fruit weighing 350–500 g produced by the hermaphrodite trees of an inbred gynodioecious strain. Parental lines are grown in isolation and only seeds from hermaphrodite trees are planted. These yield hermaphrodites and females in the ratio of 2:1. If seedlings were planted singly at the rate of 1,080 per hectare, there would be about 720 hermaphrodites and 360 females. Since there is no known means of separating the sexes in the early seedling stage, seedlings are planted in threes of which two are later eliminated. With each place reduced to a single tree about 6 months after planting when the seedlings begin to flower and the sex of each can be ascertained, the final stand is 1,020 hermaphrodites and 40 females.

The most important cultivar in South Africa is Hortus Gold, of which the fruit is borne by female plants of a dioecious strain. Seedlings of this strain segregate 1 female:1 male so there would be 540 of each in a one-hectare planting. Since 10–16 per cent of males in a dioecious planting is sufficient for pollination, seedlings are planted in threes and later reduced to a final stand of 945 females:135 males, or 12·5 per

Fig. 8.1 Breeding system of papaya, *Carica papaya*.

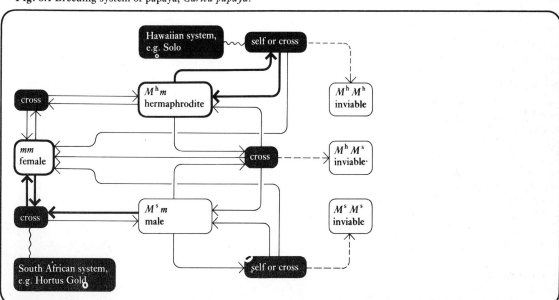

cent males.

Elsewhere, production is not so systematized and such cultivars as there are rapidly lose identity by outcrossing. Isolation of seed sources and hand pollination followed by bagging are the only certain means of keeping strains pure. Vegetative propagation, though possible, is not economically feasible.

Hamilton's (1954) study of inbreeding and crossbreeding of two Solo strains showed that inbreeding (surprisingly) has no ill effects on plant vigour.

Polyploids have been produced by treating seeds and seedlings with colchicine but none has been found useful either for fruit production or for breeding.

Breeding for fruit and latex production is seriously hampered in the West Indies and tropical continental America by two (possibly more) virus diseases and, less seriously, by several species of fruit flies.

5 Prospects

Presently, papaya breeding programmes are being carried on in Australia, South Africa, India, Trinidad and Tobago, Mexico and almost every country in Central and South America, as well as the United States of America. Investigations on reaction to papaya viruses and on proteolytic enzymatic activity of available species of *Carica* and *Jacaratia* are being conducted in Venezuela.

Some of the problems that need to be solved to enhance the probability of increased fruit and latex production are development of virus-resistant cultivars; elimination of ambisexual andromonoecious forms that tend to become female sterile at certain times of the year or show a tendency toward stamen carpellody; development of homozygous andromonoecious forms by possible elimination of the zygotic lethal factor or by *in vitro* culture of embryos; inducing a sex-linked vegetative character for eliminating unwanted sex forms in the early seedling stage; early and low bearing; selection for inflorescences of moderate length (about $7 \cdot 5 - 10 \cdot 0$ cm) bearing single fruits, to circumvent crowding with resulting misshapen fruit; selection for an ovarian cavity circular in transverse section and with an easily separable placenta; breeding of dioecious cultivars for places in which andromonoecious types are excessively sexually ambivalent; breeding to expand regions of production by hybridizing with species that are more cold tolerant (e.g. *C. pubescens*); and breeding for increased latex yield.

6 References

Badillo, V. M. (1971). *Monografia de la familia Caricaceae.* Maracay, Venezuela.

Becker, S. (1958). The production of papain – an agricultural industry for tropical America. *Econ. Bot.*, **12**, 62–79.

Hamilton, R. A. (1954). A quantitative study of growth and fruiting in inbred and crossbred progenies from two Solo papaya strains. *Haw. agric. Exp. Sta. Bull.* **38**, pp. 16.

Hamilton, R. A. and Izuno, T. (1967). A revised concept of sex inheritance in *Carica papaya. Agron. trop.*, **17**, 401–2.

Hofmeyr, J. D. J. (1967). Some genetic and breeding aspects of *Carica papaya. Agron. trop.*, **17**, 345–51.

Horovitz, S. and Jiminez, H. (1967). Cruzamientos interespecificos e intergenericos en Caricaceas y sus implicaciones fitotecnias. *Agron. trop.*, **17**, 353–9.

Storey, W. B. (1953). Genetics of the papaya. *J. Hered.*, **44**, 70–8.

Storey, W. B. (1958). Modifications of sex expression in papaya. *Hort. Adv.*, **2**, 49–60.

Storey, W. B. (1967). Theory of the derivations of the unisexual flowers of Caricaceae. *Agron. trop.*, **17**, 273–321.

Storey, W. B. (1969a). Pistillate papaya flower: a morphological anomaly. *Science*, **163**, 401–5.

Storey, W. B. (1969b). Papaya (*Carica papaya*). In F. P. Ferwerda and F. Wit (eds), *Outlines of perennial crop breeding in the tropics.* Wageningen, The Netherlands, 389–407.

Sugar beet

Beta vulgaris (Chenopodiaceae)

G. K. G. Campbell
Speen Aylesbury England *formerly*
Plant Breeding Institute
Cambridge England

1 Introduction

Crystalline sugar was a scarce luxury in the western world before the seventeenth century. Originally, all sugar came from sugar cane grown in the tropics but, at the present time, beet sugar accounts for nearly half the total world production of the refined product. Production of sugar beet is mainly in temperate climates, especially Europe, the USSR and North America, but it is also grown in the sub-tropics under irrigation. The sugar in beet is stored as sucrose in the enlarged tap-root and hypocotyl. The plants are topped at harvest, and the tops provide a valuable source of feed for livestock in areas where mixed farming is practised. Sugar is extracted from the beet in factories, using a process similar to that for cane sugar. After extracting the sugar, the residues are processed to form dried pulp which is an additional source of feed for livestock.

2 Cytotaxonomic background

Beta is an Old World genus virtually confined to Europe. It has been divided into three subgeneric taxa: sections Beta (Vulgares), Patellares, and Corallinae. The section Beta has the widest distribution and may well be the most primitive. The Corallinae occupy the eastern part of the range while the Patellares are confined to the southwest, chiefly in the Atlantic Islands and the western Mediterranean. Cultivated beets are all derived from the section Beta, although other sections contain species of limited culinary value. Useful characters, such as resistance to diseases, occur throughout the genus and attempts have been made to incorporate them into sugar beet. Interspecific hybrids involving members of different sections are more difficult to make than those derived from the same section and are less easily exploited. Systems of classification have been proposed which

vary in details (Coons, 1954; Krasochkin, 1959). Broadly, however, the section Beta comprises the following:

> *Beta vulgaris* – ssp. *vulgaris* contains many cultivars, including sugar beet, beetroot, mangolds and fodder beet; ssp. *cicla* embraces the chards; ssp. *maritima*, sea beet, is thought to be ancestral to most, if not all, cultivars.
>
> *B. atriplicifolia* – ⎫
> *B. macrocarpa* – ⎬ of little agricultural significance
> *B. patula* – ⎭

These taxa have been variously accorded specific or infraspecific rank by different authors.

Diploid chromosome numbers within the genus are $2n = 2x = 18$ (e.g. *B. vulgaris*), $2n = 4x = 36$ (e.g. *B. corolliflora*) and $2n = 6x = 54$ (e.g. *B. trigyna*). Spontaneous chromosome doubling results in the occasional occurrence of autopolyploid individuals.

3 Early history

The use of beet probably dates from prehistoric times when the leaves were almost certainly used as potherbs. Aristotle mentioned red chards and Theophrastus mentioned light green and dark green chards used in the fourth century B.C. The Romans used beet (probably *B. maritima* from the sea shore) as feed for animals and man. Called *Beta* by the Romans, it was taken from Italy to northern Europe by the barbarian invaders. By the sixteenth century it was widely used for feeding animals, particularly during the winter. Red beet or beetroot featured in Roman recipes of the second and third centuries and was recorded in English recipes of the fourteenth century. In Germany it was first described in 1557 when it was referred to as Roman beet. The crop was introduced into the USA in 1800 where it became known as garden beet.

In 1747 Marggraf noted the presence of sugar in sap from the roots of fodder beet. His pupil, Achard, laid the foundations of the improvement of the crop plant and the development of an extraction process. Achard received a subsidy from the King of Prussia and secured a ten-year monopoly for the manufacture of beet sugar. The first factory was built at Kunern, Silesia, in 1801. Silesian sweet fodder beet, in which sugar contents up to 6·2 per cent were recorded, was introduced into France in 1775. In 1811, Napoleon published the first of a series of decrees requiring beet to be grown and schools for its study to be established. This action was prompted by disruption to the supply of cane sugar from the West Indies to France caused by the British naval blockade of continental ports. Later, the import of sugar to France was restricted in order to prevent the beet industry from succumbing

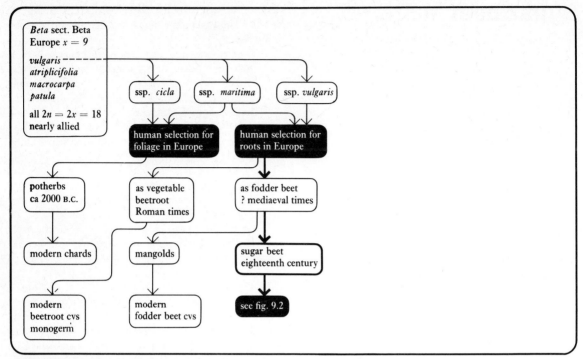

Fig. 9.1 Evolution of cultivated *Beta vulgaris*.

to competition from cane sugar produced more cheaply in the colonies.

4 Recent history (Fig. 9.2)

Conscious selection started during the nineteenth century, primarily in Germany and France, where rapid advances were made in breeding for greater sugar percentages, facilitated by the introduction of polarimetric methods of estimation. This trend was encouraged in Germany (and, later, in other countries) by the imposition of excise duty which was levied on the weight of beet. Government support proved to be essential for the development of beet industries, both in Europe and in the USA, where the first successful factory was established in 1879, in California.

Sugar beet is outbreeding and largely anemophilous (Dark, 1971). Varieties grown in the first half of the present century were multilines based on 20–30 parental stocks. The parents were maintained by sib-mating, with mass selection at each generation for root size and sucrose content. Pollination in the final generation was random, with all plants producing pollen and seed; the commercial root crop consisted of 95–97 per cent interline hybrids.

Work on polyploids began in the late 1930s, and it soon became evident that, although tetraploids were,

if anything, slightly inferior to the parental diploids, some triploid combinations were very promising. Anisoploid varieties, produced from random pollination of parental mixtures of diploid and tetraploid lines, were marketed in some European countries after the late war and are still of some importance in Europe and the UK. They are now being replaced by monogerm varieties, many of which are triploid, and thus also derived from the work on polyploidy (Savitsky, 1962; Sedlmayr, 1964; Hornsey, 1975). Wholly triploid varieties are produced with the aid of cytoplasmic male sterility (Owen, 1945).

The monogerm character is of great importance because it allows the possibility of total mechanization of crop production. It was first exploited in the USSR and the USA but has now been adopted by breeders in most other countries. A monogerm plant bears flowers singly at most nodes of the inflorescence whereas the normal form bears them in clusters. Fruits developed at a single node form an indehiscent aggregate and this is the 'seed' of agriculture. Each fruit is single-seeded so that the aggregate fruits of a multigerm variety are multi-seeded, whereas most of those from a monogerm are single-seeded. Multigerm varieties must be singled by hand in the field after emergence but this process may be avoided if monogerm varieties are planted with a precision seed-drill. The monogerm character is an

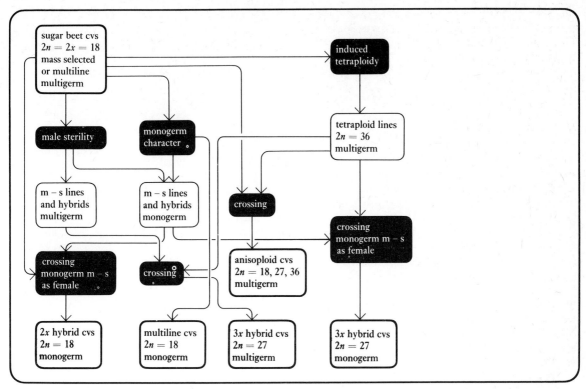

Fig. 9.2 Breeding systems in sugar beet.

inflorescence character and is determined by the genotype of the seed-parent. Hence, it is possible to use a multigerm pollinator on a male-sterile monogerm in commercial seed production (Savitsky, 1954).

Diseases of sugar beet and their control have been described by Hull (1960). In parts of the USA the curly-top virus, which is transmitted by the leaf-hopper (*Eutettix tenellus*), made beet growing unprofitable until resistant varieties were produced. Yellowing viruses transmitted by aphids occur more widely but are generally less damaging. Selection for tolerance to yellowing has met with some success and work on aphid resistance is also in progress. Varieties resistant to the leaf-spot caused by *Cercospora beticola* have been produced in the USA and parts of Europe, while resistance to root infecting fungi such as *Aphanomyces cochlioides* and *Sclerotium rolfsii* has also been developed in countries where they are important. Breeding for resistance to the cyst nematode (*Heterodera schachtii*), which has caused serious losses in Europe and the USA since the middle of the nineteenth century, has proved difficult, but some progress has been made. These and other pests of sugar beet have been described by Jones and Dunning (1972).

The original cultivars of sugar beet were selected as biennials but, as they were grown further north, the increase in day length and the lower temperatures to which the young plants were exposed after emergence, tended to cause vernalization. Plants in which flowering is initiated prematurely are described as bolters. Roots of bolters are usually smaller and more lignified than roots of normal plants; the result is depressed sugar yield and the greater lignification causes greater wear of machinery, leading to increased costs of sugar production.

Work in England showed that it was possible to select for resistance to bolting by applying artificial conditions of vernalization to plants at the four-leaf stage (Campbell, 1953). This work provided the stimulus for intensive screening of breeding material for commercial use. The bolting-resistant varieties that were produced enabled beet to be grown further north than had hitherto been possible. It also meant that the crop could be drilled earlier, thus giving the potential for heavier yields by exploiting a longer growing season.

5 Prospects

Because of continually rising costs of production, current emphasis in breeding is on the production of improved varieties with the monogerm character. In sugar beet, the first monogerm varieties to be released for commercial cultivation did not yield as well as contemporary multigerm varieties. Moreover, the use of monogerm seed with precision drilling called for greatly improved field emergence to achieve regular stands, a problem that was aggravated by the cooler seed-beds consequent upon earlier drilling. So long as seed was multigerm and each cluster had several chances of producing a viable seedling, field emergence was not a problem calling for special attention. Regularity of stand is also an important factor in regularity of beet size and shape, which has become increasingly desirable to improve the efficiency of mechanical harvesting.

The main aims of current breeding programmes are therefore to produce monogerm varieties with improved field emergence and the potential for heavier yields of extractable sugar from roots that can be harvested efficiently by machine. The trend to earlier drilling in temperate climates calls for still greater resistance to bolting, while the increased cost of chemicals used for controlling insect pests (particularly aphids) calls for increased effort to produce varieties more resistant both to aphids and to the virus diseases they transmit. The breeder must also follow closely the development of improved herbicides for use with sugar beet to ensure their selectivity with new varieties.

Although there is little published evidence on the relative success of different types of sugar beet varieties, it seems probable that triploids derived some advantage from the incorporation of two genomes of the multigerm tetraploid parent with one of the monogerm male-sterile diploid. Consequently, with genetic improvements to monogerm diploids, any advantage that the triploid may have had in the past may well not apply in the future. The popularity of triploid varieties in Europe has not been shared in the USA, however, where monogerm varieties have been predominantly of the diploid-hybrid type.

Tetraploid, anisoploid and triploid varieties suffer from the disadvantage of producing a proportion of aneuploids. Even so, many plant breeders believe that, in the long term, no particular type of variety will prove to be inherently better than any other and triploid hybrids, diploid hybrids, diploid multilines and composite varieties are known to feature in current breeding programmes. In future, therefore, it appears that the success of a new variety is more likely to be related to the amount and type of effort that has gone into selecting its components, rather than to any inherent advantages of a particular varietal type.

6 References

Campbell, G. K. G. (1953). Selection of sugar beet for resistance to bolting. *Publ. Inst. internl Rech. better. XVI Winter Congr.*, pp. 5.

Coons, G. H. (1954). The wild species of *Beta*. *Proc. Amer. Soc. Sugar Beet. Technol.*, 8, 142–7.

Dark, S. O. S. (1971). Experiments on the cross-pollination of sugar beet in the field. *J. natnl Inst. agric. Bot.*, 12, 242–66.

Hornsey, K. G. (1975). The exploitation of polyploidy in sugar beet breeding. *J. agric. Sci. Camt.*, 84, 543–57.

Hull, R. (1960). *Sugar beet diseases. Min. Agric. tech. Bull.*, 142, pp. 53, HMSO, London (2nd edn).

Jones, F. G. W. a d Dunning, R. A. (1972). *Sugar beet pests. Min. Agric. tech. Bull.*, 162, pp. 62, HMSO, London (3rd edn).

Krasochkin, V. T. (1959). [Review of the genus *Beta*.] *Bull. appl. Bot. Genet. Pl. Breeding, Leningrad*, 32, 3–36.

Owen, F. V. (1945). Cytoplasmically inherited male sterility in sugar beets. *J. agric. Res.*, 71, 423–40.

Savitsky, V. F. (1954). Inheritance of the number of flowers in flower clusters of *Beta vulgaris*. *Proc. Amer. Soc. Sugar Beet Technol.*, 8, 3–15.

Savitsky, V. F. (1962). Sucrose and weight of root in tetraploid monogerm and multigerm sugar beet populations under different mating systems. *J. Amer. Soc. Sugar Beet Technol.*, 11, 676–711.

Sedlmayr, K. (1964). Monogerme Zuckerrüben, ihre Genetik, Züchtung und Bedeutung für den Zuckerrübenbau. *Züchter*, 34, 45–51.

Note: The editor's and author's best thanks are due to Dr M. H. Arnold, Dr G. J. Curtis and Dr K. G. Hornsey of the PBI, Cambridge, for helpful advice on this chapter.

Quinoa and relatives

Chenopodium spp. (Chenopodiaceae)

N. W. Simmonds
Scottish Plant Breeding Station
Edinburgh Scotland

1 Introduction

Of the three cultivated chenopods of the tropical American highlands, only one, quinoa, makes a significant contribution to local food production. Quinoa is an annual herb cultivated for grain which is roughly comparable with wheat in nutritional quality. The flour is made into bread, biscuits, porridge, etc., or fermented to provide chicha. Cañahua is a very minor grain (Gade, 1970) and the immature infructescences of huauzontle provide a vegetable which is usually eaten fried in batter. Exceptionally, the last is eaten in Mexico as a grain (Sauer, 1950). None of the three has ever been cultivated outside its native area to any significant extent and none has been accorded much scientific study. For general reviews of economic botany see Hunziker (1943, 1952) and Simmonds (1965).

2 Cytotaxonomic background

The three species are as follows:

Name	Vernacular	2n =	Distribution
Ch. quinoa	Quinoa	$4x = 36$	Andes
Ch. nuttalliae	Huauzontle	$4x = 36$	Mexico
Ch. pallidicaule	Cañahua	$2x = 18$	Andes

The first two are certainly closely allied and may be conspecific; one strain of huauzontle examined appeared to be no more than a very large, late-maturing form of quinoa (Simmonds, 1965). Unpigmented forms of quinoa bear a general resemblance to *Ch. album*. Quinoa is extremely variable in stature, rate of maturity, habit, plant and seed pigmentation and seed size; stature and rate of maturity are negatively correlated. Variability has been little studied. Segregation of such genes as are known is disomic, suggesting allopolyploidy (Simmonds, 1971). Virtually nothing is known of the variability of huauzontle and cañahua, both now rare and declining crops. Cañahua, a small bushy herb with leafy inflorescence branches, has quite a different facies from the other two. There is no information on the cytotaxonomy of putative wild relatives.

3 Early history

All three are native American crops developed by Indian agriculturalists in pre-Columbian times. Quinoa and huauzontle, at least, were important food plants at the time of the Conquest. There are some archaeological records but they provide no real evidence as to time of evolution. The earliest record seems to be of quinoa grain in an Argentine tomb, *ca* 2000 B.P. (Hunziker and Planchuelo, 1971) but the crop must have been domesticated long before then. The ancestors of all three must have been native American *Chenopodium* spp., yet unidentified. Hunziker (1943) suggested that *Ch. hircinum* might have been the wild progenitor of quinoa.

Comparison of quinoa with wild chenopods (e.g. *Ch. album*) suggests that human selection for large seeds (strictly, fruits), non-shattering infructescences and low seed dormancy was operative; there is a striking analogy here to the evolution of the grass cereals. Quinoa seeds are generally about twice as large in linear dimensions as those of *Ch. album*. Cañahua appears to be more primitive than the other two in having small seeds which shed freely at maturity and show some dormancy.

The similarity of quinoa and huauzontle in grain characters, and the strong probability that they are conspecific, suggests that the Mexican crop represents an early migrant quinoa population selected locally for a different end-use, namely vegetable instead of grain. There is no Mexican tradition of using chenopods for grain; there, *Amaranthus* was the local analogue (Sauer, 1950).

Quinoa grain contains variable amounts of bitter (?toxic) saponin material which is washed out before consumption. Samples low in saponin occur but are rare and, perhaps surprisingly, there is no evidence of selection against saponins (Gandarillas, 1967*a*).

The breeding system of quinoa (also, presumably, of huauzontle) is distinctive. Plants are gynomonoecious (flowers mixed hermaphrodite and female) and self-pollination predominates. Natural crossing has been estimated as 2–9 per cent and there is no evident inbreeding depression on selfing (Gandarillas, 1967*b*, *c*). However, one population that carries a cytoplasmic/genetic male sterility has been identified in Bolivia, so enhanced outbreeding is sometimes favoured (Sim-

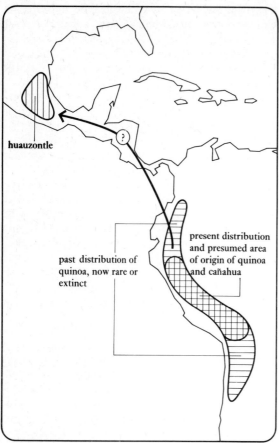

huauzontle

past distribution of quinoa, now rare or extinct

present distribution and presumed area of origin of quinoa and cañahua

Fig. 10.1 Distribution of the cultivated chenopods. (Simmonds, 1965; from *Economic Botany*, by permission).

than continued decline. Quinoa however might have a place if it could be shown to have any significant advantage over its potential competitors, the old world cereals, barley and oats; nutritionally, it is certainly quite attractive. No doubt there would be plenty of room for improvement by breeding. The evidence suggests that pure line cultivars (and/or multilines) would be feasible and the best first step in breeding.

6 References

Gade, D. W. (1970). Ethnobotany of cañahua (*Chenopodium pallidicaule*) rustic seed crop of the altiplano. *Econ. Bot.*, **24**, 55–61.

Gandarillas, H. (1967*a*). Distribucion geografica de quinuas sin saponina y granos grandes. *Sayaña*, **5**, 6–7.

Gandarillas, H. (1967*b*). Efecto de la autofecondacion sobre la quinua. *Sayaña*, **5**, 16–17.

Gandarillas, H. (1967*c*). Observaciones sobre la biologia reproductiva de la quinua. *Sayaña*, **5**, 26–9.

Hunziker, A. T. (1943). Las especies alimenticios de *Amaranthus* y *Chenopodium* cultivadas por los indios de América. *Rev. Argent. Agron.*, **10**, 297–354.

Hunziker, A. T. (1952). *Los pseudocereales de la agricultura indigena de América*. Univ. Nac. Cordoba, Buenos Aires, pp. 104.

Hunziker, A. T. and **Planchuelo, A. M.** (1971). Sobre un nuevo hallazgo de *Amaranthus caudatus* en tumbas indigenas de Argentina. *Kurtziana*, **6**, 63–7.

Sauer, J. D. (1950). The grain araranths: a survey of their history and classification. *Ann. Mo bot. Gdn.*, **37**, 561–632.

Simmonds, N. W. (1965). The grain chenopods of the tropical American highlands. *Econ. Bot.*, **19**, 223–35.

Simmonds, N. W. (1971). The breeding system of *Chenopodium quinoa*. I. *Heredity*, **27**, 73–82.

monds, 1971). Cañahua differs in having hermaphrodite, more or less cleistogamous, flowers and natural crossing is very rare (about 0·1 per cent in the glasshouse).

All three crops may be inferred to have originated in the areas to which they are still restricted and/or in which they are variable: quinoa and cañahua in the high central Andes, huauzontle in the plateau of Mexico. But the last could, conceivably, have originated by the northward migration of a quinoa population (see above and Fig. 10.1).

4 Recent history

The recent history of all three crops is one of decline. Quinoa alone retains any agricultural significance (Fig. 10.1) but very little breeding has been done.

5 Prospects

It is difficult to foresee any future for these crops other

Safflower

Carthamus tinctorius (Compositae)

P. F. Knowles
University of California Davis USA

1 Introduction

Safflower (*Carthamus tinctorius*) is an annual oilseed crop of some importance in India (600 kha), the United States and Mexico and of minor importance in Australia, Spain, Portugal and Turkey. Two oils are obtained from safflower, one poly-unsaturated, which is used in soft margarines, salad oils and surface coatings and the other mono-unsaturated, which is used primarily as a frying oil. The meal is used as an ingredient of livestock rations. Several countries of the Middle East grow very small amounts for the dried flowers, which serve as a substitute for saffron. For general reviews of the crop, see Beech (1969), Knowles (1955, 1958) and Weiss (1971).

2 Cytotaxonomic background (Hanelt, 1963; Ashri, 1973)

Cultivated safflower belongs to a group of closely related diploids ($2n = 2x = 24$) that extends from central Turkey, Lebanon and Israel in the west to northwestern India in the east. Two successful weedy species are *C. flavescens* (Turkey, Syria and Lebanon primarily) and *C. oxyacantha* (Iran and Iraq to northwestern India); the former is self-incompatible and the latter mixed self-compatible and incompatible. *Carthamus palaestinus* (desert areas from Israel to western Iraq) is self-compatible. *Carthamus gypsicolus* is found from the Caspian Sea to the Aral Sea and *C. curdicus* is reported only from western Iran. Flowers of wild species are various shades of yellow (occasionally white) and all have yellow pollen.

A large group of closely related species with $2n = 2x = 20$ occupies the Middle East from Libya to Iran. All taxa in that group have flowers that range from blue through to white and all have white pollen.

C. nitidus ($2n = 2x = 24$) will cross with cultivated safflower but the F_1 shows low meiotic pairing and is sterile. In appearance, *C. nitidus* resembles *C. leucocaulos* ($2n = 2x = 20$); both are self-compatible.

C. divaricatus, a self-incompatible diploid with $x = 11$, is found only in Libya. It resembles $x = 10$ species in general appearance but has flower colours that range from white through yellow to blue and it has yellow pollen.

There are three polyploids, all self-compatible. *Carthamus lanatus* ($2n = 4x = 44$), is presumed to be a product of a cross of $(x = 10) \times (x = 12)$ type but it has not been possible to synthesize the species from crosses of presumed parents. *Carthamus turkestanicus* ($2n = 64$) extends from Turkey to the easternmost range of the genus and into Ethiopia. *Carthamus baeticus* (also $2n = 64$) extends from Turkey westwards to Spain. Polyploidy has extended the boundaries of the genus in Europe, Africa and central and eastern Asia. Furthermore, the most successful emigrants have been the polyploids, with *C. lanatus* and/or *C. baeticus* reported to occur in California, Chile and Australia.

3 Early history

It is believed that the cultivated species had its origins in the Near East, probably from an ancestral type that gave rise to all the closely related $x = 12$ diploids. Cultivated safflower hybridizes readily with *C. flavescens*, *C. oxyacantha* and *C. palaestinus* to give fertile F_1 and F_2. Single major genes govern each of the following differences between the domesticated species and *C. flavescens*: short *vs* long rosette stage of growth; entire *vs* lobed margins; non-shedding *vs* shedding of achenes; absence of (or reduced) *vs* abundant pappus; white *vs* pigmented achenes; and green *vs* purple midveins of cotyledonary leaves (Imrie and Knowles, 1970).

The oldest evidence of safflower as a cultivated plant comes from the archeological record in Egypt. There, about 1600 B.C., safflower was grown for its flowers which showed the same variability in colour as may be found in that country now. The reddish orange florets were sewn sideways on narrow strips of papyrus or cloth to form long garlands that were wrapped about the necks and bodies of mummies. Then, or at some later date, it was discovered that the orange and red flowers could serve as a source of dye to colour cloth. This suggests that man became interested in safflower when he first found plants with orange or red flowers.

By the early nineteenth century, safflower had become one of the two most important plant sources of dye-stuffs, the other being indigo. Its importance declined, along with that of indigo, when synthetic aniline dyes were developed. The dried flowers may still be purchased in the bazaars of cities from Cairo to Tehran, being sold as colouring for foods. The pigment is carthamone, a chalcone derivative.

It is likely that safflower seed was used as a source of oil during the Roman period in Egypt and perhaps earlier. The date of its first culture for oil in India is not known, but it was probably also very early. Until the later part of the nineteenth century it was not tested as a potential oil crop in any other area.

The general picture that emerges, therefore, is of a species first domesticated in the Near East as a dye plant and subsequently very widely used for this purpose in the old world subtropics. At an early stage it was locally converted into an oil-seed. In cultivation, over a huge area and for a long time, considerable diversity developed and there is evidence (see below) of incipient genetic differentiation.

Centres of diversity (Knowles, 1969b) with predominant types in each are: 1. Far East – late, spiny, tall, red-flowered; 2. India–Pakistan – early, very spiny, short, orange-flowered; 3. Middle East – late, spineless, tall, red-flowered; 4. Egypt – variable, usually large-headed; 5. Sudan – early, very spiny, yellow-flowered; 6. Ethiopia – late, very spiny, tall, red-flowered; 7. Europe (Mediterranean) – variable.

There appears to have been introgression from the *C. oxyacantha* of northern Pakistan into the domestic safflower of that area; and from the safflower of the Sudan centre to the Egyptian centre (Knowles, 1969b). Often, when Indian types are crossed with those of other areas, sterile types appear in F_2. Interactions of recessive alleles at three loci are involved. This incipient isolation suggests that Indian populations have been separated from others for a long time (Carapetian, 1973).

4 Recent history

Following a programme of testing at the University of Nebraska, safflower was established in the 1940s on a small commercial scale in the western and northern Great Plains of the USA. However, production ceased in the 1960s, primarily because of competition from improved wheats and problems with weeds and disease. Safflower has been grown in California and Arizona since 1950, the whole area of production (60–120 kha) being grown under contract to companies interested in processing or selling the seed. Areas of production (kha) in other countries are as follows: north-western Mexico, 300; Australia, 22; Spain, 50; and Portugal, 25. As in India, safflower has been most successful on heavier textured soils that are reasonably permeable. In such soils, its roots can penetrate to depths of four metres.

The development of the crop in countries of Asia, Europe and Africa has been severely handicapped by insects, many of them adapted to survival on other Compositae including the weedy safflower species. Most serious among them is the safflower fly (*Acanthiophilus helianthi*). The safflower of the Deccan region of India shows little damage from that pest, apparently because alternate hosts are not abundant or the crop is so early that it escapes damage.

In breeding programmes safflower is handled as a self-pollinated crop, though plants are bagged to prevent outcrossing by insects. Levels of outcrossing appear to vary between 5 and 10 per cent in commercial varieties in the United States, but have been much higher in some experimental materials. Efforts to develop hybrid varieties, using a thin-hulled female parent which had a form of structural male-sterility, were unsuccessful, primarily because there were enough selfed female plants to affect yields adversely.

Much effort has been devoted to the development of disease resistance. Cultivars have been bred with resistance to wilt caused by both *Fusarium oxysporum* f. *carthami* and *Verticillium albo-atrum* and to rust caused by *Puccinia carthami*. Greater resistance to *Phytophthora* root rot is needed in California and Arizona and to *Alternaria* in the northern Great Plains.

A large germ plasm reservoir is maintained by the US Department of Agriculture at the Plant Introduction Station at Washington State University, Pullman, Washington.

5 Prospects

Disease resistance will continue to be a major objective of all breeding programmes. A higher level of resistance to root rot caused by *Phytophthora drechsleri* is needed in all areas in which irrigation is practised. Greater tolerance to foliar diseases caused by *Alternaria* spp., *Puccinia* and *Botrytis cinerea* would permit more widespread cultivation in the USA and would facilitate establishment of the crop in damper areas elsewhere.

Higher oil contents will be developed, with levels exceeding 50 per cent, both by reducing the hull and by raising the level of oil at the expense of protein. A mutant type that presumably originated in the Ganges delta has an oil with high levels of oleic acid, making it chemically like olive oil (Knowles, 1969a). The characteristic has been bred into cultivars in the USA so there are now two types in commercial production: high linoleic (poly-unsaturated) and high oleic (mono-unsaturated) types.

Attention will be given to the meal, primarily in terms of raising levels of lysine. Removal of a bitter substance, matairesinol monoglucoside (Palter and

Lundin, 1970) and of the cathartic 2-hydroxyarctin will improve safflower meal as a feed for monogastric animals and as a food for man.

Winter types are being developed in Iran and the United States that will tolerate temperatures of −15°C and which may permit the culture of safflower as a winter crop.

The University of California has released UC-26, a spineless, red-flowered type with long branches closely appressed to the main stem, for use in dried flower arrangements. The dried flowers retain their red colour long after harvest. The use of safflower as an ornamental plant will increase.

As in the recent past, most of the variety development in the United States will be done by commercial plant breeders. The US Department of Agriculture and universities will be heavily involved in germ plasm development, genetic and cytogenetic studies, cultural and weed control studies and utilization research.

6 References

Ashri, A. (1973). Divergence and evolution in the safflower genus, *Carthamus. Final Research Report. USDA P.L. 480 Project No. A10-CR-18*. Hebrew University, Rehovot, Israel.

Beech, D. F. (1969). Safflower. *Field Crop Abstr.*, **22**, 107–9.

Carapetian, J. (1973). *Inheritance, cytology and histology of genic sterility in cultivated safflower (Carthamus tinctorius)*. Thesis, University of California, Davis.

Hanelt, P. (1963). Monographische Ubersicht der Gattung *Carthamus* (Compositae). *Fedde Repert.*, **67**, 41–180.

Imrie, B. C. and Knowles, P. F. (1970). Inheritance studies in interspecific hybrids between *Carthamus flavescens* and *C. tinctorius. Crop Sci.*, **10**, 349–52.

Knowles, P. F. (1955). Safflower – production, processing and utilization. *Econ. Bot.*, **9**, 273–99.

Knowles, P. F. (1958). Safflower. *Adv. Agron.*, **10**, 289–323.

Knowles, P. F. (1969*a*). Modification of quantity and quality of safflower oil through plant breeding. *J. Amer. Oil Chem. Soc.*, **46**, 130–2.

Knowles, P. F. (1969*b*). Centers of plant diversity and conservation of crop germ plasm: Safflower. *Econ. Bot.*, **23**, 324–9.

Kupsow, A. J. (1932). The geographical variability of the species *Carthamus tinctorius. Bull. appl. Bot. Gen. Pl. Br.*, **9**(1), 99–181.

Palter, R. and Lundin, R. E. (1970). A bitter principle of safflower, matairesinol monoglucoside. *Phytochem.*, **9**, 2407–9.

Weiss, E. A. (1971). *Castor, sesame and safflower*. New York.

Pyrethrum

Chrysanthemum spp. (Compositae)

R. B. Contant
University of Nairobi Kenya

1 Introduction

Of the two principal *Chrysanthemum* species with insecticidal properties only one, *C. cinerariifolium* (often incorrectly spelled *cinerariaefolium*), is now grown commercially, the other, *C. coccineum*, having been reduced to a homestead crop in its native area.

The pyrethrum of commerce is a daisy-like perennial herb with deeply dissected leaves, grown for its flowers containing six insecticidal compounds collectively called 'pyrethrins'. The crop is planted from seed or 'splits' (divided mother plants); its economic life is 2–3 years. The pyrethrins are extracted by solvents from the hand-picked, dried, ground-up flowers and used in many formulations and in a wide range of products. Pulverized flowers are used for manufacturing mosquito-repellent smoulder coils. Pyrethrins are virtually non-toxic to all warm-blooded animals, have low persistence, possess both knock-down and killing effects, can be effectively synergized and are free from taint.

The short history of the crop (since 1840) has been characterized by dramatic changes of principal producing countries and sharp fluctuations in world production, mainly occasioned by the two world wars. Since the Second World War, the principal production centres have all been in tropical highland regions where the plant flowers for most of the year and labour costs are still low. It is now grown mainly by small farmers, for many of whom it is the only available cash crop. In 1971–72, world production of dried flowers approached 22 kt, of which Kenya produced 14·5, Tanzania 4·3, Ecuador 1·1, Rwanda 1·0 and Japan 0·4 (Casida, 1973). Further increases are expected as a result of new applications and the growing concern about the use of DDT and other chlorinated hydrocarbons and of organo-phosphorus insecticides. Synthetic pyrethrins do not at present threaten natural pyrethrum.

2 Cytotaxonomic background

The genus *Chrysanthemum*, in the widest sense, contains over 200 species, most of which are native of the Old World, with centres of diversity in China and Japan, Siberia, Tibet, central and western Europe, south-western Asia and the Caucasus and the Mediterranean.

All *Chrysanthemum* species have a basic $x = 9$. Ploidy levels vary widely between species. The highest polyploids are found in the oriental region (up to $10x$) where virtually all species exist at one ploidy level only. The European species are $2x$, $4x$ or $6x$ and several occur at more than one ploidy level. Most Mediterranean and southwest Asian species are only $2x$; some exist in both $2x$ and $4x$ forms. *Chrysanthemum maximum*, native of Spain, has an exceptionally wide ploidy range. These data indicate that eastern Asia is the oldest centre, the European centre being much younger and the Mediterranean and southwest Asian (sub-) centres probably younger still (Dowrick, 1952).

The most toxic species, *C. cinerariifolium* ($2n = 2x = 18$), is native of the Adriatic coast of Yugoslavia (Dalmatia). *C. coccineum* (also $2n = 2x = 18$) originated in the Caucasus; its taxonomic status is still controversial, partly because of variation in morphology and flower colour. The names *C. roseum*, *C. marschallii*, *C. carneum*, *Pyrethrum roseum* and *P. carneum* in the literature all apply to forms of *C. coccineum*.

Species relationships have hardly been studied. A few crosses have been made with practical aims and some spontaneous hybrids have been found (e.g. Greener, 1941; see also Contant, 1963a):

C. coccineum (carneum) × C. cinerariifolium
C. coccineum (carneum) × C. corymbosum
C. coccineum (carneum) × C. macrophyllum
C. macrophyllum   × C. cinerariifolium.

C. corymbosum is a $2x$ and $4x$ perennial of very low toxicity, native of Europe, also occurring in North Africa and the Caucasus. *Chrysanthemum macrophyllum* is a diploid perennial, native of Europe and said to have some pyrethrins in the leaves. All hybrids mentioned were completely sterile except *C. coccineum* × *C. corymbosum* which produced some seed. The hybrid *C. coccineum* × *C. cinerariifolium* showed univalents in meiosis (Tominaga, 1968).

In the absence of further data one cannot go beyond suggesting that both *C. cinerariifolium* and *C. coccineum* are relatively young species and may have evolved, each in its own region, from a low-toxicity European ancestor, possibly of *C. corymbosum* type.

3 Early history

The first mention of insecticidal pyrethrum dates from the first century A.D. (China, species unknown). *Chrysanthemum coccineum*, grown for several centuries in the Caucasus and north-western Persia, entered commercial production in 1828; a lively trade in ground flowers ('Persian powder') developed and the crop was introduced into Europe and the USA. Mass selection for flowering ability and toxic potency has probably taken place, but records are lacking. Interest declined after 1840 when the insecticidal properties of *C. cinerariifolium*, growing wild in Dalmatia, were discovered. Cultivation of the latter soon started and the crop was introduced (as seed) into France, Switzerland and the USA and from there, first as seed and then (1886) as vegetative material, into Japan. Dalmatia remained the main producer until the First World War when Japan seized the world market. In the period 1925–36 the crop spread to a dozen countries around the globe and was grown experimentally in 30 more. Of these newcomers, only Kenya, Tanganyika, Congo and Brazil had shown promise by 1939. Kenya, where seed from Dalmatia was introduced in 1928, offered strong competition to Japan from 1935 (higher pyrethrins contents and flower yields) and became the world's main producer during the second world war when Japanese exports ceased.

During this period the crop was grown mainly from seed, although there were some clonal plantings in Japan (Hokkaido). Apart from some mass-selection for abundance of flowering and (after 1936 in the USSR) some attempts to improve pyrethrins content, there was no genetic improvement prior to 1940.

4 Recent history

Chrysanthemum coccineum was taken up in the USSR during 1939–41, notably for its resistance to cold and drought. Selection for high pyrethrins content took place and crosses with *C. cinerariifolium* were made but these were sterile (see section 2). The programme was soon abandoned and the species is now grown only in its native region for domestic purposes. As an ornamental it has gained popularity in many countries and various horticultural forms have been developed.

Another drought-resistant species, *C. tamrutense* was identified as highly toxic in bioassays (Mardjanian, 1941) but seems to have received no follow-up.

The cultivation of *C. cinerariifolium* expanded greatly during the second world war in Kenya, Tanzania and Congo. A slump occurred in 1945–48 (surplus stocks, upsurge of DDT) but production recovered with new technology. Seed from Kenya provided the basis for selection in Ecuador and New

Guinea, where cultivation started in the 1950s. Congo, whose own production virtually ceased in 1960, provided improved clones and seed for a new industry in neighbouring Rwanda (1967). Improved material, mostly from Kenya, has also benefited several other countries which have started growing the crop in recent years.

As pyrethrum is strongly self-incompatible (sporophytic system: Brewer, 1968), breeding methods have been those appropriate to cross-pollinated vegetatively propagated crops. In Congo, inbreeding by various methods (cages with insects, hand-pollination) has been used for genetic studies but, for practical purposes, clonal selection prevailed, first for the production of hybrids and later (after 1950) also for direct issue of outstanding clones (advocated in the literature since 1934). In Kenya, breeding since 1946 has been based on single plant selection, clonal testing, bi-clonal crossing and testing of the hybrids at various locations, the main selection criterion being yield of pyrethrins/acre (= flower yield × pyrethrins content). Several good hybrids have been issued (Kroll, 1958). The direct issue of superior clones started only in 1962 (Contant, 1963b). The polycross progeny test was introduced to estimate the breeding value (general combining ability) of outstanding clones. This has greatly improved the choice of parents for new crosses (Parlevliet and Contant, 1970). The best hybrids (bi-clonal and synthetics) serve as sources of further clonal selections. They are also issued to farmers, in addition to clones, mainly because of increasingly serious yield reductions in clonal plantings, attributed to root-knot nematodes (*Meloidogyne* spp.) and possibly root-rot fungi.

Apart from yield of pyrethrins/acre (qualified, since 1962, by high pyrethrins content), important selection criteria are: good splitting ability, establishment, lodging resistance, resistance to bud disease. Other criteria are flower size, a favourable ratio of the pyrethrins and, recently, also nematode tolerance. High-yielding clones with 2·3 per cent pyrethrins are now being multiplied in Kenya. Clones with 3 per cent pyrethrins exist but have inadequate vigour; there is a negative relationship between the two traits. Against these achievements, new seed introductions from Dalmatia have proved disappointing. The value of introduced pollen, used on outstanding clones, is still under study.

Considering the highly outcrossing nature of pyrethrum and its short breeding history (at most 4–5 cycles), it is still essentially a wild species. Even in advanced hybrids, variability in most traits is still within the range of the wild population. The most

significant changes have been in pyrethrins content (wild material has 0·8–1·9 per cent) and in flowering adaptation to local ecological conditions, mainly temperature patterns.

5 Prospects

Reciprocal recurrent selection appears to be the most attractive breeding strategy for the near future (Parlevliet and Contant, 1970). Inbreeding for F_1 hybrid production is cumbersome because of pronounced self-incompatibility in most clones; it is not likely to be widely practised. Further work is anticipated on artificial tetraploids and triploids, mainly to increase flower yields and flower size. Japanese research has shown that triploids in particular may excel in flower yield while retaining the pyrethrins content of the corresponding diploids (Tominaga, 1959); one natural triploid clone has been grown for many years in Congo and Kenya. If successful, polyploidy would constitute the first truly evolutionary breakthrough in this almost exclusively diploid species.

Mutation induction has been suggested for breaking the presumably single-locus (Brewer, unpubl.) sporophytic incompatibility system, as a preliminary to inbreeding for F_1 hybrid production. Quick results are improbable. *In vivo* culture of plantlets from the axils of flower-shoots (Roest, unpubl.) will probably soon be applied to speed up clonal propagation and hence the release of new clones and hybrids.

No drastic change in breeding objectives is foreseen. Selection for nematode tolerance will become more prominent wherever clones are used. Increased flower size is desirable to reduce picking costs but one must guard against the risk of lowering dry matter content.

The increasing interest in natural insecticides may lead to further exploration of *C. coccineum*, *C. tamrutense* and possibly other *Chrysanthemum* (*Pyrethrum*) species. Toxic principles other than the known pyrethrins might be found. Such studies, and further interspecific hybridization and karyotype analysis (Tominaga, 1968), even if without short-term evolutionary consequence, would in any case increase the knowledge of species relationships in this genus.

6 References

Brewer, J. G. (1968). Flowering and seed-setting in pyrethrum (*Chrysanthemum cinerariaefolium*). A review. *Pyrethrum Post*, 9(4), 18–21.

Casida, J. E. (ed.) (1973). *Pyrethrum*. New York and London.

Contant, R. B. (1963a). The possible use of *Chrysanthemum* species in the genetic improvement of pyrethrum. *Proc. East African Acad.*, 85–92.

13

Contant, R. B. (1963*b*). The current position of pyrethrum breeding in Kenya. *Proc. East African Acad.*, 93–6.

Dowrick, G. J. (1952). The chromosomes of *Chrysanthemum* I: the species. *Heredity*, 6, 365–75.

Greener, B. M. (1941). [Breeding pyrethrum for insecticides]. *Proc. Lenin Acad. agric. Sci. USSR*, 6, 13–16 (Russian).

Kroll, U. (1958). The breeding of improved pyrethrum varieties. *Pyrethrum Post*, 4(4), 16–19.

Mardjanian, G. M. (1941). [On the question of toxic characters of different *Pyrethrum* species]. *Proc. Lenin Acad. agric. Sci. USSR*, 10, 26–9 (Russian).

Parlevliet, J. E. and Contant, R. B. (1970). Selection for combining ability in pyrethrum, *Chrysanthemum cinerariaefolium*. *Euphytica*, 19, 4–11.

Tominaga, Y. (1959). [Cytogenetic studies on *Chrysanthemum cinerariaefolium* and *Chrysanthemum coccineum* I: External characters and contents of pyrethrin in the polyploids of *Chrysanthemum cinerariaefolium*]. *Jap. J. Genet.*, 34, 381–5 (Japanese with English summary).

Tominaga, Y. (1968). [Cytogenetic studies on pyrethrum flowers III and IV]. *Bull. Hiroshima agric. Coll.*, 3, 93–6, 171–6 (Japanese with English summary).

Sunflowers

Helianthus (Compositae – Heliantheae)

C. B. Heiser Jr
Indiana University
Bloomington Indiana USA

1 Introduction

The genus *Helianthus* has provided man with two food plants, the sunflower (*H. annuus*) and the Jerusalem artichoke or topinambour (*H. tuberosus*). Both are native to temperate North America and were used as food plants by the Indians. The sunflower, grown for its seed, in 1970 became the world's second most important supplier of vegetable oil. The crop has been reviewed by Putt (1963). It is now fairly widely grown throughout the world, with the USSR responsible for more than two-thirds of the world's production. Argentina and the Balkan countries are other important sunflower-growing areas. The seed cake remaining after the oil is expressed is mostly used for stock feed. In addition to its use for oil, the seed is still used directly as a food for man as well as for birds. In the United States most of the sunflowers grown are for 'snacks' for humans or for feeding wild birds. The Jerusalem artichoke, grown for its tubers, has always been an extremely minor crop, but it is still grown in many places as a food for man or livestock and, in France, for the production of alcohol. The storage material in the tubers is inulin. Tubers are used for propagation. Several varieties of *H. annuus* as well as other species of the genus are sometimes grown as ornamentals.

2 Cytotaxonomic background

Helianthus contains some 70 species, divided into four sections (Heiser *et al.*, 1969). *Helianthus annuus* and 13 other species constitute the section Annui. This section comprises taprooted species, mostly annuals, confined to the western part of the United States, all of which are diploid ($x = 17$, $2n = 34$). The species most closely related to *H. annuus* is *H. argophyllus*, native to Texas. Hybrids between the two show only moderate reduction in fertility. It is possible to secure hybrids of *H. annuus* with most of the other annual species, but the fertility of these hybrids is greatly reduced. In addition to the domesticated monocephalic types, *H. annuus* comprises both weedy and wild races, which are widespread in western and central North America, as well as ornamental forms. The wild and weedy forms are much branched plants and bear much smaller heads and achenes than do the domesticated sunflowers.

H. tuberosus is placed in the section Divaricati. The 30 species of this section, all perennials, are concentrated in the eastern and central part of the United States and include diploid, tetraploid and hexaploid species. The Jerusalem artichoke is a hexaploid and, as a wild plant, is fairly common in the eastern half of the United States. Some of the wild types differ little from the domesticated form except for smaller tubers. Nearly fertile hybrids have been secured between *H. tuberosus* and several of the other hexaploid species; hybrids with low fertility have been secured in crosses with the tetraploid species; thus far it has been impossible to obtain hybrids of any of the diploid perennials with *H. tuberosus*. Hybrids, however, have been secured between *H. annuus* and *H. tuberosus* which show varying degrees of fertility.

Self-incompatibility appears to be widespread in the genus. So far as is known, only *H. agrestis*, some races of *H. argophyllus* and some varieties of the domesticated sunflower, *H. annuus*, show self-compatibility. The fact that the Jerusalem artichoke often fails to set seed in cultivation probably is the result of a single clone being grown. Also, it fails to flower when

Fig. 13.1 Evolutionary geography of cultivated *Helianthus*.

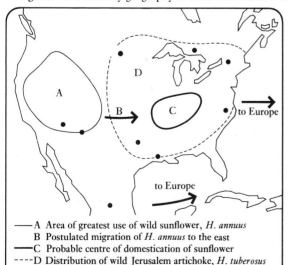

——A Area of greatest use of wild sunflower, *H. annuus*
    B Postulated migration of *H. annuus* to the east
——C Probable centre of domestication of sunflower
----D Distribution of wild Jerusalem artichoke, *H. tuberosus*
  • Sites of sunflower cultivation by Indians in historical times

grown in regions where the growing seasons are short. While considerable varietal diversity of the Jerusalem artichoke is still found in Europe, apparently nearly all now in cultivation in the United States is derived from the same clone.

3 Early history (Heiser, 1951; 1975)

Sunflower seeds were probably an important wild food source to early man in the western United States. It has been postulated that, in time, the sunflower became a camp-following weed and was introduced from the western to the central part of the country. Somewhere in the latter area the sunflower appears to have been domesticated and, as a domesticated plant, was carried both eastward and to the southwest. Achenes of the domesticated sunflower have been found in several archaeological sites in the central and eastern states, but thus far only wild sunflowers have been reported from archaeological sites in the southwest. Dates are not available for all the archaeological sunflowers, but it seems probable that domestication occurred sometime in the first millenium B.C. Possibly, the sunflower was domesticated in the central United States before the people there had acquired maize, beans and squash from Mexico. When the Europeans arrived, they found the sunflower cultivated in many places from southern Canada to Mexico. The seeds were eaten directly and used for their oil and the plant was used medicinally as well as in other ways, but nowhere did it appear to be a major crop.

In the sixteenth century the sunflower went from the Americas to Europe. Several of the herbalists indicate that the source was Peru, but this is almost certainly an error. In all probability, the first introductions were from Mexico to Spain, followed by introductions from other parts of North America. Dodoens gave the first account of the sunflower in Europe in 1568. The plant spread through Europe, grown more as a novelty than as a food plant, until it reached Russia. According to one account, the sunflower was readily adopted in Russia for, only shortly before its arrival there, the Church had laid down strict restrictions on the eating of oily foods on certain Holy days and the newly introduced sunflower was not on the list of prohibited foods. Its use for oil in Russia was suggested in 1779 and selection for high oil content began in 1860.

Samuel Champlain observed the Jerusalem Artichoke 'in cultivation' at what is now Nausett Harbor, Massachusetts, in 1605. It is clear that the Indians used it as a food plant but to what extent it was cultivated is unknown. It may have been that the Indians depended primarily upon wild plants. The first tubers to reach Europe, however, appear to have been slightly larger than those of wild plants, so it is likely that it was cultivated to some extent and that some selection had taken place. In Europe the Jerusalem artichoke had an enthusiastic reception, but it was shortly to fall into disfavour and come to be more food for swine than man. Champlain in his account of the plant mentions that the 'roots' had a taste resembling the artichoke, so there is no difficulty in explaining the adoption of the name artichoke. It had generally been assumed that the 'Jerusalem' represented a corruption of *girasole*, the Italian name for sunflower, until Salaman (1940) showed that the name *girasole* was not established until after the Jerusalem artichoke had already received that name. He suggests that 'Jerusalem' was an English corruption of Ter Neusen, from which place the tubers had been introduced into England. The history of the equally ridiculous name, topinambour, has also been traced by Salaman. In 1613 six natives of the Topinambous tribe of Brazil were brought to France and excited considerable interest. Apparently some street hawker appropriated their name for the rather newly introduced tubers in order to increase their sales value. Because of this name some botanists attributed the plant to Brazil.

4 Recent history

The sunflower is Russia's most important oil crop. Considerable areas of the country were found suitable for growing it and subsequent breeding work has resulted in greatly improved plants. Oil content has been increased from around 28 per cent to 50 per cent. Considerable headway was made with the control of fungal diseases and insect pests. Resistance to several major diseases was secured through hybridization with *H. tuberosus*. Dwarf varieties, which have now largely replaced the giant varieties throughout the world, were developed that could more readily be mechanically harvested. Sunflower breeding has been reviewed by Rudorf (1961) (see also Pustovoit, 1966).

In Argentina the sunflower became an important oil crop when that country was cut off from its olive oil source in Spain during the Spanish Civil War. During the Second World War sunflower growing was attempted in several new countries. The effort was mostly unsuccessful in England but later the sunflower did achieve some success in Canada. The sunflower has never been more than a minor crop in its homeland but presently there is a revival of interest in it there. As an oil crop in the United States it faces serious competition from the soybean, cotton seed and the peanut. Recently, there has also been some interest in the sunflower in several tropical countries.

While the sunflower has developed into a major crop, the Jerusalem artichoke continues to be of only very minor importance, a position that seems unlikely to change much in the future. Recently, however, there has been some renewal of interest in it as a food for humans in the United States. Breeding has been reviewed by Rudorf (1958).

5 Prospects

Leclercq (1969) reported the discovery of cytoplasmic male sterility in the sunflower derived from crosses of *H. annuus* with *H. petiolaris*. Then, with the finding of fertility restorer genes, the practical production of hybrid sunflowers became a reality. In trials the hybrid sunflowers have shown the same dramatic increases in yield that were found with hybrid corn. It seems likely that hybrid sunflowers will eventually largely replace the presently grown open-pollinated varieties. A number of diseases still plague the sunflower, so continued work in their control must still be undertaken.

Some of the American Indian varieties of the sunflower have been preserved by the United States Department of Agriculture, but several varieties have already disappeared. A tremendous reservoir of genes exists among the wild and weed sunflowers which occupy a wide area in North America. While these seem to be in no danger of extinction in the near future, steps perhaps should be taken to preserve adequate samples of them as well as of the older cultivated varieties.

6 References

Heiser, C. B. (1951). The sunflower among the North American Indians. *Proc. Amer. phil. Soc.*, **95**, 432–48.

Heiser, C. B. (1975). *The Sunflower*. University of Oklahoma Press, Norman.

Heiser, C. B. Smith, D. M., Clevenger, S. B. and **Martin, W. C.** (1969). The North American sunflowers (*Helianthus*). *Mem. Torrey bot. Club*, **22**, 1–218.

Leclercq, P. (1969). Une stérilité male cytoplasmique chez le Tournesol. *Ann. Amelior. Plantes*, **19**, 99–106.

Pustovoit, G. (1966). Distant (interspecific) hydridization in sunflowers. *Proc. II interl. Sunflower Conference, Morden, Manitoba, Cemeva.*

Putt, E. D. (1963). Sunflowers. *Field Crop Abstr.*, **16**, 1–6.

Rudorf, W. (1958). Topinambour, *Helianthus tuberosus* L. *Handbuch der Pflanzenzüchtung*, **3**, 327–41.

Rudorf, W. (1961). Die Sonnenblume, *Helianthus annuus* L. *Handbuch der Pflanzenzüchtung*, **5**, 89–114.

Salaman, R. N. (1940). Why 'Jerusalem' Artichoke? *J. Roy. Hort. Soc. N.S.*, **65**, 338–48, 376–83.

Lettuce

Lactuca sativa (Compositae)

Edward J. Ryder
US Agricultural Research Station
Salinas California USA
and
Thomas W. Whitaker
P.O. Box 150 La Jolla
California USA

1 Introduction

Lettuce is the major salad crop in North America and is important also in Australia and most countries of Europe and South America. Its popularity is increasing in Africa, the Middle East and Japan. Lettuce includes four salad types: crisphead, butterhead, romaine or cos and leaf. Another type, stem lettuce, eaten raw or cooked, is widely used in Chinese cookery (Thompson, 1951).

Lettuce is grown for the fresh market. It may be produced for local sale or for long-distance shipment. In the United States, Europe, and Africa, it may be grown for export. As a research organism, it is best known (cv. Grand Rapids) for the development of the germination–dormancy theory based on the conversion of phytochrome by red or far-red light (Borthwick *et al.*, 1954).

2 Cytotaxonomic background

Lactuca is a large genus of the Compositae with more than 100 species, chiefly indigenous to north temperate regions. The genus includes species with 8, 9 and 17 pairs of chromosomes (Babcock *et al.*, 1937). *Lactuca sativa* is one component of a group of four species within *Lactuca* that have nine pairs of chromosomes and are interfertile with each other. The others are *L. serriola*, *L. virosa* and *L. saligna*. All four are native to the Mediterranean basin. These species appear to be isolated within the genus by genetic barriers from other nine-chromosome species. Also, extensive tests have shown strong incompatibility barriers between the cultivated *Lactuca* species and the 17-chromosome species (Fig. 14.1).

3 Early history

Compared with corn or cucurbits, lettuce is a relatively recent addition to man's repertoire of cultivated crops. Evidently, lettuce was first cultivated about 4500 B.C. (Lindqvist, 1960). Leaves painted on the walls of Egyptian tombs, identified as those of lettuce, suggest that it was a common crop, widely known and appreciated at the time. The leaves appear to represent those of the romaine or cos type and are similar to the leaves of the lanceolate, pointed-leaf cv. Asparagus Leaf. Keimer (1924) suggests that the early Egyptians first cultivated lettuce as a seed crop for the edible oil extracted from the seeds.

Lettuce spread rapidly throughout the Mediterranean basin at an early date. Sturtevant (see Hedrick, 1919) cites numerous references to the crop in Greek and Roman literature, indicating that it was a popular and extensively used vegetable at the apogee of these civilizations. It probably spread with the Roman legions to France, England and the rest of Europe. After Columbus discovered the New World in 1492, lettuce cultivation quickly spread there. It was reported to be abundant in Haiti by 1565 and cultivated in Brazil in 1647. As early as 1806, an American seedsman listed 16 cultivars in his catalogue (Hedrick, 1919).

The origin of cultivated lettuce is uncertain, and critical tests to resolve the problem have not been devised. The *sativa-serriola* complex appears to be large, polymorphic and capable of free interchange of genes with little, if any, reduction in fertility. Through selection, *L. sativa* may have been derived directly from *L. serriola*, because nearly all the variations in cultivated lettuce are present in *L. serriola*, except the extreme forms of head formation and shape of involucre. If this theory of direct descent is not acceptable, we must account for the similarity of the two species by some other means. Lindqvist (1960) suggests three possibilities: (*a*) both species might have originated from hybrid populations that diverged into two groups, *L. sativa* cultivated by man, and *L. serriola* adapted to man-made waste habitats; (*b*) the ancestors of *L. sativa* might have been hybrids between *L. serriola* and a third species; (*c*) *L. serriola* might be a product of hybridization between *L. sativa* and some other species. Careful cytogenetic studies of the four species (*L. sativa*, *L. serriola*, *L. virosa* and *L. saligna*) will be needed before these questions can be resolved.

The emphasis of early human selection must have been on non-shattering seedheads, absence of early flowering (bolting), non-spininess, decrease in latex content and increase in seed size, as well as on the hearting character. These are the major traits separating cultivated from wild forms. Later, more formal

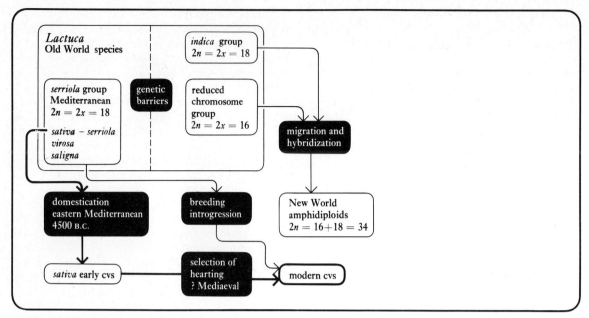

Fig. 14.1 Evolution and relationships of the lettuce, *Lactuca sativa*.

programmes continued to emphasize resistance to bolting and added resistance to disease, particularly downy mildew and lettuce mosaic.

4 Recent history

Lettuce is a highly variable crop. Not surprisingly, variation is mostly limited to the vegetative parts, which are the economic parts of the plant. There is much variation in leaf length, leaf shape, leaf colour, leaf texture, leaf size and hearting type. Perhaps the observed variation results from the origin and early development of the species in the Mediterranean, an area of many overrunning civilizations, each of which may have exerted some selection pressure on the species. Mostly, early cultivars appear to have been pointed, narrow-leaved, and non-heading. Cultivar development in Europe and, later, in the United States emphasized the head form, first of the butterhead, then of the crisphead type. European cultivars are mostly butterheads, but the crispheads predominated in the United States at an early date and now make up most of the lettuce production in that country. It is not known when the hearting character first appeared. The first definite evidence for existence of head lettuce was in 1543 (Helm, 1954).

In Europe, butterhead cultivars were developed for winter culture in the Mediterranean area and for summer culture in northern Europe. Butterhead and loose-leaf types are used for glasshouse production in France, Holland and Great Britain. Recently, crisp-head types have so increased in popularity that their production in France, Spain and North Africa is increasing, as are imports of this type into Europe from the United States.

Early breeding work in the United States stressed the development of salad types for market gardens around urban areas and for home gardens. Early in this century, crisphead lettuce became dominant because it could be grown on a large scale in the irrigated vastness of the western United States and shipped under ice to the cities of the East and mid-West. Therefore, breeding programmes emphasized size, weight and the ability to withstand the rigours of long-distance shipment in a cold, wet environment. At the turn of the century, the cultivar New York was predominant. The appearance of a disease of unknown origin, brown blight, led to the development, beginning in 1926, of a large group of resistant cultivars, the Imperial group. Continued emphasis on colour, size, weight and resistance to bolting and diseases led to the development of the Great Lakes cultivars, first appearing in 1941, which exceeded the Imperial group in many of these characteristics.

The Imperial and Great Lakes cultivars were susceptible to a new race of downy mildew (*Bremia lactucae*), which appeared in 1932. The resistant cultivars Valverde and Calmar were developed in response to this new race. Calmar and its derivatives are the latest group of cultivars to achieve dominance in the western United States. They are resistant to the races of downy mildew prevalent in the area.

Wild species of *Lactuca* have been somewhat exploited in lettuce breeding programmes. Collections of *L. serriola* have been used in a few crosses. This species was the source of the downy mildew-resistance gene incorporated into Calmar and Valverde. The cultivar Vanguard is the only known derivative of the difficult cross between cultivated lettuce and *L. virosa*. This wild species appears to have contributed genes for the dark-green colour and excellent leaf texture that characterize Vanguard and its derivatives.

Breeding methods used with lettuce are the standard ones for a self-fertilizing, polymorphic species, i.e. crosses within and between the several forms, followed by selection for various character combinations in the succeeding selfed generations. Some backcrossing has been practised. There is little evidence for heterosis. Even if hybrid vigour were shown, it could not, for technical reasons, be easily exploited. Cytogenetic studies have been numerous, but largely unexploited, because of the failure to obtain crosses between the *L. sativa–L. serriola* group and other *Lactuca* species. No biometrical studies have been reported.

5 Prospects

Several changes in the direction of breeding programmes are clear. In cultivar usage, popularity of crisphead lettuce is markedly increasing in western Europe and Japan. This trend is likely to continue and will have at least two consequences. Breeding programmes with crisphead lettuce will increase, emphasizing, outside the United States, those characteristics that accompanied the rise of the crisphead industry in the United States. There, the expanding export trade will be hampered by physiological breakdown problems, and breeding programmes will stress increased resistance to post-harvest disorders, such as russet spotting, pink rib and brown stain.

There will be continued emphasis on disease resistance and increased emphasis on insect resistance. For example, the first mosaic-resistant cultivars are now being released. One source of the mosaic-resistance gene is *L. serriola*; *L. virosa* may provide another source of resistance genes. Crosses with *L. saligna* will provide resistance to the cabbage looper, completing the list of species in the group contributing genes to the improvement of *L. sativa*.

In the vast production areas of the western United States, growers are anticipating mechanical harvesting and post-harvest handling. Mechanization will create two breeding needs: (*a*) the exploitation of such characters as head shape, head size, head density and resistance to rough treatment, all of which will facilitate machine-handling; and (*b*) uniform harvest density. The latter will require information on the genetic and genetic-environmental bases for uniform head development. Thus, there will be much emphasis on the biometrical and physiological genetics of head formation.

In northern Europe, the development of crisphead seems likely as well as of butterheads and leaf types adapted to year-round lettuce production, particularly in greenhouses in winter. This will focus breeders' attention on problems of extremes of humidity, light and temperature and of related diseases and disorders that are aggravated by stresses such as downy mildew, tipburn, botrytis and big vein.

Altogether, the increased emphasis on specific problems suggests a greater reliance on closely related species to provide specific genes. Perhaps interest in cytogenetic studies will be renewed, as well as the use of ionizing radiation and other mutagenic methods designed to increase the range of variability of the species.

6 References

Babcock, E. B., Stebbins, G. L. and Jenkins, G. A. (1937). Chromosomes and phylogeny in some genera of the Crepidineae. *Cytologia, Fujii Jub. Vol.* 1, 188–210.

Borthwick, H. A., Hendricks, S. B., Toole, E. H. and Toole, V. K. (1954). Action of light on lettuce-seed germination. *Bot. Gaz.*, 115, 205–25.

Hedrick, U. P. (1919). Sturtevant's notes on edible plants. *N.Y. Dep. Agric. ann. Rep.*, 27, 2/2, pp. 685.

Helm, J. (1954). *Lactuca sativa* in morphologisch-systematischer Sicht. *Kulturpflanze*, 2, 72–129.

Keimer, L. (1924). *Die Gartenpflanze im alten Ägypten.* Hamburg and Berlin.

Lindqvist, K. (1960). On the origin of cultivated lettuce. *Hereditas*, 46, 319–50.

Rodenburg, C. M. (ed.) (1960). Varieties of lettuce. *Varietal Desc.*, 3, *Inst. Verer. Tuinbouwgewassen*, Wageningen, Holland.

Thompson, R. C. (1951). Lettuce varieties and culture. *USDA Fmrs Bull.*, 1953.

Thompson, R. C., Whitaker, T. W. and Kosar, W. F. (1941). Interspecific genetic relationships in *Lactuca*. *J. agric. Res.*, 63, 91–107.

Whitaker, T. W., Ryder, E. J., Rubatzky, V. E. and Vail, P. (1974). Lettuce production in the United States. *USDA agric. Handb.*, 221.

15

Sweet potato

Ipomoea batatas (Convolvulaceae)

D. E. Yen
Bernice P. Bishop Museum
Honolulu Hawaii USA

1 Introduction

The sweet potato is basically a starchy vegetable, subsidiary or complementary to the *Solanum* potato in Caucasoid diets. Restricted mainly to tropical and sub-tropical regions, the tubers are of varying popularity in the markets of the United States, South America, Spain and the Mediterranean countries, Japan and New Zealand. The value of tubers and leafage as stock food was early recognized in Australia, while the tubers have been exploited in Japan as a source of industrial starch. It is in the emergent countries of the tropics, however, that the sweet potato makes its most valuable contribution as a crop, grown in mixed clonal plantings in the fields of subsistence farmers. It is often the cultigen underlying culturally more important plants in agricultural systems; for example, rice in southeast Asia and taro and yams in Oceania.

Agronomically there is a striking difference between sub-tropical and tropical practices; in the first, the plant is treated as an annual with winter storage of tubers for consumption and of 'seed' for subsequent vegetative reproduction while, in the tropics, it is virtually perennial, stem cuttings being made from standing crops in a continuous planting procedure. The latter has considerable relevance to the evolution of the crop.

2 Cytotaxonomic background

The most recent systematic revision of Convolvulaceae is by Van Ooststroom and Hoogland (1953), placing *Ipomoea batatas* in the section Batatas with the wild species *I. gracilis*, *I. tiliacea*, *I. trifida* and *I. triloba*; House (1908) had earlier grouped *I. batatas* with 25 wild species under section Batatas, subsection Aequisepalae to include *I. tiliacea*, *I. trifida*, *I. triloba* and had added *I. trichocarpa* and *I. lacunosa*. Most of these species have been the subject of cyto-

taxonomic study by Nishiyama (1971) who has suggested a new classification of *I. batatas* and its wild relatives on the basis of chromosome number and genome analysis, as follows.

Species, etc.	Synonyms	2n (x = 15)	Constitution
I. batatas			
var. *batatas*		90	BBBBBB
f. *trifida*	*I. trifida*	90	BBBBBB
var. *littoralis*	*I. littoralis*	60	BBBB
var. *leucantha*	*I. leucantha*	30	BB
I. triloba			
var. *triloba*		30	AA
f. *lacunosa*	*I. lacunosa*	30	AA
f. *trichocarpa*	*I. trichocarpa*	30	AA
f. *ramoni*	*I. ramoni*	30	AA
I. gracilis		60	—
I. tiliacea		60	—

However, general agreement on cytotaxonomic relationships has yet to be reached and Nishiyama's revision is perhaps open to some doubts, as follows. On *cytological* grounds, the range of irregularities such as multivalents and univalents at meiosis in interspecific hybrids between wild *Ipomoea* forms may be paralleled within *I. batatas* itself (Ting and Kehr, 1953; Yen, 1974, citing a cytological survey of the species representative of America, Polynesia, Melanesia and Asia by Powell and Yen, 1962–65). On *morphological* grounds, none of the wild species examined by Martin and Jones (1972) (especially *I. trifida* but also including *I. lacunosa*, *I. gracilis*, *I. tiliacea*, *I. trichocarpa* and *I. triloba*) bear sufficient resemblance to *I. batatas* to suggest direct evolutionary relationships.

3 Early history

There is now general agreement that the sweet potato is of American origin. The evidence is cross-disciplinary. The closest wild relatives are American, and Mexico has been suggested as the centre of diversity for the Batatas section of *Ipomoea* by Martin and Jones, as well as by Nishiyama; but the wild stocks of central America and northern South America have yet to be studied adequately. The comparative study of variation in cultivated populations from America, southeast Asia and Oceania shows a Vavilovian effect in that the American (and specifically western South American), population shows the widest range of variability (Yen, 1974). Archaeological remains of the tuber from Peru, including representations in pottery, ante-date any direct or indirect evidence for antiquity of the plant from any other area in the world (Towle, 1961; Yen, 1974). Engel (1970) reported tuber remains from a cave in Chilca Canyon, Peru, carbon-dated at between 8,000 and 10,000 B.C. It is not known whether these remains of sweet potato and other tubers are representative of very early agriculture or whether they are the product of plant-gathering. However, wild tuber-bearing sweet potatoes have not been found in recent times.

The evolutionary steps by which the crop originated remain uncertain. Ting and Kehr (1953) suggested hybridization between two unknown wild species followed by allohexaploidy ($2n = 2 \times 15 + 4 \times 15 = 90$). Nishiyama (1971), however, regards the sweet potato as an autohexaploid derived directly from *I. trifida* (see above). He places *I. leucantha* ($2x$), *I. littoralis* ($4x$) and *I. trifida* ($3x$ and $6x$) in a scheme resembling that of Ting and Kehr, domestication proceeding after chromosome doubling. Jones (1967) and Martin and Jones (1972) have suggested that the tuberless *I. trifida* evolved in parallel with *I. batatas* from ancient (perhaps now extinct) species, while the self-fertile diploid *Ipomoea* forms evolved more recently from earlier self-incompatible ancestors. Self-incompatibility is a characteristic of most sweet potatoes that have been investigated and Martin (1965, 1968), in reviewing the experimental results, has defined its genetic basis in terms of S-type genes with complex epistatic relationships depending on the alleles present. Subsequently, Martin (in Tai *et al.*, 1967), has addressed himself to the general problem of infertility as the result of self-pollination and hybridization; he suggested an over-riding physiological sterility caused by chromosomal aberrations and weak gamete production superimposed on the incompatibility system.

The breeding system acted as a conservative agent in the evolution of the crop. Vegetative reproduction must have had a quasi-apomictic effect; and, while the out-crossing habit tended to produce variation, the absence of conscious breeding by seed, combined with the biological barriers of sterility and incompatibility, were further impeding factors. Whatever progress there was must have depended on the discovery, selection and preservation of rare chance seedlings and somatic mutants.

There is little evidence for the much discussed extra-America distribution (or origin) of the sweet potato in pre-Columbian times, with the noted exception of Polynesia. Tuber remains from archaeological sites have been recovered from Hawaii, New Zealand and Easter Island, and structural remains from New Zealand early in prehistory indicate that storage practices were developed to convert the tropical perennial plant into an annual in response to seasonal temperate climates in which tropical cultivation practices were impossible to apply. This adaptation was the forerunner of modern practices. Yen (in Riley *et al.*, 1971) thought that the introduction of the plant from America to Polynesia occurred before the eighth century A.D.; but Brand (in Riley *et al.*, 1971) attributed it to Spanish travellers after A.D. 1500, thus returning to one of the earlier hypotheses. These and other historical issues are explored by Yen (1974).

4 Recent history

Clear records of the sweet potato in the chronicles of Columbus's first expedition to Hispaniola demonstrated the widespread adoption of the plant in America by the fifteenth century. The Caribbean region was to be the source of the earliest sweet potato stocks in Europe and, by Shakespearean times, the plant was well known in England. We are uncertain of where it was grown but, probably, the tuber was a novelty item in trade from the warmer climates of southern France and the Mediterranean. It was never, however, to achieve the economic prominence of the *Solanum* potato in Europe.

The main hypothesis for the rapid spread of the sweet potato in the sixteenth century has, on the one hand, Portuguese voyagers following the track of Da Gama, carrying the plant eastwards from the Caribbean, Brazil and Europe to Africa, India, southeast Asia and Indonesia; while, from their colonies in western South America and Mexico, the Spaniards took the plant westward to Guam and the Philippines. The confluence of these lines of introduction, together with the earlier American–Polynesian diffusion, is indicated by studies of variability in clonal populations collected largely from subsistence cultivators (Yen, 1974). The question as to whether or not the Polynesian sweet potato reached Melanesia in prehistoric times is still open.

Geographically separated populations show the influence of the founder effect, in that they tend to exhibit somewhat different ranges of variation affecting some 40 morphological and physiological (e.g. cold and disease reaction) characters. Also, there seems to have been some effect of human selection on two characters in hill regions of South America and New Guinea, namely: tolerance of cold and a preference for a long-vined growth habit correlated with soil-holding capacity on the commonly cultivated steep slopes. However, the movement of the plant into new environments has not resulted in anything like speciation or the formation of recognizably

43

differentiated races; nor are chromosome number or behaviours detectably modified in such populations.

Scientific interest in the sweet potato began late in the nineteenth century in the United States, with the cataloguing of regional collections. These catalogues are invaluable, for they record varieties now extremely rare or extinct. It was in the 1920s however that Julian C. Miller of Louisiana State University began conscious improvement by clonal selection and the investigation of hybridization and inheritance in the species. Similar programmes were soon instituted in other parts of the United States and in Japan, the Philippines and India, and Miller's breeding material was freely shared with interested breeders. Multifactorial inheritance as the genetic basis for most plant characters was early recognized; and to selection for quality and yield in clonal and hybrid populations was soon added concern for resistance to fungal and viral diseases. In the United States, cultivation was concentrated on the unfortunately-named group of varieties, the 'yam', whose tuber flesh is very sweet, soft-textured, bright orange in colour and with generally high carotene content; this group had been differentiated earlier from the 'Jersey' type which is yellow or white fleshed. At one time, the southern states produced mainly the 'yam' type, while California favoured the 'Jersey' but, by 1960, the 'yam' predominated in all sweet potato-producing areas. In Japan, however, it was the 'Jersey' type which was concentrated upon and breeding contributed improved varieties for industrial use.

5 Prospects

Breeding for disease resistance has progressed (summary by W. J. Martin, in Tai *et al.*, 1967) and may be expected to continue to do so. Other objectives, involving machine harvesting and the use of tubers in processed foods, are also likely to have considerable impact. However, it is in the developing countries of Asia, Africa, Central and South America, that a new impetus in plant breeding is already evident. The proceedings of the International Symposia on Tropical Root Crops (Tai *et al.*, 1967; Plucknett, 1970–73) are indicative of this trend. In such countries, however, the objectives of plant breeding are often difficult to define, for yield figures, areas in cultivation and even the occurrence of crop failure are often obscure where subsistence farming is common. However, there are two areas in which practical objectives are fairly well defined, namely: disease and pest resistance, (since some pathogens unimportant to temperate zone cultivation may be epidemic in the tropics) and nutritive value. In areas in which sweet potato forms a dominant part of the diet (e.g. in the Highlands of New Guinea) and protein intake is low, Oomen *et al.* (1961) have suggested the upgrading of amino-acid content in the sweet potato by breeding, since varieties exhibit considerable variability in this property.

There have been some innovations in the approaches to breeding. Jones (1965) proposed a modification of the polycross method of open pollination with the aid of insects, as an alternative to the generally applied pedigree methods. Among other advantages, this method is designed to maximize variability and to promote the expression of desirable epistatic effects. Furthermore, the conservation of variability is inherent in the procedure and this must be a major concern in breeding for subsistence agriculture, where variation appears to have functions in spread of seasonal production, in promoting dietary diversity and as a buffer against the effects of epidemic diseases. With this concern in mind, Yen (in Tai *et al.*, 1967) proposed the utilization of native selection in the selection stages of breeding programmes.

6 References

Engel, F. (1970). Exploration of the Chilca Canyon, Peru. *Current Anthropol.*, **11**, 55–8.

House, H. D. (1908). The North American species of the genus *Ipomoea*. *Ann. N. Y. Acad. Sci.*, **18**, 181–263.

Jones, A. (1965). A proposed breeding procedure for sweet potato. *Crop Sci.*, **5**, 191–2.

Jones, A. (1967). Should Nishiyama's K123 (*Ipomoea trifida*) be designated *I. batatas*? *Econ. Bot.*, **21**, 163–6.

Martin, F. W. (1965). Incompatibility in the sweet potato. A review. *Econ. Bot.*, **19**, 406–15.

Martin, F. W. (1968). The system of self-incompatibility in *Ipomoea*. *J. Hered.*, **59**, 262–7.

Martin, F. W. and **Jones, A.** (1972). The species of *Ipomoea* closely related to the sweet potato. *Econ. Bot.*, **26**, 201–15.

Nishiyama, I. (1971). Evolution and domestication of the sweet potato. *Bot. Mag.*, *Tokyo*, **84**, 377–87.

Oomen, H. A. P. C., Spoon, W., Heesterman, J. E., Ruinard, J., Luyken, R. and **Slump, P.** (1961). The sweet potato as the staff of life of the highland Papuan. *Trop. Geogr. Medicine*, **13**, 55–66.

Plucknett, D. L. (ed.) (1970–1973). *Tropical root and tuber crops tomorrow.* Proc. 2nd Intnl Symp. Tropical Root and Tuber Crops, 1970, 2 vols., Honolulu.

Riley, C. L., Kelley, J. C., Pennington, C. W. and **Rands, R. L.** (eds). (1971). *Man across the sea.* Austin and London.

Tai, E. A., Charles, W. B., Haynes, P. H., Iton, E. F. and **Leslie, K. A.** (eds). (1967). *Proceedings of the international symposium on tropical root Crops.* 2 vols, Trinidad.

Ting, Y. C. and **Kehr, A. E.** 1953. Meiotic studies in the sweet potato. *J. Hered.*, **44**, 207–11.

Towle, M. (1961). *The ethnobotany of pre-Columbian Peru.* New York.

16

Van Ooststroom, S. J. and Hoogland, R. D. (1953). Convolvulaceae. In *Flora Malesiana* I, 4, 458–88.

Yen, D. E. (1974). *The sweet potato and Oceania.* B. P. Bishop Museum Bulletin, 236, Honolulu.

Turnip and relatives

Brassica campestris (Cruciferae)

I. H. McNaughton
Scottish Plant Breeding Station
Edinburgh Scotland

1 Introduction

Storage organs, leaves and seeds are all utilized in the varying forms of *B. campestris*. The true turnips are important as forages for sheep and cattle, especially in northern Europe and New Zealand but are also eaten as a vegetable in many parts of the world. There is a range of leafy forms developed for salad and pickling purposes in China and Japan.

Oil-seed forms, both annual and biennial, are of considerable economic significance. The annual form predominates in Canada where it provides a substantial proportion of the world rapeseed crop; annual forms are also widely cultivated in India and Pakistan. The biennial form is important in Sweden where it is found to be more hardy than winter rape (*B. napus*).

2 Cytotaxonomic background (Fig. 16.1)

Brassica campestris ($2n = 2x = 20$) is polymorphic. The group has been classified into a number of subspecies (mostly former Linnaean species) on the grounds of complete inter-fertility (Olsson, 1954). It is likely that but few genes separate some of the subspecies, particularly the oriental forms.

The wild-type, ssp. *eu-campestris*, is a slender-rooted, branching annual. Truly wild *B. campestris* probably still exists today; certainly the species is a common weed both in Europe (wild rape) and in North America (field mustard).

The ssp. *oleifera*, turnip-rape, appears closest, morphologically and probably phylogenetically, to the wild-type. There are summer and winter varieties, usually referred to as var. *annua* and var. *biennis* respectively. Ssp. *rapifera*, the true turnip, is biennial. The useful part is the storage organ, technically a swollen hypocotyl. The stubble-turnip or Dutch turnip is a distinct class containing both lyrate and strap-leaved forms. There are also quick growing, early maturing varieties used as vegetables.

The ssp. *chinensis*, Pak-choi or Chinese mustard, is a leafy annual of which the young shoots, more or less blanched, are an important vegetable in China. Selection has produced some extreme forms with greatly enlarged leaf petioles with only a small fringe of lamina. Ssp. *pekinensis*, Pe-tsai or Chinese cabbage, forms distinct heads of leaves and is used as a salad vegetable in the Far East and elsewhere. Ssp. *narinosa* is a compact form with small, puckered leaves, used as a salad vegetable in China. Ssp. *nipposinica* forms rosettes of very numerous leaves. There is a strap-leaved form and a variety with highly dissected leaves (var. *laciniata*); both are used for greens or as pickled vegetables in the Orient. Ssp. *dichotoma*, Toria or Indian rape, is an annual oil-seed form, as is ssp. *trilocularis*, yellow-seeded Sarson. Indian mustard oil is generally obtained from varying mixtures of Toria and Sarson, together with Rai or wild turnip (*B.*

tournefortii), another $2n = 20$ chromosome species but one that is not inter-fertile with *B. campestris* (Olsson, 1954) (see also Chapter 19).

Brassica campestris crosses readily with *B. napus* ($2n = 4x = 38$), particularly when the latter is used as female parent. Both spontaneous and artificial amphidiploids (*B. napocampestris*, $2n = 58$) have been reported (Frandsen and Winge, 1932; McNaughton, 1973; Olsson, 1963), involving bulb-forming, leafy and oil-seed forms of the parental species. None has proved agriculturally useful but there are possibilities of introgression of desirable characters from one species into another by way of such amphidiploids (see Chapter 18).

Hybridization with *B. oleracea* is extremely difficult but may be facilitated by embryo culture (Harberd, 1969). Hybrids are highly sterile but fertility is restored in the amphidiploid *B. napus*. Artificial forms of *B.*

Fig. 16.1 Evolution of the turnip and its relatives, *Brassica campestris*. (Compare Figs. 18.1 and 19.1.)

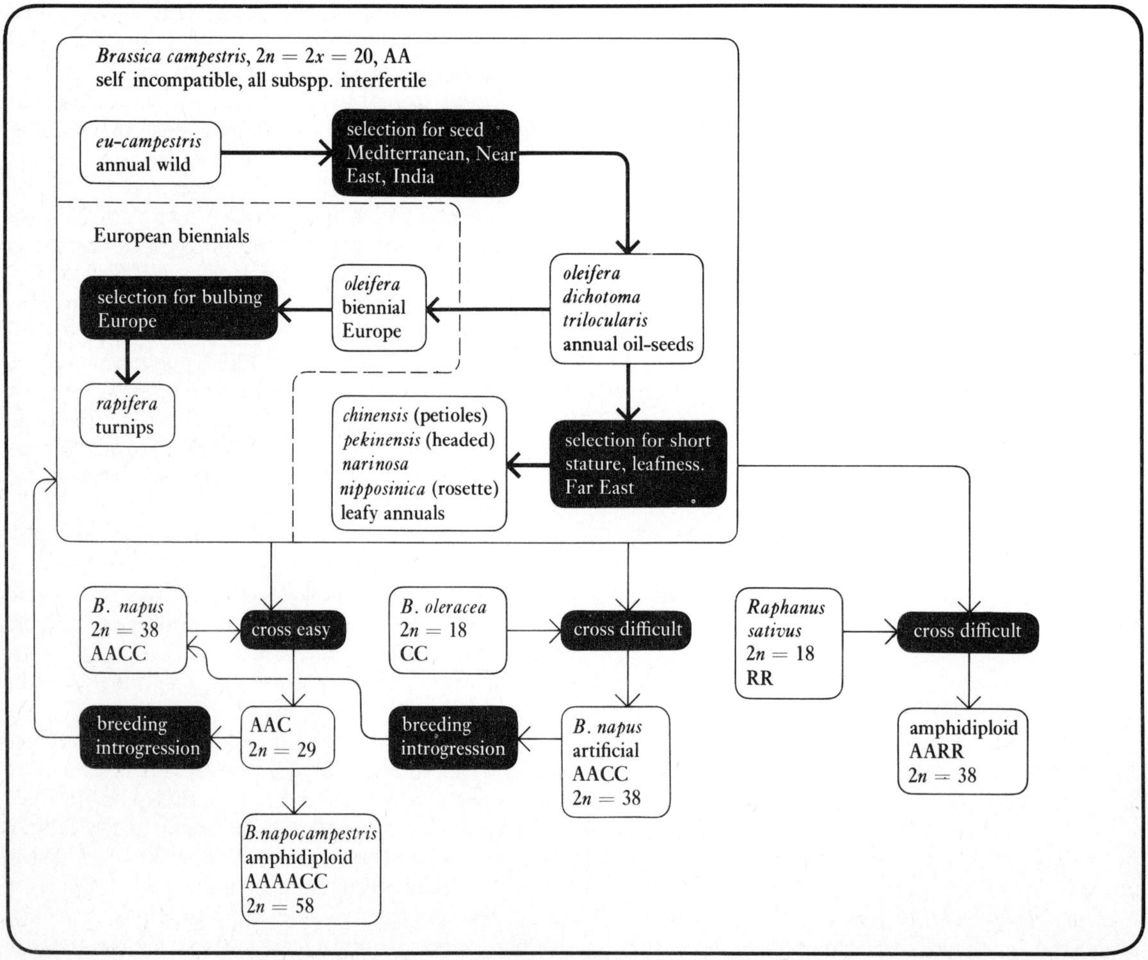

napus have been used as a 'bridge' by Japanese breeders for the introgression of disease resistance from *B. oleracea* (cabbage) into *B. campestris* (Chinese cabbage). For a discussion of *B. napus* see Chapter 18.

Brassica campestris (AA) hybridizes with great difficulty with *Raphanus sativus* (radish, RR, $2n = 18$). Amphidiploids (AARR, $2n = 38$) have been reported by Japanese workers (Hosoda, 1947); they are rather infertile and it is not known whether they have agronomic potential (Fig. 16.1).

For a treatment of *B. campestris* as a parent of the amphidiploid mustard, *B. juncea*, see Chapter 19.

3 Early history

Two main centres of origin are indicated. The Mediterranean area is thought to be the primary centre of European forms, while eastern Afghanistan and the adjoining portion of Pakistan is considered to be another primary centre, with Asia Minor, Transcaucasus and Iran as secondary centres (Sinskaia, 1928).

On grounds of comparative morphology it seems likely that subsp. *oleifera* is the basic cultivated form nearest to the wild type (Fig. 16.1 and see above). Vavilov (1926) suggested that it could have originated from *campestris* weeds of older seed crops such as flax. The time and place of domestication are unknown; somewhere in southwestern Asia in pre-Classical times seems likely, for there are old Arab and Hebrew names for the crop. According to Appelqvist and Ohlson (1972), rape seed was not cultivated by the Romans and, indeed, was used only by people who had neither the olive nor the poppy; the same authors remark that Sanskrit records show that sarson has been used in India since 2000–1500 B.C. On balance, the meagre evidence suggests multiple domestication of annual oil-seed forms from the Mediterranean to India perhaps about two millenia B.C.

Cultivation of oil-seed turnip-rape is thought to have started in Europe in the thirteenth century. It was important as a source of lamp oil until replaced by petroleum products (Appelqvist and Ohlson, 1972). Presumably, biennial oil-seed forms were the source of the turnips and all the evidence, botanical, historical and philological, points to an origin in the cooler parts of Europe. There are old Anglo Saxon, Welsh and Slav names for the crop. Turnips were known to the Romans in northern France and were probably introduced by them into Britain. In Europe, therefore, the turnip probably long ante-dates the oil-seed use of the crop.

The stubble-turnip, which has been selected for very rapid early growth from late sowing, is a recent development, originating, probably, during the great turnip era in Europe (fifteenth–eighteenth centuries). Stubble-turnips, as the name implies, are an autumn crop which were usually grown in the (rye) stubbles but there were many variations in their culture. Forms resembling modern Dutch cultivars were being grown in Britain in the early nineteenth century.

The history of *B. campestris* in the Far East is obscure. There is no evidence that it was anciently cultivated in China and, according to Appelqvist and Ohlson (1972), rape-seed reached Japan from China via Korea as recently as 200 years ago. The oriental subspecies are thought to have evolved in China, very probably from oil-seed forms, by selection for leafiness. The ssp. *nipposinica* developed in the direction of increasing leaf numbers, while, in *chinensis* and *pekinensis*, there was increase in leaf size and apical bud development, leading to head-forming characteristics. These changes were paralleled by forms of *B. oleracea* in Europe (Chapter 18).

4 Recent history

In the early nineteenth century in Britain various forms of turnips, together with swedes (*B. napus*), with which they cross readily, were reported to have been commonly multiplied side by side; as a result, it was impossible to obtain true-breeding strains. Since then, but before modern plant breeding, some progress was probably achieved by mass selection or by mild inbreeding, coupled with the realization of the need for adequate isolation distances during multiplication.

More recently, heterosis for dry matter yield of turnips has been demonstrated (Wit, 1966). *Brassica campestris* is an out-breeder with a sporophytic incompatibility system based on S-alleles. The production of F_1 hybrid varieties would, therefore, be a possibility, at least in principle. However, the difficulty of maintaining inbred lines, due to increased susceptibility to disease and reduced winter hardiness, has been a deterrent to the commercial production of hybrid stubble-turnips in Holland. Synthetic varieties, based on inbred lines of good combining ability, have, however, outyielded the best commercial cultivars (Wit, 1966).

Hybrid varieties of the leafy oriental salad vegetables (e.g. Chinese cabbage and Chinese mustard) have been available in Japan for some years; indeed, virtually all cultivars listed in seed catalogues are hybrids. Most of them have resulted from utilization of the incompatibility system.

Tetraploid stubble-turnips, mainly produced by Dutch breeders, are commercially available and are

47

reported to give dry matter yields marginally superior to the diploid strains from which they are derived. One such cultivar, Sirius, produced in Sweden by inter-crossing tetraploid strains from various sources, has given dry matter yields up to 14 per cent higher than the best commercial control; this gain is probably attributable more to heterosis than to polyploidy *per se*.

As with swede-rape, progress in breeding oil-seed turnip-rape has been mainly due to improved analytical techniques and screening methods and to an ever-increasing knowledge of the chemical components of the seed (particularly the fatty acids) and their genetic control. Strains with very low erucic acid content have been produced, particularly in Canada, where breeding and improved cultural methods have jointly resulted in increases of 40 per cent in seed yields over a 25-year period (Downey, 1971).

Winter turnip-rape tolerates late sowing better than does the *napus* analogue. It is hardier and earlier to ripen but is inferior in seed yield and oil content. In Sweden, both mass selection and pedigree selection have been tried but with poor results. The failure of pedigree selection is thought to be due to inbreeding depression in a naturally outbreeding species. Synthetic varieties offer a practical alternative (Andersson and Olsson, 1948).

5 Prospects

Poor utilization by animals of the bulbs of stubble-turnips suggests that the development of leafier forms would be a worthwhile objective. Dutch breeders are aiming for an 80:20 leaf:bulb ratio (as opposed to the existing 50:50 ratio) and good leaf retention. It is possible to combine the leafiness of ssp. *pekinensis* and *nipposinica* with the hardiness of turnip and of the biennial form of turnip-rape. *Plasmodiophora* resistance must be kept in view in such hybrids because the oriental forms are generally susceptible.

Yellow-seeded strains of oil-seed rape have been shown to have a higher oil content with more protein and less fibre in the meal than brown-seeded strains of the same genetic background. Work, particularly in Canada, is being directed towards yellow-seeded, low erucic acid combinations. Removal of undesirable glucosinolates in rapeseed meal is another breeding objective (Downey, 1971).

The breeding of hybrid varieties, difficulties notwithstanding, remains an attractive possibility. Hybrids between different cultivars of oil-seed turnip-rape, produced in Sweden, have given yields only equal to the best commercial material. However, inter-subspecific hybrids between *oleifera* and *chinensis*,

dichotoma, *narinosa* and *pekinensis* have given yields superior to the better parent. This suggests that future work should not be restricted to ssp. *oleifera* alone but should encompass the whole species (Olsson, 1955).

6 References

Andersson, G. and **Olsson, G.** (1948). Oil plants. In Åkerman, Tedin, Fröier and Whyte (eds), *Svalöf 1886–1946*. Lund, Sweden.

Appelqvist, L.-Å. and **Ohlson, R.** (1972). *Rapeseed*. London and New York.

Downey, R. K. (1971). Agricultural and genetic potentials of cruciferous oilseed crops. *J. Am. Oil Chem. Soc.*, 48, 718–22.

Frandsen, H. N. and **Winge, Ö.** (1932). *Brassica napocampestris*, a new constant amphidiploid species hybrid. *Hereditas*, 16, 212–18.

Harberd, D. J. (1969). A simple effective embryo culture technique for *Brassica*. *Euphytica*, 18, 425–9.

Hosoda, T. (1947). [On the fertility of *Raphanus-Brassica* and *Brassica-Raphanus* obtained by colchicine treatment]. *Jap. J. Genet.*, 22, 52–3 (Japanese).

McNaughton, I. H. (1973). *Brassica napocampestris* (2n = 58). I. Synthesis, cytology, fertility and general considerations. *Euphytica*, 22, 301–9.

Olsson, G. (1954). Crosses within the *campestris* group of the genus *Brassica*. *Hereditas*, 40, 398–418.

Olsson, G. (1955). [Heterosis in spring turnip rape]. *Sver. Utsädesför Tidskr.*, 65, 215–9 (Swedish).

Olsson, G. (1963). Induced polyploids in *Brassica*. In Akerberg and Hagberg (eds), *Recent plant breeding research. Svalöf 1946–1961*. New York and London.

Sinskaia, E. N. (1928). The oleiferous plants and root crops of the family Cruciferae. *Bull. appl. Bot. Genet. Pl. Breed.*, 19, (3), 1–648.

Vavilov, N. I. (1926). *Studies on the origin of cultivated plants*. Leningrad.

Wit, F. (1966). The use of inbred lines in turnip breeding. *Qualitas Pl. Mater. veg.*, 13, 305–10.

Cabbages, kales etc.

Brassica oleracea (Cruciferae)

K. F. Thompson
Plant Breeding Institute
Cambridge England

1 Introduction

This polymorphic species has provided a range of vegetables for human consumption as well as green food for animals. Within *B. oleracea* almost all parts of the plant may be modified into storage organs which are utilized by man. Numerous overlapping leaves surround the terminal bud in the cabbage; enlarged axillary buds occur in Brussels sprouts; the kohlrabi has a swollen, bulb-like stem; the marrow-stem kale has a fleshy but not bulb-like stem; and in the cauliflower (broccoli) the inflorescence and the flower buds are thickened and fleshy to give the edible curd. Although widely grown in temperate regions, these crops must still be considered to be of relatively minor economic importance.

2 Cytotaxonomic background

Brassica oleracea is a diploid with $2n = 2x = 18$. Studies of secondary association have been interpreted as showing a basic number for the genus *Brassica* of $x = 5$, making *B. oleracea* a modified amphidiploid from a cross between two primitive 5-chromosome species with subsequent loss of one pair of chromosomes. Other authors have suggested that $x = 6$ and that the constitution of the haploid set of *B. oleracea* is AABBCCDEF (Yarnell, 1956). The occurrence of up to two bivalents in kale haploids (Thompson, 1956), gives some support to the latter hypothesis. Six different chromosome types were recognised by Röbbelen (1960) on total length, symmetry of arms and especially on the shape of the heterochromatic centromere region in the three *Brassica* genomes. In each genome, different basic chromosomes had been duplicated and the haploid constitution of *B. oleracea* was given as A BB CC D EE F.

Although the wild cabbage is native to the coasts of northwestern Europe as well as to the Mediterranean, it is doubtful whether the many cultivated varieties of *B. oleracea* evolved solely from the wild cabbage, *B. oleracea* var. *oleracea*; several other wild diploid relatives may have contributed. These wild relatives have been described as different species and include *B. cretica*, *B. insularis* and *B. rupestris* (Yarnell, 1956); crosses between them, however, are interfertile and so they have, along with *B. oleracea*, been classified into one cytodeme by Harberd (1972). Some of these wild forms have characteristics quite unlike typical *B. oleracea*; for example the short, fat siliqua of *B. macrocarpa* and the corymbose inflorescence of *B. cretica*. The fixation of distinctive characters, combined with the absence of mating barriers, may have resulted from geographical isolation of populations on the sea cliffs of Europe. Perhaps the whole complex should be classified as one species, *B. oleracea*, in which morphological variability is not restricted to the cultivated types.

3 Early history

The early evolution of the different cultivated types took place in the Mediterranean area and/or in Asia Minor. The development of cultivars from wild species probably took place in several different areas of the Mediterranean; for example, the eastern Mediterranean was the evolutionary centre for the cauliflower (Helm, 1963). Cultivation of kales probably occurred first several thousand years ago, the Greeks growing them at least as early as 600 B.C.; several other types, including a heading cabbage and possibly kohlrabi, were described by writers in ancient Rome. These types would have been highly self-incompatible and almost completely cross-fertilized. As all cultivar groups of *B. oleracea* are completely cross-fertile, the maintenance of distinct types must have been difficult.

All Brassicas contain glucosinolates which, in crushed leaves, are broken down by the enzyme myrosinase to give bitter tasting and goitrogenic substances: isothiocyanates, thiocyanates, nitriles and goitrin. Selection, at an early stage in domestication, must have been for plants which were less bitter-tasting than usual. Semi-quantitative estimations of these substances by Josefsson (1967) show that the stem, petiole and lamina of wild cabbage can contain four times the quantity present in organs eaten by man; for example, cauliflower curd, cabbage heart and kohlrabi bulb. The present Mediterranean wild species are annuals growing in the warm and moist winter. As the early cultivars spread to northwestern Europe, biennial types must have been selected with greater winter-hardiness and a considerable cold

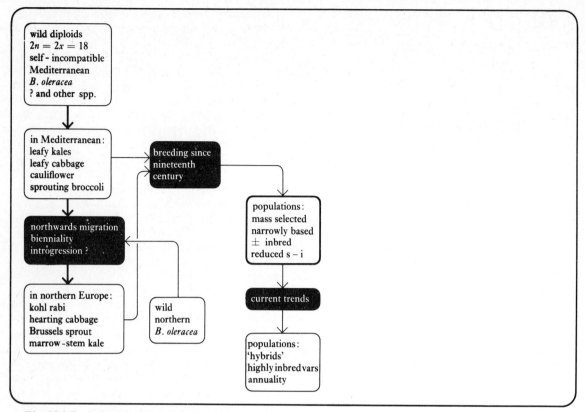

Fig. 17.1 Evolution of cultivated forms of *Brassica oleracea*.

vernalization requirement for flowering; this change probably resulted from out-crossing with native biennial or perennial wild cabbages. Less fibrous, thicker stems and more succulent storage organs were other early selection criteria.

Without doubt, leafy unbranched kales and branching, thin-stemmed kales were the earliest cultivated Brassicas. These *B. oleracea* var. *acephala* types gave rise to the forage kales, which were later fed mainly to livestock. Although an open heading cabbage, 'Tritianon', with a head 30 cm in diameter was described by Pliny, Helm (1963) considers that the real hearting cabbage, *B. oleracea* var. *capitata*, evolved in Germany. Both red and white cabbages were known about A.D. 1150 in Germany and in England by the fourteenth century (Fig. 17.1). Three types of Savoy cabbage, varying from plants with loose rosettes of crinkled leaves to others with well-formed hearts but only slightly crinkled leaves, were described in a German herbal in 1543. Savoy cabbage probably originated in Italy and spread to France and Germany in the sixteenth and seventeenth centuries.

Pliny described 'a Brassica in which the stem is

thin just above the roots, but swells out in the region that bears the leaves, which are few and slender', a description which Pease considered must refer to the kohlrabi. This vegetable was described in the sixteenth century herbals and Gerarde stated that it was grown in Italy, Spain and Germany (Helm, 1963). Boswell (1949) thought that Pliny's description applied to the cauliflower and suggested that the kohlrabi developed in northern Europe in the fifteenth century. However, Helm says that the cauliflower, in its present form, was unknown before the early Middle Ages. Three different cultivars of 'Chou de Syrie' were described in a work by an Arab botanist in Spain in the twelfth century, but it is not known whether the cauliflower was grown in Spain then. It was not mentioned by the earlier German herbalists and was only described in 1576 as *Brassica florida botrytis* and in 1583 as *Brassica cauliflora*. It was probably introduced into Italy by the Genoese from the Levant or Cyprus about 1490. It was grown for seed around the Gulf of Naples. At the beginning of the seventeenth century, it reached Germany, France and England.

Sprouting broccoli, *B. oleracea* var. *italica*, came

from the Levant, Cyprus or Crete to Italy. It was not mentioned until 1660 and was referred to as 'sprout colli-flower' or 'Italian Asparagus' in Miller's Gardener's Dictionary of 1724. As all flower buds open in sprouting broccoli (while, in the cauliflower, many are fleshy, deformed and sterile), sprouting broccoli might be expected to have preceded the cauliflower in the evolution of the species. It is not known whether sprouting broccoli did not exist until the seventeenth century or whether it was not distinguished from the cauliflower before that time.

The Brussels sprout, *B. oleracea* var. *gemmifera*, in its present form, appeared only in about 1750 in Belgium as a 'sport', but by 1800 had reached France and England and, from 1820, was known as 'Chou de Bruxelles'. Its earlier history is uncertain. Records from 1231 and 1472 in Belgium refer to 'un cent de sproq', but may have referred to *B. capitata polycephalos*, illustrated by Dalechamps in 1587, which had a large number of sprouts but lacked the characteristic terminal head (Helm, 1963).

Marrow-stem kale, *B. oleracea* var. *acephala*, is recorded with certainty only from the Vendée region of France in the early nineteenth century and its origin is not known. From France, it spread quickly to Germany and Denmark and was introduced into England about 1900. It is grown in New Zealand under its French name, chou moellier.

4 Recent history

Originally grown only in gardens, Brussels sprouts, cabbages and cauliflowers have become market garden crops in this century while the acreage of kale for autumn and winter cattle grazing reached a peak of 146 kha in 1960 in England and Wales. Since the late nineteenth century, cultivars of Brussels sprouts and cauliflowers were bred for marketing at different times of the year while breeding for improved quality, such as white curds and round heads in winter cauliflowers and small, firm Brussels sprouts, became important to meet marketing and processing requirements. Disease resistance, especially resistance to *Fusarium* wilt of cabbage in America, assumed greater importance.

Using mass selection methods, breeders achieved greater uniformity by more intense selection and cultivars were frequently developed from very few parental lines. This was probably responsible for the tendency to self-compatibility found in Brussels sprouts, cauliflowers and kohlrabi, in comparison with agricultural kales, which are morphologically less specialized and have received relatively little attention from the breeders. Thompson and Taylor

(1966) found, in this sporophytically controlled incompatibility system, that at least two of the three same pollen-recessive S-alleles were found in all varieties of *B. oleracea* examined but that their frequency was much higher and that the number of S-alleles high in the dominance series was lower in these partially self-compatible vegetable cultivars than in a commercial stock of marrow-stem kale. Self-compatibility was determined by genes outside the S-allele system but occurred only in the absence of S-alleles high in the dominance series (Thompson and Taylor, 1972).

The summer cauliflowers are the only fully self-compatible group in *B. oleracea*; Watts (1965) found little if any inbreeding depression on selfing in this crop, in contrast to the more self-incompatible autumn and winter cauliflowers. The change from outbreeding to inbreeding seems to have occurred within about 100 years, since the middle of the nineteenth century. The evolution of self-compatibility in the summer cauliflower may be paralleled by the breeding of the self-compatible winter cauliflower, Cambridge Dwarf April, which was obtained by rigorous selection for 30 years, following a cross between two plants of the winter cauliflower, St George, which is one of the most self-incompatible of its kind (Watts, 1963).

5 Prospects

With high labour costs and increasing mechanisation, growing emphasis is now being put on breeding varieties which can be harvested mechanically in one operation. This makes uniformity of heading, curding or sprout production within a cultivar much more important.

Breeders are attempting to meet the demand for uniformity combined with vigour by the production of F_1 hybrids from uniform, inbred lines, mainly by exploiting knowledge of the self-incompatibility system. However, unacceptable numbers of selfed or sibbed plants have occurred in some Brussels sprout hybrids because the parental inbreds were (indeed had to be) partially self-compatible and the pollinating honey bees prefer to visit flowers of one inbred rather than the flowers of both lines indiscriminately. Inbreds that are homozygous for S-alleles high in the dominance series should be more self-incompatible and the self-incompatibility should be less influenced by environmental factors. The production of three-way crosses and double-cross hybrids, based on single-crosses between morphologically similar but genetically unrelated inbred lines, are also being tried, as well as synthetic varieties.

51

In the agricultural kales, where uniformity is less important, a short stemmed marrowstem kale hybrid, suitable for strip-grazing with electric fencing and with less fibrous stems, is replacing taller varieties in England. Problems in producing sufficient basic seed from inbreds in the field and the labour required to produce inbred seed by bud-pollination necessitated changing the hybrid, Maris Kestrel, from a double- to a triple-cross hybrid, in which the production of hybrid seed depends on the operation of the incompatibility system. Future improvements in agricultural kales will probably be in quality, by reducing the goitrogenic thiocyanate content of the leaf and also the substance (probably S-methyl cysteine sulphoxide – Smith, Earl and Matheson, 1973) responsible for producing kale anaemia in livestock.

Future market requirements for 'mini' cauliflowers, for smaller cabbage heads and for small Brussels sprouts for canning may be met by very uniform, self-compatible cultivars in which yield is not depressed by inbreeding. The example of the cauliflowers shows this to be possible.

Calabrese, a sprouting broccoli, *B. oleracea* var. *italica*, which is an annual, flowering in the early autumn, has become very popular as a vegetable in America. With improved facilities for low temperature storage of vegetables and acceptance of frozen food by the housewife, evolution of Brussels sprouts, cabbages and cauliflowers may all be directed towards the production of annual types, the produce of which can be harvested in the late summer and early autumn and stored to avoid losses due to pigeons and hard winters. Selection for tolerance of handling and for retention of an attractive appearance after storage would be necessary. Improvements in the nutritive value of *B. oleracea* vegetables have been suggested by Schuphan (1958).

6 References

Boswell, V. R. (1949). Our vegetable travellers. *Natnl geogr. Mag.*, **96**, 145–217.

Harberd, D. J. (1972). A contribution to the cytotaxonomy of *Brassica* (Cruciferae) and its allies. *Bot. J. Linn. Soc.*, **65**, 1–23.

Helm, J. (1963). Morphologisch-taxonomische Gliederung der Kultursippen von *Brassica oleracea*. *Kulturpflanze*, **11**, 92–210.

Josefsson, E. (1967). Distribution of thioglucosides in different parts of *Brassica* plants. *Phytochem.*, **6**, 1617–27.

Röbbelen, G. (1960). Beiträge zur analyse des *Brassica* genoms. *Chromosoma*, **11**, 205–28.

Schuphan, W. (1958). Biochemische Stoffbildung bei *Brassica oleracea* in Abhangligkeit von morphologischen und anatomischen Differenzierungen ihrer Organe. I. Vegetative Organe. *Z. Pflanzenzücht.*, **39**, 127–86.

Smith, R. H., Earl, C. R. and Matheson, N. A. (1973). The probable rôle of S-methyl cysteine sulphoxide in kale poisoning in ruminants. *Biochem. Soc. Trans.*, **1**, 75–8.

Thompson, K. F. (1956). Production of haploid plants of marrow-stem kale. *Nature, Lond.*, **178**, 748.

Thompson, K. F. and Taylor, J. P. (1966). The breakdown of self-incompatibility in cultivars of *Brassica oleracea*. *Heredity*, **21**, 345–62.

Thompson, K. F. and Taylor, J. P. (1972). Self-compatibility in kale. *Heredity*, **27**, 459–71.

Watts, L. E. (1963). Investigations into the breeding system of cauliflowers *Brassica oleracea* var. *botrytis*. I. Studies of self-incompatibility. *Euphytica*, **12**, 323–40.

Watts, L. E. (1965). Investigations into the breeding system of cauliflowers *Brassica oleracea* var. *botrytis* II. Adaptation of the system to inbreeding. *Euphytica*, **14**, 67–77.

Yarnell, S. H. (1956). Cytogenetics of the vegetable crops II. Crucifers. *Bot. Rev.*, **22**, 81–166.

Swedes and rapes

Brassica napus (Cruciferae)

I. H. McNaughton
Scottish Plant Breeding Station
Edinburgh Scotland

1 Introduction

Brassica napus provides two forages and an oil-seed. Of the former, the swede remains an important animal fodder in northern Europe, Russia and New Zealand and provides a minor vegetable for human consumption; forage rape yields a leafy fodder for sheep in northern Europe and New Zealand. Of the oil-seeds, there are both annual and biennial forms. The latter (winter oil-seed rape) has predominated in the recent spectacular development of oil-seed production in the EEC which has risen from 625 kt in 1967 to 1·1 Mt in 1972. The oil finds a variety of food and industrial uses.

It is interesting to note a broad parallelism between this species and *B. campestris* (which is, after all, a progenitor of *B. napus*): both provide biennial fodder bulb crops (which are also eaten as vegetables), leafy crops (fodder in one and vegetable in the other) and oil-seeds (both annual and biennial).

2 Cytotaxonomic background

Brassica napus ($2n = 38$, AACC) has been demonstrated experimentally to be an amphidiploid of *B. campestris* ($2n = 20$, AA) and *B. oleracea* ($2n = 18$, CC) (U, 1935; and see Chapter 19, Fig. 19.1). The parental species are extremely difficult to cross artificially, due to endosperm deficiency leading to embryo abortion; occasional artificial forms of *B. napus* have, however, been successfully raised, with or without embryo culture, by crossing diploid parents, followed by colchicine treatment of progeny, and by crossing autotetraploid forms of the parents.

The spontaneous formation of *B. napus* is likely to have been an extremely rare event for, not only are there post-fertilization barriers, but differences in flower colour and form leading to discrimination by insect pollinators are likely to reduce the chances of inter-specific hybridization still further.

Brassica napus is generally self-fertile, whereas its parent species both possess effective sporophytic incompatibility systems; the chances of establishment of rare natural hybrids would presumably be enhanced if they were self-fertile from inception so the difference may reflect early natural selection for self-fertility. However, self-incompatibility is often weakened in polyploids derived from self-incompatible diploids and there may be no need to seek an evolutionary explanation. Whatever the cause, *B. napus* is tolerant of inbreeding.

Brassica napus is known to cross only with extreme difficulty with *B. oleracea* but readily with *B. campestris* (Yarnell, 1956). Natural introgression with the latter seems possible and has certainly occurred in recent plant breeding. The oriental *B. napella* has been shown to be completely inter-fertile with *B. napus* and may be regarded as a variety of it (Olsson, 1954). A separate origin seems probable, presumably from local *B. campestris* and cultivated *B. oleracea* introduced from Europe.

For a review of cytogenetic relationships in the Brassicas, see Yarnell (1956).

3 Early history

It is uncertain whether or not *B. napus* exists in truly wild form. Linnaeus recorded it in sandy coastal areas of Sweden but the plants he saw may have been escapes. If wild *napus* exists, it must be a European–Mediterranean species which originated in the area of overlap between *B. oleracea* and the much more widely distributed *B. campestris*.

A range of morphological forms, paralleling those found in *B. napus*, occur in *B. campestris*, the true turnips being equivalent to the swede and there being annual and biennial oil-seed forms of both species. On this basis, Olsson (1960) suggested that *B. napus* could have arisen several times by spontaneous hybridization of different forms of *B. campestris* and *B. oleracea*. Thus, the swede could have originated in mediaeval gardens where turnips and kale grew side by side. Another view is that the swede could have arisen by selection from an oil-seed form but this is perhaps less likely. That forage rape originated from oil-seed forms, however, seems plausible.

Rape has been recorded as an oil-seed crop in Europe at least since the middle ages but which species is, unfortunately, not known (Appelqvist and Ohlson, 1972). By the early nineteenth century, British rape (*B. napus*) and Continental rape (*B. campestris*) were recognized as being distinguishable on leaf characters. At that time, rape, almost certainly the biennial form of *B. napus*, was being used as an autumn forage for

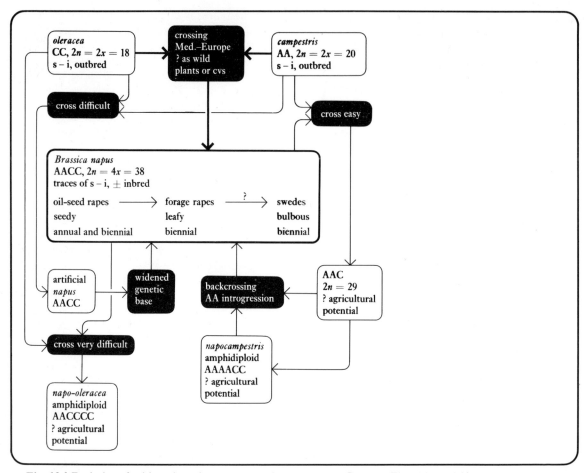

Fig. 18.1 Evolution of cultivated swedes, rapes, etc., *Brassica napus*. (Compare Figs. 16.1 and 19.1.)

sheep; sometimes it was grazed lightly and left to yield an oil crop in the following summer.

Swedes were first recorded in Europe in 1620 by the Swiss botanist Caspar Bauhin, but probably existed earlier than this (Boswell, 1949). Introduction into Britain appears to have been from Sweden around 1775–80.

In summary, *B. napus* may or may not be a cultigen. Its domestication is recent; the swedes and rapes are probably only a few hundred years old and the oil seeds may or may not be somewhat older. A multiple origin from different parental combinations of *campestris* and *oleracea* is possible, even probable. *Brassica napella* is an oriental analogue.

4 Recent history

As remarked above, *B. napus* is tolerant of inbreeding. From the nineteenth century onwards, mass and line selection, with a good deal of inbreeding, was the standard method of improvement. Land races were thus replaced by named cultivars.

Introgression between *napus* and *campestris* has been a feature of the recent history of swedes. 'Hybrid' swedes, arising from either deliberate or accidental crossing with turnips, followed by selection, were reported in Britain in the middle of the nineteenth century (Wilson, 1859). It is probable that present-day swedes owe some of their diversity to such hybridization. Swede cultivar types and breeding methods have been described by Frandsen (1958).

Artificial swedes obtained from crossing turnips with kale or kohlrabi provided useful breeding material (Namai and Hosoda, 1967). An artificial forage rape, in which marrowstem kale was one parent, has proved more acceptable to stock than a leading commercial cultivar. Some artificial oil-seed rapes have proved of value, particularly with regard to winter hardiness and seed yield (Olsson, 1963).

Artificial 'heading' forms of *B. napus* have been produced in Japan from Chinese cabbage (*B. campestris*) and cabbage (*B. oleracea*), thus introducing an entirely new morphological type into the group (Shinohara and Kanno, 1961). Artificial forms may, perhaps, best be regarded as a means whereby new variation may be introgressed into *B. napus* from the parental species. Considerable heterosis has sometimes been observed in crosses between natural and artificial lines (Sarashima, 1967).

The range of variation in forage rape is not great. Dwarf and giant cultivars and a number of intermediate forms, varying in leaf/stem ratio, are currently available. Siberian kale, Hungry Gap kale and Ragged Jack kale are distinct forms, probably best classified with forage rape; certainly they belong in *B. napus* and not with the true kales.

Recent Swedish work on forage rape has shown that the content of rhodanidogenic glucosides, known to induce goitre in stock, is under genetic control and selection for low content should be effective (Ellerström and Josefsson, 1967). Forage and silage types of rape are recognized in Sweden, the former being of higher stem edibility, the latter higher in gross yield. Selection has also been for digestibility and for resistance to *Plasmodiophora* and insect pests.

Aphid-resistant forage rapes have been developed in New Zealand from crosses with a resistant swede cultivar and have proved useful where certain viruses are prevalent. Rapes bred in New Zealand and Britain for resistance to some races of *Plasmodiophora* have had restricted success.

As to oil-seed rape forms, work at the Swedish Seed Association, from 1918 onwards, showed that it was possible, by individual plant selection, to breed cultivars with improved oil content from old land races. Later heritability studies have shown that selection for oil content in winter rape can be expected to give better results than selection for seed yield (Olsson and Andersson, 1963). Considerable advances in the breeding of oil-seed rape have taken place in recent years, largely due to improved analytical techniques and better chemical knowledge of the oil components and their genetic control. Perhaps the most important feature has been the development of a technique for extraction and analysis of oil from single seeds without impairing viability. Improved crop management and advances in plant breeding have resulted in considerable improvements in seed yield of oil-seed rape in the twenty-five year period to 1971 (Downey, 1971).

New characters have been introgressed into oil-seed rape from turnip-rape; this method is used in Japan as a regular breeding technique (Shiga, 1970).

The amphidiploid *napocampestris* (AAAACC, $2n = 6x = 58$) has been explored by several breeders but without success. Oil-seed combinations were too infertile and forage types too low in dry matter content and/or yield. However, an enormous range of plant types could be constructed thus and the group (in effect a new species) may yet be found to have agricultural uses.

5 Prospects

Two rather important trends are apparent. One is towards the breeding of F_1 hybrid varieties and the other is towards an extended use of interspecific hybridization. Thus swedes, with their exceptional potential for high-quality fodder production, are known to show yield heterosis (McNaughton and Munro, 1972) and could, in principle, be bred as hybrids by the use of the self-incompatibility system; suitable *S*-alleles can be introgressed from *campestris* relatively easily, from *oleracea* only with much more difficulty. In the oil-seed rapes, a cytoplasm-restorer male-sterility system has been identified (Shiga and Baba, 1973); whether or not it is successful, steady improvement of the crop by elimination of the several chemical disadvantages of the oil and meal for human and animal food seems certain.

Interspecific transfer of useful genes from *campestris* includes the *S*-alleles, referred to above, and *Plasmodiophora* resistance (of which turnips are a useful source). Interspecific combinations *per se* also have their uses and there is no doubt that, difficult as the cross is, *oleracea-campestris* hybrids (i.e. artificial *napus*) could materially widen the genetic base of *napus* crops; several examples have been given above. Also, AAC hybrids have been shown to have considerable potential as a rape-like crop if means of producing F_1 seed in commercial quantity could be developed (Mackay, 1973). And, though not strictly relevant to a discussion of *Brassica napus*, there seems to be considerable, but yet quite unexploited, agricultural potential in *Raphanobrassica*; many combinations of the three (A, C, R) genomes are possible. One of them (CCRR) is being developed as a rape-like crop which promises better disease resistance than rape itself (McNaughton, 1973).

Haploid plants from pollen cultures would have obvious potential for expediting the production of inbred lines, whether as varieties or as parents of hybrid varieties.

6 References

Appelqvist, L.-Å. and **Ohlson**, R. (1972). *Rapeseed.*   55

London and New York.

Boswell, V. R. (1949). Our vegetable travelers. *Natnl geogr. Mag.*, 96, 145–217.

Downey, R. K. (1971). Agricultural and genetic potentials of Cruciferous oilseed crops. *J. Am. Oil Chem. Soc.*, 48, 718–22.

Ellerström, G. and Josefsson, E. (1967). Selection for low content of rhodanidogenic glucosides in fodder rape. *Z. PflZücht.*, 58, 128–35.

Frandsen, K. J. (1958). Breeding of swede (*B. napus* var. *rapifera*). *Handb. landw. PflZücht.*, 3, 311–26.

Mackay, G. R. (1973). Interspecific hybrids between forage rape (*Brassica napus*) and turnip (*Brassica campestris* ssp *rapifera*) as alternatives to forage rape. I. An exploratory study with single pair crosses. *Euphytica*, 22, 495–9.

McNaughton, I. H. (1973). Resistance of *Raphanobrassica* to clubroot disease. *Nature, Lond.*, 243, 547–8.

McNaughton, I. H. and Munro, I. K. (1972). Heterosis and its possible exploitation in swedes. *Brassica napus* ssp. *rapifera*). *Euphytica*, 21, 518–22.

Namai, H. and Hosoda, T. (1967). On breeding of *Brassica napus* obtained from artificially induced amphidiploids. III. On the breeding of synthetic rutabaga (*B. napus* var. *rapifera*). *Jap. J. Breed.*, 17, 38–48.

Olsson, G. (1954). Crosses between *Brassica napus* and Japanese *Brassica napella*. *Hereditas*, 40, 249–52.

Olsson, G. (1960). Species crosses within the genus *Brassica*. II. Artificial *Brassica napus*. *Hereditas*, 46, 351–86.

Olsson, G. (1963). Induced polyploids in *Brassica*. In Åkerberg and Hagberg (eds.), *Recent plant breeding research, Svalöf, 1946–1961*. New York and London.

Olsson, G. and Andersson, G. (1963). Selection for oil-content in Cruciferous crops. In Åkerberg and Hagberg (eds.), *Recent plant breeding research. Svalöf, 1946–1961*. New York and London.

Sarashima, M. (1967). Studies on the breeding of artificially synthesised soiling forage rape. *Jap. J. Breed.*, 17, 26–31.

Shiga, T. (1970). Rape breeding by interspecific crossing between *Brassica napus* and *Brassica campestris* in Japan. *Jap. agric. Res. Quart.*, 5(4), 5–10.

Shiga, T. and Baba, S. (1973). Cytoplasmic male sterility in oil seed rape, *Brassica napus*, and its utilization to breeding. *Jap. J. Breed.*, 23, 187–97.

Shinohara, S. and Kanno, M. (1961). The species hybrid 'Hakuran' between common cabbage and Chinese cabbage. *Agric. Hort., Tokyo*, 36, 1189–90.

U, N. (1935). Genome analysis in *Brassica* with special reference to the experimental formation of *B. napus* and peculiar mode of fertilization. *Jap. J. Bot.*, 7, 389–452.

Wilson, J. (1859). *Our farm crops*. London.

Yarnell, S. H. (1956). Cytogenetics of the vegetable crops. II. Crucifers. *Bot. Rev.*, 22, 81–166.

Mustards

Brassica spp. and *Sinapis alba* (Cruciferae)

J. S. Hemingway
Reckitt and Colman Products Ltd
Norwich England

1 Introduction

The term 'mustard' is believed to be derived from the use of the seeds as condiment; the sweet 'must' of old wine was mixed with crushed seeds to form a paste, 'hot must' or 'mustum ardens', hence mustard. It is amongst the oldest recorded spices, with Sanskrit records dating back to about 3000 B.C. (Mehra, 1968) and an extensive literature from Greek and Roman times onwards (Rosengarten, 1969). The large use of mustard seed makes it the most important spice in the world in quantity terms: in value terms, it is exceeded only by pepper. Mustard, unusual in being a temperately-grown spice, is nowadays cheaply produced as it is a large-scale fully mechanized crop.

For use as spice, the main present-day production is in North America, in the prairie provinces of Canada and southwards into Montana and the Dakotas. Other major production centres are the UK and Denmark, with very little now being grown in former areas such as France, Germany, Italy, the Danube basin and eastern European countries. Smaller-scale production occurs in Australia and Argentina for local use. Total annual world trade for use in condiment is estimated as 150 kt, the main manufacturing centres being in the USA, UK, France, Germany and Japan.

Botanically, four species are involved, as follows:

White mustard   *Sinapis alba*
Brown mustard   *Brassica juncea*
Black mustard   *Brassica nigra*
Ethiopian mustard   *Brassica carinata*

In condiment, *S. alba* contributes a 'hot' and *B. juncea*, *B. nigra* or *B. carinata* a 'pungent' principle. *Brassica carinata* has always been a very small local peasant production in northeastern Africa. Until the 1950s, *B. nigra* was almost the sole source of pungency but, with perhaps unparalleled rapidity in any industry, let alone in an area as traditional as the spice trade, *B. juncea* replaced *B. nigra* over a single decade and now *B. nigra* cropping occupies less than 1 per cent of production. This change was mainly due to mechanization requirements, as *B. nigra* had to be hand-harvested.

Mustards are also valuable as oil crops. *Sinapis alba* was widely grown in Sweden for this purpose in the 1940–50s and *B. juncea* is a part of the *juncea–campestris* complex, one of the major oil crops of the Indian sub-continent. They are also used as salads, green manure and fodder crops, and as leaf and stem vegetables, especially in the Far East (Vaughan and Hemingway, 1959).

2 Cytotaxonomic background (Fig. 19.1)

The common nomenclature of mustards remains somewhat confused. 'White' mustard, *S. alba* is known as 'Yellow', *B. hirta* in North America; the names 'Brown', 'Black' and 'Indian' are to a degree interchangeable, and there are cultivars of 'Brown' with yellow seed coats.

Sinapis alba is the only crop species in the small genus *Sinapis*, of which the only other notable species is *S. arvensis*, charlock, a major and widespread weed. *Sinapis alba* differs from the *Brassica* mustards in having $x = 12$; it is a diploid with $2n = 24$.

Brassica nigra ($x = 8$) is one of the three basic diploid cultivated *Brassica* species which form amphidiploids in the triangular pattern elegantly described by Japanese workers, principally U (1935). The other two *Brassica* mustards are allotetraploids (Fig. 19.1).

Fig. 19.1 Cytogenetic relationships of the *Brassica* mustards.

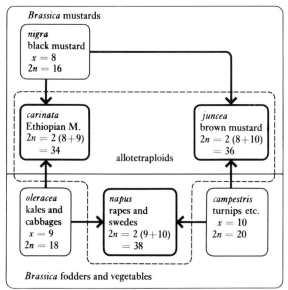

Experimental synthesis of these amphidiploids has been carried out by Frandsen (1943) and others. Considerable additional evidence on cytotaxonomic relationships in *Brassica* has recently been reported by Harberd (1972) and Vaughan and his co-workers have explored serological and protein analysis methods in further elucidation. Vaughan *et al.* (1963) suggested, from the studies of chemical and morphological variation in *B. juncea* which led to the identification of three secondary centres of diversity, that $x = 10$ *Brassicas* other than *B. campestris* (e.g. *B. japonica*, *B. pekinensis*, *B. trilocularis*) could have been parents of *B. juncea*; however these three species are sometimes treated as subspecies of *campestris*. No $x = 8$ species other than *B. nigra* is recorded. Little attention has been paid to *B. carinata*.

3 Early history (Fig. 19.2)

Sinapis alba is a wind pollinated species with a sporophytic incompatibility system similar to that studied in various horticultural Brassicas and kales. Much material, however, appears to be sib-compatible. The centre of origin is believed to be the eastern Mediterranean, and wild forms occur around most of the Mediterranean littoral, especially in the Aegean. The wild forms are typically low growing, much branched, short-season dehiscent plants, often with brown or black seeds of the form recorded as *melanosperma* (a character which has been bred out of advanced cultivars). One feature of wild material is the presence of a single seed in the siliqua beak in addition to 1–3 in each of the two locules; as such a seed is unable to germinate until the beak has rotted away, it has been suggested that this is a natural survival mechanism. The feature has virtually disappeared from cropped land-races, perhaps by unconscious human selection; in bred cultivars, the seed number per locule may rise as high as seven, though seed weight per pod is a more constant factor than seed number.

B. nigra probably originated in the Asia Minor–Iran area but was so early used as a commercial spice that it became widespread in Europe, Africa, Asia, India and the Far East. Early forms are of short-season, spreading, semi-erect growth, up to 1 m tall; deliberate selection of more erect, taller material has evidently long been practised, though no known land-race or cultivar is indehiscent.

The primary centre of origin of *B. juncea* is believed to be Central Asia-Himalayas, with migration to three secondary centres in India, China and the Caucasus but, as noted above, the strong evidence of diversity in several aspects reported by Vaughan *et al.* (1963) could

57

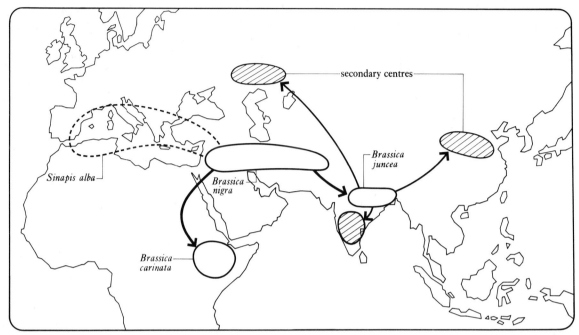

Fig. 19.2 Evolutionary geography of the mustards, *Brassica* spp. and *Sinapis alba*.

indicate independent hybridization of *B. nigra* with local *n* = 10 species (or subspecies) in those areas. Yellow-seeded forms arose in the Orient. Similarly, it is thought that *B. carinata* arose from *B. nigra* by crossing with a *B. oleracea* form in north-eastern Africa. Local selection of both species for use as vegetables must have occurred over long periods of time in many areas, particularly of *juncea* in south-east Asia and China.

4 Recent history

Land-race forms of *S. alba* and *B. nigra* were cultivated in many European countries from the Middle Ages and acquired joint origin/trade nomenclature as, for example: 'English' 'Danish' 'Dutch' 'Hungarian' White, 'Sicilian', 'Bari', 'English' or 'Russian Brown'. *Brassica juncea* was normally called 'Indian' and *B. carinata* 'Ethiopian'.

In *S. alba*, land-race forms from northern Europe tend to be of much taller, fleshier growth than those from the Danube Basin, and these more southerly forms have mainly been used in breeding for modern cultivation. In Sweden (where the aim to support domestic oil production called for high yields of seed of high oil content) a series ('Primex', 'Secu', 'Trico' successively) was successfully produced at Svalöf in the 1940–50s, a noteworthy aspect of the work being that Primex was the first crop cultivar to arise from an X-irradiation programme. Interest in tetraploid forms has waned after some breeding in Sweden and Germany. The named German varieties arising from work at Giessen and Gross-Lüsewitz (e.g. 'Hohenheimer', 'Steinacher', 'Lüsewitzer Stamm') have been bred particularly for use as green manure crops, being leafy, stemmy and late to flower; 'Gisilba' is more suited to seed production and is grown in Canada. The major centre of breeding of *S. alba* for crop yield, habit, full mechanization and for the many parameters of seed quality for condiment usage has been Norwich, England where, over the past 15 years, 'Bixley' and 'Kirby' have been successfully introduced. Cultivation in Denmark is mainly of English, German or Swedish cultivars or seedsmens' land-races; Canada still relies on merchants' stocks, though breeding has recently commenced at Saskatoon, with particular interest in oil content and quality.

Breeding of *S. alba* is mainly by line selection either after two generations of bud-pollination or 3–4 of sib-pollination, the latter method giving less inbreeding depression. The exploitation of heterosis, particularly in synthetics, is receiving attention.

Very little breeding or selection work has been attempted on *B. nigra* and none is known on *B. carinata*. As mentioned earlier, *B. nigra* was everywhere abandoned in favour of the more suitable *B. juncea* and there seems no likelihood of renewed interest in this species, ill-suited as it is to mechanical harvesting.

There is a long history of breeding *B. juncea* as an oil-seed in India, with many recorded types, such as 'Manipuri-rai', 'Gohna-sarson' and more modern pure lines such as 'RT–11', 'RL–9', etc. Similarly, in Japan and China, established vegetable and oil-seed varieties with a multiplicity of local names are recorded, and recent Japanese workers (e.g. Aoba, 1972) have begun to elucidate the inheritance of plant and seed characters. Attempts to develop *B. juncea* as an oil-seed at Svalöf ('Fiskeby-senap') were made in the 1940s and some similar work is believed to be in progress in Russia. Breeding for condiment usage, usually under the stimulus of manufacturers, has gone on during the last two decades at Giessen, Germany ('Primus'), Douai, France ('Burgonde' and 'Ekla'), Lethbridge and Saskatoon, Canada ('Lethbridge 22A' and imminent further introductions) and Norwich, England ('Trowse', 'Stoke' and 'Newton'). Lethbridge 22A, Stoke and Newton are all yellow-seeded as it has been found that this characteristic, with associated higher kernel content, aids extraction during milling. 'Ekla', 'Stoke' and 'Newton' are of greatly enhanced pungency. The work at Norwich originated from very extensive world collections showing a wide range of variability of both morphological and chemical characteristics, enabling close definition of breeding objectives for habit, yield and quality. All breeding of *B. juncea* is by means of pure lines, as this species is almost exclusively self-pollinating.

5 Prospects

Genetic studies of *B. juncea* are continuing in India and Japan and the breeding of this species for oil-seed usage is in progress at several centres in India, Pakistan and Russia. For condiment, the predominant centres are Norwich and Saskatoon; at the latter centre there is also interest in industrial usage of strains with varied fixed oil fatty acid content. *Sinapis alba* also receives attention as a potential protein crop (seed normally 30–34 per cent oil, 30–32 per cent protein) but the glucosides must be destroyed to improve palatability. *Brassica juncea* proteins have recently been studied by Mackenzie (1973).

Other than the possible exploitation of heterosis in *S. alba*, breeding methods are unlikely to change; polyploidy is disadvantageous and irradiation as an aid to producing mutants has proved too drastic a method. Very large collections of both *S. alba* and *B. juncea*, easily maintained by the long-lived seed, serve as substantial gene pools and wild material of all four mustard species is widely distributed in several continents.

6 References

Aoba, T. (1972). Inheritance of seed colour and seed coat type in *B. juncea*. *J. Yamagata Agr. For. Soc.*, **29**, 28–30.

Frandsen, K. J. (1943). The experimental formation of *B. juncea*. *Dansk Bot. Archiv.*, **11**, 1–17.

Harberd, D. J. (1972). A contribution to the cytotaxonomy of *Brassica* and its allies. *Bot. J. Linn. Soc.*, **65**, 1–23.

Mackenzie, S. L. (1973). Cultivar differences in proteins of *B. juncea*. *J. Amer. Oil. Chem. Soc.*, **50**, 411–14.

Mehra, K. L. (1968). History and ethnobotany of mustard in India, *Adv. Front. Pl. Sci.* **19**, 51–9.

Rosengarten, F. (1969). Mustard seed. In *The Book of Spices*, Philadelphia, 295–305.

U, N. (1935). Genome analysis in *Brassica*. *Jap. J. Bot.*, **7**, 389–452.

Vaughan, J. G. and **Hemingway, J. S.** (1959). The utilisation of mustards. *Econ. Bot.*, **13**, 196–204.

Vaughan, J. G., Hemingway, J. S. and **Schofield, H. J.** (1963). Contributions to a study of variation in *Brassica juncea*. *J. Linn. Soc. (Bot.)*, **58**, 435–47.

Radish

Raphanus sativus (Cruciferae)

O. Banga
Diedenweg 6 Wageningen The Netherlands
formerly Institute of Horticultural Plant
Breeding Wageningen

1 Introduction

The small short-season type of radish is a cool-weather plant. It is a popular garden vegetable and its culture is of commercial interest, especially early production under glass or large scale production in the field.

The larger radish has a wider range of temperature adaptation. Early varieties do well in cool weather, but summer varieties can stand, or even need, relatively high temperatures. The larger type is an important vegetable in Japan, Korea, China, India and other eastern countries where it may be eaten raw or cooked or preserved by storage, pickling, canning or drying. Not only the fleshy root but also the leaves and the young seed pods provide vegetables. In Europe half a century ago, the larger radish used to be eaten more than nowadays and it is still popular in parts of southern Germany.

Mougri-radish forms no fleshy roots, but is grown in southeast Asia specifically for its leaves and its young seed pods which in this form may become 20–100 cm long.

Fodder-radish, too, bears little or no fleshy root and is also grown for its foliage but this is used as fodder or green manure. Its rapid growth has given it a limited but promising popularity in western Europe during the last quarter century.

2 Cytotaxonomic background

All four types of radish belong to the species *Raphanus sativus*, with $2n = 2x = 18$. Botanically they are known as varieties *radicula*, *niger*, *mougri* and *oleifera* respectively. All four types intercross freely among each other and also with related wild species. The most important wild species are *Raphanus raphanistrum* ($n = 9$; Europe eastwards to the Volga and Mediterranean eastwards to the Caspian); *R. maritimus* and *R. landra*, which is considered to be an inland form of *maritimus* ($n = 9$; Mediterranean eastwards to the Caspian and, in Europe, along the coasts of France, Belgium, Holland and Great Britain); and *R. rostratus* (from Greece eastwards to the Caspian).

All three wild species mentioned are variable and a confused nomenclature has been further confounded by the occurrence of cultivated forms run wild. No one of the wild species can be identified as the source of the cultivated radish; quite possibly, all three contributed. The area of maximum diversity from the eastern Mediterranean to the Caspian, is the likely area of origin of the cultivars but Werth (1937) has argued for a more easterly source.

Successful crossing of *Raphanus sativus* with several *Brassica* species and with *Sinapis arvensis* suggest a fairly close relationship between the three genera, but there is no evidence that such crossing has played a role in the evolution of the crop.

3 Early history

Inscriptions on the inner walls of pyramids tell us that the *niger* type of radish was an important food in Egypt about 2000 B.C. and, from certain remarks of Herodotus, it has been inferred that it already was so by about 2700 B.C. (Becker, 1962). It is thought to have spread to China about 500 B.C. and to Japan about A.D. 700 (Sirks, 1957). This implies that radish was known in the eastern Mediterranean at least two millennia before it was known in China.

A survey has shown that variability of the crop decreases from Europe to China and from there to Japan (Sinskaja, 1928). As Europe has doubtless adopted much of the Mediterranean material, this may be taken to support the indications above but the survey results are not very decisive.

The conclusion remains that some area east of the Mediterranean was the probable source of the crop but exact definition of the area is not yet possible.

In contrast with the *niger* type, the *radicula* type must be much younger. A long white form of it first appeared in Europe at the end of the sixteenth century; its origin is unknown, but the speculation that it may have been derived from the *niger* type seems reasonable (Helm, 1957). In the eighteenth century, globular forms were developed, white at first but later also red. In the earlier phases of the evolution of the crop there was thus developed great variation in shape and colour; there were long, half-long, globular, pear-shaped, even flattened roots coloured white, red, yellow and black.

4 Recent history

Purposeful radish breeding has been practised for

some centuries, until recently by means of mass-selection or combined mass-pedigree-selection. Radishes are normally self-incompatible and insect pollinated. The incompatibility system is of the sporophytic type in which the papillae on the stigma surface form the barrier to pollen tube penetration. By disrupting the stigma surface or by applying pollen immediately to the conducting tissue of the style, it is possible to get normal seed set by selfing. Bud pollination is another, and so far more generally used, way of doing the same (Roggen and van Dijk, 1973; Dickinson and Lewis, 1973).

Male sterility has been found in Japanese radishes. Okura (1968) described a type controlled by the interaction of a recessive gene, *ms*, and S-cytoplasm (confirmed by Bonnet, 1970). Hybrid varieties can now be developed using either incompatibility or male-sterility for the control of crossing. Japanese breeders have been breeding hybrids for at least the last ten years especially for the production of *niger* types. In western countries, hybrid varieties are still in the experimental stage.

Adaptation to growing seasons of different length and character has been from the beginning an important breeding objective. Thus selection of varieties suitable for early culture in spring has led to small size, high growth rate and an annual life cycle. Both *radicula* types and the earliest *niger* types are annual; they do not need vernalizing temperatures and will ultimately flower under any photoperiod (though a long photoperiod advances flowering). They are not good for production in full summer. On the other hand, selection for summer, autumn and winter radishes has tended to produce greater sensitivity to vernalizing temperatures and a biennial character. Roots stored during the winter in a cool place flower the following season only when they have had enough cold. In late spring and summer varieties, the rogueing of early bolters must have led automatically to decreased sensitivity to daylength. (Tsukamoto and Konishi, 1959; Banga and van Bennekom, 1962).

Winter growth under poor light in glasshouses may cause difficulties which can be understood now as shortage of light in relation to temperature. Net assimilation may then be too low to allow thickening of the roots with the result that the plants ultimately bolt without having produced harvestable roots. Low temperatures provide one remedy, of course, but strains have been selected which are better adapted to such conditions (Banga and van Bennekom, 1962).

Raising the growth rate to promote earliness may, however, increase sensitivity to 'pithiness' revealed first as dry, white patches in the inner tissues (Hagiya,

1952–59). Many pithiness-resistant varieties have now been introduced, the first being A. R. Zwaan's Cherry Belle, which has spread all over the world.

Differences in resistance to pests and diseases have been reported in the USA, Europe and Japan. Lines resistant to *Fusarium oxysporum* f. *conglutinans* race 2 and *Albugo candida* have been bred by Pound and co-workers (Williams and Pound, 1963) in the USA and lines resistant to virus disease by Shimiru and co-workers (Shimiru *et al.*, 1963) in Japan.

Radishes, as noted above, are all diploid. Experimental autopolyploids have been made, especially in Japan and East Germany. They are not superior to diploids (Takii, personal communication, 1974).

5 Prospects

Most current breeding work is aimed at further adaptation to different growth conditions, at improved resistance to pests and diseases and at improved marketing characteristics. Breeding methods were, until recently, fairly simple but the production of F_1 hybrids is opening new perspectives. Already established in Japan, hybrids hold much promise but seed is still very expensive to produce.

The success of fodder-radish in western Europe as a result of recent German breeding work suggests that this form may have greater potential than superficially appears.

6 References

Banga, O. and **Van Bennekom, J. L.** (1962). Breeding radish for winter production under glass. *Euphytica*, 11, 311–26.

Becker, G. (1962). Rettich und Radies (*Raphanus sativus*) *Handbuch der Pflanzenzüchtung*, 6, 23–78.

Bonnet, A. (1970). Comportement d'une stérilité male cytoplasmique d'origine japonaise dans les variétés européenne de radis. *Eucarpia* (*hort. Sect. Rep.*), CRNA, Versailles, 83–8.

Dickinson, H. G. and **Lewis, D.** (1973). Cytochemical and ultrastructural differences between interspecific compatible and incompatible pollinations in *Raphanus*. *Proc. R. Soc., London*, B, 183, 21–38.

Hagiya, K. (1952–59). Physiological studies on the occurrence of pithy tissue in root crops. *J. hort. Ass. Japan*, 21, 81–6, 165–73; 26, 111–20, 121–5; 27, 68–77; 28, 109–14.

Helm, J. (1957). Ueber den typus der Art *Raphanus sativus*, deren Gliederung und Synonymie. *Kulturpflanze*, 5, 41–54.

Okura, H. (1968). See **Bonnet, A.** (1970).

Roggen, H. P. J. R. and **Van Dijk, A. J.** (1973). Electric aided and bud pollination. *Euphytica*, 22, 260–3.

Shimiru, S., Kanazawa, K., Kono, H. and **Yokota, Y.** (1963). Studies on breeding radish for resistance to virus diseases. *Plant Br. Abstr.*, 34, 3059.

Sinskaja, E. N. (1928). The oleiferous plants and root crops of the family Cruciferae. *Bull. appl. Bot. Genet. Pl. Br.* **19**(3), 1–648.

Sirks, M. J. (1957). Japanese genetica. *Genen en Phaenen*, **2**, 2–10.

Tsukamoto, Y. and Konishi, K. (1959). Studies on the vernalisation of growing plants of radish. *Mem. Res. Inst. Food Sci., Kyoto Univ.*, **18**, 41–7.

Werth, E. (1937). Abstammung und Heimat des Rettichs. *Angew. Bot.*, **19**, 194–205.

Williams, P. H. and Pound, G. S. (1963). Nature and inheritance of resistance to *Albugo candida* in radish. *Phytopath.*, **53**, 1150–4.

Watercress

Rorippa nasturtium-aquaticum (Cruciferae)

H. W. Howard
Plant Breeding Institute
Cambridge England

1 Introduction

Watercress is both a wild plant and a cultivated crop. As a crop it is only grown on a large scale in a small number of regions, mostly in Germany, France and the United Kingdom, because it requires a plentiful supply of spring or artesian water which must be alkaline and with a certain content of nitrates (Lyon and Howard, 1952). Watercress reproduces both by seeds and by vegetative means, roots being found at all nodes of the stem under damp conditions.

2 Cytotaxonomic background

There are two species of watercress, a diploid ($2n = 32$) and a tetraploid ($2n = 64$); a sterile, triploid ($2n = 48$) hybrid between them is also found, even being the commonest type of watercress in some districts. The tetraploid was at first thought to be an autotetraploid (Manton, 1935) but, like many other forms once considered to be autotetraploids, was later shown conclusively to be an allotetraploid (Howard and Manton, 1946). The allotetraploid was suggested to be an undescribed species, but Airy Shaw (1947) showed that it had been previously described by several authors. The two species of watercress, which have white flowers, are sometimes classified as the only two species in the genus *Nasturtium* but are now more usually included in the larger genus *Rorippa* in which all the other species have yellow flowers. The diploid is *Rorippa nasturtium-aquaticum* (or *Nasturtium officinale*) and the tetraploid is *R. microphylla* (or *N. microphyllum*). The triploid hybrid, *R. microphylla* × *nasturtium aquaticum*, has been called *R.* × *sterilis*. There is no difficulty in distinguishing the two species by clear-cut fruit and seed features (Fig. 21.1) and they differ also in other characters. These include the seed of *R. microphylla* needing light for germination whereas *R. nasturtium-aquaticum* seeds will germinate in the dark (Howard and Lyon, 1952). It has been suggested

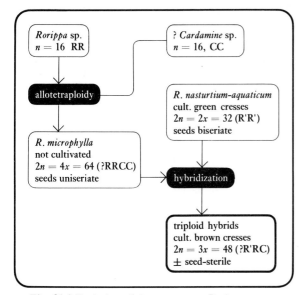

Fig. 21.1 Evolution of the watercress, *Rorippa*.

that the other parent of the allotetraploid might be a species of *Cardamine*. Both diploid and tetraploid watercress are self-compatible.

3 Early history

As pointed out by Manton (1935), there is documentary evidence for the use of watercress as a medicinal plant from the first century A.D. (Dioscorides's Materia Medica of about A.D. 77) to the nineteenth century; it disappeared from the Pharmacopaeias of London and Edinburgh in the 1809 edition. It was chiefly valued as an antiscorbutic. In addition to being a medicinal plant, there is evidence that it was commonly eaten and it is even recorded that watercress formed one of the tributes to which the paramount Kings of Erin were entitled by royal prerogative (Howard and Lyon, 1952). Although watercress was obtained from wild plants and was not cultivated until the nineteenth century, it seems probable that plants of it must often have been spread by man. This may account for the prevalence of the triploid hybrid in some areas where one or both of its parents are absent. Spread by man in recent times has also occurred, for example into New Zealand, Australia and South Africa and into several tropical or sub-tropical places where it is cultivated on a small scale by Chinese market-gardeners. As wild plants both watercress species are widespread in Europe and Asia. They are usually considered as having been introduced into North America.

Manton (1935) suggests that the first large-scale cultivation of watercress may have been in south Germany, especially near Erfurt, as early as 1750.

Large-scale cultivation in England was started at Gravesend in 1808 to supply the London market and, in France, the first watercress farm near Paris was inaugurated in 1811. The German and French watercress farmers used only the diploid species but, in England, both the diploid species (green cress) and the triploid hybrid (brown cress) were grown (Howard, 1947). The tetraploid species was not cultivated in any region.

4 Recent history

As a cultivated crop, watercress is propagated both from seed and vegetatively. Growing from seed has the advantage of starting a crop free from virus infection and is now used by most growers, a common practice being to sow one bed from seed and to use plants from this bed to plant two or more further beds. The triploid brown cress cannot be grown this way and has now virtually disappeared from cultivation. The only advantage that it had over green cress was that it was somewhat more resistant to frost. On the other hand it did not yield as well as green cress.

Repeated growing from seed has one obvious danger: unconscious selection for high seed production rather than high yield of vegetative matter. Growers claim that they are aware of this danger and have produced their own stocks which are both uniform and relatively late in producing flowers. Watercress is self-compatible and probably most seed is the result of self-pollination.

An autotetraploid green cress was produced by Howard (1952) and tested on a commercial scale. It produced very attractive samples with thicker and broader leaflets than the diploid. On the other hand it had a somewhat slower growth rate and was a little more frost tender than the diploid. It was therefore not taken up for commercial growing.

5 Prospects

Watercress is a labour-intensive crop and the cultivation and harvesting are not amenable to mechanization. The area grown may therefore decline. On the other hand the grower receives all his nitrate at no cost from the very large volume of water flowing down the bed (Lyon and Howard, 1952). There may be scope for selecting better strains of green cress but, because of the high possibility of virus infection, it is unlikely that triploid brown cress will ever again be of any importance.

6 References

Airy Shaw, H. K. (1947). The botanical name of the wild tetraploid watercress. *Kew Bull.* 1, 38–46.

Howard, H. W. (1947). Wild and cultivated watercress types. *J. Minist. Agric.*, **53**, 453–7.

Howard, H. W. (1952). Autotetraploid, green watercress. *J. hort. Sci.*, **27**, 273–9.

Howard, H. W. and Lyon, A. G. (1952). Biological Flora of the British Isles: *Nasturtium officinale* R. B. and *Nasturtium microphyllum* Boenningh ex Rchb. *J. Ecol.*, **40**, 228–45.

Howard, H. W. and Manton, I. (1946). Autopolyploid and allopolyploid watercress, with the description of a new species. *Ann. Bot., Lond.*, N.S., **10**, 1–13.

Lyon, A. G. and Howard, H. W. (1952). Science and watercress growing. *J. Minist. Agric.*, **59**, 123–8.

Manton, I. (1935). The cytological history of watercress (*Nasturtium officinale*). *Z. indukt. Abstamm.- u. VererbLehre*, **69**, 132–57.

Cucurbits

Cucumis, Citrullus, Cucurbita, Lagenaria (Cucurbitaceae)

Thomas W. Whitaker
US Department of Agriculture
La Jolla California USA,
and
W. P. Bemis
University of Arizona Tucson Arizona USA

1 Introduction

This chapter is devoted to the major cucurbit crops. Minor ones are treated in Chapter 88. The major cucurbits are usually produced in relatively small quantities for local consumption and so do not enter into production statistics in a significant way. Nevertheless, they are important items in the diets of many peoples because one or more species are elements of nearly every vegetable garden, both home and commercial. The fruits can be used for food either fresh, cooked, pickled, candied, dried or in sauces and curries. Those species with fruits having a hard, tough rind are used for containers, cutlery, musical instruments, ornaments, etc. Used fresh and uncooked the fruits are refreshing (cucumbers, muskmelons, watermelons); cooked summer squash is a pleasant vegetable dish and makes a good ingredient for low calorie diets. Baked winter squash is an excellent food, with high carbohydrate content. The seeds of several species of *Cucurbita* are relatively rich in oil and protein and make a nutritious confection when roasted and salted.

2 Cytogenetic background

The Cucurbitaceae are difficult cytologically and only recently have good methods been devised (Varghese, 1973). Little is known about cucurbit genetics, partly because of the large space requirements of individual plants and partly because crops are mainly grown in small patches. Thus, units are small even though the total acreage is substantial. They have therefore generated little research support. The relevant species are listed in Table 22.1.

Table 22.1 Major cucurbit crops.

Species	Common name	$n =$ $x =$	Native area
1 *Cucumis sativus*	Cucumber	7	India
2 *Cucumis melo*	Muskmelon, cantaloup, etc.	12	Africa, India
3 *Citrullus lanatus*	Watermelon	11	South Africa
4 *Lagenaria siceraria*	White-flowered gourd	11	Africa, the Americas
5 *Cucurbita pepo*	Summer squash, pumpkin, marrow	20	North America, north of Mexico City
5 *Cucurbita mixta*	Winter squash	20	Mexico, Central America
5 *Cucurbita moschata*	Winter squash	20	Mexico, Central America
5 *Cucurbita maxima*	Winter squash, pumpkin	20	Northern South America, Central America
5 *Cucurbita ficifolia*	Fig-leaf gourd	20	Mexico, Central America, northern South America

In the following numbered paragraphs, the crops are treated in the order of Table 22.1.

1 The cucumber, *Cucumis sativus*, is anomalous in being the only species in the genus with $2n = 2x = 14$ and in being native to India. Other species of *Cucumis* have $x = 12$ and are mostly African, primarily from southern Africa. There is more genetic information available for cucumbers than for any other cucurbit. Some 55 loci have been identified. These include genes governing plant habit, sex expression, fertility, fruit set, fruit type, fruit flavour and disease resistance (scab, virus). The genetics of sex expression in cucumber has been thoroughly explored and is being exploited in the commercial production of hybrid seed. Another interesting group of genes of wide occurrence in the cucurbits are those responsible for producing cucurbitacins, terpene compounds that produce bitter flavours in foliage and fruits. A mutant lacking cucurbitacins has made it possible for breeders to produce non-bitter cucumbers; however, the presence of cucurbitacins makes the plants resistant to some insects though more attractive to others. Another interesting mutant, useful in the production of greenhouse cucumbers, enables the fruit to develop parthenocarpically, thus increasing fruit yield where insect pollinators are likely to be scarce and ineffectual.

2 Numerous cultivars of muskmelon, *Cucumis melo*, have been examined cytologically. All have $2n = 2x = 24$ with regular meiosis and pollen fertility averaging more than 90 per cent. The economic importance of muskmelons has stimulated much plant breeding work, most of it aimed at improvements in the dessert quality of the fruit and in resistance to diseases and insect pests. Studies have also been reported on the inheritance of plant habit, sex expression, male sterility, fruit skin colour, fruit smoothness and flesh colour. The genetics of resistance to powdery mildew has been the subject of much study in *C. melo*. This disease (caused by *Sphaerotheca fuliginea*) has a devastating effect on yields of cantaloups and other muskmelons, particularly where the crop is grown in monoculture. Resistance involves two genes, a dominant (Pm^1) for resistance to race 1 of the pathogen and a partial dominant (Pm^2) for resistance to race 2. Two modifiers, in addition to Pm^2, are responsible for extreme resistance. Cultivars have been developed incorporating some, but not all, of these genes (Bohn and Whitaker, 1964).

3 For watermelon, *Citrullus lanatus*, plant breeders have collected much useful information about the inheritance of such characters as leaf shape, plant habit, male sterility, sex expression, fruit colour, fruit shape, fruit surface, seed size, seed coat colour, fruit bitterness and disease and insect resistance. Species of *Citrullus* have $2n = 2x = 22$ and *C. lanatus* is highly pollen fertile. Some of the most interesting work on the species has been done with disease resistance. Orton (1911) in a classical investigation, found that resistance to *Fusarium* wilt is determined genetically. Furthermore, he showed that resistance could be transferred from the wild form to edible cultivars. Subsequently, many cultivars resistant to *Fusarium* have been developed, but the inheritance of resistance remains unknown. A single dominant gene for resistance to at least two races of anthracnose (*Colletotrichum lagenarium*) has been reported. Much effort has been given to studies of seed coat colour and seed size in watermelons. The former evidently depends upon a system of three major genes and a specific modifier active in combination with the black seed coat phenotype. This three-gene system accounts for most of the seed coat colour phenotypes from black to tan, red and white (Poole *et al.*, 1941). For seed length, digenic inheritance is reported, short and long genotypes being recessive to medium (Poole *et al.*, 1941; Prem Nath and Dutta, 1973).

Among the interesting mutants in watermelon is a male sterile (*msg*) associated with a glabrous condition. Seed production in the original mutant was sparse but has been increased to an acceptable level by selection. Glabrous plants occasionally produce a little pollen late in the season, thus allowing propagation.

The commercial production of seedless watermelons has been an attractive prospect since 1939 when seedless polyploids were first produced (Kihara and Nishiyama, 1951). Andrus *et al.* (1971) showed that triploids are feasible, have superior consumer qualities and, in fact, are so much superior to diploids that they almost constitute a new product.

4 There is little genetic information about the bottle gourd, *Lagenaria siceraria* ($2n = 2x = 22$). The early work of Pathak and Singh (1950) showed that fruit colour was determined on a monogenic basis, as were several fruit shapes. Time of fruit maturity and seed size were polygenic. Recently, Tyagi (1973) has demonstrated heterosis for the number of pistillate flowers, number of fruits per plant, weight of fruits and number of seeds per fruit. From meiotic studies, Varghese (1973) suggested that modern forms were secondary polyploids derived from an ancestor with a basic $x = 5$; speculatively $x = 5+5+1 = 11$.

5 There are about 26 species of *Cucurbita*, five of which are domesticated cenospecies genetically isolated from each other. The wild species contain many ecospecies which are separated geographically but are genetically cross-compatible; they fall into five natural groups (Bemis *et al.*, 1970). All species have $2n = 2x = 40$ and several authors have suggested that they are polyploids based on $x = 10$; the evidence, however, is by no means clear. The centre of diversity of the genus is in the tropics near the Mexico–Guatemala border. The wild species are variable and they range from mesophytic to xerophytic and from annuals to perennials. The domesticated species are as follows:

C. pepo Includes both summer and winter type squash; cultivars of summer squash are mostly brachytic (bush) plants; has some cold resistance (although frost-sensitive).

C. moschata High quality winter squash; vine type plants; tolerant of the squash vine borer.

C. maxima High quality, soft skinned winter squash with vigorous vines and, sometimes, extremely large fruit.

C. mixta Similar to *C. moschata* but largely restricted to Mexico; used both as summer and winter squash; seeds are roasted as a confection; fruits have a large, hard, corky peduncle.

C. ficifolia Restricted to the highland, tropical regions of Mexico, Central America and northern parts of South America.

Since each of the cultivated species has desirable horticultural qualities (e.g. brachytic (bush) forms in *C. pepo*, vine borer resistance in *C. moschata*, soft rind in *C. maxima*, etc.). there has been, and continues to be, interest in interspecific crosses. The species, however, are genetically well isolated and only with considerable difficulty can characters be transferred from one to another. Whitaker and Davis (1962) listed the results of hydridization studies and only three out of the 58 reported showed some degree of fertility (always low) in the F_2. Amphidiploids of *C. moschata* × *C. maxima* have been made; fertility was enhanced but seed production was not economical (Pearson *et al.*, 1951). A number of wild species have some degree of compatibility with the cultigens and may have certain desirable horticultural characters (such as powdery mildew resistance in *C. lundelliana*). It is sometimes possible to transfer these characters from wild to cultivated species, either directly or by the use of bridging species. Some wild × cultivated amphidiploids have been made but, as the entire genome of the wild species remains intact, its undesirable characters persist and cannot be eliminated.

Curtis (1939) demonstrated heterosis in summer squash (*C. pepo*) and proposed a method of producing F_1 hybrids, nowadays the main type of cultivar in practical use. Hybrid seed is produced commercially by field emasculation of brachytic (but not vine type) plants as females. This method is economically feasible because *Cucurbita* flowers are large and male buds can be detected and removed several days before anthesis.

Genic male sterility (but not cytoplasmic) has been described in *C. pepo*, *C. maxima* and *C. moschata*. Populations are maintained by pollinating male sterile (recessive) plants with heterozygous male fertiles but the expense of field-roguing the male fertile segregates has prevented production of hybrid seed by this method.

The edible flesh of *Cucurbita* fruits is their major horticultural attribute. The seed, however, is a source of high quality vegetable oil, produced primarily in Europe. A 'naked seed' mutant of *C. pepo* eliminates the relatively heavy seed coat and results in seed containing 45–50 per cent oil.

A unique hybrid squash cultivar, Iron Cap, introduced in 1968, is an interspecific F_1 hybrid between *C. maxima* and *C. moschata*; it is itself insufficiently pollen fertile for satisfactory fruit set and requires pollen from either *C. maxima* or *C. moschata* to set well.

3 Early history

1 The cucumber is thought to be indigenous to India. The evidence, mostly circumstantial, because the cucumber has never been found in the wild state, is nevertheless highly suggestive. A tremendous array of fruit sizes, shapes, skin colours, skin ornamentations and vegetative characters are found in India and adjacent countries. Furthermore, a small, bitter cucumber, with sparse, stiff spines, *C. hardwickii*, a close relative of the cultivated kinds, grows wild in the foothills of the Himalayas. *Cucumis hardwickii* will hybridize readily with the cucumber, producing a fertile F_1, with no reduction of fertility in the F_2 (Deakin *et al.*, 1971). These observations suggest that *C. hardwickii* is either a feral form or, more likely, a progenitor of the cultivated cucumber. De Candolle (1882) thought that the cucumber had been cultivated for over 3,000 years in India. If true, this makes the crop an early domesticate, comparable in age with some of the cereals.

2 The muskmelon appears to be indigenous to Africa. It is probable that truly wild forms of *C. melo* are found only in eastern tropical Africa south of the Sahara. Wild forms reported from India are more likely to be feral escapes derived from local cultivars. *Cucumis*, except for the cucumber, is an African genus with $x = 12$. The muskmelon was evidently a latecomer to man's list of crops. Once domesticated, however, it exploded into numerous cultivars, particularly in India, which may thus be regarded as a secondary centre. Cultivars of *C. melo* were rapidly dispersed throughout Europe and, at an early date, into the Americas.

3 The watermelon, *Citrullus lanatus*, is reported to be feral in the warm parts of the globe but it is truly wild or native only in sandy, dry areas of South Africa, chiefly in the Kalahari Desert. In this area it occurs in two biochemical forms, one having bitter fruits containing cucurbitacin (E-glucoside), the other lacking this constituent. Forms with non-bitter fruits are a source of food and water for the Bushmen of the Kalahari. There are some forms of watermelon (known as citrons) which have fruits with firm, hard, greenish flesh and large seeds. Citrons should be regarded as cultivars of the watermelon, not as progenitors. They are used for stock feed.

4 While *Lagenaria siceraria*, the white-flowered gourd, is most likely indigenous to the tropical lowlands of Africa south of the equator, archaeological records indicate that it was present in the western hemisphere as early as 7000 B.C., probably as a cultivated plant. Schweinfurth has identified remains of *L. siceraria* from Egyptian tombs dated at about 3300–3500 B.C. Thus the white-flowered gourd has a long history of cultivation in both hemispheres documented by well-preserved remains of shards (broken pieces of the tough rind of the fruit), seeds, peduncles and an occasional intact fruit. The widespread occurrence of gourd remains in many archaeological sites of the western hemisphere is perhaps a result of the importance of the crop in the day-to-day affairs of primitive societies. The shells were (and still are) used as containers for liquids, for food storage, as eating utensils and for musical instruments, etc.

5 *Cucurbita*, containing the squashes, pumpkins and gourds, has been associated with man in the Americas for at least 10,000 years (Whitaker and Cutler, 1971). The five cultivated species (*C. pepo*, *C. moschata*, *C. maxima*, *C. mixta* and *C. ficifolia*) were selected by American Indians long before the Discovery in 1492 and they were staple items of their diet. It is well established that the advanced civilizations of the Americas (Aztec, Inca, Maya), prior to the European influx, were founded upon a food complex of maize, beans and squash.

The fruits of most wild species of *Cucurbita* are of a size and colour to attract the attention of primitive man but they have hard, tough rinds and excessively bitter flesh. The seeds, however, are non-bitter, tasty and nutritious. We assume that early man, in sampling the fruits of wild *Cucurbita* species for seeds, found mutants that lacked the bitter principle and thus began the long process of selection that, presumably, resulted in the modern domesticated species.

In Mexico and Guatemala there is a group of wild species of *Cucurbita* closely allied, in breeding and other criteria, to the cultivated forms. These include *C. lundelliana* and *C. martinezii* which are to some extent cross-compatible with the cultivars. They are vigorous, lush plants bearing small fruits with tough rinds, bitter, coarse, stringy flesh and abundant seeds. The evidence that the complex of five cultivated species derived from these small, bitter gourds is suggestive but by no means decisive.

4 Recent history

Recent work with cucumber has stressed disease resistance, mainly to scab, downy mildew, powdery mildew and cucumber mosaic virus. Sources of resistance to scab have been found and incorporated into commercial cultivars. Tolerance to downy and powdery mildews have been identified and utilized but cultivars with satisfactory resistance or tolerance to cucumber mosaic virus have not yet been developed.

The muskmelon, *C. melo*, includes fruits popularly known as cantaloups, honey dews, casabas and persians. Muskmelons are grown throughout the warmer parts of the world but are of peculiar economic importance in the USA, especially in the states of Arizona, California, New Mexico and Texas. In these areas, muskmelons are grown on large acreages for distant markets. Much effort has therefore been devoted to developing cultivars with superior consumer qualities such as improved flesh colour and texture, increased soluble solids and fruits that will withstand severe shipping stresses, so as to arrive at distant markets in good condition.

Most of the recent breeding work with watermelon has been directed towards producing cultivars with good consumer qualities, tough rinds for long-distance transport and multiple disease resistances. The cultivar, Charleston Gray, is a good example of this trend.

The economic potential of *Lagenaria*, the white-flowered gourd, is small. Virtually no research has been done to improve the food or ornamental value of the crop.

The major objectives of the plant breeder in *Cucurbita* has been to improve the quantity and quality of the fruit flesh of the winter types (used mature) and the earliness and productivity of the fruit of the summer types (used immature). The primary breeding method has been to make selections within intraspecies crosses. Little significant genetic information has come from this work; *Cucurbita* plants, particularly the indeterminate vine types, demand much space.

5 Prospects

1 Current trends in cucumber production are concerned mainly with two issues: (*a*) the genetic manipulation of sex expression designed to find inexpensive means of producing hybrid seed; and (*b*) breeding of a plant-type adapted to 'once over' harvest. Good progress is being made with both projects. Gynoecious (female, male sterile) lines have been developed to serve as mother plants, producing fruit from which the F_1 seed is harvested. The gynoecious lines are planted in isolated blocks with the desired pollinators. Breeders are also attempting to develop cultivars with short internodes and large numbers of fruits that can be harvested mechanically and are suitable for processing.

2 New breeding techniques for *Cucumis melo*, the muskmelon, have been devised (Andrus and Bohn, 1967; Bohn and Andrus, 1969). A search for biological means of controlling insect pests of the crop has revealed collections within *C. melo* tolerant of the melon aphid (*Aphis gossypii*); this character is controlled by a single dominant gene (Bohn *et al.*, 1973).

3 Efforts are being made to breed watermelons, *Citrullus lanatus*, that have smaller fruits and are more prolific than those currently available; small fruits are easier to handle than large ones and are thus less liable to damage.

4 *Lagenaria* has recently provoked considerable attention. There has been a remarkable revival of interest in the gourds for use as ornamentals and in handicraft. They are made into bird-feeders, planters, musical instruments, animal figures, ornaments, etc. This activity constitutes a small but thriving trade for growers and craftsmen and it seems likely to continue, as more people with leisure time adopt the hobby. A Gourd Society has been organized with its own publication, *Gourd Seed*. This new interest in gourds has been combined with a renewed search for novel cultivars.

5 In *Cucurbita*, F_1 hybrids have been shown to be more desirable than open pollinated cultivars but (as mentioned above) the production of vine-type F_1 hybrids is not economic. However, ethylene-producing compounds (such as ethephon, 2-chloroethyl-phosphonic acid), when sprayed on seedlings of normally monoecious *Cucurbita* plants, induce a gynoecious (male-sterile or female) condition which persists for at least ten flowers after treatment. By judicious management of the chemical, the production of hybrid seed is possible (Robinson, *et al.*, 1971).

The transfer of single chromosomes from a wild species into the genome of a cultivated form has been demonstrated (Bemis, 1973). This procedure may prove valuable in utilizing desirable traits from wild species without incurring their disadvantages.

One of the most exciting developments in the utilization of *Cucurbita* is the possible domestication of the wild xerophytic species. These species are perennial and produce abundant fruit with a limited supply of water. The fruit is bitter but the seeds contain 30–35 per cent high quality vegetable oil and 30–35 per cent protein. Seed yields of 1·7–3·3 t/ha have been projected but have yet to be realized in large-scale cultivation. If these projections prove reliable, new sources of oil and protein of great potential for arid countries would be in prospect.

6 References

Andrus, C. F. and Bohn, G. W. (1967). Cantaloup breeding: shifts in population means and variability under mass selection. *Amer. Soc. hort. Sci. Proc.*, **90**, 209–22.

Andrus, C. F., Seshadri, V. S. and Grimball, P. C. (1971). Production of seedless watermelons. *USDA tech. Bull.*, **1425**, pp. 12.

Bemis, W. P. (1973). Interspecific aneuploidy in *Cucurbita*. *Genet. Res.*, **21**, 221–8.

Bemis, W. P., Rhodes, A. M., Whitaker, T. W. and Carmer, S. G. (1970). Numerical taxonomy applied to *Cucurbita* relationships. *Amer. J. Bot.*, **57**, 404–12.

Bohn, G. W. and Andrus, C. F. (1969). Cantaloup breeding: correlations among fruit characters under mass selection. *USDA tech. Bull.*, **1403**, pp. 21.

Bohn, G. W., Kishaba, A. N., Principe, J. A. and Toba, H. H. (1973). Tolerance to melon aphid in *Cucumis melo*. *J. Amer. Soc. hort. Sci.*, **98**, 37–40.

Bohn, G. W. and Whitaker, T. W. (1964). Genetics of resistance to powdery mildew race two in muskmelons. *Phytopath.*, **54**, 587–91.

Curtis, L. C. (1939). Heterosis in summer squash (*Cucurbita pepo*) and the possibilities of producing F_1 hybrid seed for commercial planting. *Amer. Soc. hort. Sci. Proc.*, **37**, 827–8.

Cutler, H. C. and Whitaker, T. W. (1961). History and distribution of the cultivated cucurbits in America. *Amer. Antiquity*, **26**, 468–85.

Deakin, J. R., Bohn, G. W. and Whitaker, T. W. (1971). Interspecific hybridization in *Cucumis*. *Econ. Bot.*, **25**, 195–211.

De Candolle, A. (1882). *Origin of cultivated plants*. New York (reprint, 1959).

Kihara, H. and Nishiyama, I. (1951). An application of sterility of autotriploid to the breeding of seedless watermelons. *Seiken Ziho*, **3**, 5–15.

Orton, W. A. (1911). The development of disease resistant varieties of plants. *IV Conf. Int. Genetic Paris, C.R. et Rapp.*, 247–65.

Pathak, G. N. and Singh, S. N. (1950). Genetical studies in *Lagenaria leucantha* (*L. vulgaris*). *Ind. J. Gen. and Pl. Breed.*, **10**, 28–35.

Pearson, O. H., Hopp, R. and Bohn, G. W. (1951). Notes on species crosses in *Cucurbita*. *Amer. Soc. hort. Sci. Proc.*, **57**, 310–22.

Prem Nath and Dutta, O. P. (1973). Inheritance of some seed characters in *Citrullus lanatus*. *Ind. J. Hort.*, **30**, 388–90.

Poole, C. F., Grimball, P. C. and Porter, D. R. (1941). Inheritance of seed characters in watermelons. *J. agric. Res.*, **63**, 433–56.

Robinson, R. W., Whitaker, T. W. and Bohn, G. W. (1970). Promotion of pistillate flowering in *Cucurbita* by 2-chloroethylphosphonic acid. *Euphytica*, **19**, 180–3.

Tyagi, I. D. (1973). Heterosis in bottle gourd. *Ind. J. Hort.*, **30**, 394–9.

Varghese, Benny M. (1973). *Studies on the cytology and evolution of South Indian Cucurbitaceae*. Ph.D. Thesis, Kerala Univ., India.

Whitaker, T. W. and Cutler, H. C. (1971). Pre-historic cucurbits from the Valley of Oaxaca. *Econ. Bot.*, **25**, 123–7.

Whitaker, T. W. and Davis, G. N. (1962). *Cucurbits. Botany, cultivation and utilization*. New York.

23

Yams

Dioscorea spp. (Dioscoreaceae)

D. G. Coursey
Tropical Products Institute
London England

1 Introduction

Yams are a staple food crop in many tropical countries: global production has been estimated as around 20 Mt/yr. The greater part derives from West Africa, Nigeria providing about half the world total. Yams are also important in parts of southeast Asia and Oceania (especially Melanesia); and in the Caribbean and neighbouring parts of tropical America (Coursey, 1967). They are essentially crops of subsistence agriculture although recently there has been some semi-commercial production for local markets in West Africa and the Caribbean. International trade, other than some trans-frontier or inter-island regional trade, is very limited.

Yams are primarily a starchy staple. Nevertheless, their protein content is appreciable and contributes significantly to dietary value. Apart from food yams, interest has recently developed in several other *Dioscorea* spp. as sources of steroidal sapogenins for pharmaceutical purposes (Coursey and Martin, 1970; Anon., 1972).

2 Cytotaxonomic background

The genus *Dioscorea*, including some 600 species, is by far the largest in the Dioscoreaceae. The family was formerly classified in the Liliales (Burkill, 1960; Coursey, 1967) but has now been separated to form, together with the Trichopodiaceae and Roxburgiaceae, the separate Order Dioscoreales (Ayensu, 1972; Schubert in Anon., 1972).

The Dioscoreaceae show some primitive characters, of both monocotyledons and dicotyledons, suggesting that they were amongst the earlier Angiosperms. The ancestral Dioscoreaceae may thus have appeared in the late Triassic or early Jurassic, in what is now southeast Asia, as suggested by Axelrod (1970, cited by Coursey, 1975) for the origin of the Angiosperms. Burkill (1960) also indicates the emergence of

the ancestral proto-Dioscoreaceae in southeast Asia, although at a later date. Separation of Old and New World species by the formation of the Atlantic Ocean at the end of the Cretaceous was followed by divergent evolution. There are consistent differences in vascular anatomy (Ayensu in Anon., 1972) while only Old World Dioscoreas contain alkaloidal toxins (Willaman *et al.*, 1953, cited by Coursey, 1967). Desiccation of the Middle East in the Miocene separated African from Asiatic species, although subsequent divergence has been slight.

The evolutionary separation of Old and New World groups is reflected in their cytology. The Old World species have chromosome numbers based on $x = 10$ (except for the aberrant temperate section Borderea and the phytogeographically related genus *Tamus*, both based on $x = 12$). The primitive section Stenophora has $2n = 2x = 20$, but in many tropical species related to the food yams, $2n = 4x = 40$ (Burkill, 1960; Coursey, 1967). The New World Dioscoreas and the related *Rajania* have $x = 9$, most being tetraploids or hexaploids ($2n = 36, 54$) (Martin and Ortiz, 1963, 1966, cited by Coursey, 1967; Ayensu and Coursey, 1972).

Within Old World yams, high degrees of polyploidy occur. Chromosome counts of 50, 60, 70, 80, 90, 100, 120 and 140 have been reported (Coursey, 1967). Most of these are from cultivated forms, but some of the highest counts are from wild material, indicating that the relationship between cultivation and polyploidy stressed by Burkill (1960) is by no means exact. Among cultivated yams, aneuploids ($2n = 38, 52, 55, 66$ and 81) have been reported and, rarely (Sharma and De, 1956, cited by Coursey, 1967), variation in chromosome number has even been recorded within a single individual.

Only a few species are economic plants of any significance. Food yams are derived mainly from the twelve listed in Fig. 23.1. The two first-named Asian, and the first African, species are by far the most important. Some 50 or 60 other species are gathered or cultivated to a very limited degree (Coursey, 1967; Alexander and Coursey, 1969). The species exploited for pharmaceutical purposes are quite distinct (Fig. 23.1). Interest has concentrated largely on six, although others have been utilized (Coursey and Martin, 1970; Martin in Anon., 1972; Martin, 1969, cited by Coursey, 1972).

3 Early history

The domestication of the yams in Asia, Africa and tropical America took place entirely separately, different species being involved. Intercontinental contacts

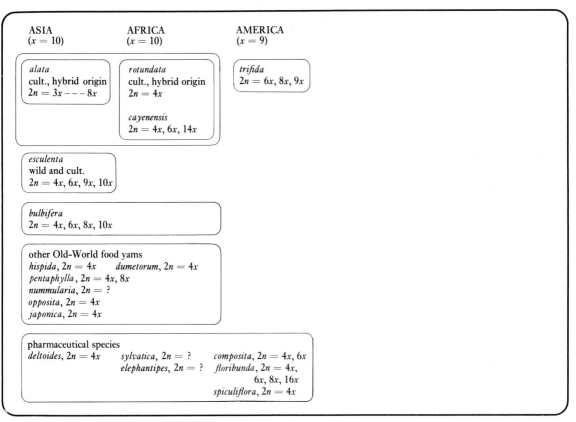

Fig. 23.1 Relationships of the cultivated yams, *Dioscorea*.

are comparatively recent. It has been suggested that yam-based agricultural systems evolved only through external cultural influences, subsequent to the initiation of grain-crop agriculture (Burkill, 1924, 1951; Murdock, 1959; Baker, 1962, cited by Coursey, 1967). Culture contacts between grain-using intrusive and non-grain-using aboriginal populations probably took place in Asia, Africa and northern South America, around 5000 B.P., contributing to the present form of yam-based agriculture (Alexander and Coursey, 1969) but the beginnings of the evolution of yams as crop plants lie earlier in history, within the aboriginal cultures.

In Asia, the cultigen *D. alata* appears to have arisen under human selection from wild forms related to *D. hamiltonii* and *D. persimilis* in the north-central parts of the southeast Asian peninsula. *Dioscorea esculenta* is indigenous to the same area and cultivated forms appear to have been developed by selection from the wild (Burkill, 1924; 1951, cited by Alexander and Coursey, 1969; Coursey, 1975). The major centre of diversity of cultivars of both species is now in Papua-New Guinea but the lesser diversity

in other areas may have resulted from erosion of a former greater variability.

These domestications were probably ancient. Evidence exists of vege-cultural civilizations in southeast Asia, at radiocarbon dates of 10 000 B.P. or more, associated with mesolithic Hoabinhian ceramics. Yams had evolved far enough as domestic plants to have been taken across the Pacific in the Polynesian migrations (which left the Asiatic mainland around 3500 B.P.) and to have been associated, even earlier, with sophisticated premetallic cultures in Melanesia. Both species were taken to Madagascar and, probably, to the East African littoral by later Malaysian migrations (*ca.* 1500 B.P.) but did not influence African yam domestication (Coursey, 1972, 1975). The domestication of the warm-temperate species, *D. opposita* and *D. japonica*, may be associated with the Jomon culture in Japan, contemporary, and having affinities with, the late Hoabinhian.

The evolution of African yams as crop plants was initially independent of external influences: 'to the African himself is entirely due the invention of *D. cayenensis* as a crop plant' (Burkill, 1939, cited by

Ayensu and Coursey, 1972). Of the two major species, *D. cayenensis* is indigenous to much of forest zone Africa while *D. rotundata*, a cultigen, like its Asiatic counterpart *D. alata*, is so close to *D. cayenensis* that it has been regarded as a sub-species (Miège, 1968; Ayensu, 1970; cited by Ayensu and Coursey, 1972). These African species belong, like *D. alata*, to section Enantiophyllum of *Dioscorea*. *Dioscorea rotundata*, the most important of the African yams, appears to be of hybrid origin. One parent is undoubtedly *D. cayenensis*, a forest species, adapted to a short dry season and having limited tuber dormancy. The other must be a savanna species; *D. praehensilis* bears the greatest resemblance to some *D. rotundata* forms, although *D. abyssinica* and *D. togoensis* may also have contributed.

Hybrids may often have arisen spontaneously in the ecotonal forest/savanna mosaic but were unsuited to the local ecology and did not persist until the advent of Mesolithic man who, with his concepts of 'protection' of food plant resources, created new ecological niches which favoured survival of the archetypal *D. rotundata*. His recognition of this superior food resource led to the many, eventually clonal, cultivars known today (Ayensu and Coursey, 1972; Coursey, 1975). As in Asia, this initial phase appears to have been ancient: major cruces were probably the contraction of the forest-savanna ecotonal environment caused by climatic change and the approximately simultaneous diffusion of the proto-Negro people into the West African savanna, *ca.* 11 000 B.P. (Coursey, 1975).

Little is known of New World yam domestication; yams were of secondary importance in pre-Columbian times, perhaps because of the availability of cassava. *Dioscorea trifida* is clearly an Amerindian domesticate, distribution of cultivars suggesting domestication on the borders of Brazil and Guyana, followed by spread through the Caribbean (Ayensu and Coursey, 1972).

Yams are primarily Old World domesticates. A close parallel exists between those of Asia and Africa, botanically and in cultural importance (Coursey, 1972). In all continents, they are domesticates of the tropical/ equatorial 'non-centres' defined by Harlan (1971): regions, diffuse in both space and time, where crops developed by a slow, widely spread process of evolution towards symbiosis with man. These contrast sharply with the subtropical Vavilovian centres where grain crops were domesticated in closely defined geographical areas during the so-called 'Neolithic Revolution'. Within these 'non-centres', yams of tropical, rather than those of equatorial, regions are inherently best adapted to exploitation by man: the

more pronounced dry season being associated with the development of larger, more dormant tubers. Yams are thus crops of the forest fringes and savannas, rather than of the more humid areas (Barrau, 1965; 1970; Burkill, 1951; cited by Coursey, 1972; 1975).

It has been proposed on ethnographic grounds (Lomax and Berkowitz, 1972) that, at very early times, 'a continuous ring of gardening cultures once linked Oceania to Africa in the tropical latitudes'. These cultures have survived only at the West African and Melanesian extremities and among relict peoples of mainland Asia and the Philippines, the massive 'incursion of higher cultures', subsequent to the 'Neolithic Revolution', having virtually eliminated older cultural patterns in the areas between. It is within these gardening cultures that yam domestication began, with a ritualization of man's relationship with his originally wild food plants. This permitted the emergence of forms suited to man's requirements but possibly ill-adapted to survival in the wild (Burkill, 1953; Barrau, 1970; Coursey and Coursey, 1971; cited by Ayensu and Coursey, 1972). Criteria of domestication (cf. non-shattering rachis in cereals) are hard to define: loss of defence mechanisms such as toxicity, deep-burying tuber habit and spininess; reduction of sexual fertility; and high polyploidy are indicative (Coursey, 1975).

Many species of *Dioscorea* have been used in traditional medicines in both Africa and Asia since ancient times, anticipating modern pharmaceutical use. Others have been used as sources of poison.

4 Recent history

Recent evolution of yams as crop plants began when the paleo-colonial Lusitano–Iberian expansion into the tropics around 500 B.P. brought the Indo-Pacific area, Africa and America into regular contact. Asiatic and, later, African yams were extensively used for the victualling of ships; through this agency, *D. alata* was taken to West Africa while the Atlantic slave trade resulted in both this species and the African *D. rotundata* and *D. cayenensis* being taken to the Caribbean. Early historic records of *D. alata* in West Africa and of African yams in the Americas exist (Burkill, 1938; cited by Coursey, 1967). These transfers have resulted in relatively few cultivars (mainly those superior in quality and storage characters) being taken outside their continents of origin. The Asiatic *D. esculenta* was only taken to Africa and America within the last century and is still little grown outside the Indo-Pacific. Jamaican and Puerto Rican forms of *D. cayenensis* are often superior to those in Africa, indicating that transdomestication has taken place.

There has been virtually no transfer of food yams from America to the Old World or from Africa to Asia (Burkill, 1960; Coursey, 1967; 1975), except experimentally in the last decade.

Food crops of tropical subsistence agriculture have, until recently, been badly neglected by agricultural science and yams more so than most. Experimentation has been spasmodic and no serious breeding has yet been undertaken.

Although spontaneous hybridization must have contributed to the ancestry of some yams, propagation has been entirely vegetative, each cultivar being a single clone. Improvement must have been far more often by selection of somatic mutants rather than of sexual seedlings. Vegetative propagation has led to reduced sexual fertility, resulting from limited production and incomplete opening of flowers and failure of pollen release. The polyploidy of many preferred cultivars may also contribute to infertility. In *D. rotundata*, at least, a considerable proportion of the better clones are male. Infertility appears most extreme in the Asiatic *D. alata* and *D. esculenta* where seed set is almost unknown. Occasional seed setting has been recorded in botanical gardens where the plants have become more mature; in farming practice, in which small tuber fragments are planted annually, juvenility of form is artificially maintained. Seed set in *D. rotundata* takes place rather more readily and the difficulties of obtaining fertile seed may be less than has been suggested. Occasional attempts to breed food yams have been made over the last two decades, notably in Nigeria (Waitt, 1959–61, cited by Coursey, 1967). Little success has been achieved, though this may be due to lack of support. (But see Sadik and Okereke (1975) for recent developments).

The drug Dioscoreas emerged as crop plants since the late war owing to developments in biochemical industry which created a demand for steroidal saponins for manufacture of pharmaceuticals (Feuell, in Coursey, 1967). Much of the demand has been met from wild yams from arid regions of South Africa and Mexico. Initially destructive, gathering has now been disciplined by legal sanctions into more systematic collecting, with precautions to ensure regeneration; while attempts are also being made to cultivate some species (Martin and Gaskins, 1968, cited by Coursey and Martin, 1970; Martin in Anon., 1972). This progress towards domestication constitutes a recapitulation over decades of what took place over millenia with the food yams (Coursey, 1972). The drug yams are still essentially wild species, and seed freely, but little breeding has been undertaken.

5 Prospects

It is often said that yam production is declining. There has been a decline over recent decades, *relative* to cassava and rice, critical factors being high labour requirements and consequent costs of production. Nevertheless, yams are often so strongly the preferred staple that, bearing in mind population increases, there will not have been an *absolute* decline and demand will probably continue, even at increasing prices.

Plans for the development of food yams as crops need to be aimed: first, towards improved clones for the small farmer for growing under virtually horticultural conditions at relatively high cost; second, towards attempts to modify the crop the better to adapt it to mechanized agriculture. Parameters to be considered in improvement programmes have been indicated (Coursey and Martin, 1970; Ayensu in Anon., 1972): Ayensu and Coursey (1972) have emphasized the need to eliminate staking, which accounts for about a third of production costs.

The variation already known among yam cultivars is so vast that much can be done simply by selection and dissemination of the best existing clones. For more radical improvement, breeding will be required, necessitating study of the reduced sexuality of many clones to overcome low fertility. Many of the most important yams today are of species producing one or few large tubers, most appropriate to the manual farmer. Others, such as *D. esculenta* and *D. trifida* which form many smaller tubers, should be more adaptable to industrial agriculture; the latter species is sexually highly fertile. The introduction into cultivation of other species from the wild needs serious investigation. It may well be mainly chance that led millenia ago to the ennoblement of ancestors of the present economic species. A much larger proportion of the variability within the genus should be tapped, either by direct ennoblement or by hybridization with cultivated forms.

Few, if any, species are in danger of extinction, but serious genetic erosion is occurring among cultivated yams. Twenty years ago, it was estimated (Torto, 1956, cited by Coursey, 1967) that 80 per cent of Ghana's crop was provided by two clones. Since then, much remaining variation has probably been lost. There is urgent need to collect such forms before further erosion occurs. Even apparently inferior clones may have value for use in different ecosystems or for particular breeding characteristics.

The future of the drug Dioscoreas is uncertain. As with edible yams, production systems are labour-demanding, which may inhibit future development as competing sources of steroids become available.

24

6 References

Alexander, J. and **Coursey, D. G.** (1969). The origins of yam cultivation. In Ucko and Dimbleby (eds.), *The domestication and exploitation of plants and animals.* London.

Anon. (1972). *Primer simposio internacional sobre Dioscoreas. Publ. Esp. Inst. Nal. Invest. For., Mexico, 8,* pp. 130.

Ayensu, E. S. (1972). *Dioscoreales.* In C. R. Metcalfe (ed.), *Anatomy of the Monocotyledons,* 6. Oxford,

Ayensu, E. S. and **Coursey, D. G.** (1972) Guinea yams. *Econ. Bot.,* **26**, 301–18.

Burkill, I. H. (1960). The organography and the evolution of the Dioscoreaceae, the family of the yams. *J. Linn. Soc. (Bot.),* **56**, 319–412.

Coursey, D. G. (1967). *Yams.* London.

Coursey, D. G. (1972). The civilizations of the yam: interrelationships of man and yams in Africa and the Indo-Pacific region. *Archaeol. Phys. Anthropol. Oceania,* 7, 215–33.

Coursey, D. G. (1975). The origins and domestication of yams in Africa. In J. R. Harlan (ed.) *Origins of African plant domestication.* The Hague,

Coursey, D. G. and **Martin, F. W.** (1970). The past and future of yams as crop plants. *Proc. 2nd. intnl Symp. trop. Root Crops, Hawaii,* 1, 87–90; 99–101.

Harlan, J. R. (1971). Agricultural origins: centres and non-centres. *Science, N.Y.* 154, 468–74.

Lomax, A. and **Berkowitz, N.** (1972). The evolutionary taxonomy of culture. *Science, N.Y.* 177, 228–39.

Sadík, S. and **Okereke, O. U.** (1975). A new approach to improvement of yam, *Dioscorea rotundata. Nature, Lond.,* **254**, 134–5.

Tung

Aleurites spp. (Euphorbiaceae)

F. Wit

Hamelakkerlaan 38 Wageningen
The Netherlands *formerly*
Foundation for Agricultural Plant Breeding
Wageningen The Netherlands

1 Introduction

Tung oil or wood oil is a high-quality rapid drying oil pressed from the seeds of two trees, *Aleurites fordii* and *A. montana*. The two species yield similar oils which are not differentiated in commerce. *Aleurites fordii*, a small deciduous tree, naturally occurring in central and western China, produces 90 per cent of the oil. Outside China, it is now grown commercially in several warm temperate and sub-tropical regions. *Aleurites montana* is a medium to large deciduous tree, indigenous to the subtropical regions of southern China and northern Indo-China. It has been successfully introduced into some tropical countries.

Tung trees are propagated by seeds or by grafted seedlings. They start fruiting after 3 to 6 years, are in full production by 10 to 12 years and may have a productive life of 30 years or more. The fruits are allowed to dry after falling to the ground and are then collected, either by hand or by machine. They easily split into one-seeded sections, the seed remaining partly enclosed by a hull. *Montana* fruits usually contain three seeds, *fordii* fruits five seeds. Fruits of *montana* tung are less easy to decorticate mechanically as they have a harder hull than those of *fordii*.

World production of tung oil is about 100 kt. The major producers are China, Argentina, the United States and Brazil. Tung oil is resistant to weather and water and owes its superior quality to the high content of elaeostearic acid. It is widely used in the manufacture of finer paints and varnishes for protecting both wood and metal surfaces. There are many special applications (Godin and Spensley, 1971).

2 Cytotaxonomic background

There are three related species, all trees yielding drying oils, but commercially not important. The oils of

A. cordata (Japan, Formosa) and *A. trisperma* (Philippines) contain less elaeostearic acid than tung oil; the oil of *A. moluccana*, the candlenut tree (indigenous to Malaysia), has none.

The somatic chromosome number is $2n = 2x = 22$ in all species except in *A. moluccana* where $2n = 4x = 44$. Spontaneous hybrids between *A. fordii* and *A. montana* have been found in China and have also been produced experimentally. The hybrids showed vegetative heterosis but fruit and seed setting were very low (Webster, 1950; Wit, 1969). In the USA sterility appeared to be associated with meiotic irregularities. A fertile backcross clone showed regular chromosome pairing but had 12 pairs of chromosomes instead of 11 (Draper, 1966). This indicates the possibility of chromosome additions. Hybrids also have been obtained between *A. fordii* and *A. cordata*. Crosses of *A. montana* with *A. moluccana* and *A. trisperma* failed.

3 Early history

The first known record of the use of tung oil in paints occurs in the *Book of Poetry*, compiled by Confucius more than 24 centuries ago. *Aleurites fordii* must have been grown in central China since very early times. Its probable native habitat is between 26° and 33°N.

Aleurites montana is indigenous to the subtropical areas south of 25°N. Nothing is known about the early evolution of the crop. Within the relatively unselected seed-propagated species no varieties were developed in China.

A. fordii is monoecious and produces in mixed inflorescences staminate and pistillate flowers. Although self-pollination prevails, most trees are more or less heterozygous. *Aleurites montana* trees have a tendency to flower either predominantly male (non-bearing trees) or female (bearers) and to produce seed by cross-pollination. They are not obtained true to type from seeds. Both species are insect-pollinated.

The first shipment of oil came to the United States in 1869. It was not until the end of the century, however, that western technologists discovered how to utilize fully the excellent drying qualities. China could not meet the growing demand which ensued, so the consuming countries tried to make their industries less dependent upon Chinese exports by introducing the crop to their own countries or overseas territories. Introduction was attempted in nearly all areas where it might have a chance of success. Many efforts failed but *A. fordii* is now planted commercially in: a narrow belt north of the Gulf of Mexico in the USA (introduced in 1904); in the Soviet Union (1926);

Fig. 24.1 Evolution and relationships of tung, *Aleurites fordii* and *A. montana*.

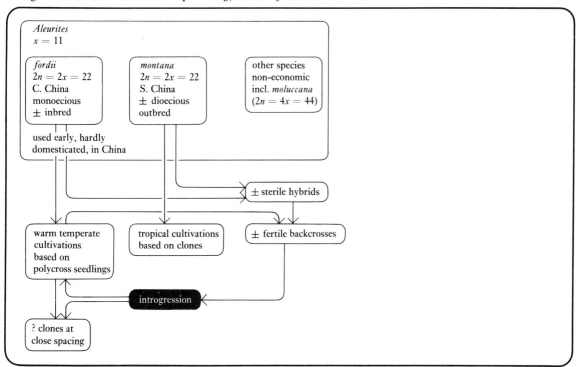

Argentina (1928); Brazil (1930); Paraguay and Malagasy. At least a part of the material introduced into Latin America came from the United States. *Aleurites montana* is grown successfully in Malawi (1930) and Brazil (1930) and experimental plantations have shown some promise in India (1918), Ngwane (1923), Indonesia (1930), Guinea and Zaïre (1933).

4 Recent history (Fig. 24.1)

Conscious plant breeding started with *A. fordii* in the USA, where the unselected introductions appeared to vary widely in size, tree type, yield and fruit characters. In an extensive selection and breeding programme, several hundred outstanding trees were traced in commercial orchards from 1938 onwards. Their seedling progenies were screened and a few of the most promising mother-trees so selected were also tested as buddings. In no case did the budded trees outyield the seedlings. A very few seedling progenies appeared to be sufficiently uniform for commercial planting. These were released as named varieties, the buddings of the mother-trees being maintained for seed production (Potter, 1968). From 1958, commercial plantings were almost exclusively of these seedlings.

In later years much emphasis was placed on breeding a late flowering variety, losses due to late spring frosts having been found to be serious in many years. The flowers of *A. fordii* are borne on terminal shoots of the previous season; they open with the first warm weather in spring. *Aleurites montana* produces female flowers on the end of the shoots of the current season, after shoot growth has been made. In a programme for the breeding of late-blooming types *A. fordii* was hybridized with *A. montana*. Because the flowering period of some of the hybrids is intermediate between the two parent species, such trees are more likely to escape damage from spring frosts than trees of *A. fordii*. Some of the backcross hybrids appear to be inherently productive (Merrill and Sitton, 1968).

Breeding in *montana* tung started in the 1930s in Indochina, Indonesia and Malawi. The unselected seedling plantations were found to consist of 40–60 per cent of predominantly male trees which bore little or no fruit (Webster, 1950; Wit, 1969). The productive bearers formed all or most of their flowers in female or mixed clusters. Buddings of such trees flowered as predominantly female (both on male rootstocks and on female ones) and they outyielded their highly heterogeneous seedling families. In the first instance, vegetative propagation of superior mother-trees, therefore, offered the best possibility of establishing profitable plantations. The criterion by which

a clone was selected was the yield per unit area of dry seed of the budded progeny. Trees of which the oil did not come up to specification or which owed their high yield to exceptional size, were eliminated before clone trials. In the final choice between clones yielding equally high crops the oil content of the seed became an important factor (Foster, 1963).

5 Prospects

The economic outlook for tung is not as bright as it was a few decades ago. Tung oil is increasingly being replaced, in various fields of utilization, by substitutes made from cheaper drying oils and synthetics. Prices have been declining. No increased demand for the oil is to be expected in the near future. However, the annual consumption has remained constant over the past decade and a number of users continue to rely on tung oil (Kilby, 1970). In order to protect this market from the competition of cheaper substitutes, the tung grower should strive for higher yields and lower costs. Increased production efficiency can be obtained with the aid of improved varieties.

The methods of breeding are not likely to undergo great changes. Both species may move up a rung on the evolutionary ladder from unselected seedlings to seedlings from selected mother-trees, to clones, to improved seedlings, to secondary clones and so on. Until very recently, *A. fordii* lingered in the phase of selected seedlings. It may now enter the clonal phase as it has been shown that the smaller yields of budded trees are closely associated with their smaller size. By closer planting, buddings can very well surpass seedlings in per-area-yield and new hybrids (e.g. late flowering types) can be utilized as buddings long before homozygous seedling offspring can be obtained.

Montana tung entered the phase of clonal selection some decades ago and the breeders soon realized that the best genotypes were not likely to be found in the first cycle of selection. Following the controlled selfing of superior mother-trees and after the crossing of trees which complement each other in desirable characters, improved seedling progenies are produced. From these, the best trees are propagated by buddings and some of them may be expected to give improved clones. At the same time, these seedling progeny tests will give information on the heritability of economic characters and on the combining abilities of the mother-trees. This should make possible the informed choice of parents for the next cycle of crosses.

Budded clones have not given yields as high as their mother-trees, presumably as a result of the reduction in vigour due to budding and the effect of variable seedling rootstocks. The possibility is by no means excluded that families might be bred consisting solely of bearing trees able to compete with buddings. This would make budding unnecessary and lower the cost of planting.

Meanwhile there is the possibility of improvement by way of better rootstocks. Recent experiments with semi-dwarf clones showed that spacings much closer than those conventionally used resulted in increased yields over 12 harvests (Hill, 1966; Spurling and Spurling, 1973). Therefore it would be advisable to include rootstock and spacing tests in clonal and variety trials. Apparently, closer planting can improve the economics of tung growing.

6 References

Draper, A. D. (1966). Cytological irregularities in inter-specific hybrids of *Aleurites*. *Proc. Amer. Soc. hort. Sci.*, **89**, 157–60.

Foster, L. J. (1963). The tung oil trees. *World Crops*, 220–6.

Godin, V. J. and Spensley, P. C. (1971). Oils and oil seeds. *TPI Crop and Product Digests, 1*. Tropical Products Institute, London.

Hill, J. (1966). Close planting in *montana* tung (*Aleurites montana*). *Trop. Agric., Trin.*, **43**, 11–18.

Kilby, W. W. (1970). The American tung nut industry. *Agric. Sci. Rev.* **8**, 29–35.

Merrill, S., Jr. and Sitton, G. (1968). Responses of tung clones to chilling and warming. *Proc. Amer. Soc. hort. Sci.*, **93**, 224–34.

Potter, G. F. (1968). The domestic tung industry to-day. *J. Am. Oil Chem. Soc.*, **45**, 281–4.

Spurling, D. and Spurling, A. T. (1973). A close planting trial with tung (*Aleurites montana*). *Trop. Agric., Trin.*, **50**, 347–8.

Webster, C. C. (1950). The improvement of yield in the tung oil tree. *Trop. Agric., Trin.*, **27**, 179–220.

Wit, F. (1969). Tung trees. In Ferwerda and Wit (eds.), *Outlines of perennial crop breeding in the tropics*. Wageningen, 495–507.

Rubber

Hevea brasiliensis (Euphorbiaceae)

P. R. Wycherley
Kings Park and Botanic Garden
West Perth Western Australia
formerly Rubber Research Institute of
Malaysia Kuala Lumpur Malaysia

1 Introduction

Natural rubber consists mainly of *cis*-1, 4 polyisoprene, $(C_5H_8)_n$, of molecular weight, $0 \cdot 7$ to $1 \cdot 4$ million. The principal source is *Hevea brasiliensis* cultivated in southeast Asia ($93 \cdot 5\%$) and West Africa ($5 \cdot 2\%$), total $3 \cdot 4$ Mt in 1973 (compare synthetic elastomers, $7 \cdot 3$ Mt). Latex, an aqueous suspension of about 40 per cent dry matter which contains over 90 per cent hydrocarbon, flows from concentric cylinders of latex-vessels in the phloem on 'tapping', that is excising 'bark' with a curved blade from a sloping cut. Yield is sustained under moderate continuous exploitation and the bark grows again. The tree's economic life is about 30 years. Latex is processed simply with modest equipment.

The expense of purifying rubber from alternative species that could be mechanically harvested (e.g. *Parthenium, Taraxacum*) led to their abandonment; although dependence on manual collection is an economic limitation with *Hevea*, yet the need for 'tappers' makes work in populous countries. Moreover, 'stimulation', by application of growth regulators, yields more rubber by processes physiologically akin to more intensive tapping but at lower cost.

Hevea brasiliensis was introduced into the Old World free from South American Leaf Blight (SALB) caused by *Microcyclus* (= *Dothidella*) *ulei*, which has severely limited *Hevea* monoculture throughout the neotropics. This accounts largely for the present distribution of production by (in descending order): Malaysia, Indonesia, Thailand, Sri Lanka (Ceylon), South India, the rest of Asia, Liberia, Nigeria, the rest of Africa, Brazil and other American countries.

2 Cytotaxonomic background

Hevea occurs naturally throughout the Amazon Basin

and in parts of Matto Grosso, Upper Orinoco and the Guianas (Schultes, 1970). The nine species listed below are now recognized. Important varieties and synonyms are given, with ranges and habitats. The first three species yield commercially acceptable latex, some of the others are too resinous or even anticoagulant. The sequence is roughly in descending order of use in production and breeding.

(1) *H. brasiliensis*: southern half of range of genus, usually on well drained soils, sometimes subject to light seasonal flooding.

(2) *H. benthamiana*: mainly north of the Amazon, on low sites, often heavily flooded.

(3) *H. guianensis* (var. *lutea*; var. *marginata*): almost whole range of genus, on well-drained soils.

(4) *H. pauciflora* (= *H. kunthiana*; var. *coriacea* = *H. confusa*): from the Guianas through the Rio Negro to the Upper Amazon; *H. pauciflora* to the east, the more widespread and abundant var. *coriacea* to the west; on strongly drained soils.

(5) *H. spruceana*: middle and lower Amazon, often on heavily flooded sites.

(6) *H. microphylla*: mid and upper Rio Negro, often flooded for long periods.

(7) *H. nitida* (= *H. viridis*; var. *toxicodendroides*): upper Amazon, on well drained sites (*H. nitida* sometimes lightly flooded); var. *toxicodendroides* on quartzitic hills, Colombia.

(8) *H. rigidifolia*: upper Rio Negro, on well drained soils.

(9) *H. camporum*: sandy savanna between headwaters of southern tributaries.

Most diversity is about the Rio Negro and tributaries from its confluence with the Amazon into Colombia, although *H. brasiliensis*, *H. spruceana* and *H. nitida* var. *toxicodendroides* are only on the fringe of this area and *H. camporum* is absent.

Putative hybrids representing most combinations of the more common species occur naturally. The geographic range of each species overlaps with one or more of the others. Experimental crosses reveal no genetic barriers. Hybridization and introgression

Fig. 25.1 Distribution of *Hevea*.

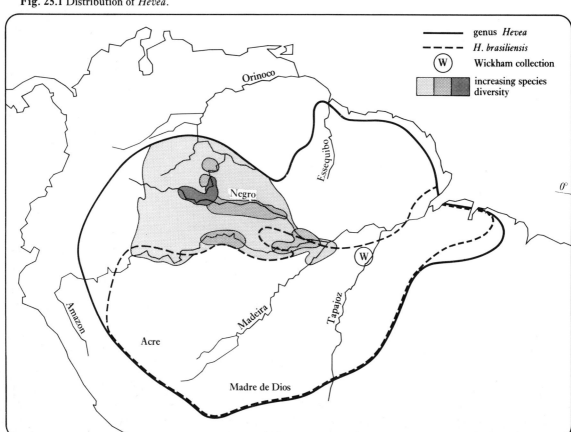

abound. Speciation seems to be based on ecological adaptation, the short range of the weakly-flying insect pollinators (e.g. midges and thrips) and the lack of coincidence in flowering between some species or even, rarely, within species.

With the exception of one triploid clone of *H. guianensis* ($2n = 3x = 54$), all reliable counts of chromosomes in *Hevea* are $2n = 2x = 36$. No counts have been reported for *H. camporum* and it is uncertain if authentic *H. microphylla* and *H. nitida* have been examined. Allotetraploidy on a base of $x = 9$ has been suggested for this and several related genera (Majumder, 1964).

Male flowers open slightly earlier than females in the same inflorescence, but there is considerable overlap within a tree. Fruit set is less than 1 per cent of the female flowers in nature and about 4 per cent of those artificially pollinated. Each fruit contains three seeds which mature in five months. There are usually two flowerings each year, the main flowering following leaf change stimulated by a dry spell.

There is a wide range in fertility. A few trees are virtually sterile whether selfed or crossed. Some are self sterile only. Putative male sterility is probably complete sterility. The direction of reciprocal crosses makes little difference, except for some parents poor as females because of susceptibility to leaf diseases.

3 Early history

Early explorers noted various uses of rubber from several wild sources (Schultes, 1956). Cooked seed of *Hevea* are regularly eaten by aborigines in the northwest of the range, although it is famine fare elsewhere. Hypotheses to the effect that Indians carried seeds on their travels or planted trees to produce more seed for food fail for lack of evidence; there is no reason to think that *Hevea* was ever planted, either for latex or for seed until H. A. Wickham tried near Santerem about 1872; nor that latex or seed were harvested except from wild trees until the coming into bearing of trees raised in Malaysia and Sri Lanka from seed exported from Brazil in 1876. Seed is seldom edible or viable for more than three weeks unless specially packed or kept at 5°C. Prior to the 1900s, the effects of man, if any, on *Hevea*'s evolution were limited and conjectural, e.g. eliminating ecological barriers between species by clearing or drainage.

In 1876 Wickham collected 70,000 seeds of *H. brasiliensis* from near Boim on the Rio Tapajoz and from the well-drained undulating country towards the Rio Madeira. This area produced excellent wild rubber. Except for the acceptable *H. guianensis*, all other species (in particular, the undesirable *H.*

spruceana) are distant. About 2,800 seedlings were raised at Kew, England, and 2,397 of them were despatched during 1876–7, mainly to Sri Lanka; a few went to Malaysia, Singapore and Indonesia. As far as is known, virtually all the trees cultivated in the Old World are descended from this one collection. Subsequent introductions have had little impact yet on plant breeding in Asia and none on the material actually planted commercially.

During the 20 years after 1876 there was haphazard seed multiplication of this material. Commercial planting began in the late 1890s when H. N. Ridley had devised a practical method of exploitation, when other crops were failing in Malaysia and when the world demand for rubber was rising. Seed was in short supply and all available was planted with little or no selection. The industry was therefore founded upon an unselected sample of wild genotypes.

4 Recent history (Wycherley, 1969)

Variations in trunk-girth and yield were investigated from 1910 onward in Indonesia, Malaysia and Sri Lanka. Selection of seed from high yielding mother-trees was initiated but discarded from commercial practice when, in 1917, bud-grafting on seedling rootstocks was developed in Sumatra. The first clones were multiplied from individual high yielding seedlings in commercial plantings in all three countries. Hand-pollination was developed in 1920 (also in Sumatra) and, within a few years, was adopted by all *Hevea* breeders as the main means of producing seedling populations for the selection of new clones. About 1930, some proprietary breeders planted seed orchards with the best parents emerging from their breeding programmes to provide large scale sources of improved seedlings, mainly for immediate commercial planting, but also to provide, directly or indirectly, populations for clone selection.

Results of clone selection were indeed striking. The best of the first clones nearly doubled the average yield; and the best of the first generation of bred clones nearly doubled it again.

Although most clones now under trial are of the second or even third bred generation, few clones of advanced breeding have yet been planted commercially. Test procedures were, and are, lengthy, extending up to 30 years between pollination and practical impact. Until recently the oriental breeding policy has been to cross 'the best with the best', with strong emphasis on precocious yield in selection. All this has been within 'Wickham' material of *H. brasiliensis*. Resistance to leaf diseases endemic in the Old World (*Colletotrichum* (= *Gloeosporium*), *Oidium* and *Phytophthora*) has been eroded, especially if the original selection was made where the disease was not prevalent. Concurrently, with rising yields, storm damage has increased. Hence wind-fastness (which is negatively correlated with yield) has been added to the list of breeding objectives.

Crown-budding (three-component trees) may overcome the defects of simple buddings (two-components) but it is not yet widely used. On the seedling rootstock is grafted the trunk or panel clone on which, in turn, the top or crown is grafted. This enables selection of the trunk clone for yield and of crown clones for resistance to storm-damage and locally prevalent leaf diseases. Single component clonal trees such as self-rooted marcots and mist-propagated cuttings failed, due to their lack of taproots and hence adequate anchorage.

While these developments were taking place in southeast Asia, plantation rubber and *Hevea* breeding had taken a different course in the Americas. During the first 20 years of this century plantations of *H. brasiliensis* in the Guianas failed due to South American Leaf Blight (SALB) (Holliday, 1970). Plantings of local material and of Asian selections at Fordlandia near Boim (1927) and Belterra (1934) on the Rio Tapajoz were also severely damaged by SALB. So too eventually were plantings in other parts of Brazil and throughout Central America. In 1937 the Ford Motor Company initiated breeding and crown budding to combine high yield and SALB resistance by combined genetical and horticultural means. This programme was taken over by official Brazilian institutions in 1946. The Firestone Rubber Company (whose main producing plantations are in Liberia) and government research have carried on breeding programmes in Guatemala.

These have been almost entirely recurrent back-crossing to high yielding oriental *H. brasiliensis* of SALB-resistant selections of *H. brasiliensis* (especially from the Upper Amazon, Acre and Madre de Dios) and of *H. benthamiana* (in particular, clone Ford 4542 from the Rio Negro). These have reached the third generation with generally progressive deterioration in resistance, aggravated by the appearance of new pathological races of the fungus, of which at least four have been differentiated. No selection of *Hevea* is resistant to all four, which do not yet, however, all occur at any one place.

5 Prospects

Biometrical analysis and inferences from the general pattern of the results of breeding and selection indicate that commercially important characters are polygenically determined and that variation is predominantly additive (Gilbert *et al.*, 1973).

Rubber is an outbreeder and vigour and yield suffer inbreeding depression. The physical bases of yield have been identified as vigour of growth (efficiency in assimilation), the number of rings of latex vessels in the bark and the duration of flow before the cut vessels are sealed by coagulated latex. The correlation of storm losses with yield is probably due partly to excessive leafiness in some very vigorous selections but mainly to unfavourable partition associated with long flow, especially in precocious high yielders. In future the choice of parents and the selection of new cultivars may take more conscious account of these physiological characters.

More rigorous screening for resistance to leaf diseases has been initiated. Resistance to the different strains of SALB and to the various diseases in Asia and Africa must be combined from several sources, which implies reduced dependence on strict backcross or conventional 'between highest yielder' programmes.

A few non-Wickham *Hevea* were sent to the Old World from the New between 1873 and 1947; most were lost, including those planted in unsuitable regions, and others were destroyed for fear of genetic contamination. Collections to broaden the genetic base (and especially to introduce SALB resistance) were sent in 1947 and 1951 to Liberia and Malaysia respectively. These countries received budwood of SALB-resistant Brazilian clones in 1953. Since then there have been several despatches of seed and clones from America to the main rubber growing and breeding countries. These include all species except *H. camporum*, although the survivors of some species may be few or hybrids. These imported collections and the early oriental selections are maintained in mixed plantings and provide a pool of genetic variation for future incorporation.

Other innovations, which have not yet made impact, include induced mutations, artificial polyploids, a search for genetic dwarfs and attempts to culture haploid tissues.

6 References

Gilbert, N. E., Dodds, K. S. and Subramaniam, S. (1973). Progress of breeding investigations with *Hevea brasiliensis*. V. Analysis of data from earlier crosses. *J. Rubb. Res. Inst. Malaya*, **23**, 365–80.

Holliday, P. (1970). South American leaf blight (*Microcyclus ulei*) of *Hevea brasiliensis*. *Phytopathological Papers*, **12**, Commonwealth Mycological Institute.

Majumder, S. K. (1964). Chromosome Studies of some Species of *Hevea*. *J. Rubb. Res. Inst. Malaya*, **18**, 269–65.

Schultes, R. E. (1956). The Amazon Indian and evolution in *Hevea* and related genera. *J. Arn. Arb.*, **37**, 123–48.

Schultes, R. E. (1970). The History of taxonomic studies

in *Hevea. Bot. Rev.*, **36**, 197–276.

Wycherley, P. R. (1969). Breeding of *Hevea. J. Rubb. Res. Inst. Malaya*, **21**, 38–55.

Cassava

Manihot esculenta (Euphorbiaceae)

D. L. Jennings
Scottish Horticultural Research Institute
Dundee Scotland

1 Introduction

Cassava is a perennial shrub which produces a high yield of tuberous roots in one to three years after planting. Among the tropical staples it produces exceptional carbohydrate yields, much higher than those of maize or rice and second only to yams (de Vries *et al.*, 1967). World production is about 92 million tons, of which a third is produced in South America, a third in Africa and the remainder in Asia and various tropical islands. It is mainly grown by peasant farmers, for many of whom it is the primary staple, but it is also a cash crop, being used to produce industrial starches, tapioca and livestock feeds. For a long time research on the crop was neglected but its exceptional capacity to produce high yields of calories is now recognized and the crop is receiving appropriate attention from breeders and agronomists.

2 Cytotaxonomic background

The genus *Manihot* is a member of the Euphorbiaceae; it has two sections, the Arboreae, which contains tree species and is considered the more primitive, and the Fruticosae, which contains shrubs adapted to savanna, grassland or desert. *Manihot esculenta* (cassava) belongs to the latter; it is a cultigen, unknown in the wild state. The genus occurs naturally only in the western hemisphere, between the southwest USA (33°N) and Argentina (33°S). It shows most diversity in two areas, one in northeastern Brazil, extending towards Paraguay, and the other in western and southern Mexico. An early classification by Pax included 128 taxa but is not satisfactory. Recently, Rogers and Appan (1973) have used a computer-aided method to delimit biological species, which they call 'closed gene pools'. They define 98 species and separate one species into a new genus called *Manihotoides*. All species can be intercrossed, but they show evidence of being reproductively isolated in nature.

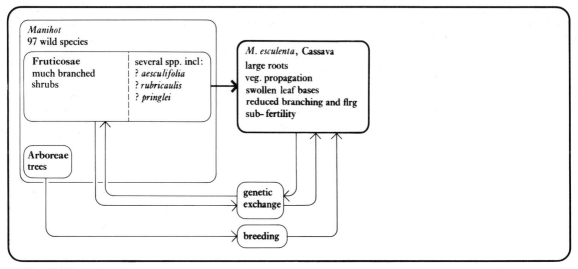

Fig. 26.1 Evolution of cassava, *Manihot esculenta*.

All species so far studied have 36 chromosomes and show regular bivalent pairing but, in both cassava and *M. glaziovii* (sect Arboreae), studies of pachytene karyology have given evidence of polyploidy. Thus, first, there are three nucleolar chromosomes, which is high for true diploids and, second, duplication occurs for some of the chromosomes. Magoon *et al.* (1969) suggest that *Manihot* species are segmental allotetraploids derived from a combination of two diploid taxa whose haploid complement had six chromosomes in common but differed in the other three.

3 Early history

Several writers have suggested that the ancestors of cassava were among the first plants to be used as food when man first migrated southwards into Central and South America, and there seems good evidence that cassava flour was important in the trade of north-western South America in the second and third millennia B.C. (Lathrap, 1973; Reichel-Dolmatoff, 1965). The evidence is too tenuous to decide whether present-day cassavas have descended from a single species or from several. The work of Rogers and Appan (1973) indicates that *M. aesculifolia*, *M. rubricaulis* and *M. pringlei* are its closest relatives, each having a more or less erect growth habit and tuberous roots; *M. pringlei* is particularly interesting (and unusual among the wild species) in having a low content of the poisonous cyanogenic glycosides. It seems likely that the variability of cultivated forms has been enhanced by hybridization with several wild forms; most of the wild species have a propensity for colonizing disturbed areas and there would have been ample

opportunity for gene exchange in areas adjacent to cultivation. Indeed, Harlan (see Rogers, 1963, 1965) has suggested that gene exchange produced hybrid swarms from which both new cultivated and wild forms were derived. Selection in one direction by man and in the other by nature would have provided the kind of disruptive selection which Doggett considers to have been so potent a force in sorghum evolution. It would have produced diversity both between and within the wild and cultivated forms and could have enhanced reproductive isolation.

Under domestication, selection was for large roots, for more erect and less branched growth and for the ability to establish easily from stem cuttings. These pressures evoked correlated responses in several other characters. Branching, for example, occurs when inflorescences are formed, so the less branched plants selected were inevitably less floriferous. Selection for the ability to establish rapidly from cuttings probably favoured genotypes the stems of which carried adequate food reserves; hence the presence of swellings at the leaf scars, a characteristic feature which now separates *M. esculenta* taxonomically from its nearest relatives. The virtual discontinuation of seed propagation tended to encourage sterility, so much so that Jennings (1970) suggested interspecific crossing to restore fertility.

No forms completely devoid of cyanogenic glycosides are known. Any recessive mutant gene which conferred an incapacity for producing glycosides would have little chance of becoming homozygous in an allotetraploid like cassava. Forms with very low contents of glycoside have evolved locally but the bitter ones are still preferred by some tribes, especially where predators are a problem. Processes for fermenting or heating the roots to remove the poison prior to eating must have been developed at an early stage, and are a tribute to the ingenuity of the early cultivators.

4 Recent history

Cassava spread rapidly from South America in post-Columbian times. It arrived on the west coast of Africa, via the Gulf of Benin and the river Congo, at the end of the sixteenth century and on the east coast, via the islands of Réunion, Madagascar and Zanzibar, at the end of the eighteenth century. Cultivation spread inland from both sides. The crop arrived in India about 1800. Its importance in the Old World increased only slowly until the middle of the nineteenth century when it became widely grown as a famine reserve on account of its tolerance of drought, locust attack and poor husbandry. Although propagated almost entirely by cuttings, self-sown seedlings frequently became established and the more successful of them were given names and fixed by vegetative propagation. This gave diversity to the crop both in South America and in its new habitats; indeed, the process possibly proceeded faster in Africa, where virus degeneration of old stocks was more serious and the superiority of virus-free seedlings consequently more conspicuous.

Controlled breeding did not begin until the 1920s. Interspecific crossing and backcrossing was started in Java about that time and in East Africa in 1937. In both places, the emphasis was on hybrids between cassava and the tree species *M. glaziovii* (Ceara rubber) and *M. dichotoma* (Jequie Manicoba rubber), probably because these two species had already been introduced from South America as sources of rubber and so were readily available to the breeders. Resistance to virus diseases was the main objective of breeding in Africa and *M. glaziovii* and *M. melanobasis* (a closely related shrub type, possibly misnamed) proved the best sources of resistance genes. But this early work also illustrated the value of interspecific crossing to broaden the genetic base; thus, *M. glaziovii* also contributed genes for improved vigour, drought resistance and resistance to bacterial disease; and *M. melanobasis* contributed genes for modifying the leaf canopy and increasing the protein content of the roots. Breeding work elsewhere placed greater emphasis on local adaptation, low cyanogenic glycoside content and suitability for industrial starch or tapioca. Usually, extensive controlled crossing was attempted among large numbers of parents and the outstanding selections were propagated vegetatively. However, this method has limitations for large-scale breeding because

potential parents often flower only sparsely and, though the fruit capsules are trilocular, their fertility is so low that, on average, only one viable seed is obtained for every two flowers pollinated. (In this respect, however, cassava is better than rubber.)

5 Prospects

Recognition of cassava's great potential as a source of food calories has led to a major increase in the resources made available to improve the crop. The change in the scale of operation has brought with it a change in breeding methodology, with a shift of emphasis towards methods designed to improve populations in the first instance, followed later by selection of outstanding clones in a range of contrasting environments (Hahn *et al.*, 1974). The separation of male from female flowers on the plant and the marked protogyny facilitate these methods; thus the use of blocks of parent material and the removal of male flowers from some of the breeding lines prior to flowering allows the application of various cyclical breeding procedures (such as recurrent selection or half-sib progeny-testing) which are currently used successfully in other outbreeders such as maize. The objectives of breeding are changing only slightly, though some change may arise from the needs of mechanization, which requires more compact roots that neither spread too far laterally nor penetrate too deeply into the soil.

An important development is the realization that cassava's many wild relatives contain reservoirs of unexplored germplasm for crop improvement (Rogers and Appan, 1970). Species from the whole of the Fruticosae section of the genus can be crossed readily with cassava and, even with the tree species, crossing presents few problems. There seems no reason why the ecological range of cassava could not be greatly extended by crossing with species adapted to drought (from areas of low rainfall or sandy soil), lower temperatures (higher altitudes) or acid soils. Unfortunately, the roots of most of the wild species are particularly high in cyanogenic glucosides; the emphasis is now on selection of clones with minimal levels, following medical evidence of chronic toxicity resulting from regular intake of even small quantities.

6 References

de Vries, C. A., Ferwerda, J. D. and Flach, M. (1967). Choice of food crops in relation to actual and potential production in the tropics. *Neth. J. agric. sci.*, 15, 241–8.

Hahn, S. K., Howland, A. K. and Terry, E. R. (1974). Cassava breeding at the International Institute of Tropical Agriculture. *Proc. 3rd intnl. Symp. Trop. Root Crops, Ibadan, Nigeria.*

Jennings, D. L. (1970). Cassava in Africa. *Field Crop Abstr.* 23, 271–8.

Lathrap, D. W. (1973). The antiquity and importance of long-distance trade relationships in the moist tropics of Pre-Columbian South America. *World Archaeology,* 5, 170–86.

Magoon, M. L., Krishnan, R. and Bai, K. V. (1969). Morphology of the pachytene chromosomes and meiosis in *Manihot esculenta. Cytologia,* 34, 612–24.

Reichel-Dolmantoff, G. (1965). *Colombia: ancient peoples and places.* London.

Rogers, D. J. (1963). Studies of *Manihot esculenta* and related species. *Bull. Torrey bot. Club,* 90, 43–54.

Rogers, D. J. (1965). Some botanical and ethnological considerations of *Manihot esculenta. Econ. Bot.,* 19, 369–77.

Rogers, D. J. and Appan, S. G. (1970). Untapped genetic resources for cassava improvement. *Proc. 2nd intnl. Symp. Tropical Root and Tuber Crops, Hawaii,* 72–5

Rogers, D. J. and Appan, S. G. (1973). *Manihot, Manihotoides* (Euphorbiaceae). *Flora Neotropica.* Monograph 13, New York.

Castor

Ricinus communis (Euphorbiaceae)

Dharampal Singh
UP Institute of Agricultural Sciences
Kanpur India

1 Introduction

Castor is an important oil crop of the spurge family. The oil, the viscosity of which changes little with temperature, is considered to be one of the best lubricants for high speed aero-engines. Its unique chemical properties give it a wide variety of uses and make it an oil of great industrial importance. The finest quality is used for medicinal purposes.

Castor is widely grown in tropical, sub-tropical and temperate countries. The most important producers are Brazil, India, the USSR and China. Its cultivation has recently been taken up in the USA and some east European countries. Of a total world production of about 723 kt (1970), about 38 per cent was contributed by Latin America followed by the Far East (22 per cent) and Africa (6 per cent).

For a review of economic botany, see Kulkarni (1959), Purseglove (1968) Weibel (1947) and Zimmerman (1958).

2 Cytotaxonomic background

The genus *Ricinus* is monotypic. Hilterbrandt (1935) recognized four sub-species: *persicus, chinensis, africanus* and *zanzibarinus.* The first is considered to be the most productive and has no caruncle; *chinensis* has a small caruncle and the last two have large ones.

All the sub-species and forms of castor are diploid with $2n = 2x = 20$. It is reported to be a secondary balanced polyploid with a basic number $x = 5$. On the basis of haploid studies Narain (1974) thought that the species had arisen as tetraploid from a diploid progenitor ($2n = 10$), now extinct. If so, the event must have been ancient for no basic number as low as $x = 5$ is now known in the Euphorbiaceae. All subspecies and forms inter-cross freely and are fully fertile. There is however a report suggestive of incipient subspeciation between two strains, one each from China and Africa.

3 Early history

Ricinus is now pantropical in distribution and weedy or escaped forms abound. There is some disagreement as to whether truly wild plants exist or not; if they do it is probably in eastern Africa. The area of origin of the crop is, accordingly, unsettled, some authors favouring India, others Africa. Vavilov assigned castor to the Abyssinian primary centre, with a secondary centre of diversity in the Near East.

Early records of the use of the crop come from Egypt and India (in the great Sanskrit medical work, the *Susruta Ayurveda*). It is thus presumably of ancient domestication but no dates can be given. In Europe, it was reported to have been cultivated by Albert Magnus, the bishop of Ratisbon, in the middle of the thirteenth century and was well known as a garden plant in the sixteenth century. The oil came into use for medicinal purposes in the eighteenth century but the small quantities of oil and seeds then required for European medicines were mostly obtained from Jamaica and India.

Castor is extremely variable in habit (size and longevity), colour, amount of waxy bloom, flowering habit, inflorescence type, fruit and seed characters. According to Narain (1974) the primitive type was probably a tree, unpigmented and with a waxy bloom; it had a large, lax inflorescence, spiny dehiscent fruits and strongly carunculate seeds. It does not seem possible to infer which characters changed in the early phases of domestication. Certainly, long term selection has tended towards an annual, dwarf, herbaceous habit and indehiscent and sparsely spiny fruits that mature more or less synchronously.

The waxy bloom is of some ecological interest (Harland, 1947). Waxy plants are adapted to hot, dry environments, non-waxy plants to cooler and moister ones. The character is a feature of local ecotypic adaptation in the Andes.

4 Recent History

Castor breeding work had its beginning in the early twentieth century in the USA and USSR. Breeding procedures depend on the mating system of the crop. Castor is normally monoecious but outbred, with pistillate flowers on the upper 30 to 50 per cent of the raceme and staminate flowers below. There are, however, several variants which include racemes with: (*a*) staminate and pistillate flowers interspersed; (*b*) 70 to 90 per cent pistillate flowers; (*c*) 100 per cent pistillate flowers; and (*d*) a few hermaphrodite flowers. These expressions are somewhat influenced by environment but genetically stable lines, little sensitive to environment, have been developed. Two types of femaleness, N-pistillate and S-pistillate, are recognized. In the former there is a recessive sex-switch gene, *f*, which determines femaleness. Female stocks (*ff*) can be maintained by sib-mating (*ff* × *Ff*). The S-pistillate type was obtained by selection within sex-reversal variants which start female and then revert to monoecism. The genetics of sex expression and the use of pistillate lines in hybrid seed production have been described by Shifriss (1960) and Zimmerman and Smith (1966).

The frequency of natural outcrossing has been reported to vary from 5 to 50 per cent but a recent report from Texas gave about 90 to 100 per cent outcrossing in some dwarf varieties.

In the USA, castor improvement work was started at Oklahoma in 1902 and at the Brooklyn Botanic Garden in New York in 1918. Since the 1940s, the main objective has been to breed varieties high in yield and oil content and suitable for mechanical harvesting; dwarfness, reduced branching, indehiscent capsules and resistance to capsule dropping were important selection criteria. Successful varieties have been grown since 1947.

An important recent development has been the production of hybrid seed through the use of pistillate lines. N-pistillate type lines which segregate 1*Ff* (monoecious):1 *ff* (female) are grown in a crossing block. The monoecious plants are rogued and the remaining plants crossed with a selected monoecious inbred (male) line. The pistillate line is maintained by sibbing.

Work on the improvement of castor is also in progress in Brazil, USSR, Italy and France to develop early maturing varieties suitable for mechanical harvesting. In India, the main breeding objective has been to develop varieties with high seed yield and oil content. Maturity is important, varieties with a period of 120–150 days, suitable for double cropping and of uniform maturity and non-shattering habit being sought. Mass selection and pedigree methods have been largely employed but emphasis has recently been laid on the production of hybrid varieties through the use of N-pistillate lines. The first commercial hybrid was developed in Gujrat State in 1970.

5 Prospects

Castor breeding has already reduced the plant from a tree (up to 10 m tall) to a herb (at about 1·5 m), with concomitant reduction in branching. The next step will be still smaller, unbranched plants adapted to very dense populations. More breeding for disease and pest resistance can also be foreseen; several resistances are already known; for example, the strong waxy bloom

85

which repels *Empoasca* and jassids.

Hybrid varieties are already well established and their use will no doubt spread. There is, however, need for biometrical studies to aid the development of breeding plans.

Castor breeders have no recourse to related species but a vast range of variability is available.

6 References

Harland, S. C. (1947). An alteration in gene frequency in *Ricinus communis* due to climatic conditions. *Heredity*, 1, 121–5.

Hilterbrandt, V. M. (1935). The plant resources of the world as initial material in plant breeding. The castor plant. *Lenin Acad. Agri. Sci. Inst. Plant Industry, Moscow and Leningrad*, 6, 55–70.

Kulkarni, L. G. (1959). *Castor*. Indian Central Oilseeds Committee, Hyderabad.

Narain, A. (1974). Castor. In J. B. Hutchinson (ed.), *Evolutionary studies in world crops. Diversity and change in the Indian subcontinent*. Cambridge, 71–80.

Purseglove, J. W. (1968). *Tropical crops. Dicotyledons*. 1, 180–6.

Shifriss, O. (1960). Conventional and unconventional systems controlling sex variations in *Ricinus. J. Genet.*, 57, 361–88.

Weibel, R. O. (1947). The castor oil plant in the United States. *Econ. Bot.*, 3, 273–83.

Zimmerman, L. H. (1958). Castor beans – a new oil crop for mechanized production. *Advanc. Agron.*, 10, 257–88.

Zimmermann, L. H. and Smith, J. D. (1966). Production of F_1 seed in castor beans by use of sex genes sensitive to environment. *Crop. Sci.*, 6, 406–9.

Oats
Avena spp. (Gramineae-Aveneae)

J. H. W. Holden
Scottish Plant Breeding Station
Edinburgh Scotland

1 Introduction

Oats are one of the major temperate cereals, ranking fourth behind wheat, maize and barley. World figures for 1972 give a production of 51 Mt from 31 Mha. Oats, as a group of species, has a wider range of ecological adaptation than wheat and barley. Traditionally, it has been the dominant cereal in temperate latitudes with moist maritime climates and, on the other hand, an important farm crop in Mediterranean climates. The oat kernel is rich in protein (about 16 per cent) and fat (about 8 per cent) and both fractions are of high nutritional quality. Oat straw is of value as a roughage for ruminants. The grain (when dehusked) has long been a human staple food and (when husked) a high energy diet supplement for farm animals. During the last decade, world production has remained fairly constant and yields have risen steadily but areas sown to the crop have declined sharply in several countries, due partly to a decline in the use of oats for stock feeding on the farm and partly to the high husk content which reduces the attractiveness of oats as a raw material for processing and compounding when compared with other cereals.

2 Cytotaxonomic background

Since the publication by Maltzew (1930) of his classic monograph on the systematics of the Euavenae, little was added to our knowledge of the systematics or evolution of the genus until recently, when a series of remarkable finds have been reported of new and apparently good, genetically isolated species. Indeed, it is now clear that other distinct entities remain to be identified before the evolution of the cultivated hexaploids is understood and that differentiation between genetic entities at the same or different ploidy levels is frequently obscured by marked morphological similarities. The present review should therefore be regarded as an interim assessment only.

Avena species occur at three ploidy levels; diploids ($2n = 2x = 14$), tetraploids ($2n = 4x = 28$) and hexaploids ($2n = 6x = 42$). All are regular bivalent formers. Cultivated forms with non-shedding grains occur in each group: *strigosa-brevis* diploids (fodder oats of negligible agricultural significance); tetraploid *abyssinica* (confined to Ethiopia); and, most important, as hexaploids of the *sativa-byzantina-nuda* group.

A summary of present knowledge on species relationships is given in Fig. 28.1. This summary is based on genome affinities (Holden, 1966; Rajhathy and Sadasivaiah, 1969; Ladizinsky, 1971; Rajhathy and Baum, 1972; Ladizinsky, 1973; Rajhathy and Thomas, 1974) and on comparative morphology and ecology (Ladizinsky and Zohary, 1971; Zohary, 1971). Thickness of connecting lines indicates closeness of genetic relationship but the positions of the lines do not necessarily indicate evolutionary pathways.

Clearly, either *magna* or *murphyi* must be regarded as the tetraploid ancestor of *sterilis* and cytogenetic studies implicate *strigosa* as a genome donor to the hexaploids. The occurrence of at least one other diploid species, presently unknown, can be predicted as a common ancestor to *magna*, *murphyi* and *sterilis*, contributing the characteristic spikelet morphology of

these three species, which is lacking in all diploids known to date. Finally, the diploid donors of the third genome of *sterilis* and the second genomes of *magna* and *murphyi* are as yet unknown.

The *barbata* group of tetraploids is essentially autopolyploid in origin, derived directly from the *strigosa* diploids (Holden, 1966). It does not appear, from either cytogenetic or morphological evidence, to have contributed to the evolution of *sterilis* but rather to have evolved independently as a distinct group of wild, weedy and cultivated forms.

While present evidence indicates the independent evolution of at least two tetraploid groups, it is not necessary to postulate more than a single origin to account for variability in the hexaploids. All are completely interfertile. The two major cultivated species, *sativa* and *byzantina*, are separated taxonomically by the position at which the rachilla fractures during threshing (a difference governed by one or two major genes). *Nuda* forms are doubtless free-threshing variants derived from *sativa* or *byzantina* and have a characteristic distribution range, eastwards from the centre of origin into China. All three types lack a dispersal mechanism and can exist only in cultivation. The two serious candidates for the position of hexaploid

Fig. 28.1 Relationships in *Avena*. Cultivated forms in *large italics*
Thickness of connecting lines indicates degree of relationship, not necessarily evolutionary connection.

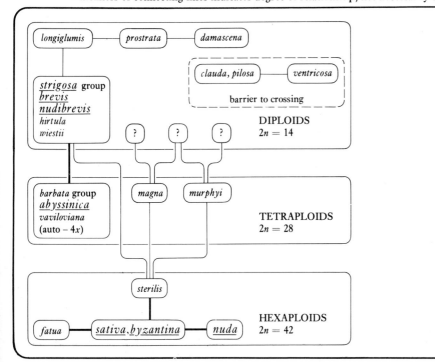

progenitor are *sterilis* and *fatua*. Both have seed dispersal mechanisms; in *sterilis* the unit is the spikelet (minus glumes); in *fatua*, it is the single grain and both have effective seed dormancy systems. However, their ecological preferences and distributions are quite different. *Sterilis* occurs as a major component of the herbaceous flora around the shores of the Mediterranean and through Asia Minor to the slopes of the Zagros mountains in the east. It forms dense stands in the (often relic) oak park-forest belt in relatively stable primary habitats. Geographically, it overlaps its putative tetraploid and diploid ancestors. It is also an aggressive pioneer of disturbed land and abandoned agriculture and a troublesome weed of crops. *Fatua* on the other hand is virtually confined to crops and disturbed land on the margins of cultivation. Self supporting populations in primary habitats are not known and *fatua* should be regarded as a specialized weedy type, derived either directly from weedy ecotypes of *sterilis* or, perhaps, from *sativa* or *byzantina*. *Sterilis* may therefore be regarded as the basic hexaploid type from which the others evolved, in the sequence given in Fig. 28.1.

3 Early history

Excavations of early sites in the Fertile Crescent (6000–7000 B.C.) have revealed emmer, einkorn, barley and pulses but not oats. The first oats appear about 1000 B.C. and then in central Europe (Helbaek, 1959). Oats are generally regarded as a secondary crop that evolved in western and northern Europe from the weed oat components of the primary grain crops, wheat and barley. Farming experience shows that oats grow best in cool, moist, maritime climates and it is argued that, as cultivation of einkorn and emmer moved westwards and northwards from the gene centre, they became progressively less well adapted and were superseded by the weed oat component (Fig. 28.2). Recent studies amply demonstrate the aggressiveness and adaptability of *A. sterilis* and support this view of the origin of the cultivated hexaploids. The critical genetic changes (a solid, non-shedding grain base and loss of dormancy) that occurred at this time would have been at a strong selective advantage in cultivation.

By the end of the first century A.D. hexaploid *sativa–byzantina* oats appear to have become established as a major crop in Europe, although its importance relative to wheat and barley is not clear.

Fig. 28.2 Evolutionary geography of the cultivated oats, *Avena*.

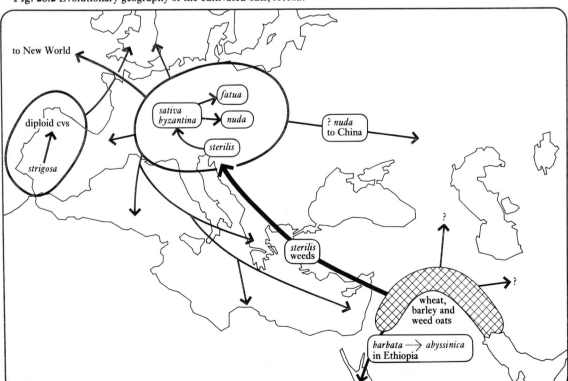

In parallel with these developments, and quite independent of them, it seems likely that cultivated diploid forms of the *strigosa* group evolved, possibly in Iberia. Relics of this culture are still to be found today on the poorest soils of the northern and western fringes of the European land mass (Fig. 28.2). Although diploid oats were probably widely cultivated as dual purpose grain-fodder varieties in early times, they were replaced by the early hexaploid forms carried northwards and westwards by migrating peoples.

There are occasional references in the literature to the growing of Pilcorn, a naked oat, in the British Isles in mediaeval times, for example, by John Gerard in his *Herball* (Coffman, 1961). This is thought to have been a form of naked diploid *nudibrevis*, since lost to cultivation, or of hexaploid *A. nuda*. The culture of naked oats (*A. nuda*) has been practised since early times in China and the mountainous parts of southwestern Asia, and seems to have been restricted to this area. The unpopularity of *nuda* in Europe is traditionally attributed to poor storage characteristics and poor germination of seed grain. The latter may well be connected with the relatively exposed embryo and its liability to damage during threshing.

The cultivated tetraploid, *abyssinica*, is described by Zohary (1971) as a 'tolerated man-dependent weed' restricted to Ethiopia; it probably evolved from weedy *barbata* types introduced into Ethiopia with barley from the Middle East. It is sown, harvested and consumed as an integral part of the barley crop.

Thus the various diploid and tetraploid cultivated types, while of great interest and of considerable importance in the areas where they were grown, are of little agricultural importance today and played little part in the development of oats as a world wide crop. The large grained *sativa-byzantina* hexaploids, on the other hand, had become well established in Europe in time for them to play their part in the colonization of North America, Argentina and Australasia. It is only in very recent times that they have returned to the Middle East as crops from Europe.

The European 'landraces' of the seventeenth, eighteenth and nineteenth centuries were almost certainly heterogeneous mixtures of homozygotes with the capacity for some adaptive response to the conditions of their new habitats. The names of the early oat varieties of North America indicate their origins in Russia, Poland, Sweden, Finland, France, England, Scotland and Ireland (Coffman, 1961). Similarly, the colonization of temperate regions of the southern hemisphere was accompanied by the introduction of the landraces characteristic of the country of origin of the colonists. The genetic variability of these landraces, in permitting an adaptive response to environmental change, was doubtless an important factor in the establishment of a successful culture of old world crops in new places.

4 Recent history

It is probable that spring and winter (autumn sown) oats existed in European landraces before conscious selection began. Hence there were two basic types on which conscious selection was first practised in the latter half of the nineteenth century.

First conscious attempts to improve oats were made, probably more or less contemporaneously, in several European countries at the end of the eighteenth and the beginning of the nineteenth centuries and similar, and equally successful, attempts followed in America by the middle of the nineteenth century. In all cases, the material for selection was the variation, often considerable, in the landraces grown at the time. The principles of selection varied from broadly based mass selection to rigid single-plant pedigree schemes. Experience, rather than theory, revealed that the most effective improvement followed single-plant selection and landraces were supplanted by the first pure line varieties. Consequently, genetic variation declined and further progress by simple selection came to a halt. The outcome of this phase of oat improvement was the widespread use of higher yielding, more uniform spring and winter varieties and a reduction in the *sativa-byzantina* gene-pool.

The next phase can be said to have started with the use of controlled hybridization to generate new variability, a practice (first attributed in oats to Patrick Sherriff, a Scottish farmer, in the 1860s) which marked the beginning of deliberate breeding work. This approach was rapidly adopted by many breeders in Europe and North America. Parents were selected on phenotype with the object of combining desirable complementary characters in the progeny. Many notable varieties were produced which persisted in cultivation for many years.

Subsequent breeding, until recent times, has been based on inter-crosses between the original pure line varieties, maintained in breeders' collections, and their derivatives. In general, sources of new genetic variation have been obtained through exchange of cultivars between different countries and continents. In particular, major-gene resistances to fungal diseases (rusts, smuts, mildew) generally lacking in *sativa* varieties, have been transferred from *byzantina* types. Exceptionally, and more recently, breeders have outcrossed to *ludoviciana* (a weedy *sterilis* type) for resistance to eelworm and powdery mildew. Generally, though, the genetic base of oat breeding has been narrow, resting primarily on European landraces and on the pure lines selected from them. There has been little attempt to exploit the gene pool of the related wild and weedy forms of *sterilis* and *fatua* and none at all to explore the resources of putative tetraploid and diploid ancestral species.

Breeding methods for the management of segregating generations following artificial hybridization have generally been of the pedigree or population type, appropriate to the detection and selective multiplication of desirable segregates in an inbreeder. During the past forty years there has been widespread use of the backcross technique for the transfer of particular major-gene loci, chiefly in connection with breeding for disease resistance. Attempts have been made by means of irradiation and chemical mutagens, to create new variations for use in breeding but success has been limited. Oat breeding since the turn of the century may therefore be described as conservative in approach, conventional in its methods and limited in its objectives.

5 Prospects

World production of oats has shown a slight upward trend in the period 1961–71 (FAO 1973) but, for several traditional western European producers and for Canada, USA and Argentina, areas under the crop have dropped considerably (up to 50 per cent in the UK).

The explanation seems to lie in the competitive disadvantage of oats, as a cash crop, compared with wheat and barley, and this is due principally to the high percentage of husk (up to 25 per cent by weight). Differences in gross yields of grain/hectare for the three cereals in these countries are not sufficient to account for the decline.

Sporadic attempts have been made in the past to breed improved free threshing varieties of *nuda* but with little success. While yield increases were achieved, the varieties so produced were all *nuda* in type, with large, pendulous, multiflorous spikelets and rather weak straw, serious lodging hazards in wet and windy climates.

The breeding of modern *sativa* types with free threshing grain would doubtless reverse the declining area under the crop. While some modest projects with this aim have been started in recent years, it is clear that a much greater effort is needed if the considerable problems involved are to be overcome. Given a successful outcome, the natural advantages of oats over wheat and barley, in terms of animal nutrition (qualitative and quantitative superiority in protein and fat),

then become exploitable attributes. From this point of view, the potential of wild populations remains to be determined. Isolated researches into variation in protein and fat content in *Avena* and examples of the successful manipulation of chemical composition of other crops all point to the possibility of reviving the culture of oats as a high energy crop of high nutritional value.

Resistance to the major fungal pathogens is certain to remain a major breeding objective but new approaches (other than the traditional major gene resistance of transient effectiveness) and new genetic resources are required. Again, *sterilis* with its wide range of wild and weedy ecotypes would appear to be the most likely source of new variation. Recent breeding work has done much to improve the strength of the culm and the ability of the plant to stand until full maturity. However, it would be advantageous if this process continued further so as to produce ultra short-strawed oat varieties, preferably with more compact panicles, similar to the modern varieties of wheat and barley.

6 References

Coffman, F. A. (ed.) (1961). Oats and oat improvement. *Agronomy*, 8, Wisconsin.

Helbaek, H. (1959). Domestication of food plants in the Old World. *Science*, N.Y. 130, 365–72.

Holden, J. H. W. (1966). Species relationships in the Avenae. *Chromosoma (Berl.)*, 20, 75–124.

Ladizinsky, G. and **Zohary, D.** (1971). Notes on species delimination, species relationships and polyploidy in *Avena*. *Euphytica*, 20, 380–95.

Ladizinsky, G. (1971). Chromosome relationships between tetraploid ($2n = 28$) *Avena murphyi* and some diploid, tetraploid and hexaploid species of oats. *Can. J. Genet. Cytol.*, 13, 203–9.

Ladizinsky, G. (1973). The cytogenetic position of *Avena prostrata* among the diploid oats. *Can. J. Genet. Cytol.*, 15, 443–50.

Maltzew, A. I. (1930). Wild and cultivated oats. Section *Euavena*. *Bull. appl. Bot. Genet. Plant Breeding (Leningrad)*, *Suppl.*, 38, 1–522.

Rajhathy, T. and **Sadasivaiah, R. S.** (1969). The cytogenetic status of *A. magna*. *Can. J. Genet. Cytol.*, 11, 77–85.

Rajhathy, T. and **Baum, B. R.** (1972). *Avena damascena*: a new diploid oat species. *Can. J. Genet. Cytol.*, 14, 645–54.

Rajhathy, T. and **Thomas, H.** (1974). Cytogenetics of oats (*Avena*). *Genet. Soc. Canada*, misc. *Publ.*, 2, pp. 90.

Zohary, D. (1971). Origin of south-west Asiatic cereals. Wheats, barley, oats and rye. In P. Davis *et al.* (eds.) *Plant life in southwest Asia*, Edinburgh, 235–63.

Millets

Eleusine coracana, Pennisetum americanum
(Gramineae)

J. W. Purseglove
East Malling Research Station
Kent England

1 Introduction

There are two important millets, treated here, and
four minor ones (for which see Chapter 88). The former
are warm country cereals which have peculiar merits;
the grain of one, *Eleusine*, is capable of prolonged
storage, while the other, *Pennisetum*, finds its main
agricultural niche in places too arid to support
sorghum or maize.

Eleusine coracana, finger millet (also known as
African millet, koracan and, in India, ragi), extends in
Africa from Nigeria eastwards to Eritrea and southwards to southwest Africa and Natal. It is a staple
food in parts of eastern and central Africa, particularly
in northern and parts of western Uganda and northeastern Zambia, and in India, with the greatest area in
Mysore. Some 2–2·5 Mha are grown in India and 0·4
Mha in Uganda. The small grains, only 1–2 mm in
diameter, are usually reddish-brown in colour, although
a white form is known. They are ground into flour,
which is made into a thick porridge. They may also be
germinated, dried and ground (i.e. malted) for making
into beer. The great merit of finger millet is that it can
be stored for long periods of up to ten years or more
without deterioration or weevil damage and so is
excellent for storage against times of famine.

The second main species is bulrush millet, *Pennisetum americanum* (*P. typhoides*, *P. glaucum*). This
species (also known as pearl millet, spiked millet, cat-
tail millet and, in India, bajra) is an erect annual, 0·5–
4m tall. It is usually grown as a rain-fed crop in semi-
arid regions of Africa and India, although it is sometimes irrigated in the latter country. It is the most
important crop in the Sahel zone approaching the
Sahara in tropical Africa; in the Sudan zone to the south
it is of equal status with sorghum. The shorter maturing cultivars can be grown with less rainfall than sorghum and the northern limit is around the 250 mm

isohyet. Late-maturing cultivars become more important as the rainfall increases. The crop is grown to a
lesser extent in eastern and central Africa. According
to Deshaprabhu (1966), bulrush millet in India
occupies the fourth place after rice, sorghum and
wheat, with a total average annual area of about 10·7
Mha, and with the greatest amount in Rajasthan.

The great merit of bulrush millet is that it can be
grown in low rainfall areas on poor, sandy soils and that
it will give economic, albeit low, yields on soil too poor
and too worn out to support most other cereals. It
stores well but, unfortunately, the ripening crop is very
susceptible to bird damage. The grain, about 4 mm
long, which may be white, yellow, grey or light blue,
may be eaten in the same way as rice, or it may be
ground and made into porridge or cakes. The green
plant provides a useful fodder and it is sometimes
grown for this purpose.

2 Cytotaxonomic background

Eleusine is predominantly an African genus with six of
its nine species confined to tropical Africa. One species,
E. tristachya, is South American. Apart from this
species, the highly successful annual weed, *E. indica*,
which is now pantropical, and the cereal, *E. coracana*,
the genus has a very limited distribution, mainly in
eastern and northeastern tropical Africa. The species
can be divided into annuals and perennials and Phillips
(1972) states that, within each group, the differences are
often small and, among the annuals in particular,
introgression is frequent, with intermediates occurring.

The cultivated *E. coracana* is a tetraploid with
$2n = 4x = 36$; there are diploids ($2n = 18$) among its
wild relatives. For many years it was considered to have
originated from *E. indica* either in Africa or India.
Kennedy-O'Byrne (1957) separated the African tetraploid form ($2n = 36$) from the pantropical diploid
E. indica ($2n = 18$) as a distinct species which he
named *E. africana*. Phillips (1972) later made this a
subspecies of *E. indica*, namely subsp. *africana*.
Eleusine indica subsp. *indica* is widespread throughout
the tropics and subtropics and, in Africa, occurs
mainly in West Africa, along the coast of east and southeast Africa and in Madagascar. *Eleusine indica* subsp.
africana is confined to Africa and is found mainly in
the uplands of eastern and southern Africa. It occurs
commonly as a weed in the fields of *E. coracana* with
which it hybridizes freely, forming a range of intermediates. Its seeds are about half the size of those of
the cultivated species and are shed about three weeks
before the latter mature.

Pennisetum americanum is a diploid with $2n = 2x =$ 91

14 and belongs with *P. purpureum* (Napier grass) in that part of the genus with $x = 7$; other species have $x = 9$.

Stapf and Hubbard (1934) recognized 18 species of cultivated bulrush millet but Bor (1960), Purseglove (1972) and others have recognized but a single collective species, all the forms of which cross readily with each other, as well as with related wild forms which occur as weeds of cultivation or grow in the area. The cultivated races have broad-tipped persistent spikelets, usually with protruding grains, whereas, in the related wild species, the spikelets are narrower, pointed and readily deciduous and the grains are smaller and entirely enclosed by the lemma and palea. Thus the wild and weedy races have an efficient dispersal mechanism, which is absent in the cultivated forms, which would have been selected for freedom from shattering.

In summary, therefore, both the major millets are African and both are capable of genetic exchange with related wild forms living in the same area.

3 History

The two major millets have remarkably similar histories of which geographical aspects are summarised in Fig. 29.1.

Two groups of cultivars of *E. coracana* are recognized: (*a*) African highland types, bearing a resemblance to *E. indica* subsp. *africana*, with long spikelets, long glumes, long lemmas and grains enclosed within the florets: (*b*) Afro-Asian types, with a close resemblance to *E. indica* subsp. *indica*, with short spikelets, short glumes, short lemmas and the mature grains exposed distally. Both are tetraploids with $2n = 36$. Both Kennedy-O'Byrne (1957) and Mehra (1963*a*, *b*) consider it probable that both types originated from subsp. *africana* and that the African

Fig. 29.1 Evolutionary geography of the African millets, *Pennisetum americanum* and *Eleusine coracana*.

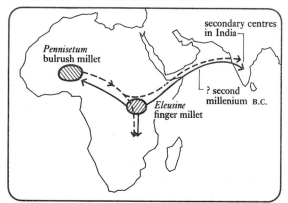

highland forms were selected from large-grained mutants. Mehra believes that Afro-Asian cultivars then evolved from African highland forms by mutations for short glumes and exposed grains. On the basis of this hypothesis, Purseglove (1972) considered that Uganda or a neighbouring territory could well have been the centre of origin. The crop is certainly of very ancient cultivation in the area, as is shown by the part it plays in religious and tribal ceremonies and by the many customs connected with its cultivation.

It is postulated that *E. coracana* was taken at an early date to India, probably over 3,000 years ago. Sorghum and bulrush millet could have been taken about the same time. The crop was not known in ancient Egypt, but it has a Sanskrit name in India. It seems likely that it was taken up the seaward edge of the Arabian peninsula along the Sabaean Lane. India became a secondary centre of evolution of the crop. There has been no spread of finger millet in recent times.

Finger millet is mainly self-pollinated, but it is reported that about 1 per cent cross-pollination by wind takes place, which accounts for contamination by *E. indica* subsp. *africana* when the two grow together. Numerous cultivars are recognized in Africa and India and it is usual to plant mixtures.

Line breeding and selection from single plants produced few worth-while results in Uganda. Sastri (1952) reports that a number of improved cultivars have been obtained by breeding in India, including 'EC.593', a general purpose type evolved in Madras. Further work is required on the improvement of the crop both in India and Africa.

There seems little doubt that despite the name, *Pennisetum americanum*, tropical Africa is the home of bulrush millet. It probably originated in western tropical Africa, where the greatest number of cultivated and related wild forms occur. Chevalier (1934) states that the cultivated bulrush millets in the Sahel zone of West Africa are systematically very near some of the spontaneous and subspontaneous forms and he supposes that the cultivated species originally arose in this region. The crop was taken at an early date, probably at least 3,000 years ago, to East Africa and thence to India, where, at the periphery of distribution, a second centre of variability developed and improved cultivars were evolved.

The spikelets are markedly protogynous, most of the styles having started to dry up before the pollen is shed, so that the crop is mostly cross-pollinated. As might be expected, the crop is very variable and most fields have a mixture of types. When there is some overlapping of female and male phases in the same inflorescence, or in different inflorescences on the same plant, some self-pollination can occur and this can be used for the production of inbred lines for the breeding of hybrid seed, as is done with maize. Deshaprabhu (1966) reports that male sterile lines are now available in India for the production of hybrid seed and that the more diverse the geographical origins of the inbreds the greater is the degree of hybrid vigour on crossing. F_1 hybrid seed has to be replaced for each crop. The production of 'synthetic varieties' is also being attempted, involving six or more parental combinations. Murty (1969) states that hybrids are now available which give 3,286 kg of grain per hectare compared with 1,519 kg/ha for the local cultivars.

In summary, it is clear that the two major millets, though botanically quite distant from each other, have many similarities. Both are of central African origin; both evolved by local selection for large grains and non-shattering heads from wild forms which persist in the area; both exist in variable populations capable of continued genetic exchange with their wild relatives; and both migrated early to India where secondary centres of diversity became established.

4 References

Bor, N. L. (1960). *The Grasses of Burma, Ceylon, India and Pakistan.* Oxford.

Chevalier, A. (1934). Étude sur les Prairies de l'Ouest Africain. *Rev. Bot. appl. Agric. trop.*, 14, 17–48.

Deshaprabhu, S. B. (1966) (ed). *The wealth of India: Raw Materials*, 7, New Delhi.

Kennedy-O'Byrne, J. (1957). Notes on African Grasses: XXIX: A New Species of *Eleusine* from Tropical and South Africa. *Kew Bull.*, 1, 65–72.

Mehra, K. L. (1963a). Differentiation of the cultivated and wild *Eleusine* species. *Phyton*, 20, 189–98.

Mehra, K. L. (1963b). Considerations of the African origin of *Eleusine coracana. Curr. Sci.*, 32, 300–1.

Murty, B. R. (1969). New hybrids of Bajra. *Indian Fmg*, 19, 13–5.

Phillips, S. M. (1972). A survey of the genus *Eleusine* (Gramineae) in Africa. *Kew Bull.*, 27, 251–70.

Purseglove, J. W. (1972). *Tropical Crops: Monocotyledons.* London.

Sastri, B. N. (1952) (ed.). *The Wealth of India: Raw Materials*, 3, New Delhi.

Stapf, O. and Hubbard, C. E. (1934). *Pennisetum.* In D. Prain (ed.), *Flora of Tropical Africa*, 9, London.

Barley
Hordeum vulgare (Gramineae – Triticinae)

J. R. Harlan
University of Illinois Urbana Ill USA

1 Introduction

Barley is an important cereal, domesticated from wild races found today in southwestern Asia. It is a short-season, early maturing grain with a high yield potential and may be found on the fringes of agriculture because it can be grown where other crops are not adapted. It extends far into the arctic, reaches the upper limits of cultivation in high mountains and may be grown in desert oases, where it is more salt tolerant than other cereals. Barley is a cool-season crop; it can tolerate high temperatures if the humidity is low, but is not suited to warm-humid climates. It is little grown in the tropics except in cool highlands, as in Mexico, the Andes, and East Africa. Major production areas are: most of Europe, the Mediterranean fringe of North Africa, Ethiopia, the Near East, USSR, China, India, Canada and the USA.

In order of importance, barley is used (*a*) for animal feed, (*b*) for brewing malts, (*c*) for human food. In most cultivars, the grain is tightly encased in husks, which increase the roughage and lower nutritional quality, but which improve malting quality. World production is about half that of maize and 40 per cent that of wheat, but the grain is largely consumed locally and little appears in world trade.

2 Cytotaxonomic background

Barley is a self-pollinating diploid with $2n = 2x = 14$. (Anon., 1968; Nilan, 1971; Takahashi, 1955). Tetraploids have appeared spontaneously but are a negligible part of the crop. The wild and weed races are usually designated *Hordeum spontaneum* but, biologically, they belong to the same species as the cultivated races. Hybrids between wild and cultivated forms are easily made and occur naturally where the two are found together. Hybrid plants are fully fertile; the chromosomes pair well, and segregation is normal. In the spontaneous forms, the spikes fragment at maturity and the grains fall while, in domesticated races, the spike is tough and the grains persistent. The difference is controlled by either one of two tightly linked 'brittle' (Bt and Bt_1) genes. Wild-type is dominant.

All the truly wild forms of *Hordeum* are two-rowed, that is, of the three spikelets at each node of the ear, the two lateral ones are female-sterile and only the central one develops a grain. Under domestication, six-rowed races appeared in which all three spikelets produce grains. There are two genes involved, both with multiple allelic series, but a single recessive mutation (vv) is adequate to cause a two-rowed barley to become six-rowed. Six-rowed genotypes with fragile ears are known, but do not appear to be truly wild plants and are probably derived from six-rowed cultivars.

In naked barley, the husks do not adhere to the grain, which falls free on threshing. The naked grain is much more acceptable as human food and such cultivars are preferred where barley is a major part of the human diet. The character is controlled by a single recessive gene (n).

Barley has been crossed with several other species of *Hordeum*, but the hybrids are highly sterile or anomalous and no other species is known to be involved in barley evolution.

3 Early history (Fig. 30.1)

Barley was one of the earliest crops domesticated in the Near East (Harlan, 1968; Harlan and Zohary, 1966). Archaeological remains of what appear to be wild forms with fragile ears were found in considerable quantities at Mureybat on the Euphrates in Syria dating, perhaps, to 8000 B.C. Similar remains were found at Beidha in the southern Jordan Highlands dated about 6800 B.C. and at Jarmo in the Zagros foothills in Iraq at about the same age. It is not possible to determine if these remains were from wild harvests or from early attempts to grow the crop.

Cultivated barley was found at Ali Kosh (Iran), Ramad (Syria), Jericho (Palestine) and at Catal Hüyük and Hacilar (both in Turkey) at time ranges between 6000 and 7000 B.C. (Renfrew, 1969). Barley usually appears with remains of other primary crops of the region such as emmer wheat, einkorn wheat, flax, pea, vetch and lentil.

All the earliest barley finds were of two-rowed, covered (i.e. non-naked) sorts but both six-rowed and naked types had appeared by 6000 B.C. (Helbaek, 1966; Renfrew, 1969; Murray, 1970).

Agriculture was first developed in the Near East, on dry land and at elevations sufficient for adequate rainfall. The arts of irrigation soon developed and several sites have been identified at which irrigation was evidently practised by 6000 B.C. Indeed, crops may have been watered considerably earlier at the oasis of Jericho. At any rate, barley was a basic crop for the early irrigated agriculture, not only of Mesopotamia but of ancient Egypt as well. Our first evidences of agriculture in Egypt are from the Fayum and date to about 4500 B.C. It is quite likely that cultivation began much earlier than this but the evidence has been buried under silts deposited by the annual floods of the Nile.

Early barley remains from the city states of Mesopotamia and dynastic Egypt are much more abundant than those of wheat and the earliest literatures in cuneiform, linear B and Egyptian hieroglyphic all

Fig. 30.1 Distribution of wild *Hordeum spontaneum* and of early archaeological sites for cultivated barley, *H. vulgare* (after Harlan and Zohary, 1966. Copyright by the American Association for the Advancement of Science).

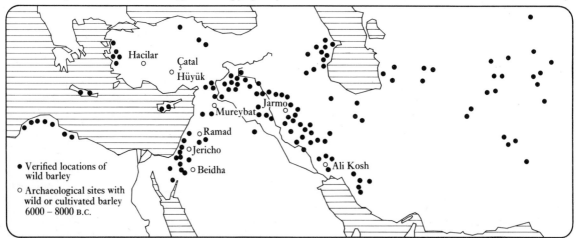

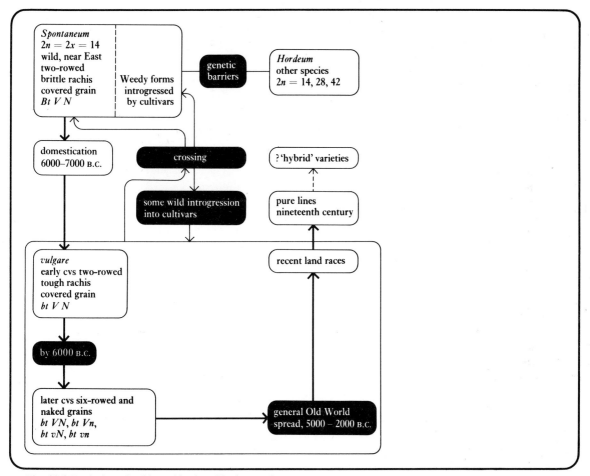

Fig. 30.2 Evolution of cultivated barley, *Hordeum vulgare*.

suggest that barley was more important than wheat for human food at that time. The Sumerians had a god for barley but not one for wheat. Toward the end of the third millennium B.C., the irrigated lands of southern Mesopotamia began to salt up and wheat production there declined sharply. A near monoculture of barley was established by about 1800 B.C., since it was the only crop that could tolerate the high salt content. Even so, the records indicate a sharp decline in yields about that time (Jacobsen and Adams, 1958).

Barley was the most abundant grain of the ancient Near East and the cheapest. It was the standard fare of the poor, the ration of the soldier, serf and slave, and the staff of life for the Greek peasantry. It had the reputation of being a strong food and was awarded to victors in the Eleusinian games. Ancient gladiators were called *hordearii* or 'barley-men' because they trained on barley. The shift to wheat as human food came in classical times but, as late as first century A.D.,

Pliny wrote: 'Barley bread was much used in earlier days, but has been condemned by experience, and barley is now mostly fed to animals . . .', although beer was highly touted for its healthful properties.

According to archaeological evidence, barley reached Spain in the fifth millennium B.C. and the lower Rhine a little later. It spread eastward to the Indus where it was grown in the third millennium B.C. and reached China late in the second. At some time, not now known, it reached the highlands of Ethiopia where it became one of the dominant crops.

The essentially simple view of the domestication of the crop outlined above may be summarized by saying that it originated in a limited area from a two-rowed ancestor; that three recessive mutants were of key importance; and that, as a corollary and contrary to earlier views, six-rowed 'wild' barleys (such as *Agriocrithon* types) reflect introgression from cultivars. The situation is taxonomically much simpler than was once supposed and there is no need to postulate multiple domestication. For reviews of earlier interpretations, see Takahashi (1955) and Nilan (1964).

Cultivated barleys, in general, are freely inter-fertile and there is no evidence of substantial genetic differentiation. There is, however, a body of breeding experience, much of it unrecorded, to the effect that there may be subtle, but yet unanalysed, differentiation between two-row and six-row types. Long ago, Harlan *et al.* (1940) showed that, inexplicably, the yielding ability of $2R \times 6R$ and $6R \times 2R$ combinations was inferior to that of crosses within groups; and it is common experience (noted by Harlan *et al.*, 1940) that inter-group crosses often generate morphologically peculiar or otherwise uninterpretable segregates. Here one recalls Takahashi's studies of the distribution of *bt* alleles and other genes which are at least suggestive of geographical differentiation related to row number (review in Takahashi, 1955). It may be that acceptable expression of the six-row character has demanded so much adjustment of the genetic background as to generate an incipient differentiation between the groups. The matter has not had the study it deserves.

4 Recent history

Barley was taken to the New World by Columbus on his second voyage. It did not succeed in Hispaniola but, early in the sixteenth century, it was introduced to the highlands of Mexico where it continues today as a minor crop, sometimes sown between rows of maize. In Peru, it had considerable success among the Indians because of its short season and cold tolerance which adapted it to the cold highlands. It also filled a special need as fodder for guinea pigs; whole plants are often cut green and sold in the markets for the purpose. By the beginning of the eighteenth century barley was being grown in the Spanish colonies from California to Chile at appropriate elevations.

A series of introductions is documented for the early seventeenth century in New England and Virginia. The crop expanded slowly as settlers moved westward. Wheat was more in demand for early subsistence farming in North America and barley was relatively little grown until the cities developed an appreciable demand for meat and dairy products. In the USA the area harvested did not exceed 0·8 Mha until 1880. Because of its early maturity, barley was particularly suited to the western plains of Canada and production increased enormously as these were settled. The USA and Canada remain the only countries of the New World with really significant productions from a global point of view.

Barley production has increased sharply in recent decades as the demand for livestock feed has grown. In the 20 years from 1950 to 1970, world production increased from about 59 Mt to 150 Mt. Increases of three- to six-fold were registered in Denmark, France, West Germany, the UK, the USSR and Canada. A great deal of the barley grown in western Europe today is used to fatten cattle. The rise in production of malting barleys has been, comparatively, much less.

The general trend follows that of the other major cereals. At the end of the late war, industrial nations converted some of their munitions plants to the production of nitrogenous fertilizers. For the first time in history, fixed nitrogen became both abundant and cheap. It then was possible and practical to apply nitrogenous fertilizers to enormous areas of crops of which the cereals were among the most responsive. Spiralling increases in yields have continued since, centred primarily in industrial nations with a substantial potential for fertilizer production, but later extending to other countries through export and local production. A sharp yield 'take-off' occurred in all industrial countries about 1945, except in Japan which had already achieved significant per hectare increases in yields early in the twentieth century.

Up to the middle or late nineteenth century, depending on locality, all barleys must have existed as highly heterogeneous 'land races', mixtures of more or less inbred lines with a significant admixture of hybrid segregates, the products of a low level of random crossing in earlier generations. Many barleys exist in this state today but, over the past 100 years or so in advanced agricultures, the land races have been virtually wholly displaced by pure-line cultivars. These were, at first, selections from land races (such as Gull, Hanna and Arctic in Europe and Mission in California); later, they have come from successive cycles of crosses between established pure lines, sometimes of diverse geographical origins. The result has surely been a marked narrowing of the genetic base in many, probably all, advanced agricultures, partially concealed by a great diversity of cultivar names. There has, therefore, been a fairly profound change in the genetic structure of barley populations. There has also been a recent morphological change in that barleys have generally become shorter in stature and thus not only more efficient biologically but also tolerant of high-N fertilizer régimes.

5 Prospects

For feed barleys the current trends parallel those of wheat and rice. The direction is toward more intensive culture with greater inputs in fertilizers, tillage, seeding rates, herbicides and so on. Cultivars have been and are being developed suitable for such altered conditions; for example, short-strawed or semi-dwarf types that are less vegetative, more resistant to lodging, that are fertilizer-responsive and have greater photosynthetic efficiency. There is a search for yield genes and heterotic combinations that will increase yield, improve test weight and provide a wider range of adaptation. The trend in this direction is already evident in world production statistics, and yields per hectare have increased substantially in recent years.

Great attention is being given to improved feed quality. Genes that influence both quantity and quality of protein have been discovered and the inter-actions between the two are being investigated. The digestibility of protein and efficiency of conversion from grain to animal gain is considered critical. Simple increases in grain protein do not always result in a better feeding efficiency. Animals can be fattened on a full barley ration but a number of nutritional problems need to be overcome in order to develop truly high quality feed grain cultivars. The trend, however, is conspicuously in the direction of much improved feed barleys for livestock nutrition. Some selections from Ethiopia have been especially useful in quality improvement.

Major emphasis is being placed on genetic resist-ance to the most serious diseases of barley. Among the most common and damaging are powdery mildew (*Erysiphe*), rusts (*Puccinia*), smuts (*Ustilago*), net blotch (*Pyrenophora*), spot blotch (*Helminthosporium*), leaf blotch (*Septoria*), scald (*Rhynchosporium*), barley yellow dwarf virus (BYDV) and barley stripe mosaic virus (BSMV). General resistance and tolerance are sought as well as genes conferring specific resistance to individual races of pathogen. Resistance to aphids (greenbugs) has been found and incorporated into some cultivars. Ethiopian barleys have been parti-cularly useful in supplying resistance to leaf diseases. Conditions on the high plateau of Ethiopia are gener-ally favourable to leaf disease of all sorts and the local races of barley have responded over the millennia by developing high levels of tolerance or resistance.

A considerable emphasis is being given to problems of adaptation, for example: greater hardiness for winter barleys; special cultivars for cool, wet regions; early short-season drought resistant cultivars for more difficult environments; and wide-range adaptation for general use. If the barley acreage is to expand substantially, it will be necessary to exploit the crop's unique characteristics of earliness, drought resistance, salt tolerance, vigorous root system, high tillering capacity and so on.

Trends in malting barleys are somewhat along the same lines but tempered by special demands for quality and uniformity. The specifications for good malting quality are very different from those for good feed barley and selection, in some respects, is in opposite directions. Two-rowed cultivars are pre-ferred in Europe and in some parts of the USA, although highly acceptable six-rowed barleys have been produced. Uniformity is important for malting barleys but is not critical for feed and breeding methodology may differ accordingly.

Barley has, for some time, been an important experi-mental subject for basic genetic research. As a self-pollinating diploid that can be easily grown, it has some special advantages. The genes often have very neat, specific effects; the chromosomes are relatively long and have been fairly well mapped. Trisomic and translocation stocks have been built up that permit interesting experiments in chromosome engin-eering. Some composite cross populations (mixtures of crosses between many cultivars) were made in Idaho and California many years ago (Harlan, Martini and Stevens, 1940) and continue to provide excellent material for population–genetic studies as well as useful breeding stocks. The importance of this work on composite crosses far transcends barley; unfor-tunately there is no general review of the subject but basic references will be found in papers in Nilan (1971).

A number of proposals for producing hybrid barley have evolved from such basic research. It is too early to tell if any of the schemes will be suitable for com-mercial production, but the trend toward increasingly intensive culture and ever higher yields is well established. Rising standards of living and rising expectations around the world have enormously in-creased demand for more meat in the human diet and this can only be supplied by greater production and more efficient use of feed grains.

Although Ethiopian barleys have not shown much promise outside Ethiopia, as cultivars they have been an immensely valuable source of genetic resistance to diseases and of improved nutritional quality. High resistance to powdery mildew, loose smut, leaf rust, net blotch, septoria blotch, scald, yellow dwarf virus and stripe mosaic virus has been found in them. Not only is resistance common in these materials but many lines have multiple resistances and some lines have resistance to all races of a disease tested. Protein as high as 18 per cent and lysine levels of 4·4 per cent of protein have been found in Ethiopian barleys.

These discoveries have helped to reinforce concern for genetic erosion and the possible loss of indigenous genetic resources. Ethiopia has not been thoroughly collected and there is a danger that important and useful populations will be discarded in the future and replaced by selected cultivars developed in plant breeding programmes. There are other parts of the world in which sampling is inadequate and it is to be hoped that efforts will be made to salvage these genetic resources before they disappear. The world barley collections are more complete and better maintained than those of many other crops but there is no way to replace materials once they are lost. International efforts are under way toward assembly, conservation and utilization of genetic resources on a global scale.

6 References

Anon. (1968). Barley: origin, botany, culture, winter-hardiness, genetics, utilization, pests. *USDA Agriculture Handb.*, **338**, pp. 127.

Harlan, J. R. (1968). On the origin of barley. *USDA Agriculture Handb.*, **338**, 9–31.

Harlan, J. R. and Zohary, D. (1966). Distribution of wild wheats and barley. *Science*, N.Y., **153**, 1074–80.

Harlan, J. V., Martini, M. L. and Stevens, H. (1940). A study of methods in barley breeding. *USDA tech. Bull.*, **720**, pp. 25.

Helbaek, H. (1966). Commentary on the phylogenesis of *Triticum* and *Hordeum. Econ. Bot.*, **20**, 350–60.

Jacobsen, T. and Adams, R. M. (1958). Salt and silt in ancient Mesopotamian agriculture. *Science*, N.Y., **128**, 1251–8.

Murray, J. (1970). *The first European agriculture.* Edinburgh.

Nilan, R. A. (1964). The cytology and genetics of barley, 1957–1962. *Res. Studies, Wash. State Univ.*, **32**(1), pp. 278.

Nilan, R. A. (ed.) (1971). Barley Genetics II. *Proc. 2nd internatl Barley Genet. Symp.*, Pullman, Washington, USA, pp. 622.

Renfrew, J. M. (1969). The archaeological evidence for the domestication of plants; methods and problems. In P. J. Ucko and G. W. Dimbleby (eds) *The domestication and exploitation of plants and animals.* Chicago, 149–72.

Takahashi, R. (1955). The origin and evolution of cultivated barley. *Advanc. Genet.*, **7**, 227–66.

Rice

Oryza sativa and *Oryza glaberrima*
(Gramineae-Oryzeae)

T. T. Chang
International Rice Research Institute
Los Baños Philippines

1 Introduction

Rice equals wheat in importance as a staple food for man. It feeds a large proportion of the people inhabiting the densely populated areas of the humid tropics and subtropics. The rice endosperm is highly digestible and nutritious although the protein content is relatively low (7 per cent).

Rice is cultivated from 53°N latitude to 35°S. In 1971–72, world production of rough rice was estimated at 310 Mt. China and India are the leading producers. International trade takes up less than eight per cent of the world production. Thailand and the USA are the leading exporters.

Asian rice, *Oryza sativa*, is planted on a much larger area than African rice, *O. glaberrima*, which it is rapidly replacing. The two species show small morphological differences but hybrids between them are highly sterile.

Chang (1964) and Nayar (1973) have given general treatments of the biosystematics and cytogenetics of the genus.

2 Cytotaxonomic background

Recent taxonomic revisions have reduced the number of species recognized in *Oryza* to 20. Some rice workers, however, do not endorse all the revisions. Chang (1970) has listed the 20 species and their distributions. *Oryza sativa*, *O. glaberrima* and their wild relatives are diploids ($2n = 24$), while more than half the wild rices are tetraploids ($2n = 48$) (Table 31.1).

The '*sativa* complex' includes *O. sativa* and its wild relatives (*O. rufipogon* and *O. nivara*) as well as *O. glaberrima* and its relatives (*O. barthii* and *O. longistaminata*). Members of this complex carry genome A which is subdivided into A, A^b, A^{cu}, and A^g to indicate partial sterility and minor pairing aberra-

Table 31.1 Species of *Oryza*, chromosome numbers, genome symbols and geographical distributions

Species name (synonym)	$x = 12$ $2n =$	Genome group	Distribution
O. alta	48	CCDD	Central and South America
O. australiensis	24	EE	Australia
O. barthii (O. breviligulata)	24	A^gA^g	West Africa
O. brachyantha	24	FF	West and central Africa
O. eichingeri	24, 48	CC, BBCC	East and central Africa
O. glaberrima	24	A^gA^g	West Africa
O. grandiglumis	48	CCDD	South America
O. granulata	24	—	South and Southeast Asia
O. latifolia	48	CCDD	Central and South America
O. longiglumis	48	—	New Guinea
O. longistaminata (O. barthii)	24	A^bA^b	Africa
O. meyeriana	24	—	Southeast Asia, southern China
O. minuta	48	BBCC	Southeast Asia
O. nivara (O. fatua, O. rufipogon)	24	AA	South and Southeast Asia, southern China, Australia
O. officinalis	24	CC	South and Southeast Asia, southern China, New Guinea
O. punctata	48, 24	BBCC, BB(?)	Africa
O. ridleyi	48		Southeast Asia
O. rufipogon (O. perennis, O. fatua, O. perennis subsp. balunga, O. perennis subsp. cubensis)	24	AA, $A^{cu}A^{cu}$	South and Southeast Asia, southern China, South America
O. sativa	24	AA	Asia
O. schlechteri	—	—	New Guinea

tions in hybrids between members of different subgenomes (intermediate between the primary and secondary gene pools of Harlan and de Wet, 1971). The A genome shows partial homology with genome B (*O. minuta*, *O. eichingeri* and *O. punctata* are BBCC) and with genome C of *O. officinalis* (International Rice Research Institute, 1964).

Earlier postulates as to the putative ancestor of *O. sativa* include the following: (*a*) *O. officinalis*, (*b*) 'O. fatua' (referred to here as spontanea forms of *O. sativa*), (*c*) 'O. perennis' (an ambiguous name often

used to include *O. rufipogon*, *O. nivara* and 'O. fatua' of Asia, or *O. longistaminata* and *O. barthii* of Africa, or even both Asian and African wild relatives of the two cultivated species); and (*d*) 'O. perennis subsp. balunga' (now designated as the Asian race of *O. rufipogon*). Chang (1964) and Nayar (1973) have reviewed extensive literature related to this problem. Studies of the origin of *O. glaberrima* have pointed to either the rhizomatous *O. longistaminata* (formerly called *O. barthii*) or the annual wild race, *O. barthii* (until recently widely known as *O. breviligulata*) as the putative ancestor.

3 Early history

Direct evidence of the early evolution of the cultivated rices is fragmentary and often controversial. Although many workers have supposed that the Indian subcontinent is the ancestral home of *O. sativa*, the earliest archaeological evidence from India goes back only to 2500 B.C. On the other hand, rice glumes belonging to the neolithic period in China have been dated to 2750–3280 B.C. and the recorded history of rice cultivation in that country goes back to the third millennium B.C. A few workers have considered southeast Asia as the centre of origin; recently excavated pottery from Thailand dated to 3500 B.C., bears imprints of rice glumes. In general, therefore, the archaeological evidence suggests considerable antiquity but is indecisive as to time and place.

Since there are certainly two cultivated species of rice, the Asian and the African, it is natural to enquire if there is any evolutionary connexion between them. In as much as *Oryza* must be presumed monophyletic, the two cultigens must have, ultimately, a common ancestor but there is no agreement as to what it might have been or whether it still exists. The cosmopolitan 'O. perennis' complex has been proposed as the probable common progenitor (IRRI, 1964; Oka, 1974). Whatever the solution of this problem, it seems clear that the two cultigens represent two independent and parallel domestications. The pan-tropical and -subtropical distribution of the wild relatives of the two cultivated species in Africa, south and southeast Asia, Oceania and Australia, and Central and South America strongly suggests a common progenitor which existed in the humid zone of the Gondwanaland continents before their breakup and drift.

Among the wild relatives of *O. sativa*, the perennial and weakly rhizomatous, *O. rufipogon*, is widely distributed over southern and southeast Asia, southern China, Oceania and South America, usually in deepwater swamps. A closely related annual wild form, *O. nivara*, is found in the Deccan plateau of India, many parts of southeast Asia and Oceania. The habitats of *O. nivara* are ditches, water holes and edges of ponds. Morphologically similar to (and sometimes indistinguishable from) *O. nivara* are the very widely distributed spontanea forms of *O. sativa* ('*O. fatua*'), which represent numerous intergrading hybrids between *O. sativa* and its two wild relatives. Throughout southern and southeast Asia, the spontanea rices are found in canals and ponds adjacent to rice fields and in the rice fields themselves.

The wild and all the above weed races have long awns (often pigmented), a high frequency of natural cross-pollination, light and shattering spikelets and strong seed dormancy, traits which enable them to persist as weeds in spite of the hot, dry season alternating with the monsoon. Because of extensive hybridization, typical specimens of *O. rufipogon* and *O. nivara* are now rarely found and, as rice agriculture has become

more intensive, many wild populations have disappeared from their known habitats.

While the ultimate progenitor of *O. sativa* was, no doubt, a perennial form (morphologically similar to *O. rufipogon*), domestication most likely started from an annual which may have resembled *O. nivara*. Similarly, *O. glaberrima* is most likely to have derived from the annual *O. barthii*. Although annual wild forms appear to be the immediate progenitors of both the Asian and African cultivated rices, the associated weed races (the spontaneas in Asia and *O. barthii* in Africa) have undoubtedly played a significant role in the development of the more recent cultivars (IRRI, 1964; Harlan, 1969). Recent samples of the weed races indicate that considerable introgressive hybridization has taken place and that gene flow has been largely from the cultivated rices to the wild forms. The principal barriers to gene flow between cultivated

Fig. 31.1 Evolution of the cultivated rices, *Oryza sativa* and *O. glaberrima*.

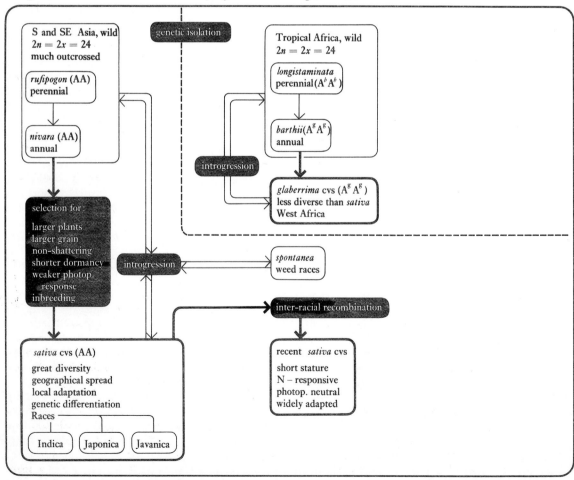

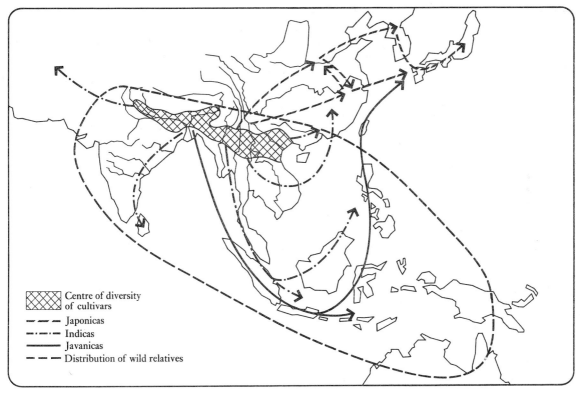

Fig. 31.2 Evolution and spread of the geographical races of *Oryza sativa*.

and wild forms are hybrid sterility, non-viability or weakness, some of which are controlled by complementary or duplicate genes (Oka, 1974).

A number of morphological and physiological changes occurred in the evolution of *O. sativa*. Larger leaves, longer and thicker culms and longer panicles resulted in a larger plant size. There were also increases in the number of leaves and in their rate of development, in the number of secondary panicle branches and in grain weight, in the rate of seedling growth and tillering capacity and in the synchronization of tiller development and panicle formation. There was a slight increment in the net photosynthetic rate of individual leaves and the period of grain filling lengthened. Concomitantly, there were decreases in (or losses of) pigmentation, rhizome formation, ability to float in deep water, awning, shattering, duration of grain dormancy, photoperiod response and sensitivity to low temperatures. There was also a decline in the frequency of cross-pollination so that the crop became more inbred than its wild ancestors.

Many, perhaps all, these changes were called forth by semi-natural selection imposed by the facts of cultivation in diverse climates, soils and seasons and under diverse cultural practices (Oka and Morishima, 1971; Chang, 1976).

The continuous distribution of *O. rufipogon*, *O. nivara* and the spontanea forms of *O. sativa* in a belt 2,000 miles long from the foothills of the Himalayas to the Mekong region (Chang, 1975) strongly suggests a diffuse origin. Since the domestication of a crop is not necessarily confined to the centre of diversity of its wild relatives, the area of greatest diversity of cultivated forms may provide a more useful clue to the centre of domestication. On this basis, along with information from philology, palaeo-climatology and ethnology, the area including northeastern India, northern Bangladesh and the triangle adjoining Burma, Thailand, Laos, Vietnam and southern China appears to be the primary centre of domestication (Fig. 31.2). From this region, rice was introduced into the Yellow River valley of China where the temperate race (keng or Japonica) evolved. From China, rice was introduced to Korea and, later, to Japan (300 B.C.). The tropical race (Indica or sen) was introduced into the Yangtze River valley about A.D. 200. The tall, large and bold-grained bulu varieties (Javanica) of Indonesia appear to be more recent derivatives from the tropical continental forms. The bulu varieties spread from Indonesia to the Philippines, Taiwan and Japan.

Geographical isolation followed by natural and human selection resulted in some hybrid sterility between the three variety groups and even between some members within the Indica group. This sterility has been ascribed to duplicate genes, chromosome structural changes and genetic imbalance (IRRI, 1964; Engle *et al.*, 1969).

The introduction of *O. sativa* into Europe and Africa was rather recent. Rice was introduced from Europe into South and Central America. The crop was first planted in the USA during the seventeenth century from Malagasy seed.

Physiological and anatomical studies indicate that rice is a semi-aquatic plant. Though some workers have considered that upland culture preceded lowland culture, the reverse appears more probable. Observation of the very large collection at the International Rice Research Institute (IRRI) suggests that the greatest diversity of plant characters and the more primitive cultivars are found among varieties adapted to lowland culture. 'Dry land' rices, by contrast, often possess one or more of the advanced features: glabrous leaves and glumes, heavy grains, long and non-shattering panicles and thick roots.

Compared with *O. sativa*, *O. glaberrima* exhibits less diversity and its distribution is limited to tropical West Africa (Portères, 1956; Chang, 1975). *O. glaberrima* probably originated about 1500 B.C. or later. Its primary centre of diversity is in the swampy area of the upper Niger, and it has two secondary centres which lie to the southwest, on the Guinea coast (Portères, 1956). In the field, *O. glaberrima* is often grown mixed with its annual weed race. Africans collected both species, as well as the rhizomatous *O. longistaminata*, for food. *Oryza glaberrima* is more closely related to *O. barthii* than it is to *O. longistaminata*. The weed race is sometimes called *O. stapfii*.

4 Recent history

Recorded history rarely mentions man's efforts to improve the rice plant. But, long before the advent of science, man undoubtedly made full use of natural variability in the crop, spontaneous mutations, natural hybrids and introductions from foreign lands. The introduction of the early maturing Champa varieties from central Vietnam into China in the eleventh century was largely responsible for the practice of double cropping in south China. The significant change in Thailand, from predominantly bold-grained rices between the fourteenth and eighteenth centuries to long-grained varieties 200 years later, illustrates the

relative ease with which rice varieties could be replaced.

Comparative trials of native varieties were begun in Japan about 1893 and the earliest report of rice breeding by hybridization came from the same country where, in 1906, the hybrid variety Ominishika was developed; by 1913, 20 varieties of hybrid origin were commercially grown. Progress achieved through the combined efforts of breeders, agronomists and soil scientists, was remarkable in that, in only 30 years, rice cultivation spread northwards as far as 45° latitude and the northernmost island of Hokkaido became the highest yielding region in Japan.

Hybridization of rice in the United States began about 1922 when breeders crossed varieties introduced from Japan, China, India and the Philippines. These programmes emphasized nitrogen responsiveness, smooth hulls and stiff straw. Varieties selected or bred in the USA later spread to South America and Australia and there attained local dominance.

Mass and pedigree selection methods were the mainstay of breeding programmes in tropical Asia before 1930. The first variety of deliberate hybrid origin was bred in Indonesia about 1913. After the late war, tropical and Japanese varieties were crossed extensively under a region-wide Indica–Japonica hybridization project, but hybrid sterility and extreme differences in ecological adaptation between the parents precluded any major breakthrough. The modest progress attained in improving rice yields during the periods 1934–1938 and 1956–1960 was achieved largely through pure-line selection and, to a more limited extent, by hybridization but was not sufficient to cope with the increase in population in tropical Asia (Parthasarathy, 1972).

On the subtropical island of Taiwan, in 1931–43, continuous cycles of disruptive selection over the two crop seasons in crosses among Japanese material led to the development of the ponlai varieties. These were non-seasonal, high yielding and adaptable over a wide area inclusive of the tropics, subtropics and even several temperate areas. Later, in the mid 1950s, the development of the semi-dwarf Taichung Native 1 signalled a breakthrough in rice breeding. An Indica type of rice was developed that, because of its short straw and profuse tillering, outperformed the Japonica type in nitrogen-responsiveness and yielding ability. The success of Taichung Native 1 stimulated the wide adoption of semi-dwarfs in India during 1965–66 (Huang *et al.*, 1972).

Shortly after the International Rice Research Institute began work in 1962, breeders, geneticists, physiologists and agronomists directed their efforts towards the identification of an improved plant type that would markedly raise the yield level of tropical rices above $2 \cdot 0 – 3 \cdot 5$ t/ha when given 30–40 kg of nitrogen/ha. A cross between the tall tropical variety Peta (from Indonesia) and the subtropical semi-dwarf Dee-geo-woo-gen (from Taiwan) produced the semi-dwarf IR8.

The IR8 selection established records of grain yield (up to 11 t/ha) and nitrogen response (up to 150 kg N/ha) at several locations in tropical Asia during 1966–68. IR8 and many other improved semi-dwarfs from several national breeding programmes or from IRRI combined most of the desired features in the improved plant type that was sought: plant stature of about 100 cm; erect, relatively short, dark green leaves; high tillering; stiff culms; early maturity and photoperiod insensitivity; nitrogen responsiveness; and a high ratio of grain to straw.

The wide adaptation of IR8, IR20 and IR22 made it possible for the semi-dwarfs to become major varieties in Brazil, Colombia, Peru, Ecuador, Cuba, Mexico, Indonesia, Malaysia, Philippines, India, Pakistan, Bangladesh and South Vietnam. In 1972/73, semi-dwarfs occupied a large part of the area planted to high-yielding varieties (10 per cent of the world total; 15 per cent in tropical Asia). The semi-dwarf habit, which is controlled by a single recessive gene, has been incorporated in many recent varieties by different national centres in the tropics (IRRI, 1972). The photoperiod insensitivity of the semi-dwarfs has made it possible for South Korea to grow the nitrogen-responsive Tongil (IR667-98) in a short temperate growing season.

Rice breeding research at IRRI is also aimed at improving grain quality and incorporating high levels of resistance to the major pests and diseases; the work includes the transfer of a dominant gene for resistance to the grassy stunt virus from wild *O. nivara* (Beachell *et al.*, 1972).

Current breeding plans depend greatly upon inter-racial recombination. Thus, new varieties can derive: resistance to pests and diseases, grain quality and dormancy and growth vigour from the tropical Indicas; short stature, nitrogen response, photoperiod insensitivity and high harvest index from the semi-dwarf Indicas; and tough leaves, slow senescence, tolerance of low temperatures, and resistance to bacterial leaf blight from the Japonica varieties.

5 Prospects

Most of the semi-dwarf varieties came from primary crosses. Complex crosses and recurrent selection will be used more widely to recombine multiple sources

of resistance to various pests and to enhance nutritive quality. Wider crosses and random mating of hybrids, along with a judicious use of induced mutations or protoplast fusions, have promise for obtaining unusual recombinations or breakup of tight linkages. Further rises in grain yield hinge on an increase in the size of sink and in photosynthetic capacity.

So far, the semi-dwarfs are grown mainly in areas of the tropics where water control is effective, where the soil is free of toxicity or deficiency problems and where there are no climatic constraints. To assist farmers in less favoured areas, varietal adaptation requires tolerance of deep water and photoperiod sensitivity in ill-drained areas, drought resistance in upland or rain-fed lowland fields, tolerance of adverse factors in problem soils, and tolerance of low temperatures at high elevations. Incorporating the desired features into improved varieties for specific environments requires both multi-disciplinary evaluation and the utilization of diverse germ plasm. Such concerted efforts are being undertaken by IRRI in cooperation with several national research centres.

It was fortunate for rice researchers that, until recently, great genetic diversity still existed in nature and was available for conservation. Many national institutions and international organizations are collaborating with IRRI in collecting indigenous varieties. During 1972–74 more than 12,000 seed samples were assembled from 12 countries in tropical Asia (Chang *et al.*, 1975). The evaluation and utilization of such diverse germ plasm promise new gene-pools for the further improvement of *O. sativa*. The conservation of *O. glaberrima* and its wild relatives in Africa remains to be vigorously pursued, however.

6 References

Beachell, H. M., Khush, G. S. and Aquino, R. C. (1972). IRRI's international program. In *Rice breeding*, IRRI, Los Baños, Philippines, 89–106.

Chang, T. T. (1964). Present knowledge of rice genetics and cytogenetics. *Intnl Rice Res. Inst. tech. Bull.*, 1, pp. 96.

Chang, T. T. (1970). Rice. In Frankel, O. H. and E. Bennett (eds), *Genetic Resources in Plants*. Oxford and Edinburgh, 267–72.

Chang, T. T. (1975). Exploration and survey in rice. In Frankel, O. H. and J. G. Hawkes (eds), *Crop genetic resources for today and tomorrow*, Cambridge, 159–65.

Chang, T. T. (1976). The rice cultures. In *The early history of agriculture*. Royal Society of London (in the press).

Chang, T. T., Villareal, R. L., Loresto, G. C. and Perez, A. T. (1974). IRRI's role as a genetic resources centre. In Frankel, O. H. and J. G. Hawkes (eds), *Crop genetic resources for today and tomorrow*, Cambridge, 457–65.

Engle, L. M., Chang, T. T. and Ramirez, D. A. (1969).

The cytogenetics of sterility in F_1 hybrids of indica×
indica and indica×javanica varieties of rice (*Oryza
sativa*). *Philipp. Agric.*, **53**, 289–307.

Harlan, J. R. (1969). Evolutionary dynamics of plant
domestication. *Jap. J. Genet.*, **44** (Suppl. 1), 337–43.

Harlan, J. R. and De Wet, J. M. J. (1971). Toward a
rational classification of cultivated plants. *Taxon*, **20**,
509–17.

Huang, C. H., Chang, W. L. and Chang, T. T. (1972).
Ponlai varieties and Taichung Native 1. In *Rice Breeding*,
IRRI, Los Baños, Philippines, 31–46.

International Rice Research Institute (1964). *Rice
genetics and cytogenetics*, Amsterdam, pp. 274.

International Rice Research Institute (1972). *Rice
breeding*, Los Baños, Philippines, pp. 738.

Nayar, N. M. (1973). Origin and cytogenetics of rice. *Adv.
Genet.*, **17**, 153–292.

Oka, H. I. (1974). Experimental studies on the origin of
cultivated rice. *Proc. 13th intnl. Congr. Genet.*, **1**, 475–86.

Oka, H. I. and Morishima, H. (1970). The dynamics of
plant domestication: cultivation experiments with wild
Oryza populations. *Evolution*, **25**, 356–64.

Parthasarathy, N. (1972). Rice breeding in tropical Asia
up to 1960. In *Rice breeding*, IRRI, Los Baños, Philippines,
5–29.

Portères, R. (1956). Taxonomie agrobotanique des riz
cultivés, *O. sativa* et *O. glaberrima*. *J. Agr. trop. Bot.
appl.*, **3**, 343–84, 541–80, 627–700, 821–56.

Sugarcanes

Saccharum (Gramineae-Andropogoneae)

N. W. Simmonds
Scottish Plant Breeding Station
Edinburgh Scotland

1 Introduction

The sugarcanes are large perennial grasses which are
propagated vegetatively by means of stem pieces.
The first ('plant') crop is generally taken about a
year after planting and thereafter annual 'ratoon'
crops are taken until replanting falls due. Traditionally
the object of hand labour, sugarcane cultivation is
becoming increasingly mechanized. Sucrose is extracted by crystallization from juice expelled from the
canes by crushing rollers. World production is rising
fairly steadily. In 1972 it was 44 Mt (in comparison
with beet sugar, 32 Mt).

The sugarcanes are tropical by origin and cultiva-
tions are all in warm countries: India, Cuba and
Brazil are major producers and there are significant
productions also in the southern USA, the West
Indies, several central and South American countries,
Egypt, Natal, Java, China, Taiwan, Philippines and
Australia.

For a general treatment of sugarcane evolution and
cytology see Stevenson (1965).

2 Cytotaxonomic background

The sugarcanes and their relatives are members of the
tribe Andropogoneae with a basic $x = 10$. Members
of the genus *Saccharum* themselves are high polyploids and no diploids ($2n = 2x = 20$) are known.
Great intraspecific variability of chromosome number
and much aneuploidy are characteristic, as follows
(Price, 1963, 1965):

S. robustum $2n = 60$, 80 are frequent but putative
interspecific hybrids with numbers in the range
63 to about 200 occur. Wild species, distributed
from Borneo to the New Hebrides with much
variability in New Guinea. Daniels (1973) has
proposed that this is a 'hybrid species' derived from
natural crossing between *S. spontaneum* and *Miscanthus floridulus* in the New Guinea area.

S. spontaneum A great variety of numbers in the range $2n = 40 - 128$ occurs. Nodes occur at $2n = 64, 80, 96, 112, 120$, with the first centred in India, the last in Africa and the middle three nodes in southeast Asia from New Guinea to Taiwan. A very variable wild species distributed from Africa to the Solomon Islands and Japan, with a centre of diversity in India.

S. officinarum $2n = 80$, with a few, probably hybrid, exceptions. These are the cultivated 'noble' canes distributed throughout the tropics.

S. sinense $2n = 82 - 124$. These are old cultivated canes, natural hybrids of *officinarum* and *spontaneum*; chromosome numbers broadly correspond with expectations based on putative parentage. *Saccharum barberi* is included here.

To add to the cytotaxonomic complexities of the group, wild *spontaneum* × *robustum* hybrids have been recognized with $2n = 80 - 101$ and also intergeneric *Saccharum* × *Miscanthus* hybrids with very high numbers in the range $2n = 114$ – about 205. Formal taxonomy and nomenclature are confused but the usage above is probably generally acceptable.

3 Early history

The evidence points to the essential early steps in sugarcane evolution having taken place in or near New Guinea. On grounds of morphology and chromosome number, *S. robustum* ($2n = 80$) is the likely primary ancestor and human selection would have been for a chewing plant and in favour of clones having a sweet juice and low fibre content. The New Guinea area is still the primary centre of diversity of 80-chromosome *officinarum* clones so derived. (This seems secure whatever view one takes of the origin of *S. robustum* – see reference to Daniels (1973), above.) No date can be given for the process. Subsequent evolution followed migration of *officinarum* ('noble') canes northwestwards to continental Asia where they hybridized with local *S. spontaneum* to produce the hybrid clones grouped under *S. sinense*. These, the 'thin' canes, throve in the seasonal monsoon climates of the northeast India–southern China area and became the foundation of local syrup production. The noble canes also migrated eastwards across the Pacific and, in the course of time, somatic mutation has contributed materially to diversity, here and near the centre of origin. The crop was known to Europeans in classical times only by repute and reached the Mediterranean after the Moorish conquest of Spain. Early European-controlled cultivations of cane in the tropics used the clones available, primarily the Creole cane (under many names, a *sinense* derivative with $2n = 81$) and, later, a few noble canes (notably the cultivar, Bourbon, from the Pacific).

4 Recent history

Cane breeding began in Java and Barbados more or less simultaneously at the end of the nineteenth century and there are now about 25 breeding programmes in existence. The original stimulus was the need for new varieties which had the sugar-producing potential

Fig. 32.1 Evolutionary geography of the sugarcanes, *Saccharum*.

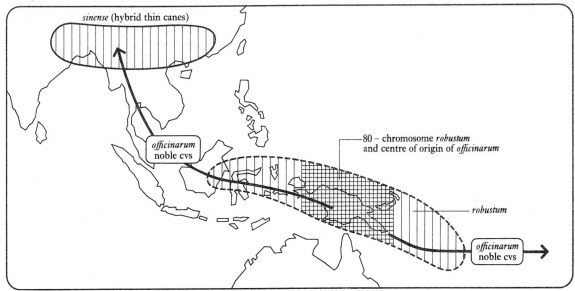

of the noble canes and the hardiness, disease resistance and ratooning capacity of the hybrid *sinense* types. The first interspecific crosses were made as early as 1893 in Java but success with this approach did not arrive until the 1920s. In the interval, some effort went into breeding the noble canes themselves and a number of useful varieties emerged (e.g. in Barbados) and persisted in declining acreages down to the Second World War.

The key event in cane breeding was the production in 1921 in Java of the greatest cane of all, POJ 2878, the first of the so-called 'nobilized' canes, now generally outclassed, but present in nearly every modern pedigree. These canes are all essentially derivatives of *S. spontaneum* backcrossed to noble types.

Bremer's cytological studies in the 1920s (summarized by Bremer, 1961–63) revealed a very curious situation which is even now not fully understood.

Essentially, the cross noble (female, $2n = 80$) × wild *spontaneum* (male) yielded progeny having the somatic complement of the female parent (80 chromosomes) plus the gametic number of the male. The reciprocal cross, however, gave the expected number. This behaviour implies some sort of restitution on the female side and also intense selection for the products of restitution, since selfs or crosses among nobles yielded only 80-chromosome progeny; exceptions to this latter statement are now known but the contrast between intra- and inter-specific crosses (with *spontaneum*) remains, and the cytology of the restitution process is yet uncertain (though mechanisms have been proposed).

The fact of female restitution is, however, abundantly confirmed; its effect is to increase chromosome numbers and to hasten phenotypic convergence on the recurrent parent. Restitution occurs for two generations but not for the third, so there is a limit to

Fig. 32.2 Evolution of the sugarcanes, *Saccharum*.

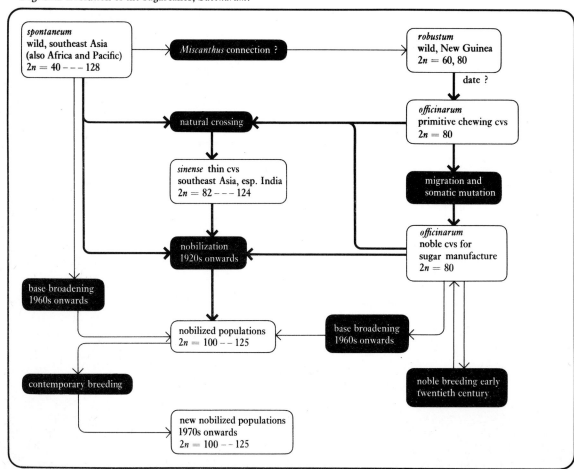

increase of chromosome numbers. Actual pedigrees (e.g. Bremer, 1962; Price, 1963) are complicated by the presence of some natural hybrid parents, by uncertainties and by some aneuploidy but the broad pattern is clear. A simplified example is (N – noble, S – *spontaneum*):

NN (80) × SS (64)　　　→NNS　　(80+32 = 112) (S% = 29)
NN (80) × NNS (112)　→NN(S)　(80+56 = 136) (S% = 12)
NN (80) × NN(S) (136)→NN(S)　(40+68 = 128) (S% = 6)

The proportions of wild chromosomes in the backcross generations are somewhat over half what would be expected in the absence of restitution. In practice, it turned out that canes with chromosome numbers in the range 100–125 and about 5–10 per cent wild contribution were what was wanted, and modern canes throughout the world derive from extensive recombination among materials of this kind.

Sugarcanes are wind-pollinated and show inbreeding depression. Breeding methods are those characteristic of outbred clonally propagated crops. The progeny of promising parents are explored in numerous crosses and potential new varieties isolated as clones. Large numbers of seedlings are generally raised, despite some practical problems connected with flowering and fertility. Biometrical genetic studies (Brown, Daniels and Latter, 1968, 1969) indicate that economic characters usually have a fairly high additive component ('quasi-additive' might be a better description in a high polyploid), in reasonable justification of the usual policy of exploiting proven parents in a fairly wide range of crosses. Criteria of selection are, essentially, sugar yield (cane yield × sugar content – but see below) qualified by various disease resistances, ratooning capacity, freedom from flowering and good plant habit in relation to harvesting.

Besides a large number of natural and experimental interspecific hybrids within *Saccharum* itself, many intergeneric hybrids have been recorded, between *Saccharum* and: *Erianthus, Miscanthus, Miscanthidium, Narenga, Imperata, Sorghum, Sclerostachya* and *Zea*. They have not had any significant practical use.

Flowering deserves comment (Coleman, 1969). Sugar canes are day-length sensitive and flower in response to light periods somewhat greater than 12 hours (12·5–14 hrs, depending on species and clone). The amount and time of flowering therefore depends upon clone, on growth cycle in relation to season and on latitude. Cultivated canes generally flower most abundantly (and potentially twice a year) in middle tropical latitudes, least near the equator and near the tropics. Forms of *S. spontaneum*, with a relatively long day-length requirement often do not flower at all at low and middle latitudes. In cultivation, flowering is

undesirable because it checks growth and may lead to 'pithiness' and late-tillering; at middle latitudes intense selection against flowering is therefore practised. For the cane breeder, flowering is, of course, essential; it is promoted by careful choice of site for breeding station (latitude and temperature are significant), supplemented by photoperiod control for refractory clones.

5 Prospects

The general pattern of cane breeding – selection of new clones from large populations of nobilized seedlings with chromosome numbers in the range 100–125 – seems well established and most unlikely to change. Biometrical sophistication, however, allied to computerized data handling, is surely yielding a steady improvement in efficiency (see papers by Brown *et al.*, 1968–71). Specialized techniques such as mutation-induction in buds or tissue cultures may have a useful, though essentially minor, place; non-flowering mutants of otherwise desirable canes that flower too much have already been induced. Criteria of selection also seem to be essentially stable, though there is an increasing emphasis on juice quality, time of maturity and aptitude for mechanical harvesting. On juice quality it turns out that the most profitable cane is not the one with highest sugar yield but rather that which, at high yield, has a sweet juice, thus minimizing costs of harvesting non-sugars; in practice, profitability is proportional to $Y(Q-a)$ instead of simply to YQ (Y, cane yield; Q sugar per cent).

By far the most important recent event, from the point of view of the continued evolution of the crop, has been the general realization that the genetic base is very narrow (Arcenaux, 1967). A few clones recur in nearly every pedigree and modern canes are founded on perhaps 20 nobles and fewer than 10 *spontaneum* or *spontaneum* derivatives. In the past decade, accordingly, a great deal of effort at many stations has gone into widening the genetic base. Some 30 *spontaneum* lines have been introduced to breeding programmes and a mass selection approach to noble improvement has been initiated in Barbados. By repeating the recent history of the sugar canes with these new materials (i.e. by generating noble × (noble × (noble × *spontaneum*)) populations) it will surely be possible to break out of the charmed circle of few original nobilizations.

As to genetic conservation, there have been many recent expeditions and will no doubt be more. Collections are maintained as living museums and there seems little prospect of prolonged seed storage being either practicable or very useful. Given more expeditions and replicated maintenance, the situation seems satisfactory. International collaboration is outstanding and the excellent Sugar Cane Breeders Newsletter has been circulating since 1956.

6 References

Arcenaux, G. (1967). Cultivated sugarcanes of the world and their botanical derivation. *Proc. Int. Soc. Sugar Cane Tech. Congr., Puerto Rico* (1965), **12**, 844–54.

Artschwager, E. and Brandes, E. W. (1958). Sugarcane. *USDA Agric. Handb.*, **122**, pp. 307.

Bremer, G. (1961–63). Problems in breeding and cytology of sugarcane. *Euphytica*, **10**, 59–78, 121–33, 229–43, 352–42; **11**, 65–80; **12**, 178–88.

Brown, A. H. D., Daniels, J. and Latter, B. H. D. (1968, 1969). Quantitative genetics of sugarcane, I–III. *Theor. appl. Genet.*, **38**, 361–9; **39**, 1–10, 79–87.

Brown, A. H. D., Daniels, J. and Stevenson, N. D. (1971). The mass selection reservoir and sugarcane selection. *Theor. appl. Genet.*, **41**, 174–80.

Coleman, R. E. (1969). Physiology of flowering in sugarcane. *Proc. Int. Soc. Sugar Cane Tech. Congr., Taiwan.* (1968), **13**, 92–100.

Daniels, J. (1973). Origin of *S. officinarum* and *S. robustum*. *ISSCT Sugarcane Breeders Newsl.*, **32**, 4–13.

Price, S. (1963). Cytogenetics of modern sugar cane. *Econ. Bot.*, **17**, 97–106.

Price, S. (1965). Cytology of *Saccharum robustum* and related sympatric species and natural hybrids. *USDA agric. Res. Serv. tech. Bull.*, **1337**, pp. 47.

Stevenson, G. C. (1965). *Genetics and Breeding of Sugar Cane*. London.

Stevenson, N. D., Brown, A. H. D. and Latter, B. D. H. (1972). Quantitative genetics of sugarcane IV. *Theor. appl. Genet.*, **42**, 262–6.

Rye

Secale cereale (Gramineae-Triticinae)

G. M. Evans
Department of Agricultural Botany
University College of Wales
Aberystwyth Wales

1 Introduction

Rye (*Secale cereale*), in common with the other temperate cereals, wheat, barley and oats, is considered to have originated in southwestern Asia. Although primarily a grain crop it is extensively used in many areas as a forage, particularly useful for early spring grazing. As a grain, it is somewhat in decline, the world crop being less than any of the other major cereals. In 1972 the total production was estimated at only 28·2 Mt compared with 35·5 Mt in 1961. Nevertheless, rye is still an important food plant in many areas of northern and eastern Europe and some regions of the USSR. In its capacity to produce an economical crop in areas of cold winters and hot dry summers, it is superior to the other temperate cereals.

Much of the world's supply of rye is consumed in the form of rye bread (black bread) and rye crispbread or biscuits. It is also used to some extent in animal feed and for the production of rye starch. A rather specialized use in Canada and the USA is for the production of rye whisky.

2 Cytotaxonomic background

Apart from artificial polyploids, all *Secale* species are diploid ($2n = 2x = 14$). There are however, numerous reports of accessory or B chromosomes in both the wild and cultivated outbreeders *S. montanum* and *S. cereale*. The taxonomy of *Secale* is confused. Roshevitz (1947) recognised 14 species but it is questionable whether, in fact, all these should be given specific rank. Some taxonomists recognize fewer, even as few as five, species. Two distinct aggregates of species (or subspecies) can readily be separated as being important in the evolution of the cultivated forms. First, there is the group of annual weeds such as *S. ancestrale*, *S. dighoricum*, *S. segetale* and *S. afghanicum* which cytologically resemble each other and cultivated rye.

In fact there is no reason why these should not be included as subspecies of *S. cereale*. This group is virtually confined to agricultural lands, the weedy types being widespread in cereal crops in north-eastern Iran, Afghanistan and Transcaspia (Zohary, 1971). Second, there is an aggregate of wild perennial races widely distributed from Morocco eastwards through the Mediterranean countries and the plateau region of central and eastern Turkey to northern Iraq and Iran. These have sometimes been separated into distinct species but are most probably best described as variants of a single species, *S. montanum*. Again, members of this group are cytologically similar to each other and completely interfertile. They differ from the *S. cereale* complex by two major interchanges (reciprocal translocations) involving three pairs of chromosomes.

There are two other species which have, from time to time, been implicated in the evolution of cultivated rye. These are *S. vavilovii*, an annual self-pollinating species of limited geographical distribution, and *S. sylvestre*, also an annual inbreeder but widely distributed from central Hungary to the steppes of southern Russia. It is generally agreed that *S. sylvestre*, although morphologically distinct from *S. montanum*, has a similar karyotype. The status of *S. vavilovii* is still in doubt. It is not at all clear whether the chromosome arrangement is the same as that of *S. cereale* and therefore different from *S. montanum* by two reciprocal translocations involving three pairs of chromosomes (Stutz, 1972) or whether it differs from *S. montanum* by one reciprocal translocation and from *S. cereale* by two (Khush, 1962).

3 Early history (Fig. 33.1)
In general, the evidence confirms the original hypo-thesis of Vavilov (1917) that rye is a classical example of a secondary crop. Having first arisen as a weed of wheat and barley it would have been introduced as a crop in its own right at a somewhat later date. It is highly likely that the immediate progenitors of the cultivated form would have come from one or more of the weedy races (named above) which are chromosom-ally identical to it. It is also reasonable to assume that these weedy races would themselves only have evolved with the development of agriculture in western Asia. There is also general agreement that the original ancestor of these weedy races (and hence of cultivated rye itself) was *S. montanum*. It is perennial, outbreed-ing and widely distributed.

How the change, from truly wild *S. montanum* to the agriculturally dependent *S. cereale* complex, occurred is still not clear. If no other distinct intermediate

species were involved (as indeed has been proposed by Khush and Stebbins (1961) and by Zohary (1971)), then the question arises as to how a major structural rearrangement (involving a double translocation) could have become established in the first place, bear-ing in mind the initial handicap of reduced fertility which would certainly have been incurred. Some adaptive superiority of the original structural heter-ozygote seems highly likely (Riley, 1955; Khush and Stebbins, 1961). Eventually, structural homozygosity for both translocations became established though it is not easy to imagine how. The partial fertility barrier between the new chromosome arrangement and the old would serve as an isolating mechanism thus preventing the swamping of the evolving weedy races of *S. cereale* by their progenitor, *S. montanum*.

Stutz (1972) proposed, alternatively to the above, a stepwise evolution of *S. cereale* from *S. montanum*. The annual weedy races were, he thought, derived from the introgression of *S. montanum* into *S. vavilovii*. This annual inbreeder was in turn derived from the annual *S. sylvestre* as a consequence of chromosome trans-location. *S. sylvestre*, itself capable of a degree of cross pollination, would have had the perennial outbreeder *S. montanum* as its progenitor. The main weakness of

Fig. 33.1 Evolutionary relationships of rye, *Secale cereale*.

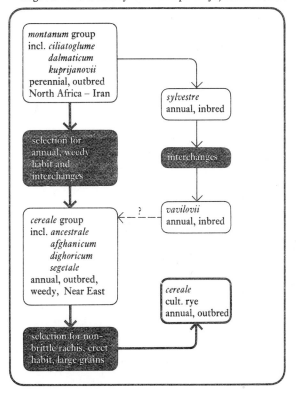

109

this hypothesis is that it implies the evolution from the cross-fertilized *S. montanum* of self-fertilization (*S. sylvestre* and *S. vavilovii*) followed by cross-fertilization once more in *S. cereale*. An attractive feature, however, is that the fixation of the double translocation is more easily explained because it is assumed to have occurred in a self-pollinating species (*S. vavilovii*).

The most likely place of origin of the weedy races of rye is somewhere in the area central and eastern Turkey, north-western Iran and Armenia. This area of maximum genetic diversity coincides in part, at least, with a high degree of variability of the perennial *S. montanum*. The northward spread of wheat and barley cultivation from the more fertile areas of the Jordan valley and what is now Israel to this plateau region would have penetrated one of the natural habitats of *S. montanum* and some of its relatives. The cold, harsh climate of this area is not ideal for the two main cereals and it is not unreasonable to assume that under such conditions rye could have become established as a cornfield weed. There is no doubt that the weedy annual races (with brittle or semi-brittle rachises) were, and still are, successful colonizers. A second area showing considerable genetic diversity of these annual types exists further east centred on Turkestan, Afghanistan and north eastern Iran.

Once established as a weed of the larger grained cereals, it is not difficult to imagine the kind of selection, most of it unconscious, that would have been exerted. A more upright culm would result from selection for competitive ability. There would also have been unconscious selection for genes controlling non-shattering of the ears. Cultivations prior to sowing would have tended to destroy seedlings originating from seeds dropped by shattering spikes the previous year. Seeds of non-shattering genotypes would have been harvested with the main cereal crop and subjected to the same cultural procedures during the next season. At the same time there would have been selection for larger grain, since even the most primitive winnowing procedure favours the retention of grains approaching the size of those of wheat and barley.

The exact site of domestication of the crop must remain in doubt. Probably, it was domesticated at several locations independently but, presumably, within the general area defined above. The timing of the event (or events) is also difficult to define with any accuracy because there is little archaeological information. A reasonable estimate for domestication is about 3000 B.C. and, for introduction into Europe, about 2000–2500 B.C. (Khush, 1963). The latter probably occurred by two separate routes, northwards and westwards through the Caucasus from the primary centre and through central Asia from the secondary centre further east.

4 Recent history

The spread of rye cultivation in Europe during Roman and post-Roman times is not well documented. Archaeological evidence is scarce; a comprehensive account of what there is is given by Helbaek (1971). Rye was probably grown originally in the same areas as the other temperate cereals but its tolerance of low rainfall, cold winters and poor light soils made it particularly suitable to large areas of northern and eastern Europe and parts of European Russia so that, by the end of the eighteenth century, it had become the major cereal of the region. In spite of the poor baking quality of the flour most people in the region used it for making bread, the so called 'black bread' of eastern Europe. In addition, the straw (often 2 m long) was valued for thatching. It has been estimated that, even as late as the early part of this century, rye bread was the main cereal food of a third of the population of Europe. Since then, however, it has been gradually replaced by wheat.

By the time conscious plant breeding began, in the later part of the nineteenth century, many locally adapted 'land races' had arisen and the early plant breeders relied heavily on them as their immediate source of variability.

Breeding methods have, naturally, been much influenced by the outbreeding nature of the crop. Rye has a gametophytic two-locus incompatibility system. (Lundquist, 1956). Early breeding techniques could best be described as forms of simple recurrent selection. With recently improved knowledge of the genetic structure of outbreeding populations, more sophisticated methods have appeared. Rye shows inbreeding depression but inbred lines of acceptable vigour can be isolated and used in the construction of synthetic varieties, following suitable progeny tests for combining ability. The objectives of rye breeding (unlike those of wheat and barley breeding) have not been dominated by aspects of disease resistance. Improvement of grain yield, protein content and quality together with cold tolerance and shorter straw, have been the aims of recent breeding. A disease which has caused some trouble from time to time is Ergot (*Claviceps purpurea*). The poisonous sclerotia, which totally replace the grain and occasionally get into flour, have been reported as causing hallucinations in people and abortion in farm animals. The utilization of rye as a forage has led to the breeding of varieties bred solely for the purpose; emphasis is then placed on characteristics other than grain production, total dry matter yield, growth in winter and early spring and digestibility of the herbage being important.

5 Prospects

Conventional breeding methods are still likely to dominate rye improvement in the immediate future. Greater emphasis will undoubtedly be placed on exploring heterotic effects, either through synthetic varieties based on a few inbred genotypes, or by breeding hybrid varieties. Systems of pollination control, based on cytoplasmic male sterility and restorer genes, are certainly available but whether it will ever become economic to produce hybrid seed remains to be seen. From an evolutionary standpoint, these methods could perhaps tend to increase the uniformity within cultivars at the expense of restricting the genetic base.

Of greater potential evolutionary importance are some other current developments. First there are auto-tetraploids ($4x = 28$) (review in Kranz, 1973). These have been used on a small scale, even for grain production, for some years and this use is likely to develop. In common with other synthetic autopolyploids, there is a decrease in fertility which, however, in rye is compensated by increased grain size and protein content. Response to selection for fertility has been slow but some progress is being made. Second, there is the potential of interspecific and intergeneric hybrids. Attempts to breed perennial rye for forage production are not new. Hybrids of *S. cereale* and the perennial *S. montanum* have shown considerable promise in terms of dry matter production but it remains to be seen whether high yield can be fixed in genotypes which are both perennial and sufficiently fertile.

Intergeneric hybrids of rye and wheat (Triticale) are treated elsewhere in this book (Chapter 35).

6 References

Helbaek, H. (1971). The origin and migration of rye, *Secale cereale*: a palaeo-ethnobotanical study. In P. H. Davis *et al.* (eds), *Plant life of south-west Asia*. Edinburgh, p. 265.

Khush, G. S. (1962). Cytogenetic and evolutionary studies in *Secale*. II. Interrelationships of the wild species. *Evolution*, 16, 484–96.

Khush, G. S. (1963). Cytogenetic and evolutionary studies in *Secale*. III. Cytogenetics of weedy ryes and origin of cultivated rye. *Econ. Bot.*, 17, 60–71.

Khush, G. S. and Stebbins, G. L. (1961). Cytogenetic and evolutionary studies in *Secale*. I. Some new data on the ancestry of *S. cereale*. *Amer. J. Bot.*, 48, 721–30.

Kranz, A. R. (1973). Wildarten und primitivformen des Roggens (*Secale*). *Fortschr. PflZücht.*, 3, 1–60.

Lundquist, A. (1956). Self-incompatibility in rye. I. Genetic control in the diploid. *Hereditas*, 42, 293–348.

Riley, R. (1955). The cytogenetics of the differences between some *Secale* species. *J. agric. Sci.*, 46, 277–83.

Roshevitz, R. J. (1947). A monograph of the wild, weedy and cultivated species of rye. *Acta Inst. bot. Acad. Sci. U.S.S.R.*, ser. 1, 6, 49–58.

Stutz, H. C. (1972). The origin of cultivated rye. *Amer. J. Bot.*, 59, 59–70.

Vavilov, N. I. (1917). On the origin of cultivated rye. *Bull. appl. Bot.*, 10, 561–90.

Zohary, D. (1971). Origin of south-west Asiatic cereals: wheats, barley, oats and rye. In P. H. Davis *et al.* (eds); *Plant life of south-west Asia*. Edinburgh, p. 235.

Sorghum

Sorghum bicolor (Gramineae–Andropogoneae)

H. Doggett
International Crops Research Institute
for the Semi-Arid Tropics (ICRISAT)
Hyderabad India *on secondment from*
the International Development Research
Centre of Canada

1 Introduction

Sorghum and bulrush millet (*Pennisetum typhoides*) are the major cereals of rain-fed agriculture in the semi-arid tropics. These may be defined as tropical areas where rainfall exceeds evapotranspiration for two to seven months in the year, is usually less than 1,200 mm, and where there is a long dry season. Sorghum is grown on the heavier, bulrush millet on the lighter, soils and, in the Old World, they occupy between them a substantially greater area than maize, forming the staple food of some 400 million people. The estimated world area of sorghum in 1972 was 40 Mha the largest areas being in India (16 Mha) and Africa (10·3 Mha). In the developing world, there were 33 Mha of sorghum, 33 Mha of millet and 51 Mha of maize (FAO figures).

Sorghum grain is used for stockfeed in the New World, in Japan and in Europe but provides human food in India and food and beer in Africa. The straw is fed to livestock and is used for fencing, while the plant bases provide fuel for cooking. The crop may be grown for forage and the modern Sudan grasses in the USA have been developed from wild sorghum. Under drought stress, hydrocyanic acid may be produced in the foliage, which can then poison livestock. Some sorghums have sweet juicy stems which may be chewed or used to produce syrup. The broomcorns are sorghums with short rhachises and long panicle branches.

For general treatments of the crop, see Doggett, (1970), Wall and Ross, (1970).

2 Cytotaxonomic background

Stapf, in the *Flora of Tropical Africa*, separated the Sorghastrae as one of the sixteen sub-tribes of the Andropogoneae. This sub-tribe contains as main genera *Sorghum, Sorghastrum* and *Cleistachne*. Within the genus *Sorghum* are the sections Sorghum, Parasorghum, Chaetosorghum and Heterosorghum, all with chromosome numbers based ultimately on $x = 5$ (see below). These represent stages in the development of the Sorghastrae, spreading east and west through the tropics. *Sorghastrum* arose while there were still links between America and Africa and Parasorghum while there were still links between Africa, Asia and Australia and possibly also with America, since *S. trichocladum* occurs in western Central America. All these genera and sections are therefore old. *Sorghum* section Sorghum has a polymorphic diploid population ($2n = 2x = 20$) in tropical Africa (which should probably be referred to as *S. arundinaceum*) and a distinctly different diploid population in southeast Asia, Indonesia and the Philippines, *S. propinquum*. The latter has strongly developed rhizomes, a distinctive leaf shape and very small seeds. Although hybrids of these two diploids are fully fertile, the present populations are probably sufficiently different in morphological characters and distribution to be considered distinct species. In between the distributions of *S. arundinaceum* and *S. propinquum* and overlapping with them, the tetraploid *S. halepense* ($2n = 4x = 40$) occurs. This species occupies a continuous area from southern and eastern India through Pakistan and Afghanistan across Asia Minor to the Levant and the Mediterranean littoral. One form is widely known as Johnson Grass. *Sorghum halepense* almost certainly arose from the doubling of the chromosomes in a cross between the *S. arundinaceum* and *S. propinquum* groups. The controversy as to whether *S. halepense* is auto- or allotetraploid arose because fertile crosses are produced by *S. arundinaceum* and *S. propinquum*. The conclusion reached depends on whether or not these two are to be regarded as distinct. Certainly, very fertile forms of *S. halepense* occur, their fertility being due to a low frequency of quadrivalents (three or fewer per cell), their orientation at metaphase in a manner leading to equal separation, and a very low frequency of trivalents and univalents.

The distribution of section Sorghum overlaps those of Parasorghum and Heterosorghum, as well as those of the genera *Cleistachne* and *Sorghastrum*. Many attempts to cross members of section Sorghum with the other genera and sections have been unsuccessful and it seems certain that the origin of the cultivated crop lies within *Sorghum* section Sorghum. It was developed primarily from the wild *S. arundinaceum* of Africa but has probably been influenced by some introgression from *S. propinquum* and *S. halepense* in the areas in which these species occur (Doggett, 1970).

A natural cross between a diploid sorghum and *S. halepense* in Argentina gave rise to Columbus grass, *S. almum*, a tetraploid resulting from restitution in the diploid parent.

The diploid chromosome number of *Sorghum* is $2n = 20$ and, since the number for the sub-tribe Sorghastrae is $x = 5$, *Sorghum* is presumed to be, ultimately, of tetraploid origin. It is hard to find cytological evidence for homologous segments in the chromosomes of the haploid and *Sorghum* behaves in all respects as a good diploid.

Crop plants (such as sorghum) derived from wild types which cross-pollinate freely, diverged by a process of disruptive selection. Man was selecting for cultivated characters (mainly non-shattering heads, large seeds and heads, threshability and suitable maturity) while natural selection favoured wild-type characters. Thoday (1972) has shown that divergence into poly-

morphic populations occurs under disruptive selection, without isolation and, indeed, in the presence of considerable gene flow between the components of the population. Doggett and Majisu (1970) stressed the importance of the interaction between wild and cultivated sorghums and the gene flow between them, through the medium of intermediate forms surviving in the intermediate habitats provided by recently abandoned cultivations. Many isolation mechanisms have contributed to the diversification of the crop, in addition to the polymorphisms created by disruptive selection, but Harlan and de Wet (1972) have provided a pretty demonstration of the essential correctness of the conclusion that disruptive selection was the major influence in the development of the sorghum crop. Systematists have found the greatest difficulty in separating the cultivated sorghums into discrete groups when numerous plant characters and large numbers of plants are used in the classification. How-

Fig. 34.1 Evolution of sorghum, *Sorghum bicolor*.

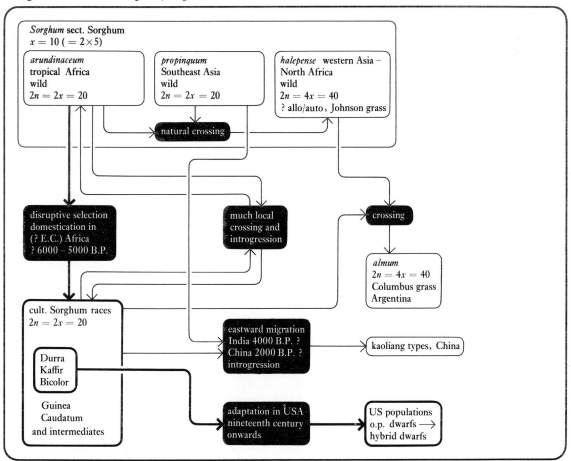

ever, Harlan and de Wet, using only the few cultivated characters listed above but excluding maturity period, were able to divide the cultivated sorghums into five basic races and a series of intermediate forms between all combinations of these races. Natural selection has kept a continuous distribution of plant characters but man has developed by selection populations polymorphic in respect of the cultivated characters. Harlan and de Wet's basic cultivated races are Bicolor, Guinea, Caudatum, Kafir, and Durra. Their designation of the intermediates between the five basic races as hybrid races is probably premature. Some may, indeed, be of hybrid origin but others may come from the overlapping areas of the polymorphic distributions. More remains to be learnt about the cultivated races and their relationship to the wild forms. Thus, a natural cross of cultivated × wild sorghum (*verticilliflorum*) in Uganda produced three of Harlan and de Wet's main races in the F_3 generation (Doggett and Majisu, 1968). Harlan *et al.* (1973) have stressed the essential similarity of the evolution of the sorghum crop to that of many other cereals.

3 Early history

There are no established facts from which to derive the history of the crop but one can infer a probable sequence of events. There is, however, much room for difference of opinion (De Wet and Huckabay, 1967: Doggett, 1970). Centres of origin are less favoured than they used to be (Harlan, 1971) but crop development requires people who are beginning, or have begun, to practise agriculture. There is no doubt that cultivated sorghums were developed in Africa: it is debatable whether they were developed in Ethiopia and the surrounding countries, or in West Africa by the 'Nuclear Mande' (Murdock, 1959) or, again, in a long belt, a non-centre, across Africa. The main area of variation of the crop lies in eastern central Africa (Doggett, in Hutchinson, 1965). Considerable crop diversity is to be expected in a mountainous region such as Ethiopia, where numerous rather isolated habitats occur but where opportunities for hybridization between small isolated populations must offer from time to time. However, the diversity of both the wild and cultivated sorghums in this region is widespread rather than localized in the mountainous areas. Indeed, Harlan noted a lack of botanical diversity in the Ethiopian cultivated forms, many of which belong to Durra, the most highly developed of the cultivated races, or to the Durra–Bicolor intermediate type. This he regarded as an argument against Ethiopia as the area of origin of the crop but there seems to be no intrinsic reason why a crop plant should not reach its most

developed form in its centre of origin, where it has been under selection for the longest time. Certainly, there is much diversity of head and grain type (including the occurrence of high lysine forms, which are eaten green) and it is probable that the excellence of these Durras resulted in their being favoured to the exclusion of other forms.

Probably, agriculture was brought to Ethiopia from the Middle East centre of crop development by the Cushites, a people of Caucasoid origin and Hamitic stock (Murdock, 1959), perhaps carrying with them wheat and barley. The Ethiopian highlands provide a suitable environment for these crops, and have a healthy, malaria-free climate. There, the Cushites developed a range of crops, including finger-millet, teff and sorghum. They must have tried many grasses (few can have looked as unpromising as teff in terms of seed size) and wild sorghum would have been an obvious choice in that region. The date of origin can only be guessed at around 3000 to 4000 B.C.

The developing crop was then carried across North Africa to West Africa, the Sahara being much better watered 5000 years ago than it is now. Crop improvement would have continued on the journey and it is not surprising to find evidence of continued improvement, north of the Congo forests from Ethiopia and the Sudan to West Africa, since some people would have settled along the route. These would have been the people who picked up *Pennisetum violaceum* and developed bulrush millet. The technology, as well as the developing crop, travelled to the west. When the improved sorghums and the technology of crop improvement reached West Africa, work began on the local wild plants, especially in the region of the upper Niger. Cultivated sorghum was brought here at an early date and crops such as *glaberrima* rice, *Brachiaria* millet, the two *Digitaria* (fonio) millets and cow-pea were developed, in part, at least, by the Mande people.

The Guinea race was diversified in West Africa and Durra in the Ethiopia–Sudan area. Cultivated sorghum meanwhile moved down into East Africa, where many of the sorghums with white, high quality grains have the Guinea race characteristic of the grain turning 90° between gaping involute glumes. This character association may well have been present in the original material which moved into East Africa as well as in that which moved to West Africa. Agriculture was practised at an early date, at least in the highland areas of Kenya and probably also in those of Uganda and Tanzania.

Meanwhile, the Bantu people, who originated in the Cameroons, worked along the southern edge of the Congo forests with the aid of the Malaysian food plants, principally coco yams and bananas, which had originally enabled them to pass through the forest belt from the north. At the eastern end of the forest belt they came in contact with sorghum which provided the basic provision for their rapid expansion into the dry savanna country to the south, where the race Kafir can be associated with them from Tanzania southwards. The Caudatum race is associated with the Nilotic and Nilo-Hamitic people in Uganda and western Kenya, rather than with the Bantu, and it is also important in the Sudan, Chad and Nigeria. Race Bicolor has an uncertain position; it is low yielding, with generally poor grain quality and is widely scattered. It may have been preserved for its sweet, juicy stem in Africa and Asia and for its broom-making potential in the Mediterranean area (Harlan and de Wet, 1972).

Race Durra was developed in Ethiopia and the Sudan and moved out to the Near East and to India. These sorghums are often associated with Hamitic peoples and Islamic culture.

The development of the main cultivated races is associated with the wild varieties of sorghum, as Snowden (1936) noted. Thus the wild variety *arundinaceum* is associated with the race Guinea in West Africa, *aethiopicum* with bicolor and durra and *verticilliflorum* with races Kafir and Caudatum. Independent origins of the cultivated sorghums from the wild forms have been proposed (Snowden, 1936) but it seems intrinsically more probable that the early cultivated forms of the crop were transferred from place to place at the same time as the new technology. Undoubtedly, there has been, and still is, much interaction between the wild and cultivated forms, and the wild ones must have imposed their stamp on the cultivated kinds, occurring frequently as weeds in the crop. Equally, the cultivated races influence wild sorghums in the areas in which they occur, and the latter act as a reservoir for some of the racial characteristics of the former.

It is possible that the wild sorghums were carried along as weeds of the crop and diversified in that way but also possible that much variation existed in wild *S. arundinaceum* before the arrival of the cultivated plants. Probably, both situations occurred: the wild variety *arundinaceum* in West Africa may well have been there before the arrival of the primitive crop; but wild *verticilliflorum* could well have developed and travelled as a crop weed since it shows affinities with all the cultivated races.

Outside Africa, the crop was not in contact with wild diploid forms until it reached southeast Asia, where the diploid *S. propinquum* occurs and this population may have contributed to the characteristic appearance of the kaoliangs of China. There is certainly some genetic exchange possible between tetraploid *S. halepense* and the crop, but the extent to which this species has influenced the crop is uncertain (and possibly slight).

Sorghum most probably moved to India from East Africa, initially overland, but movement along the coast must have occurred at an early date, as the dhow trade between the two countries via Arabia is ancient. The first reliable archaeological record of the crop in India comes from spikelets mixed with potting clay at Ahar, Rajasthan, in $1725 + 140$ B.C.; possibly, however, an older picture on a potsherd from Mohenjo-Daro, 2300–1750 B.C. is of sorghum. The crop therefore probably reached India soon after 2000 B.C.

Sorghum must have been carried to Arabia from East Africa at an early date and thence it would have spread into the Persian Gulf with the dhow traffic and so to the Near East.

The spread along the coast of southeast Asia and so to China was probably rather later, perhaps around the beginning of the Christian era. However, an identification from the Chinese neolithic period, if confirmed, would put the arrival there much earlier and would suggest that sorghum was moved overland along the silk road.

Grain sorghum first went to America from West Africa with the slave trade, when guinea corn and probably chicken corn were introduced. Brown and white durra were introduced from North Africa about 1874, milo about 1880, feterita in 1906 and hegari in 1908. The kafirs were introduced from South Africa about 1876. The crop soon spread into Central and South America (where it is assuming growing importance in the semi-arid areas) and high altitude types are becoming increasingly popular in parts of Mexico.

4 Recent history

Sorghum improvement was pioneered in the USA, the crop being developed there as a stock feed. A series of mutants affecting both height and length of maturity were selected by farmers and adaptation steadily improved. Giant milo had given rise to standard milo and dwarf yellow milo by 1906 and, in many parts of the USA, 'milo' or 'milo maize' are still general names for sorghum. Early white milo was grown in 1911 and double dwarf yellow milo appeared in 1918. Milos belong to the race Durra. A similar series developed in the Kafirs, blackhull giving rise to Dawn (a dwarf form) and also to Club kafir. The basis of sorghum improvement in the USA has been the kafir × milo cross and natural crosses selected by farmers before 1914 included types such as header milo which could be harvested by a mechanical header.

This first period (that of selection of natural mutants and crosses by farmers) was followed by deliberate plant breeding work. Segregates from kafir × milo, kafir × feterita and from three-way crosses involving hegari, were released as varieties in the 1920s. These included Chiltex, Bonita, Beaver and Wheatland. The last two cultivars had straight, erect peduncles and Wheatland could be harvested by standard wheat machinery thus establishing the future for the large scale growing of 'combine' sorghums. Wheatland gave rise to a number of other varieties, including Martin which was resistant to *Periconia* stalk-rot and which has a drying-peduncle character that makes it easy to combine. Martin occupied the greater part of the US sorghum acreage before the release of hybrids. In this same period, kafir × milo crosses gave rise to many varieties, including 7078 and Combine kafir 60, the latter possessing resistance to chinch bug, a pest of wheat and barley in the Great Plains. Waxy starch types were developed and exploited during the Second World War and some industrial starch is still produced from sorghum.

Hybrid vigour was recognized early in certain crosses but years of work were needed before a cytoplasmic male-sterile suitable for the commercial production of hybrid seed had been produced. Combine kafir-60 was sterilized and 7078 proved to be a good restorer, giving rise to the widely grown hybrid 610. Many other hybrids and parent lines were soon developed. Hybrid seed distribution to growers began in 1957 and almost all the US sorghum acreage was covered by hybrids four years later. The mean yield for 1954–56 of 12·8 q/ha over 4·6 Mha increased to a mean of 27·5 q/ha over 4·9 Mha for the period 1962–64. The 1972 figure was 38·1 q/ha over 5·5 Mha. Hybrid development represented another stage in the exploitation of the cross between Kafirs from South Africa and milos (race Durra) from North Africa, the Kafirs providing the male-sterile parents when the appropriate genome is in milo cytoplasm and race Durra furnishing the restorer lines.

The sorgos were developed for syrup production as a small farmer activity and some sorgos can be used for sugar production, although the process is not economic at present in the USA.

Elsewhere, outside the developing world, sorghum improvement in South Africa led to the production of good combine type hybrids (many coming from the USA) and *Striga*-resistant varieties. Here, the market was primarily for a brewing grain to make 'kafir beer'.

In the developing world, where the need for sorghum grain as a human food is greatest, progress has been much slower. In colonial days, emphasis was laid on cash crops to develop viable cash economies through export sales of produce and very little effort went into the improvement of food plants such as sorghum. Effort was also often wasted in 'selecting within the local material', (which the indigenous people had already been doing for generations) since this procedure had proved very successful in Africa in the recently introduced Upland cotton. Latterly, breeding programmes in Africa and India have met with some success. In India, the hybrid programme has produced high yielding material, especially CSH 1 and CSH 5 and some promising new varieties are being released; but resources available for national programmes in the developing world are still very inadequate. Problems are great, especially insect pests (such as shoot-fly, stem-borer and midge), birds and the witchweed, *Striga*. Occasional hybrids from the USA, such as NK 300, prove productive over a wide range of conditions but, generally, such hybrids have stock-feed grain types of poor quality and they lack the necessary resistances against pests and *Striga*, which do not occur in the sorghum areas of the USA. Types resistant to birds have persistent testas containing tannins, which are bitter and which may reduce the digestibility of the grain proteins. In many parts of Africa, birds prevent the cultivation of more palatable types.

5 Prospects

Greater attention is being given in the USA to grain quality, directed purely at the better utilization of the grain in stock-feed and partly at improvement in protein quality by utilizing high lysine genes (with human requirements also in view). Studies of crop physiology have begun and there is no apparent reason why much more efficient sorghum plants should not be developed since the crop has a C-4 photosynthetic pathway and a good translocation system. The crop should have at least as good a production potential as maize. The genetic base is being broadened by a programme to convert part of the world collection to dwarf, photoperiodically-insensitive types. Disease resistances are receiving more attention, especially to downy mildew (Wall and Ross, 1970).

In the developing world, the establishment of the International Crops Research Institute for the Semi-Arid Tropics (ICRISAT) holds out hopes that a properly funded, concentrated research effort may greatly improve the sorghum crop in due course. The remarkable progress made in the USA in establishing an important place for sorghum in the home of maize, using a rather narrow genetic base, holds out great promise for the future of sorghum as the major cereal of the semi-arid tropics. Grain quality could be much improved by breeding, especially for ripening under rainy conditions. Better levels of resistance to pests and *Striga* can be obtained. In the developing world, these will have to be coupled with better agricultural practices if substantial advances are to be made.

There are no restrictions on height imposed by mechanical harvesting and although, in theory, very dwarf plants should be capable of a better harvest index, in practice, the best grain yields come from plants about two metres tall. Genetic male-steriles have made possible recurrent selection in random-mating populations, a procedure which holds out promise of steady crop improvement. The development and study of an adequate world collection will also make available to breeders a better range of material to exploit. Hybrids show a valuable ability to perform relatively well under adverse conditions; and, now that genetic steriles are available (Rao and House, 1972), synthetic varieties are a possibility where seed production facilities are poor.

Population pressure in the semi-arid tropics will require the development of sorghum to the greatest possible extent; the potential seems to be there but sufficient R and D inputs over a long enough period to achieve results will be essential.

6 References

De Wet, J. M. J. and Huckabay, J. P. (1967). The origin of *Sorghum bicolor*: distribution and domestication. *Evolution*, 21, 787–802.

Doggett, H. (1970). *Sorghum*. London.

Doggett, H. and Majisu, B. N. (1968). Disruptive selection in crop development. *Heredity*, 23, 1–23.

Harlan, J. R. (1971). Agricultural origins: centers and non-centers. *Science, N.Y.*, 174, 468–74.

Harlan, J. R. and De Wet, J. M. J. (1972). A simplified classification of cultivated sorghum. *Crop Sci.*, 12, 172–6.

Harlan, J. R., De Wet, J. M. J. and Price, E. G. (1973). Comparative evolution of the cereals. *Evolution*, 27, 311–25.

Hutchinson, J. B. (1965) (ed.). *Essays on crop plant evolution*. Cambridge.

Murdock, G. P. (1959). *Africa, its peoples and their culture history*. New York.

Rao, N. G. P. and House, L. R. (1972) (eds). *Sorghum in the seventies*. Oxford and New Delhi.

Snowden, J. D. (1936). *The cultivated races of sorghum*. London.

Thoday, J. M. (1972). Disruptive selection. *Proc. R. Soc. Lond. B.*, 182, 109–43.

Wall, J. S. and Ross, W. M. (1970) (eds). *Sorghum production and utilization*. Connecticut.

Triticale

Triticosecale spp. (Gramineae–Triticinae)

E. N. Larter
Plant Science Department University of Manitoba Winnipeg Canada

1 Introduction

Triticale is a small-grain cereal which represents the first successful attempt to synthesize a new crop species from intergeneric hybridization. As the common name indicates, triticale is the result of the combination of the two genera, wheat (*Triticum*) and rye (*Secale*). In the early stages of development of triticale, breeders envisaged the synthesis of hybrids (amphidiploids) between wheat and rye as a means of combining the desirable winter hardiness of rye with the agronomic and commercial properties of wheat.

Morphologically, triticale resembles wheat both in plant-type and kernel characteristics. The main difference lies in its greater vigour relative to wheat and its larger spike and kernel size. Depending upon the varieties (or species) of wheat and rye used in its synthesis, triticale can be of either spring or winter-habit and can be used as either a grain or forage.

2 Cytotaxonomic background (Fig. 35.1)

Both hexaploid ($2n = 6x = 42$) and octoploid ($2n = 8x = 56$) forms of triticale are being developed. The $6x$ forms result from the hybridization of tetraploid wheats ($2n = 4x = 28$) with rye ($2n = 2x = 14$), with subsequent doubling. Similarly, the $8x$ triticales arise from the combination of hexaploid wheats ($2n = 6x = 42$) with rye. Because octoploid triticales are more easily made than hexaploids, they received more attention during the early years. With the improvement of embryo culture and chromosome doubling techniques, however, the hexaploid form has proved more useful because it is more seed fertile. Tetraploid forms ($2n = 4x = 28$) have also been experimentally developed (Krolow, 1973) but, in their present state, are agronomically inferior to forms of higher ploidy.

Concurrently with the breeding work just outlined, there has been much cytological study of reproductive behaviour for reviews of which see Scoles and Kaltsikes

(1974) and Tsuchiya (1974). Generally, triticale has exhibited varying degrees of aneuploidy, depending upon the wheat–rye parentage and also upon the level of ploidy. Hexaploid triticale, on average, exhibits greater cytological stability than other forms.

In recent years, the development of hexaploid triticale as a new crop has centred on the production of 'secondary' hexaploids (see Fig. 35.1). They are superior to primary hexaploids and the improvement is attributable to modified cytoplasm carried by the octoploid parent (Larter and Hsam, 1973; Hsam, 1974). Moreover, the combination of the two forms (the octoploid, genomically AABBDDRR, and the hexaploid, genomically AABBRR), can result in viable hexaploids in which R-chromosomes are replaced by D-chromosomes of common wheat. In other words, the 'pivotal' genome AB from the tetraploid wheats remain unaltered while the D–R genomes become reconstructed through substitution. Secondary hexaploids selected from these recombinants have been found to possess desirable agronomic traits from both the D and R genomes (Gustafson and Zillinsky, 1973). Recently, the identification of each of the rye chromo-somes by heterochromatic banding has enabled workers to distinguish the rye chromosomes that have been replaced by D-genome wheat chromosomes in a number of hexaploid lines (Darvey and Gustafson, 1975.).

3 Early history

Although natural hybrids between wheat and rye have frequently been reported, most were sterile and were consequently lost. The first report of the spontaneous occurrence of a partially fertile hybrid dates back almost 85 years when Rimpau (1891) described partially fertile tillers on an otherwise sterile plant. The frequency with which such hybrids occur however is too low to be of practical value and it was not until the improvement of embryo culture and chromosome doubling techniques in the 1930s that intensive work began.

One of the first reports was that of Müntzing (1939). He began work in 1932, using an amphidiploid (8x) which arose as a restitution product from a wheat (6x) × rye cross. He crossed this hybrid with the early Rimpau strain from Germany and concluded that

Fig. 35.1 Evolution of triticale, *Triticosecale*.

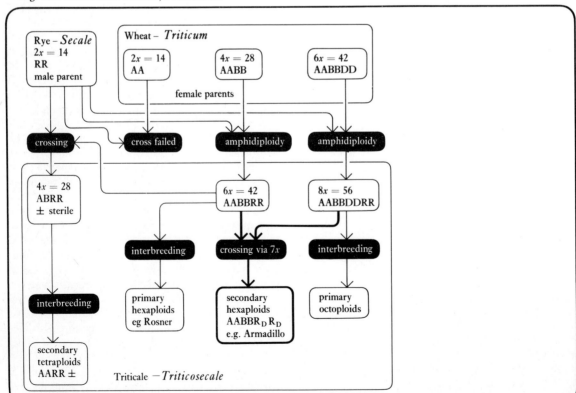

superior triticales could be developed through conventional plant breeding methods. Indeed, by 1950, 18 years after the initiation of his programme, Müntzing's best triticale lines were already yielding 90 per cent of the bread wheats grown at that time and their fertility and seed quality were steadily improving. Similarly, Pissarev (1966) who began work in Russia in the early 1940s also reported progress in octoploid material.

Meanwhile, work with primary hexaploids indicated that they were superior in meiotic stability and fertility relative to octoploid forms (Kiss, 1966; Sanchez-Monge, 1958). As a result, there was renewed interest in the crop but at the hexaploid rather than at the octopoid level. The first large-scale programme in North America was initiated at the University of Manitoba, Winnipeg, Canada in 1954 under the direction of Drs. L. H. Shebeski and B. C. Jenkins. The early stages of the programme were devoted to the introduction, synthesis and evaluation of primary hexaploids under Canadian conditions. Intercrossing of promising lines began in 1958 and, by 1967, the grain yields of at least a few lines were equal to those of the then standard Canadian bread wheat varieties. Meanwhile, feeding trials indicated that triticale grain was nutritionally equal to wheat and barley and preliminary commercial-scale milling tests showed that the grain could be used as human food. Various forms of experimental breakfast cereal were made having a characteristic 'nutty' taste and good quality bread was produced from the flour by using modified mixing and fermentation times (Lorenz et al., 1972).

4 Recent history

The first variety of triticale to be released for commercial production in Canada was the variety Rosner, a product of a double cross involving four hexaploids: three from *T. durum* (4*x*) and one from *T. persicum* (4*x*) (Larter et al., 1970). Rosner represented a marked improvement in straw strength, fertility and earliness in comparison with earlier triticales. Moreover, the same cultural practices as applied to wheat could be used for Rosner. Its adaptability however, was limited because of partial sensitivity to day-length; and there was also need for improved yield performance.

The initiation of a triticale programme by the Centro Internacional de Mejoramiento de Maiz y Trigo (CIMMYT) in 1965 provided the necessary stimulus to develop triticale on an international basis. Using the Canadian material as the initial breeding stock, lines were quickly isolated (under the short-day conditions of Mexico) that were both day-length insensitive and highly fertile (Zillinsky and Borlaug, 1971). These

lines were designated Armadillo and they represented a major advance. In 1971, the Canadian International Development Agency (CIDA) approved major financial support for an international triticale developmental programme involving CIMMYT and the University of Manitoba. The main objective is the development of triticale as a human food.

5 Prospects

The use of the Mexican Armadillo strains as parents in the Canadian programme has resulted in a marked improvement in yield. Results from advanced trials in Canada in 1973 showed that 60 per cent of all lines tested outyielded Rosner and 2 per cent outyielded the highest yielding Canadian wheat variety. Similarly, Dr F. Zillinsky, reported 1973 CIMMYT results which showed triticale as outyielding wheat for the first time since the initiation of the programme in Mexico in 1965 (pers. comm.). Gustafson et al. (1972) reported that yields of some triticales exceeded 6700 kg/ha in California, USA, a return which closely approached that of the highest yielding wheat varieties. Similar reports from workers in other parts of the world indicate that progenies derived from the Armadillo type material are more widely adapted than the later-maturing, light-sensitive strains previously used.

From the standpoint of quality, triticale has consistently exhibited higher contents of protein and essential amino acid than wheat. Villegas et al. (1970) have reported extensive analyses and found a wide range upon which selection could be based. Feeding trials in Canada have shown that certain lines of triticale support higher growth rates in rats, mice and voles than other cereal grains (with the exception of oats). The quality of triticale as a forage and grazing crop has received less attention than its potential as a grain. However, in certain regions of the United States fall-sown triticales have been used successfully for grazing and have exhibited excellent recuperative properties.

The genetic base of triticale is still very narrow. However, co-ordinated international breeding programmes are now being initiated which will result in the utilization and testing of a range of germplasms native to various regions of the world (Larter, 1974). In this way, the genetic base will be widened and the spectrum of adaptation can be expected to increase rapidly during the next decade.

6 References

Darvey, N. L. and **Gustafson, J. P.** (1975). Identification of rye chromosomes in wheat-rye addition lines and triticale by heterochromatic bands. *Crop. Sci.*, 15, 239–43.

Gustafson, J. P., Qualset, C.O., Prato, J. D., Puri, Y. P., Isom, W. H. and Lehman, W. F. (1972). Triticale in California. *Calif. Agr.*, 26, 3–5.

Gustafson, J. P. and Zillinsky, F. J. (1973). Identification of D-genome chromosomes from hexaploid wheat in a 42-chromosome triticale. *Proc. 4th intnl. Wheat Genet. Symp.*, *Columbia, USA*, 225–31.

Hsam, S. L. K. (1974). A study of the application of nucleo-cytoplasmic relationships to the improvement of hexaploid triticale. *Ph.D. Thesis, University of Manitoba, Winnipeg, Canada*, pp. 194.

Kiss, A. (1966). Neue Richtung in der Triticale-Züchtung. *Z. Pflanzenzücht.*, 55, 309–29.

Krolow, K. D. (1973). 4x-Triticale production and use in triticale breeding. *Proc. 4th intnl. Wheat Genet. Symp.*, *Columbia, USA*, 237–43.

Larter, E. N., Shebeski, L. H., McGinnis, R. C., Evans, L. E. and Kaltsikes, P. J. (1970). Rosner, a hexaploid Triticale cultivar. *Can. J. Plant Sci.*, 50, 122–4.

Larter, E. N. (1974). A review of the historical development of triticale. In Tsen (ed.) *Triticale: first man-made cereal*, American Association of Cereal Chemists, Minnesota, USA, 35–52.

Larter, E. N. and Hsam, S. L. K. (1973). Performance of hexaploid triticale as influenced by source of wheat cytoplasm. *Proc. 4th intnl. Wheat Genet. Symp.*, *Columbia, USA*, 245–51.

Lorenz, K., Welsh, J., Norman, R. and Maga, J. (1972). Comparative mixing and baking properties of wheat and triticale flours. *Cereal Chem.*, 49, 187–93.

Müntzing, A (1939). Studies on the properties and the ways of production of rye-wheat amphiploids. *Hereditas*, 25, 387–430.

Pissarev, V. (1966). Different approaches in *Triticale* breeding. *Proc. 2nd intnl. Wheat Genet. Symp.*, Lund. *Hereditas Suppl.*, 2, 279–90.

Rimpau, W. (1891). Kreuzungsprodukte landwirtschaftlicher Kulturpflanzen. *Land. Jbuch.*, 20, 335–71.

Sanchez-Monge, E. (1958). Hexaploid triticale. *Proc. 1st intnl Wheat Genet. Symp.*, *Winnipeg*, 181–94.

Scoles, G. and Kaltsikes, P. J. (1974). The cytology and cytogenetics of triticale. *Z. Pflanzenzücht.*, 73, 13–43.

Tsuchiya, T. (1974). Cytological stability of triticale. In Tsen (ed.), *Triticale: first man-made cereal*, American Association of Cereal Chemists, Minnesota, USA, 62–89.

Villegas, E., McDonald, C. E. and Gilles, K. A. (1970). Variability in the lysine content of wheat, rye and triticale proteins. *Cereal Chem.*, 47, 746–57.

Zillinsky, F. J. and Borlaug, N. E. (1971). Progress in developing triticale as an economic crop. *CIMMYT Res. Bull.*, 17.

Wheats

Triticum spp. (Gramineae-Triticinae)

Moshe Feldman

Department of Plant Genetics Weizmann Institute of Science Rehovot Israel

1 Introduction

'In the sweat of thy face shalt thou eat bread'
(Genesis, 3, 19).

Ever since man's first successful attempts, some 10,000 years ago, to produce food in southwestern Asia, the history of cultivated wheat and that of human civilization have been closely interwoven. In the course of domestication, the wheat plant has lost the ability to disseminate its seeds effectively and is now completely dependent on man for dispersal. But man has fostered this cereal to such an extent that it is now the world's foremost crop plant. Domestication of wheat has entailed that of other edible plants and enabled man to produce food in large quantities. This, in turn, has led to community settlement, population increase and rapid cultural evolution.

An enormous amount of variation has developed in the crop; so far, some 17,000 different varieties have been produced. The plant is high yielding in a wide array of environments, from 67°N in Norway, Finland and Russia to 45°S in Argentina but, in the subtropics and tropics, its cultivation is restricted to higher elevations. The world's main wheat-producing regions are southern Russia, the central plains of the United States and adjacent areas in Canada, the Mediterranean basin, north-central China, India, Argentina and south-western Australia.

Most modern varieties belong to hexaploid wheat, *Triticum aestivum* var. *aestivum*. Because of the high gluten content of the endosperm this 'common' wheat, and especially its harder grained varieties, is highly valued for bread making. The sticky gluten protein entraps the carbon dioxide formed during yeast fermentation and enables a leavened dough to rise. Durum wheat, *T. turgidum* var. *durum* is the main modern tetraploid type; it is mainly grown in relatively drier regions, namely in the Mediterranean basin, in India, the USSR and in low rainfall areas of the great plains of the United States and Canada. Its large, very hard grains yield a low-gluten flour suitable for macaroni and semolina products. There are today no economically important diploid wheats.

A large proportion of man's essential nutrients is contained in the wheat grain. These are: carbohydrates (60–80%, mainly as starch); proteins (8-15%, which contain adequate amounts of all essential amino acids except lysine, tryptophane and methionine); fats ($1 \cdot 5$–$2 \cdot 0\%$); minerals ($1 \cdot 5$–$2 \cdot 0\%$); and vitamins such as the B complex and vitamin E.

In addition to its high nutritive value, the low water content, ease of transport and processing and good storage qualities have made this crop the most important staple food of more than one billion people or 35 per cent of the world's population. During the last two decades, the global wheat area has increased by 25 per cent, reaching about 213 Mha in 1972. During the same period, production has increased by more than 100 per cent, mainly due to wider use of fertilizers and to improved varieties. In 1972, total crop was 350 Mt and accounted for more than 20 per cent of the total food calories consumed throughout the world.

Table 36.1 classification of cultivated wheats and closely related wild species.

Species	Genomes	Wild hulled	Cultivated hulled	Cultivated free-threshing
Diploid ($2n = 14$)				
T. speltoides	S (= G?)	all	--	--
T. bicorne	S^b	all	--	--
T. longissimum	S^1	all	--	--
T. tauschii (= Ae. squarrosa)	D	all	--	--
T. monococcum	A	var. boeoticum (wild einkorn)	var. monococcum (cult. einkorn)	--
Tetraploid ($2n = 28$)				
T. timopheevii	AG	var. araraticum	var. timopheevii	--
T. turgidum	AB	var. dicoccoides (wild emmer)	var. dicoccum (cult. emmer)	var. durum var. turgidum var. polonicum var. carthlicum
Hexaploid ($2n = 42$)				
T. aestivum	ABD		var. spelta var. macha var. vavilovii	var. aestivum var. compactum var. sphaerococcum

2 Cytotaxonomic background

The tribe Triticeae (= Hordeae) is economically the most important group of the family Gramineae. It has given rise to cultivated wheats, ryes and barleys. Wheats (*Triticum*) and ryes (*Secale*), together with the genera *Aegilops*, *Agropyron*, *Eremopyron* and *Haynaldia*, form the subtribe Triticinae. This subtribe is relatively young and therefore hybridization between entities of different genera can still take place and result in either direct exchange of genetic material or amphiploidy. Since polyploid wheats contain genomes which have been derived from *Aegilops* as well as *Triticum*, the present article follows a current trend and places both taxa in the single genus, *Triticum* (Morris and Sears, 1967).

The pioneering cytogenetic work of Sakamura, Sax, Kihara and others (Riley, 1965; Morris and Sears, 1967) have shown that the various species of *Triticum* form a polyploid series based on $x = 7$ and consisting of three different ploidy levels: diploids ($2n = 2x = 14$), tetraploids ($2n = 4x = 28$), and hexaploids ($2n = 6x = 42$) (Table 36.1).

The wild diploid species, some of which have contributed to polyploid wheats, are presumably monophyletic in origin though they have diverged considerably from each other. This divergence is particularly evident in the morphologically well defined seed dispersal units of the species and their specific ecological requirements and geographical distributions. Cytogenetic data have corroborated the taxonomic classification by showing that each diploid contains a distinct genome (Kihara, 1954). The related chromosomes of the different genomes show little affinity with each other and do not pair regularly in interspecific hybrids, thus leading to complete sterility and isolation of the diploid species from each other.

The polyploid species are a classic example of evolution through amphidiploidy. They behave like typical genomic allopolyploids; that is, their chromosomes pair in a diploid-like fashion and the mode of inheritance is disomic. The allopolyploid nature of the *Triticum* polyploids has been verified from cytogenetic analysis of hybrids between species of different ploidy levels. Each polyploid species can be identified as a product of hybridization followed by chromosome doubling.

Since the different genomes are closely related (Morris and Sears, 1967), polyploid wheats are segmental rather than typical genomic allopolyploids. The diploid-like behaviour of polyploid wheats is due to suppression of pairing of homoeologous chromosomes (i.e. related chromosomes of different genomes) by a specific gene. In hexaploid *T. aestivum*, this gene

121

is located on the long arm of chromosome 5 of genome B and known as the 5B^L gene (Riley, 1965). Plants deficient for this gene behave like segmental allo-polyploids; their homoeologous chromosomes pair and form multivalents.

The development of this diploidizing mechanism has been critical for the evolution of the polyploid wheats and, indeed, for their domestication. By restricting pairing to completely homologous chromosomes, the diploidizing gene ensures regular segregation of genetic material, high fertility and genetic stability. Synthetic polyploid wheats which do not contain this gene are partially sterile. Another asset of the mechanism is that it facilitates genetic diploidization under which existing genes in double and triple doses can be diverted to new functions. Furthermore, in segmental allopolyploids, permanent heterosis

between homoeoalleles (i.e. homologous genes in different genomes) can be maintained.

Three groups of polyploids are recognized (Zohary and Feldman, 1962). Species in each group have one genome in common and differ in their other genomes. Polyploids of group A share the genome of diploid wheat, *T. monococcum*; those of group D share the genome of *T. tauschii* (= *Ae. squarrosa*); and those of group C^u the genome of *T. umbellulatum* (= *Ae. umbellulata*). In basic morphology and particularly in the structure of the seed dispersal unit, the polyploids of each group resemble the diploid donor of the common pivotal genome. The cultivated polyploid wheats (Table 36.1) belong to group A and the arrowhead-shaped dispersal unit of wild diploid wheat, *T. monococcum* var. *boeoticum*, can be recognized in all the wild polyploids, while the non-brittle ear of culti-

Fig. 36.1 Evolutionary relationships of the wheats, *Triticum*.

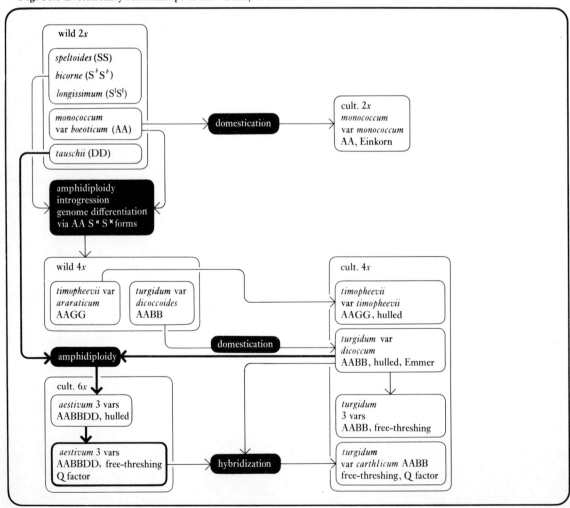

vated *T. monococcum* var. *monococcum* reappears in all the cultivated polyploids. Group A polyploids comprise the tetraploids, *T. turgidum* (AABB) and *T. timopheevii* (AAGG), and the hexaploid *T. aestivum* (AABBDD).

Hybrids between the two tetraploid species exhibit partial asynapsis and high sterility. In other words, the B genome of *T. turgidum* is non-homologous with the G genome of *T. timopheevii*. The two tetraploid taxa could have arisen independently from crosses between wild forms of *T. monococcum* and two distinct diploid species or could have a monophyletic origin. Hexaploid *T. aestivum* (AABBDD) contains two genomes homologous with the A and B genomes of *T. turgidum*. Hence *T. aestivum* has arisen from hybridization, between *T. turgidum* and a diploid species having genome D.

The identification of the diploid donors of the B, G, and D genomes has been the subject of intensive cytogenetic studies in the last 50 years. The donor of the D genome of hexaploid wheat has been identified as *T. tauschii* (Morris and Sears, 1967). Synthetic hexaploids have been produced from crosses of different varieties of *T. turgidum* with *T. tauschii* and resemble certain established hexaploids. Hybrids between synthetic and natural hexaploids are usually fully fertile.

The AB genomes of hexaploid wheat could have been donated by either wild or, more plausibly, by cultivated *T. turgidum*. While distribution of the wild tetraploid *T. turgidum* var. *dicoccoides*, overlaps with that of *T. tauschii* to only a very limited extent (if at all), in western Iran and eastern Turkey, cultivated *T. turgidum*, especially var. *dicoccum*, has been grown throughout the distribution area of *T. tauschii* at one time or another. Furthermore, because of the different mode of seed dissemination of wild tetraploid wheat and of *T. tauschii*, the amphidiploid which could originate from these two species would have been unable to disseminate its seeds effectively and would therefore have been quickly eliminated. It is thus most likely that the hexaploid originated in wheat fields after the spread of cultivated *T. turgidum* into the distribution area of *T. tauschii*. Such an origin could also explain why hexaploid wheat, in contrast to most other cereal crops, has no wild relatives.

Judging from their tremendous variability, hexaploid wheats were probably formed recurrently by numerous hybridizations involving different genotypes of tetraploid varieties and various races of *T. tauschii*. The donor of the D genome grows today within and at the edges of wheat fields in Iran and Armenia. Natural hybrids between tetraploids and *T.*

tauschii can be found there and they usually set some seeds with $2n = 42$ chromosomes, due to the formation of unreduced gametes. *T. tauschii* also has ample geographical contacts with both wild and cultivated *T. timopheevii* (AAGG) but no natural hexaploid deriving from these two species is known.

In contrast to the D genome donor, that of the B genome has so far defied conclusive identification. Morphological, geographical and cytological evidence has been used to implicate *T. speltoides* (genome S) or a closely related species (Riley, 1965). However, recent cytogenetic data (Kimber, 1974) indicate that *T. speltoides* may have donated the G genome of *T. timopheevii* rather than the B of *T. turgidum*. An incomplete homology between S and G is accounted for by the supposition that G has undergone some modification at the tetraploid level.

On the assumption that the B and G genomes are monophyletic, B could have differentiated from the G of wild *T. timopheevii* or from an earlier tetraploid of constitution AASS. Accordingly, both B and G would have to be considered as modified S genomes. The occurrence of a modified genome or genomes side by side with a stable one (i.e. a genome which is very similar to that of an existing diploid), is characteristic not only of polyploids of group A but of all polyploids of the genus *Triticum*. Such a constitution is believed (Zohary and Feldman, 1962) to have resulted from hybridization between initial polyploids sharing one genome and differing in one or two additional ones. Such hybridizations are eased by the shared genome which acts as a buffer and ensures some fertility in the resulting hybrids. In such hybrids, the two differential genomes, brought together from different parents, can exchange genetic material and become assimilated to each other. Accordingly, in the polyploids of group A, the initial amphidiploid (AASS) could have exchanged chromosome segments with other amphidiploids or with diploids such as *T. longissimum* (genome S^l), *T. bicorne* (genome S^b) or others. As a result of such introgressions, the S genome could have become modified in different directions, giving rise to G and B.

T. speltoides (genome S) is in contact with wild *T. monococcum* (i.e. var. *boeoticum*) in eastern Turkey, northwestern Iraq and western Iran. In that area, there are numerous mixed populations of the two species, which contain also wild *T. timopheevii* and, more sporadically, wild *T. turgidum*. However, the main distribution of wild *T. turgidum* var. *dicoccoides* is southwest of this *speltoides–monococcum* area; it occurs in northeastern Israel, northwestern Jordan and southern Syria, where it does not form mixed popula-

123

tions with *T. speltoides* but does grow locally with *T. longissimum*.

In contrast to the diploids, which are genetically isolated from each other and have undergone divergent evolution, the polyploids are characterized by convergent evolution, due to the facts that they contain genetic material from two or three different diploid genomes and can (by way of hybridization and introgression) exchange genes with each other, resulting in numerous genomic recombinations. Polyploidy is, therefore, of evolutionary significance here mainly because it has facilitated the formation of a superstructure that combines the various genetic materials of the isolated diploids and allows them to recombine. Moreover, polyploidy, reinforced by the diploidizing genetic system and by a predominant inbreeding, has proved to be a very successful genetic system. The evolutionary advantage of the polyploids over the diploids is obvious and is reflected in the very wide morphological and ecological variation. No wonder, therefore, that cultivated polyploid wheats exhibit a wide range of genetic flexibility and can adopt themselves to a great variety of environments.

3 Early history

The earliest known grains of domesticated wheats date to approximately 7500–6500 B.C. Together with barley and pulses, they have been found in prehistoric sites of the 'fertile crescent' (which comprises the mountain chains flanking the plains of Mesopotamia and the Syrian desert) and also in Anatolia and the Balkans. It is only natural that the domestication of wheats should have taken place in the fertile crescent since this is the centre of their wild progenitors' geographical distribution (Harlan and Zohary, 1966). Wild diploid wheat or 'einkorn', *T. monococcum* var. *boeoticum*, occurs naturally in open herbaceous oak park-forests and steppe-like formations in south-western Iran, northwestern Iraq and southeastern Turkey. In this area and beyond it, in Transcaucasia, Syria, central and western Turkey and Greece, wild einkorn is also very common in secondary habitats such as roadsides and field edges. Wild tetraploid or 'emmer' wheats, *T. timopheevii* var. *araraticum* and *T. turgidum* var. *dicoccoides*, occupy primary habitats similar to those of the wild diploid but are less continuous in their distribution. *Triticum timopheevii* grows in western Iran, northwestern Iraq, eastern Turkey, Armenia and Transcaucasia, while *T. turgidum* is distributed in southern Syria, northwestern Jordan and northeastern Israel. Recently, *T. turgidum* has also been found in certain sites in the southernmost area of wild *timopheevii*, locally in mixed populations with the latter.

Wheats were preadapted for domestication. The large seed size, which aids rapid germination, made them attractive to the ancient collector. Their annual habit, through which they escape the dry season, also made them amenable to dry farming. Their system of predominant self-pollination could have aided in the fixation of desirable mutants and of recombinants resulting from rare outcrossing events. While these grasses occupy poor, thin, rocky soils in their natural sites, they respond well when transferred to richer habitats.

One of the characters which was of no pre-adaptive advantage for cultivation is the wild mode of seed dissemination. Wild wheats are characterized by a brittle spike which disarticulates into arrowhead-shaped spikelets when mature. While these seed dispersal units facilitate self-burial in the soil (hence protection during the long dry summer and successful germination after the first rain) they must have proved a nuisance to the ancient farmer who had to collect most of the spikelets from the ground or cut the culms before the grains were mature. No wonder, therefore, that types with non-brittle heads were unconsciously selected from the earliest stages of domestication. The loss of self-propagation was followed by the loss of self-protection. The wild forms have tightly closed glumes resulting in a 'hulled' grain after threshing. Only few of the cultivated wheats of today retain this feature. Mutations which spread under domestication led to loosely closed glumes which release the naked 'free-threshing' grain.

Both the wild, brittle diploid wheat, *T. monococcum* var. *boeoticum* and the cultivated, non-brittle var. *monococcum* are hulled types (Table 36.1). The earliest known carbonized grains of brittle diploid wheat occur in the prehistoric settlement of Tell Mureybit, northern Syria (eighth millenium B.C.) where the cereal was apparently collected rather than farmed. (For palaeoethnobotanical data see Renfrew, 1973; Helbaek, 1966.) At somewhat later sites (e.g. Ali Kosh in Iranian Khuzistan, 7500–6750 B.C.; Jarmo in Iraqi Kurdistan, *ca.* 6750 B.C.; Cayönü Tepesi in southeastern Turkey, *ca.* 7000 B.C.; and Hacilar in west-central Anatolia, *ca.* 7000 B.C.), brittle diploid wheat occurs side by side with a non-brittle type and is gradually replaced by it. Towards the end of the seventh millenium there are also records of non-brittle *T. monococcum* in the southern Balkans. It is significant that no remains of either wild or cultivated *T. monococcum* have been found in irrigated settlements in the lowlands of the Mesopotamian plains or in the Nile valley, to which wheat farming spread in the fifth millenium. During that millenium, cultivated einkorn (*T. monococcum*) spread to central and western Europe through the Danube and Rhine valleys. In subsequent eras (Bronze and early Iron Ages) einkorn wheat attained a wide distribution in Europe and the Near East. In modern times it occurs only as a relic crop in mountainous parts of Yugoslavia and Turkey where it is used mainly for fodder. *Monococcum* wheats show considerable resistance to cold and to rust diseases.

The wild tetraploids, *T. timopheevii* var. *araraticum* and *T. turgidum* var. *dicoccoides* (Table 36.1), are morphologically indistinguishable. Judging from present-day geographical distribution of the two species, the carbonized grains, spikelets, and clay impressions found at Jarmo (*ca.* 6750 B.C.) and at Cayönü Tepesi (*ca.* 7000 B.C.) could belong to either. While the brittle form of *T. timopheevii* gave rise to only a restricted number of non-brittle cultivars, all of which are cultivated in Armenia and Transcaucasia, that of *T. turgidum* is the predecessor of most cultivated tetraploid and hexaploid wheats. The non-brittle, hulled tetraploid wheat, *T. turgidum* var. *dicoccum* (cultivated emmer), occurs in prehistoric villages of the Near East as early as 7500 B.C. Together with cultivated einkorn, the non-brittle emmer spread rapidly to all seventh millenium farming areas of the Near East, including valleys not penetrated by *T. monococcum* cultures. Brittle forms of both diploids and tetraploids were possibly harvested in the wild state throughout the fertile crescent long before actual farming began. Cereal farming itself may have originated in areas adjacent to, rather than within, the regions of greatest abundance of wild forms. Examples of prehistoric settlements peripheral to, rather than within, natural distribution areas are Ali Kosh in Iran, Tell es Sawwan in the plains of Mesopotamia, Tel Ramad in Syria, Jericho in Israel and Beidha in Jordan.

Emmer was the most prominent cereal in the early farming villages of the Near East. From the mountainous areas of the fertile crescent it was transferred to the lowlands of Mesopotamia in the sixth millenium B.C. and, during the fifth and fourth millenia, to Egypt, the Mediterranean basin, Europe, central Asia and India. It was taken to Ethiopia some 5,000 years ago by the Hamites. In all these regions it remained the principal cereal until the first millenium B.C., when it was replaced by the more advanced free-threshing tetraploid form, *T. turgidum* var. *durum*. Today, emmer is grown on a limited scale in Ethiopia, Iran, Transcaucasia, eastern Turkey and the Balkans.

The *durum* wheats presumably originated from cultivated emmer by an accumulation of mutations that reduced the toughness of the glumes to the point at which free-threshing was attained. Most other naked grain tetraploid wheats (Table 36.1) are probably of relatively recent origin, deviate from var. *durum* in only single characters and share the genetic system which determines the free-threshing habit. *T. turgidum* var. *carthlicum* is the only tetraploid known to carry a different gene complex determining free-threshing. This complex, known as the Q factor, is also present in hexaploid wheats (Morris and Sears, 1967). The narrow distribution of var. *carthlicum* in Transcaucasia may indicate that this variety originated relatively recently, presumably by hybridization of an unknown tetraploid with a hexaploid of the *aestivum* group.

Hexaploid *T. aestivum*, probably originated and entered cultivation only after the more or less simultaneous domestication of diploid and tetraploid forms. It appears in archaeological data from the seventh millenium B.C. The earliest finds, attributed to club wheat, *T. aestivum* var. *compactum*, are at Tell Ramad, Syria (about 7000 B.C.). Finds from the sixth millenium, identified as ancestral forms of free-threshing hexaploids, have been unearthed at Tepe Sabz in Iranian Khuzistan, at Tell es Sawwan in Iraq, at Çatal Hüyük in central and Hacilar in west-central Anatolia, and at Knossos on Crete. Between 6000 and 5000 B.C. *T. aestivum* penetrated, together with cultivated emmer, into the irrigated agriculture of the plains of Mesopotamia and western Iran and, in the fifth millenium, into the Nile basin. In the fifth millenium, *T. aestivum* appears also in finds from the central and western Mediterranean basin. Compact forms of hexaploid wheat were cultivated in central and western Europe at the end of the fourth millenium, where they are found associated, together with einkorn and emmer, with the first traces of agricultural activities.

While tetraploid wheats, in keeping with their Near Eastern origin, are adapted to mild winters and rainless summers, the addition of the central Asiatic *tauschii* (D) genome must have contributed to the adaptation of hexaploid wheats to more continental climates. This could have greatly facilitated the spread of bread wheat into central Asia and, by way of the highlands of Iran, to the Indus valley, where it appears at the beginning of the third millenium B.C. The earliest recorded evidence in China is from the middle of the third millenium.

Triticum aestivum was introduced to the New World in 1529 when the Spaniards took it to Mexico, and to Australia in 1788.

The hexaploid varieties (Table 36.1) *spelta*, *vavilovii* and *macha* are hulled while vars. *aestivum*, *compactum*,

125

and *sphaerococcum* are free-threshing and thus considered to be more advanced. The compound genetic locus (the Q factor) determining the naked grain trait is located on chromosome 5 of genome A. It could have arisen from the q gene of the hulled varieties by a series of mutations; five doses of the q of *spelta* have the same effect as two doses of Q (Morris and Sears, 1967). The primitive status of hulled hexaploids is supported by the observation that artificial hybridizations between almost all tetraploid wheats, either free-threshing or hulled, with all known races of *T. tauschii*, give rise to hulled types. Only var. *carthlicum*, which contains the Q factor of free-threshing hexaploids, yields *aestivum*-like naked grain forms when crossed with *T. tauschii*. *Aestivum* and *compactum*-like types appear as segregants from crosses between Iranian and European *spelta*, indicating that these forms of *spelta* may contain compound loci of q which can recombine in hybrids yielding duplicate or even triplicate loci. The effects of these multiple loci resemble those of Q.

These genetic data, which show that the hulled hexaploid wheats are more primitive than the free-threshing forms, are not in accord with archaeological chronology. While *aestivum* was abundant in the prehistoric Near East from the sixth millenium onwards, there are, so far, no indications of the cultivation of *spelta* or other hulled forms before 2000 B.C. Moreover, in central Europe, *spelta* appears about 1,000 years later than compact forms of free-threshing wheat. These discrepancies between genetic and archaeological data raise some difficulties in understanding the early history of the hexaploids. On the assumption that the first hexaploids were hulled, their absence from the prehistoric remains of the Near East indicate that they were not taken into cultivation in that area, possibly because they did not have any advantage over cultivated emmer. However, *spelta* is grown today in extreme environments such as the high plateau of west-central Iran, very close to the area of contact between *dicoccum* and *T. tauschii*, and this cultivation is possibly of ancient origin. Spelt wheat was either brought to Europe relatively late (*ca.* 2000 B.C.) and replaced the compact free-threshing type known as *T. antiquorum* then grown by lake dwellers in the upper Rhine region, particularly at high altitudes where temperatures were extreme; or it could have rearisen in the Rhine valley as a result of a cross between a hexaploid *compactum*-like form and tetraploid *dicoccum*, both of which were grown in that area. Such a cross could yield hexaploid progenies lacking the Q factor of the free-threshing *compactum* parent.

Spelt wheat is cultivated today in several areas of central Europe and in the plateau of western Iran.

The economically more advantageous free-threshing hexaploids may have competed with or replaced hulled tetraploid emmer at a time when free-threshing tetraploids had not yet evolved. Var. *aestivum* is economically by far the most important hexaploid and is now grown on a world-wide scale. It has given rise to *compactum* and *sphaerococcum* through mutations. Var. *compactum* is grown today in certain restricted areas of Europe, the Near East, and the northwestern United States; *sphaerococcum* is grown to some extent in India and central Asia and is known from India as early as the third millenium B.C.

To sum up: both diploid and tetraploid wheats were apparently first taken into cultivation in the Zagros area and *T. turgidum* at the watershed of the Jordan River in the southwestern part of the fertile crescent; hexaploids originated southwest of the Caspian Sea. With migration into new areas, cultivated wheats encountered new environments to some of which they responded with bursts of variation and formation of many endemic forms. Several such secondary centres of variation were described by Vavilov, namely the Ethiopian plateau and the Mediterranean basin for tetraploids and Afghanistan (Hindu-Kush area) for hexploids. Transcaucasia is a secondary centre for tetraploid as well as hexaploid types. Such secondary centres of diversity are valuable to wheat breeders as gene pools additional to those existing in the primary centre of origin.

4 Recent history

Modern wheat cultivars have developed through three main phases of selection: subconscious selection by the earliest food grower simply by the processes of harvesting and planting; mainly deliberate selection in the polymorphic fields of the primitive farmer; and scientifically planned modern breeding. The main attainments of the first phase were non-brittle spikes, simultaneous ripening of grains, rapid and synchronous germination, perhaps also erect versus prostrate culms and, to some extent, the free-threshing trait. Through the expansion of wheat culture into new areas during that phase, an additional achievement was wider adaptation to different environments.

Primitive farmers selected and planted the grains most desirable for their specific needs. Selection pressure was therefore exerted consistently but in different directions by different farmers. These efforts resulted in increased yield, larger seed size, better flour quality and adaptation to a wider range of climatic and farming regimes. Since numerous genotypes were grown in mixtures in the field, the genetic material at the disposal of the primitive farmer would have been enhanced by the production of occasional recombinants.

Under modern breeding, the wheat field has become genetically uniform and is no longer conducive to spontaneous gene exchange. On the other hand, gene migration on a large scale has been promoted by world-wide introduction services. Massive scientific screening aids in disclosing desirable genes and modern methods for manipulating these genes and transferring them from one genetic background to another are available. Hybridizations have been confined mainly to intraspecific crosses and diploid and tetraploid gene pools have been little utilized for improvement of the hexaploids.

Today's selection techniques can achieve the objectives of the primitive farmer with much greater certainty. High yielding cultivars owe their performance to genetic increase of the number of fertile florets per spikelet and, sometimes, length or density of the spike, reduction of shattering and resistance to rust and other diseases. High-yielding cultivars respond to boosting by agrotechnical practices. Thus they perform well under high doses of artificial fertilizers without lodging. The most striking examples are the semi-dwarf (90–120 cm) or dwarf (60–90 cm) strains which are now replacing the conventional taller (120–140 cm) varieties. The genes determining dwarfness were first introduced from the economically unimportant Japanese cultivar Norin 10 into the northwestern American winter wheat cultivar Brevor by the American breeder, O. Vogel. The release in 1961 of the semi-dwarf cultivar Gaines has opened a new era in wheat breeding. Under the leadership of N. E. Borlaug of the International Maize and Wheat Improvement Centre in Mexico, the same dwarfing genes were transferred to Mexican spring wheat and, in the course of 'the green revolution' the resulting semi-dwarf Mexican varieties were taken to India, Pakistan, Iran and the Mediterranean basin. Today, the dwarfing genes are being very widely incorporated into existing cultivars.

One of the modern breeding aims is the production of early maturing varieties which can, by virtue of a short life cycle, escape adverse climatic conditions and certain diseases. In breeding for grain quality, the main achievements have been improvements of milling and baking characteristics. Certain modern cultivars are easily milled because the pericarp and seed coat are only loosely attached to the endosperm. Flour yield is particularly high in varieties with short, almost spherical, grain. To date, less progress has been made in raising the nutritional value of the grain and further efforts are needed, in particular to increase the protein content and to remedy deficiencies in amino acids composition.

5 Prospects

Wheat germ plasm has so far not been utilized to the fullest extent because of our yet incomplete understanding of the genetic architecture of the group. The large number of chromosomes and the occurrence of duplicate loci make polyploid wheat unsuitable for analysis and breeding by conventional genetic methods. A technique which promises to overcome these difficulties makes use of the aneuploid lines produced by E. R. Sears in the bread wheat variety Chinese Spring. With the aid of these lines and of substitution lines (in which the missing chromosome pair is replaced by a homologous pair from a known donor variety) genes can be assigned to specific chromosomes, linkage maps can be constructed and genetic differences between varieties or even species can be established (Morris and Sears, 1967; Sears, 1969; Law, 1968). Moreover, by using these lines, entire chromosomes or chromosome segments can be precisely transferred from one variety to another. These transfers affect only single chromosomes and no reshuffling of the entire complement occurs as is the case in conventional hybridization. The utilization of wild wheats has only just begun. A simple way to use them is to add their entire genomes to tetraploids or hexaploids, that is, to produce synthetic amphidiploids. The new crop, triticale, produced from different cultivars of *T. turgidum* ($2n = 28$) and rye, *Secale cereale* ($2n = 14$), is one promising result of such an approach (see Chapter 35). Another example is the resynthesis of hexaploid wheat from *T. turgidum* crossed by different strains of *T. tauschii*.

Since entire genomes introduce also unwanted genes, a more elegant and precise way of transferring genetic material from wild to cultivated forms, makes use of substitution lines. One obstacle to such transfer is the suppression of pairing between homoeologous (i.e. related) chromosomes in the presence of the diploidizing gene on chromosome 5B. This difficulty can be overcome either by transferring the desired chromosomal segment from wild to cultivated material with the aid of induced translocations or, more readily, by eliminating or suppressing the 5B gene altogether (Feldman, 1968; Sears, 1972).

Attempts are also being made to raise yields through heterosis (Wilson, 1968). The basic tools for producing F_1 hybrid wheat, namely male-sterile lines and fertility restoration genes, are available. High seed set under field conditions has already been reported for several hybrid combinations. However, the mating system of cultivated wheats is not adapted to open pollination. Most of the pollen is shed before the flower opens and the amount reaching stigmas of male steriles may often be insufficient.

To the current trends must also be added the present emphasis on gene conservation. Attempts are being made to collect and preserve genetic resources for eventual future use, not only from wild relatives but also from primitive or little used cultivated varieties. Much of this material is at present awaiting evaluation.

6 References

Feldman, M. (1968). Regulation of somatic association and meiotic pairing in common wheat. *Proc. 3rd intnl Wheat Genet. Symp.*, Canberra, 169–78.

Harlan, J. R. and Zohary, D. (1966). Distribution of wild wheats and barley. *Science, N.Y.*, **153**, 1074–80.

Helbaek, H. (1966). Commentary on the phylogenesis of *Triticum* and *Hordeum. Econ. Bot.*, **20**, 350–60.

Kihara, H. (1954). Considerations on the evolution and distribution of *Aegilops* species based on the analyser-method. *Cytologia*, **19**, 336–57.

Kimber, G. (1974). The relationships of the S-genome diploids to polyploid wheats. *Proc. 4th intnl Wheat Genet. Symp.*, Columbia, 81–5.

Law, C. N. (1968). Genetic analysis using inter-varietal chromosome substitutions. *Proc. 3rd intnl Wheat Genet. Symp.*, Canberra, 331–42.

Morris, R. and Sears, E. R. (1967). The cytogenetics of wheat and its relatives. In Quisenberry and Reitz (eds), *Wheat and wheat improvement*, Madison, USA, 19–87.

Renfrew, J. M. (1973). *Palaeoethnobotany – the prehistoric food plants of the Near East and Europe.* London.

Riley, R. (1965). Cytogenetics and evolution of wheat. In J. B. Hutchinson (ed.), *Essays on crop plant evolution*, Cambridge, 103–22.

Sears, E. R. (1969). Wheat cytogenetics. *Ann. Rev. Genet.*, **3**, 451–68.

Sears, E. R. (1972). Chromosome engineering in wheat. *Stadler Symp.*, **4**, 23–38.

Wilson, J. A. (1968). Problems in hybrid wheat breeding. *Euphytica*, **17** (suppl. 1), 13–33.

Zohary, D. and Feldman, M. (1962). Hybridization between amphidiploids and the evolution of polyploids in the wheat (*Aegilops-Triticum*) group. *Evolution*, **16**, 44–61.

Maize*

Zea mays (Gramineae–Maydeae)

Major M. Goodman
North Carolina State University
Raleigh NC USA

1 Introduction

Maize, rice and wheat are the three leading food crops. Unlike wheat and rice, most maize is consumed indirectly as feed for livestock rather than directly as human food. In parts of Latin America, however, most maize is grown for human consumption, and the diet consists largely of maize.

The chief maize-producing regions include the United States (especially the Corn Belt region of the north central states), which produces almost half of the world's total, southeastern Europe (especially the USSR, Romania, Yugoslavia, Hungary, and Italy), China, Brazil, Mexico, South Africa, Argentina, India and Indonesia. Only about 5 per cent of the US crop is exported, but that is over half of the world's total exports. Argentina exports about 40 per cent of her total production for second place, while southeastern Europe and South Africa are other major exporters. Western Europe and Japan are the major importers.

2 Cytotaxonomic background

Maize (*Zea mays*), teosinte (*Euchlaena mexicana* or *Zea mexicana* or even *Zea mays* ssp. *mexicana*), and tripsacum (numerous *Tripsacum* species) are the three New World members of the tribe Maydeae. Teosinte (see Wilkes, 1967), a weedy annual, is a close relative of maize. It shares the same chromosome number, $2n = 20$, and usually has rather similar chromosome morphology; a rare perennial form of teosinte with 40 chromosomes has apparently become extinct outside botanical gardens. The two taxa differ most in the structure of their female inflorescences and in their chromosome knob patterns, chromosome knobs being

* Paper No. *4439* of the Journal Series of the North Carolina Agricultural Experiment Station. Supported in part by NIH research grant number GM 11546 from the National Institute of General Medical Sciences.

especially dark staining sites best identifiable at pachytene. These knobs are inherited in Mendelian fashion, the knobs of maize being much less often at terminal and subterminal positions. Morphologically, teosinte plants often resemble maize. The stems are usually less robust and individual plants, unless depauperate, tend to have more tillers. Whereas the terminal (male) inflorescences are similar, the teosinte tassel does not possess the prominent central spike which is characteristic of maize. The female inflorescences of maize usually are one to three (rarely as many as ten) short lateral branches terminating in ears which have 8 to 24, or rarely more, rows of perhaps fifty kernels each. The female inflorescences of teosinte are also lateral branches (which may be further branched), each of which may terminate in a two-rowed spike of perhaps a dozen hardened fruit-cases. Each fruit-case holds a single seed enclosed by an indurated glume. The larger lateral branches may bear terminal staminate spikes. The ear of maize is enclosed by numerous husks (modified leaves) of the lateral branch and the kernels (technically single-seeded fruits or caryopses rather than seeds) adhere to the tough cob or rachis. The teosinte lateral spike is very loosely enclosed by a few husks, the rachis of the spike becoming very fragile upon maturity, and the fruit-cases (the seeds of which often have variable dormancy) disseminating easily. Maize, with neither natural seed dispersal nor seed dormancy, is wholly dependent upon man for its propagation.

Tripsacum species (see Randolph, 1970) are perennials with chromosome numbers in multiples of $x = 18$. *Tripsacum* appears to be more closely related cytologically and morphologically to the genus *Manisuris* (tribe Andropogoneae) than to maize or teosinte. Vegetatively, the various *Tripsacum* species are quite variable. In plant size they vary from that of wheats, or smaller, to beyond that of the larger types of maize. Their inflorescences differ from those of maize and teosinte in that male and female flowers are borne separately, but in tandem, in terminal spikes. The female flowers occur on the lower parts of the inflorescence, with the male flowers developing above them. The seeds are embedded in virtually cylindrical, indurated rachis segments, which break apart at maturity. Although *Tripsacum* also has chromosomal knobs, its general genetic (and chromosomal) structure is different from that of maize (Galinat, 1973). It has alleles of maize genes, although not always on the same chromosomes. Chromosome shapes and knob positions differ greatly from those of maize.

Teosinte crosses readily with maize and the descendents are fertile. Since these two species, like *Tripsacum*, are cross-(wind-)pollinated and grow sympatrically in Mexico and Guatemala, reciprocal introgression between them is possible and doubtless occurs. *Tripsacum* can be crossed with maize (with difficulty, except for *T. floridanum*) but the offspring show varying degrees of sterility. Through backcrossing, small portions of the *Tripsacum* genome can be incorporated into that of maize.

The Oriental Maydeae (*Coix*, *Sclerachne*, *Polytoca*, and *Chionachne*) are usually acknowledged to be but distantly related to maize (Mangelsdorf, 1974), although it has occasionally been speculated that *Coix* (Job's tears), which has knobbed chromosomes in multiples of $x = 5$, is more closely related to maize than are the other Oriental genera. It has been suggested that the $x = 10$ Maydeae arose as a result of amphidiploidy between two $x = 5$ species, such as the $x = 5$ species of *Coix* and *Sorghum*. Other suggestions for the origin of the Maydeae are based upon derivation from the Andropogoneae (see Weatherwax, 1954; Mangelsdorf, 1974). Since the Maydeae seem heterogeneous and since the separation between the Andropogoneae and the Maydeae rests solely upon the singe character monoecy (vs. dioecy), it may well be that the Oriental Maydeae, and the American Maydeae each arose from more than one Andropogonoid ancestor.

There appear to be three current hypotheses concerning the origin of the American members of the Maydeae. One is that maize, teosinte and *Tripsacum* are descendants of a common ancestor (Weatherwax, 1954), the last-named having differentiated earlier than maize and teosinte. A second hypothesis is that maize is derived from teosinte. This suggestion has arisen sporadically over the years, having been advocated by Longley, Beadle and, more recently, by Galinat (1973), Iltis and others. Current work involves identifying a probable ancestral race of teosinte, the minimum number of loci differentiating the two taxa and the linkage relationships of these loci. Galinat (1971) and Mangelsdorf (1974) have cautioned that there is no archaeological evidence to support the idea that teosinte is the ancestor of maize; indeed, almost all the earliest known maize is both polystichous (many-rowed) and soft-glumed, as well as having paired spikelets. Both Mangelsdorf and Weatherwax (1954) have emphasized that these traits appear to be evolutionarily more specialized in teosinte than in maize. For these reasons, among others, Mandelsdorf (1974) suggested a third hypothesis, namely that teosinte originated from maize.

Studies initiated by Galinat and Mangelsdorf many

years ago to test the hypothesis that teosinte arose as a hybrid between maize and *Tripsacum* have recently been reported by Galinat (1971, 1973). Earlier studies had shown that the arrangements of loci in maize and teosinte are quite similar, with linkage between several of the distinguishing loci. In *Tripsacum*, in contrast, the same loci are often found on different chromosomes. These findings do not support the hypothesis, which formed one part of the Mangelsdorf and Reeves 'tripartite hypothesis' concerning the origin and evolution of maize. Neither do they support the hypothesis, suggested by Galinat and his colleagues, that *Tripsacum* arose as a result of hybridization of *Zea* (a genus which, in Galinat's opinion, includes teosinte) and *Manisuris*, a member of the tribe Andropogoneae (Mangelsdorf, personal communication). Mangelsdorf (1974) regards unpublished electron microscope studies by Galinat, Barghoorn and Banergee as conclusive proof that teosinte did not arise as a result of corn-*Tripsacum* hybridization. These studies showed in brief, that the spinules of the pollen grains of maize and teosinte are uniformly distributed, while those of *Tripsacum* are clumped. Maize-

Tripsacum hybrids are intermediate, as are derivatives of such hybrids and backcrosses which contain as little as a single *Tripsacum* chromosome (Mangelsdorf, personal communication). Similar studies have not been conducted with *Manisuris* and may not be possible.

3 Early history (see Fig. 37.1)

More is known about the evolution and domestication of maize than is known about the evolution of the tribe Maydeae as a whole. Pollen samples identified as belonging to maize, teosinte or their common ancestor have been dated at 60,000 to 80,000 years ago in drill cores collected in Mexico City in the mid-1950s. In addition, the evolution of maize under domestication has been documented in a series of publications by Mangelsdorf, MacNeish and co-workers dealing with the archaeology of the southwestern USA and Mexico. Their discoveries at Bat Cave in New Mexico in the late 1940s and at Tehuacán, Mexico in the early 1960s are particularly revealing. At the earliest Tehuacán levels, dated about 5000 B.C., they found very small cobs (little larger than an ordinary pencil eraser), some

Fig. 37.1 Evolution of maize, *Zea mays*.

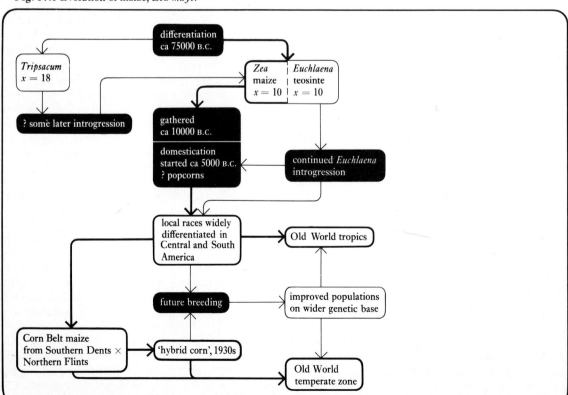

of which they considered 'wild maize' (agreement is far from unanimous on this point but the specimens are clearly very primitive). Some of these cobs have long, soft glumes, supporting to some extent the second part of the 'tripartite hypothesis': that modern maize is descended from pod corn, the most common form of which is determined by a series of alleles at the *Tu* (tunicate) locus on chromosome 4. In the later levels, larger cobs with firmer glumes were uncovered; it was hypothesized that these cobs represented maize which had introgressed with teosinte, from experimental evidence that teosinte introgression resulted in greater induration of the rachises and lower glumes of the cobs. This evidence has been considered to support the third part of the 'tripartite hypothesis': that teosinte contributed to the evolution of modern maize by widespread introgression.

In South America, archaeological evidence on domestication is relatively scanty and is mostly limited to the dry coastal areas of Peru. The earliest materials date to about 1000 B.C. Complete ears dated at about 500 B.C. are clearly similar to Andean races still found in Peru and Bolivia and are quite distinct from current or archaeological Mexican maize.

At the earliest levels of domestication, it appears that kernel size was small; thus it is believed that the earliest maize was a popcorn. Later, larger-kernelled types of maize appear. On the basis of the variability of currently grown maize races, these appear to differ from the popcorns in the constitution of the endosperm of the grain. The endosperm is the portion of the maize kernel surrounding the embryo and enclosed by the pericarp, the external layer of the kernel. Flint maize differs from popcorn mostly in having larger kernels but, unlike popcorn, there is often a small amount of soft, floury, opaque tissue near the centre of the kernel. The floury corns have mostly soft, opaque, floury endosperm tissue, perhaps with a thin layer of evenly distributed hard endosperm around the surface. These types of maize have smooth, either rounded or pointed, kernels. Two other common types of maize, dent corn and sweet corn, do not. Dent corn has a central core of soft starch which shrinks more in drying than the surrounding hard endosperm, resulting in a dented appearance of the kernels. In sweet corn, the carbohydrates are largely stored as sugars rather than starch, so the kernels are heavily wrinkled and translucent after drying.

The size of the kernel depends mainly upon the size of the endosperm. The smallest popcorns have kernels measuring only several mm in length, width and thickness. The largest Cuzco Gigante kernels are about two cm in length and width and perhaps one cm

in thickness. Colouration is found in the endosperm (white, yellow, orange), in the aleurone (the thin outer layer of the endosperm, which may be colourless, purple, red, lemon-yellow or brown) and in the pericarp (colourless, various pinks, reds, yellows, browns and purples). The aleurone and/or pericarp may be patterned (stippled, speckled, dotted, striped, streaked, etc.). Although most commercial maize is either yellow or white, flint or dent, much of the maize cultivated by American Indians is highly coloured and also floury in texture.

From agrobotanical studies of variability (summarized by Brandolini, 1970; Mangelsdorf, 1974), it appears that Mexico and/or lowland Central America (Fig. 37.2) is the centre of variability for commercially important dent types. These forms have spread around the tropics since A.D. 1500. Derivatives of the Mexican dents apparently spread into the southern United States shortly before, or after, colonization of that region. In the mid-1800s these dents were crossed with flints indigenous to the northern USA by midwestern farmers, who often replanted poor stands of late-maturing southern dents with the early-maturing flints. Among the descendants of these crosses were the Corn Belt dents, upon which most of the world's corn production is now based. The Corn Belt dents are well

Fig. 37.2 Geographical distribution of principal races of maize, *Zea mays* (much simplified).

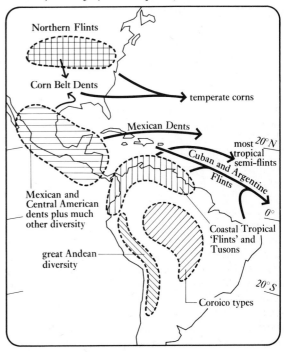

adapted to southern Europe, where they have been widely used in recent years.

The Cuban and Argentine Flints, or Catetos, were apparently scattered along the Atlantic coast from the Caribbean to Argentina after 1600. These yellow to orange flints have had several different suggested origins but there is neither archaeological nor historical evidence for their occurrence anywhere within the Atlantic coastal area until the 1800s, while fairly clear descriptions of other types of maize for the same region date back to about 1500. The true flints which are used throughout the tropics and subtropics usually trace to this source of germplasm. In contrast, the flints of the temperate zones often trace their origin back to the northern flint and flour corns of the USA, whose precise origins are equally obscure. These do not resemble closely either the 8-rowed Mexican or Guatemalan races which are occasionally cited as being ancestral to them.

The northern edge of South America and the Caribbean appears to encompass the centre of diversity for the Coastal Tropical Flints, a group of semi-flints, which, together with their close relatives, the Tusóns, a group of cylindrical flinty-dents from the same region, appear to have arisen as a result of crosses between dents of Mexican origin and indigenous flints from somewhere in the Caribbean region. The time and place of origin of this type of maize is uncertain.

Several other centres of variability (Fig. 37.2) exist in Latin America but materials from them have been of less world-wide importance than those described above. In northwestern Mexico there are a number of long-eared, flinty and floury races of maize. These often have low row numbers and ears that taper at both ends. At high elevations (ca 2000 m) in central Mexico a series of maize varieties with conically shaped ears and narrow kernels predominates. The plants are heavily pubescent, have few tassel branches and tend to root lodge badly. The 'shoepeg' types of maize, once popular in the southern USA, are thought to trace their origin to these corns.

Highland Guatemala is a centre of variability for a series of long-eared flint corns, which appear to extend as far south as the northern Andes. At mid to high elevations in the central Andes is found the greatest single source of variability in maize in the form of the greatest arrays of kernel, cob and plant colours and kernel sizes. While floury endosperm predominates, flints are also present. Dents are rare and apparently introduced. Ears are generally grenade-shaped, with size varying inversely with altitude. East of the northern and central Andes, in the Amazon basin and surrounding lowlands, a single type of maize pre-

dominates. Typically, it has long, narrow ears with bronze-coloured, floury kernels. Its most distinctive feature, however, is the modification of the usual paired arrangement of kernel rows so that kernels from what would ordinarily be adjacent rows are positioned alternately in the same row, somewhat as bricks are overlapped in a wall. This results in a lower row number (as low as half that which would occur without such overlapping) and an increase in ear length.

The development of these centres of variability in the New World parallels in many ways the development of American Indian civilizations and the spread of intensive agriculture. The Mexican and lowland Central American dents appear to have been associated with the Mayan civilization, while the conical corns from higher elevations in central Mexico appear to have been associated with the Aztecs and their predecessors. The variability of maize in the central Andes correlates with the extensive agricultural development of the Incas and their predecessors. The long-eared Guatemalan and North Andean flints may have been spread by the Chibchan culture. It is not known whether the semi-flints and flinty-dents of the Caribbean islands and the northern edge of South America were spread by the Spanish or were already present at the time the first Europeans arrived.

It seems apparent that the major portion of the variability now found in maize developed before 1500; however, several of the most widely grown races, notably Corn Belt Dent, developed later. The large differences between the types of maize found in the various centres of variability have led several students to suggest the hypothesis that the domestication of maize occurred independently in different regions from different types of wild maize. Mangelsdorf (1974) has further postulated that six still-living races of maize represent products of such independent domestications. While conclusive archaeological evidence for this hypothesis has not appeared, the earliest known South American maize, dated at about 1000 B.C., differs substantially from any recovered at Tehuacán. Since 1500, secondary centres of variability have developed, especially in regions in which maize has undergone widespread use, such as the north central USA and southeastern Europe. In many respects, most maize from the Caribbean region and the northern coast of South America also represents secondary variability as a result of the trading activities of the early colonists.

The agricultural practices of American Indians at the time of Columbus were described by Weatherwax (1954). The most common form was the milpa system. Forest was cleared by the slash and burn method still used in many tropical areas today. Crops, which usually included maize, squash and beans (thus insuring a relatively balanced diet), were grown on such land for about three years. As yields decreased and weed problems increased such plots were abandoned, for perhaps a decade or more, before being cleared again.

Fertilizers were rarely used, except in New England, where a fish was added to each hill of corn, and in Peru, where guano was collected from coastal islands. In Peru, terracing and irrigation reached levels not encountered elsewhere in the New World. The terracing systems constructed at that time are still used by Peruvian Indians today.

In the dry areas of the Atacama desert of northern Chile, and in the south-western part of what is now the USA, specialized methods were used to grow maize. Thus, in northern Chile, sunken field farming was practised. Sand was cleared down to a level at which soil moisture was present, then crops were planted. Such areas usually averaged about 0·2 ha, although some exceeded 1 ha. In the southwestern USA, the Hopi Indians, even today, dig widely spaced holes several metres apart, plant 10 to 12 kernels of maize per hole, cover it shallowly until it germinates and continue adding soil around the plants as they grow.

In general, the American Indians who practised maize agriculture appear to have been very careful in the selection and maintenance of their varieties, even when growing several varieties in the same area. Since maize seeds usually lose viability in about three to five years, maintenance of varietal distinctness required continued diligence, especially with a wind-pollinated crop.

The areas in which maize growing was most important at the time European colonization began differ greatly from those in which most of its production is centred today (see Weatherwax, 1954; Fig. 18). The crude tools available to the American Indians did not enable them to farm the grassy plains of the central USA or the pampas of southeastern South America. In those regions, agriculture, if practised at all, was largely limited to the flood-plains of streams.

4 Recent history

From the time of the colonization of the Americas until the mid-1800s, there was little, if any, formal breeding of maize. The European settlers accepted local American Indian varieties or planted similar varieties from neighbouring settlements and seed was saved from crib-run material. In Europe, tropical varieties, apparently introduced from the Caribbean, were widely used in the south and southeast.

In the 1800s maize growers in the USA began to show their products at various fairs and exhibitions. This was the beginning of the corn show era, which ended early in the 1900s. In the corn shows, emphasis was placed upon uniformity of the sample of ears entered and conformity to an 'ideal' type of Corn Belt dent ear. The development of the Corn Belt dents from crosses between the southern dents and northern flints was sketched above. The southern dents were tall, late maturing, soft-kernelled, non-tillering, and many-rowed, frequently with tapering ears. The northern flints were largely eight-rowed, early maturing, short in stature, tillering, and prolific (many-eared), usually with cylindrical ears. Largely as a result of the influence of the corn shows, the Corn Belt dents were selected toward singly- and cylindrically-eared, tillerless plants. The emphasis of the shows on uniformly large ears resulted in widescale elimination of prolific and tillered plant types which tended to produce more but smaller ears. The publicity which accompanied such shows ensured widespread use of specific open-pollinated varieties of corn across the US Corn Belt and abroad. Toward the end of this era (about 1875–1910), several relatively independent experiments were conducted which were later to have dramatic effects upon corn growing throughout the world. The first of these were the observations, by Charles Darwin in England and by Beal at Michigan State, of hybrid vigour (heterosis) in varietal crosses of maize. Such heterosis often resulted in yield increases of about 20 per cent. Somewhat later, the concept of yield-testing the various corn show entries to determine whether the judges' rankings corresponded to the relative yields of the offspring seems to have been developed, first in Iowa. Again, differences of the order of 20 per cent, and sometimes greater, were found. As might be expected, the judges' choices were often poor yielders, as were seedstocks supplied by commercial seedsmen. At about the time that yield-testing began, interest developed in inbreeding corn. The resulting inbred lines were weak and poor yielding, but East at Connecticut and Harvard, and Shull at Cold Spring Harbor, New York, found that crosses between two inbred lines often produced uniform, F_1 (single-cross) hybrids which were superior in yield to any open-pollinated varieties of the time. The early inbred lines were so unproductive that their use for seed production seemed impracticable. It remained for D. F. Jones at Connecticut in 1918 to report a cross of two different single-crosses (e.g. $(A \times B) \times (C \times D)$) which resulted in an economically feasible, reasonably uniform, and highly productive 'double-cross' hybrid. Within fifteen

133

Maize

years, double-cross hybrids became economically important and, by the early 1950s, essentially all the maize grown in the US Corn Belt was from double-cross hybrid seed.

The three leading sources of inbred lines for double-cross hybrids were three open-pollinated varieties. Reid's Yellow Dent, developed by the Reid family of Illinois in the late 1800s, won a prize at the 1893 Chicago World's Fair. It later spread to Iowa and Indiana and was widely used in subsequent corn breeding studies, one widely used strain of it being Funk's Yellow Dent. A second source of useful inbreds was the variety Krug, developed by George Krug of Illinois, from a combination of a strain of Reid's and another variety. The superiority of Krug to other open-pollinated varieties was discovered as a result of yield-tests conducted in the early 1920s and the variety was immediately incorporated into the breeding programme of Lester Pfister, one of the early commercial producers of hybrid corn seed. The third source of useful inbred lines was the variety Lancaster Sure Crop, developed by Isaac Hersey in Lancaster County, Pennsylvania. This variety's high yields impressed F. D. Richey of the US Department of Agriculture, who incorporated it in early inbreeding programmes. For many years, inbreds from Krug and Lancaster were frequently crossed with inbreds of Reid origin in the production of commercial hybrids (Wallace and Brown, 1956).

After the late 1950s, more and more US Corn Belt farmers have planted single-cross, rather than double-cross, hybrid seed. Because the single-cross seed must be produced on an inbred line, the kernels are less uniform and more expensive than double-cross seed. However, under most conditions, the best single-cross hybrids outyield the best double-crosses. What has made single-cross seed economically feasible are the extensive breeding programmes which have been almost exclusively aimed at the development of new inbred lines by selfing and subsequent selection of the descendants of crosses between older, élite, inbred lines. This type of selection programme has resulted in a new generation of inbreds which are not only themselves higher yielding than were their predecessors but which also yield better in hybrid combinations. It has also resulted in a marked loss of variability in breeding materials, which may increase the vulnerability of our maize crop and hamper the progress of future breeding programmes. Outside the US Corn Belt, where production of seed on inbred lines is difficult, or in the Corn Belt when poorly productive inbred lines must be used, three-way crosses are now being widely grown. These are crosses between a

single-cross (female line) and an inbred male (e.g. $(A \times B) \times C$, where A and B may themselves be closely related to each other).

Production of hybrid seed requires crossing of inbred lines (or single-crosses) in isolation from other maize and detasseling of female plants to prevent self-pollination. Initially, detasseling was done by hand prior to pollen shedding. In the late 1940s, however, a stable type of cytoplasmic male sterility was discovered in Texas material derived from the variety Mexican June. Thus, by backcrossing inbred lines to this source of sterility, it was possible to obtain male-sterile inbred lines for use in seed production and avoid at least a portion of the detasseling effort by blending seed produced by such lines with normal seed. Soon, dominant restorer genes were discovered which restored male fertility to plants carrying the Texas source of cytoplasmic male sterility. In 1956, a US patent on the use of such restorer genes was issued (Mangelsdorf, 1974). By incorporating restorers into inbred lines used as males, the seed produced would be heterozygous and plants developing from them would shed normal pollen. These discoveries came at a time when labour costs were increasing and labour supplies decreasing; hence industry quickly adopted them. By the late 1960s, essentially all US production was based upon male-sterile lines using the Texas form of cytoplasmic sterility, except for a few experimental hybrids. This further restriction on genetic variability produced in the USA (and in a number of other countries) became very apparent in the summer of 1970, when a mutant form of the southern leaf blight fungus, *Helminthosporium maydis*, Race T, spread northwards across the USA at a rate of about 150 km per day. It attacked all hybrids containing Texas cytoplasmic male sterility.

One of the major changes in maize breeding over the past 25 years has been the emphasis, chiefly by researchers trained in the area of quantitative genetics, on development of improved maize populations, rather than on the immediate production of improved inbred lines. Various breeding methods such as mass, recurrent and reciprocal recurrent selection procedures have been used to improve population performance before extraction of inbred lines is begun. Perhaps the outstanding single example of such work, on a practical scale, is that of Jenkins, Sprague, Russell and others at Iowa State University with the Stiff Stalk Synthetic and the outstanding inbred lines extracted from it.

5 Prospects

One recent discovery promises to be extremely important from the standpoint of human and animal

nutrition. In addition to being relatively low in protein content, maize is low in two essential amino acids, lysine and tryptophane. However, Mertz, Nelson and their co-workers at Purdue University found that the *opaque–2* mutant increased significantly the amounts of these amino acids in maize kernals. Shortly thereafter several other mutations were found to have similar effects (including one in *Sorghum*). Some of these genes have been incorporated into a few commercial hybrids, despite problems with kernel quality and appearance. Tests on children suffering from acute protein malnutrition indicate that some of these high lysine corns can be used as a complete diet for such children, with subsequent recovery. The possible importance of such materials in areas where the diet is largely maize can readily be seen, despite the fact that Katz *et al.* (1974) have recently indicated that the quality of essential amino acids in maize can be, and often is, at least in Latin America, improved by processing. Specifically, they indicate that cooking maize in an alkaline solution, such as boiling in a lime (or calcium hydroxide) water solution, enhances its nutritional quality with respect to protein, despite a lowering of overall nutrient content.

A change in breeding methods, in part connected with emphasis upon population improvement rather than the production of inbred lines, appears to be developing. A few maize breeders now make crosses between individual plants, while at the same time selfing those plants. Others cross plants on to testers (single-crosses, inbred lines, etc.), while at the same time selfing the individual plants. However, rather than just intercrossing the selfed seed of the best crosses, as has been done for years with various recurrent selection schemes, a few breeders, especially those interested in rapid production of material for new markets, have increased the selfed seed of the parents of the best test crosses and used such seed actually to produce a hybrid between two such slightly inbred lines or between an established inbred or single-cross and such a rapidly developed, slightly inbred line. While such rapidly developed hybrids lack uniformity, the rapidity with which they can be developed and the flexibility they allow the breeder may compensate for lack of uniformity, which is often not critical. The production of experimental hybrids traditionally takes at least seven seasons and usually more: five seasons of selfing (with selection among and within progenies), at least one season of test crossing, then one season of inter-crossing the best lines with themselves and established inbreds. In contrast, the procedure outlined above takes only two to four seasons to arrive at experimental hybrids.

Such rapid breeding systems may become increasingly important, not simply because of the speed and economy with which they can be implemented, but because of necessity. Most cultivated cereals have been plagued for decades, if not for centuries, by diseases (rusts, wilts, blights, smuts, etc.) which have required constant, almost full time, attention from their breeders. Until recently this has not been the case with maize. However, various forces have acted over centuries to reduce the variability, not of maize as a species (there is still much variability available, even if rarely used), but of maize as a cultivated plant important in world commerce. These forces have ranged from the hill planting of corn and the influence of corn shows, both of which selected against tillering and prolificacy, to the production of single-cross hybrids by the few dominant commercial producers of hybrid seed. Most of the world's maize production is based upon derivatives of the Corn Belt dents. These in turn were derived from crosses between only two of the more than 200 races of New World maize. Furthermore, it appears that over 70 per cent (and this figure seems to be increasing) of the hybrid corn seed produced in the US is currently based on no more than a half dozen inbred lines (and, unfortunately, even this figure seems to be decreasing), some of which were derived from the same base population (Committee on Genetic Vulnerability, 1972). The situation in Europe is perhaps not so extreme, but much of the production there is based upon derivatives of the same materials, and there has often been actual exchange of inbred lines with the USA.

Use of sources of tropical germplasm in temperate (long day) environments to increase available genetic diversity has been dramatically hindered by the photoperiodic response of such material. Typically, it requires a day length of 12–14 hr to produce tassels and ear shoots; hence tropical material does not flower in temperate regions until late fall, at plant heights of about 6 m. Those tropical materials which do appear to mature normally in temperate regions are usually extremely early under conditions to which they are adapted, with the usual limitations on yield and plant strength that extreme earliness implies. Until breeders make more effort to use tropical materials with normal maturity (under short day conditions), little progress can be expected from efforts to incorporate tropical materials into breeding programmes outside the tropics. Limited investigation indicates that the photoperiodic response in maize is reasonably simply inherited, but it seems to have effectively excluded many possibly useful sources of variability for the past fifty years.

135

Actually, knowledge of the variability available outside the USA is limited, despite extensive collecting and despite the fact that impressive descriptions of the races of maize have been published. Many, perhaps most, of the 12,000 Latin American collections assembled under the sponsorship of the Rockefeller Foundation and the National Academy of Sciences of the USA during the late 1940s and 1950s may still be available. Much effort is being devoted to salvaging these collections, particularly by Dr Mario Gutiérrez of CIMMYT in Mexico; this effort merits more support. From these collections, about 250 races have been described but few have been carefully studied (Sprague, 1975) and those few men who have been involved in such studies are either in or near retirement or occupied by administrative or other duties. Our knowledge of maize in other areas of the world is even more fragmentary. One area of inquiry showing promise of clarifying the evolution of the races of the crop is the study of chromosome knobs by Dr Barbara McClintock of the Carnegie Institution, Cold Spring Harbor, New York, and her colleagues, Kato of Mexico and Blumenschein of Brazil. They are attempting to interpret maps of the frequencies of specific chromosome knobs of different sizes to determine direction(s) of migration of such knobs. Similar work is being attempted in several laboratories using electrophoretically detectable isozyme frequencies.

Finally, in the area of plant breeding, it is apparent that development of breeding materials needs to be coordinated more closely with agronomic studies of the best methods of growing maize. While results for maize will likely not be as spectacular as the Green Revolution with rice and wheat, much of the yield increase which has occurred since the introduction of hybrid corn has been due not only to genetic improvement of yield *per se* but also to improved standability, introduction of fully mechanized harvesting, better fertilization practices and higher population densities. Improved biochemical screening techniques, perhaps ultimately combined with tissue culture, will probably eventually lead to laboratory- (rather than field-) conducted selection for such traits as disease- and insect-resistance.

6 References

Brandolini, A. (1970). Maize. In Frankel, O. H. and E. Bennett (eds), *Genetic resources in plants—their exploration and conservation.* Philadelphia, 273–309.

Committee on Genetic Vulnerability of Major Crops (1972). *Genetic vulnerability of major crops.* National Academy of Sciences, Washington, D.C., pp. 307.

Galinat, W. C. (1971). The origin of maize. *Ann. Rev. Genet.*, 5, 447–78.

Galinat, W. C. (1973). Intergenomic mapping of maize, teosinte, and *Tripsacum. Evolution*, 27, 644–55.

Katz, S. H., Hediger, M. L. and Valleroy, L. A. (1974). Traditional maize processing techniques in the New World. *Science, N.Y.*, 184, 765–73.

Mangelsdorf, P. C. (1974). *Corn its origin, evolution and improvement.* Cambridge, Mass.

Randolph, L. F. (1970). Variation among *Tripsacum* populations in Mexico and Guatemala. *Brittonia*, 22, 305–37.

Sprague, G. F. (ed.). (1976). *Corn and corn improvement.* 2nd edn, Agronomy Society of America, Madison, Wis.

Wallace, H. A. and Brown, W. L. (1956). *Corn and its early fathers.* East Lansing, Mich.

Weatherwax, P. (1954). *Indian corn in old America.* New York.

Wilkes, H. G. (1967). *Teosinte: the closest relative of maize.* Bussey Institution, Harvard University, Cambridge, Mass.

Temperate grasses

Lolium, Festuca, Dactylis, Phleum, Bromus
(Gramineae)

M. Borrill
Welsh Plant Breeding Station
Aberystwyth Wales

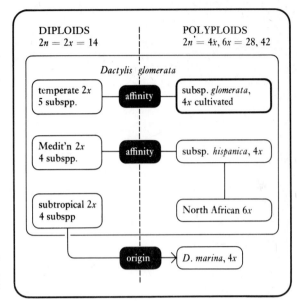

Fig. 38.1 Relationships in the *Dactylis glomerata* complex.

1 Introduction

The rôle of herbage grasses is to provide a source of digestible energy and other nutrients for the ruminant animal. Grown intensively, an average conversion of up to 3 per cent of the incoming light energy is possible, with up to 5 per cent during the most productive part of the growing season, resulting in a potential dry matter production of over 20 t/ha–yr under British conditions (Cooper, 1970). Fully utilized, this level of production would provide over 50 G cal digestible energy per hectare, enough for the requirements of over four dairy cows each giving about 15 kg of milk per day.

This level of production is rarely attained but sown (as distinct from natural) grassland is of great agricultural importance in certain areas within the temperate zone; for example, in western Europe, North America and New Zealand. It is perhaps well to recall that natural and semi-natural grasslands still dominate the world scene. Sown grass land is a relatively recent agricultural concept.

Seven species are to be considered, as follows:

Table 38.1 Temperate grasses.

Common name	Species	$2n =$
Perennial ryegrass	*Lolium perenne*	$2x = 14$
Italian ryegrass	*Lolium multiflorum*	$2x = 14$
Meadow fescue	*Festuca pratensis*	$2x = 14$
Tall fescue	*Festuca arundinacea* var. *genuina*	$6x = 42$
Cocksfoot	*Dactylis glomerata*	$4x = 28$
Timothy	*Phleum pratense*	$6x = 42$
Smooth brome	*Bromus inermis*	$8x = 56$

Three phases can be recognized in their evolution: as wild plants in Europe and western Asia; as camp-followers adapting themselves in varying degree to the disturbed habitats created by early agriculture; and, very recently, as cultivars adapted by modern plant breeding to intensive agriculture.

2 Cytotaxonomic background

Cocksfoot (*Dactylis glomerata*, Fig. 38.1) is a good example of a polyploid complex. It consists of a wide ranging Eurasian assemblage of tetraploids, amongst which enclaves of diploids occur. Wild hexaploids are found in parts of North Africa. Fertile hybrids are easily obtained: between ecotypes within species; between tetraploid subspecies; and, to a large extent, between the diploids. The main barrier to crossing is between diploid and tetraploid, in which 1 per cent triploids are formed and about the same number of tetraploid hybrids by heteroploidy, usually by the functioning of unreduced (diploid) gametes (Carroll and Borrill, 1965). Introgression from diploid to tetraploid can occur by direct crossing or by way of the partially female-fertile triploids. In the diploids, bivalent pairing is usual, whereas the tetraploids show enough quadrivalent formation to indicate autopolyploidy.

The common widely distributed subsp. *glomerata* was the only agriculturally important component of the complex until recently, when breeders started to use diploid subspecies and *D. marina* (Borrill *et al.*, 1972).

Tall fescue (*F. arundinacea* var. *genuina*, $6x$) is one of several chromosome races of *F. arundinacea*. An interpretation of relationships between these races and

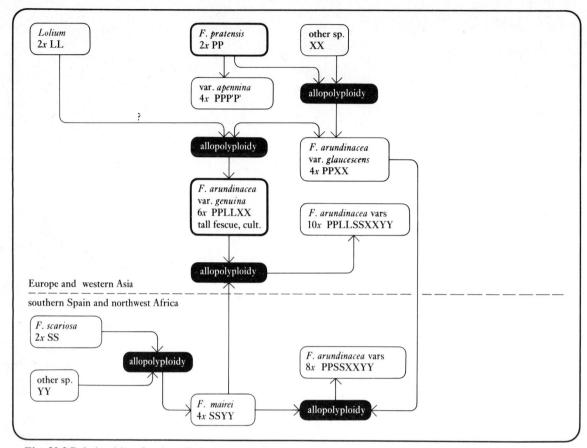

Fig. 38.2 Relationships of various *Festuca* spp. and *Lolium*.

other species of broad-leaved fescues is given in Fig. 38.2. There are three points to be made. First, as in cocksfoot, the common wild fescues with the greatest geographical distribution are polyploid but in *Festuca* they are high polyploids ($6x$ and $10x$). Second, unlike cocksfoot, the fescues have few diploids and this makes for difficulty in analysing the parentages of polyploid forms such as the $6x$ tall fescue. Cytological analysis is complicated by a high incidence of bivalent pairing in the polyploids, suggestive of some genetic control. The third point concerns low crossability and hybrid infertility. This occurs not only in hybrids between chromosome races but also between species of the same chromosome number. Furthermore, within $6x$ tall fescue, although ecotypes from different parts of the geographical range can be crossed easily, some combinations give infertile hybrids.

The relationship of *Lolium* to *Festuca* must be considered (Figs. 38.2 and 38.3), because the three outbreeding *Lolium* species (Fig. 38.3), which are interfertile diploids, hybridize with $2x$ meadow fescue, with

$6x$ tall fescue and with other 'wild' fescue species to about the same extent as do the fescues with one another. Hybrids are largely infertile. The outbreeding *Lolium* species are very closely related to $2x$ *F. pratensis*. Cultivated tall fescue (*F. arundinacea* var. *genuina*, $6x$) is a fertile allohexaploid having $2x$ *F. pratensis* as one parent and probably containing a genetic contribution from *Lolium*.

The very important ryegrasses, *L. perenne* and *L. multiflorum* (Fig. 38.3), are diploid and they cross freely, giving fully fertile hybrids. The inbreeding *Lolium* species are probably derived from the outbreeding group. There are no wild polyploids in *Lolium* (Terrell, 1966). However, many fertile auto- and allotetraploids have recently been produced in ryegrass breeding (see below). Numerous artificial hybrids have been produced between meadow fescue, tall fescue and the ryegrasses. Most probably, the wild polyploid derivatives of ryegrass are to be found amongst the broad leaved fescues. Artificial 'intergenerics' have recently been bred (see below).

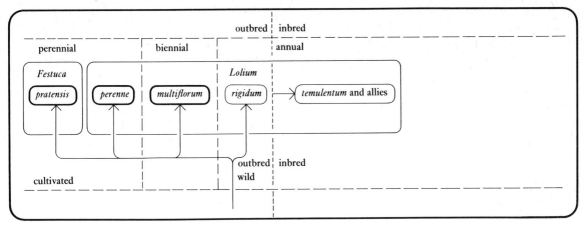

Fig. 38.3 Relationships of *Lolium* and *Festuca pratensis*; all $2x = 14$.

Timothy (*Phleum pratense*, $6x$) belongs to a polyploid series containing diploid and tetraploid species. The evidence suggests that it is an autohexaploid of diploid *P. bertollonii* (previously known as *P. nodosum*). The group has been reviewed by Nath (1967).

The main features of smooth brome grass (*Bromus inermis*, $8x$) are summarized in Fig. 38.4. The genus *Bromus*, like *Festuca*, contains many complex, polyploid series. *Bromus inermis* belongs to section Bromopsis, containing wild species with chromosome numbers ranging from $2x$ to $8x$. Little is known about their relationships. Following its introduction to North America, smooth brome, an aggressive species, has tended to replace native American species of sect. Bromopsis, especially *B. pumpellianus*. Hybridization studies have been made (Elliott, 1949; Hanna, 1961)

Fig. 38.4 Evolution of introduced *Bromus inermis* in North America.

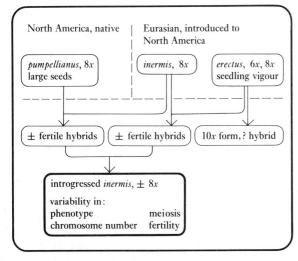

between: *B. inermis*, $8x$; *B. pumpellianus*, $8x$; and material of the introduced *B. erectus* complex ($6x$, $8x$, $10x$, though the last is probably itself hybrid). An objective of these studies was to incorporate into *inermis* the larger seeds of *pumpellianus* and better seedling vigour of *erectus*. The three species are related, crossable and give quite fertile hybrids. Collections of *B. inermis* in North America show varying meiotic irregularity and aneuploidy. This is probably a result of introgression from *B. pumpellianus* and the *B. erectus* complex (Fig. 38.4). Quite a high degree of meiotic irregularity may be tolerated under natural conditions in a long lived, cross-pollinated species such as *B. inermis*, which can spread aggressively by rhizomes.

3 Early history

The main subdivisions of the grasses, including the groups under consideration here, had probably evolved by the end of the Pleistocene. Ice movements must have been associated with plant migration, opportunities for hybridization and bursts of evolutionary variation. By about 8000 BC the ice sheet was retreating from western Europe, to be succeeded by climax vegetation of mixed oak forest. The temperate herbage grasses now used as sown species, then occupied forest and forest margin habitats; by contrast, the temperate climax grasslands (such as prairie) have yielded no species of importance in cultivated pastures.

Western Europe is a main centre of variation of the tribe Festuceae (which includes five out of the seven species considered) and, in this area, they have had a long association with human settlement (Hartley and Williams, 1956). *Bromus inermis* (tribe Bromeae) is a species of temperate-zone forests in Eurasia. The activities of man as a herdsman and, later, as a cultivator, starting with 'slash and burn' methods, made inroads into the forest from early Neolithic times. The destruction of these forests coincided with the beginning of settled agriculture (Godwin, 1965). The woodland grasses colonized new habitats and expanded into cleared areas. The extent to which such grasses were sought by livestock and evolved further in association with the grazing animal is uncertain (Hartley and Williams, 1956). The evolution of European grassland is discussed by Scholz (1975).

Lolium perenne was perhaps associated (though unintentionally) with agricultural practices from early times and was the first herbage grass to become a crop plant. It has an obvious preference for nitrogen-rich, disturbed habitats. It is reasonable to suppose that it followed the westward migration of Neolithic farmers into Europe (Butzer, 1971), reaching Britain in prehistoric times. It seems to be more a weed of cultivation and less a forest species than the others. At present its survival probably depends on man. Its geographical origin is obscure.

4 Recent history

Measured on the time-scale of most crops, herbage grasses have a very short history of deliberate cultivation. The earliest recorded is Italian ryegrass (*L. multiflorum*) which has been cultivated under irrigation since the late twelfth century in Piedmont and Lombardy. From here it was introduced to other parts of western Europe but did not reach Britain until 1831, when it was imported from Hamburg by Lawson of Edinburgh. It seems that the deliberate use of native grasses began in western Europe in the middle seventeenth century. In Britain, for example, local material of *L. perenne* was grown in Oxfordshire in 1677 and harvesting of seed of the species became a common practice among English farmers in the late eighteenth century. William Pacey was one of the best known; contemporary accounts (1795) refer to his ryegrass and he was also interested in cocksfoot.

Settlers of many nationalities carried the native species of Europe to other continents. Italian and perennial ryegrass, meadow fescue, tall fescue, cocksfoot and timothy were introduced into North America by the middle of the eighteenth century. Increased awareness of the agricultural value of individual species followed and the reputation so gained led to seed (especially of cocksfoot and timothy) being sent to Britain, where they gave impetus to the growing of local native races in the early nineteenth century (Beddows, 1953).

Smooth brome (*Bromus inermis*) was imported into North America in 1880 from Hungary, where it had developed a good reputation as a pasture grass during the preceding 30 years. It satisfied a requirement for drought resistance in the USA, and winter hardy stocks for use in the north were later imported from Siberia. The species is important in eastern Europe and Russia.

Some differentiation of species introduced into new lands must have occurred because: initial samples must have been small and unrepresentative; local selection pressures must have sometimes been disruptive; and, in the case of *Bromus inermis*, hybridization with local species occurred.

Seeds of local races were collected and grown without knowledge of the breeding system or of the need for isolation of seed stocks. Even so, some consistency of type could have been maintained by cyclic sowing and harvesting under constant management systems, as exemplified by twelfth century Italian ryegrass, by the American timothy variety 'Shilby' (grown on one farm from 1855 to 1940) and by the 'Irish perennial' ryegrass which was still widely grown some 25 years ago. A consequence of this system was unintentional selection of high seed yield at the expense of leafiness and persistency.

Deliberate breeding was pioneered in Britain by R. G. Stapledon and his colleagues beginning in 1919. Little was then known about the relative values of different species for livestock production. Local races were collected and evaluated, leading to greatly increased emphasis on ley farming, the replacement of permanent pastures (containing mixed grasses, herbs and legumes) by sown leys comprising grass–legume mixtures of proven yielding ability. It was found that the performance of an ecotype was related to the selection pressure to which it had been subjected and the idea followed that certain growth forms were valuable in relation to particular agronomic objectives; for example, densely-tillered, prostrate ryegrass from old 'fattening pastures' gave high yields under intensive grazing and erect-growing cocksfoots from hedgerows and woodland margins were suitable for conservation.

Breeding systems were also studied. In these outbred grasses, inbreeding depression and heterosis were very evident especially in diploids. Selfing of *Lolium*, for example, is quickly evidenced by the appearance of dwarf, chlorophyll–deficient and other mutant seedlings. With this knowledge, Jenkin (1931) formulated breeding plans based on progeny tests and synthetic varieties. The object was to discover superior individuals from similar ecotypes and growth forms, capable of inter-breeding true to type (phenotypically consistent, narrow range of heading dates) and of giving high yields of leafy herbage and good seed production.

Cultivars produced by these methods outperformed any others available at the time and some are still in use. Essentially similar methods were adopted by breeders in many parts of the world; for example, with tall fescue, cocksfoot and smooth brome in America (Hanson and Carnahan, 1956).

The phase of adaptation of the wild grasses to the environmental demands of settled agriculture reached a conclusion in these synthetic varieties, which are essentially modified versions of local ecotypes. That some evolutionary change occurred as a result of breeding is certain but it was probably small in relation to preceding changes called forth by selection for adaptation to the new environments created by settled agriculture. The degree of change, however, depends upon the species concerned. Thus, cross-fertilized *Lolium* species have evolved towards fitness in artificial habitats, associated with N–rich arable cultivations. Ryegrass is an aggressive species in these circumstances but is unfitted to compete with native species outside them. Deliberate selection for yield under higher N–levels would accentuate this trend. In the other cultivated species, breeding for yield and seed production has not changed the cultivars enough to reduce their fitness outside the farm systems under which they are usually grown. They can still colonize habitats occupied by wild relatives with which they will hybridize in hedgerows, woodland margins and waste places.

5 Prospects

During the last 15 years, new breeding objectives and methods have become apparent, namely: the improvement of conversion efficiency by the ruminant by improving grass quality; and the production of new hybrid grasses (often artificial polyploids), usually combining desirable features of two or more species.

Quality (including intake and digestibility factors) can be estimated chemically and has been for about 10 years. High heritability of digestibility has been shown in several species, including cocksfoot, ryegrass and smooth brome. Rapid responses to selection have been obtained, both for digestibility and for other quality features such as water-soluble carbohydrate content. Such changes can best be regarded as close adaptation to the highly specialized environment of intensive agriculture.

The current trend with the greatest evolutionary significance, however, is towards polyploidy and interspecific hybridity. Induced polyploidy has three aspects: as a means of producing autopolyploid cultivars, as a method of genetic bridging and as a

means of restoring fertility in otherwise useful but sterile hybrids.

Autotetraploid cultivars have been developed in *L. perenne* and *L. multiflorum* and are reproductively isolated from the diploids. 'Gigas' characters associated with increased cell size (and, often, fewer tillers and leaves) are evident. There may sometimes be some loss of vigour, but certain $4x$ ryegrasses are often as vigorous and agriculturally useful as the diploids, or even more so. More generally, however, comparison of autotetraploids with their diploid progenitors shows that polyploidy *per se* confers no overall advantage.

An example of the genetic bridging function is provided by the use of induced tetraploids of *Dactylis glomerata* subsp. *lusitanica* ($2x$) to extend the growth-season of tetraploid cultivars; this is, in effect, an example of introgression and the potential of the technique for grass improvement must be very considerable (Borrill *et al.*, 1972).

Examples of amphidiploids which actually extend the range of species available to agriculture are provided by the *Lolium–Festuca* group. These include: allotetraploid derivatives of *L. perenne* × *L. multiflorum* and of these species crossed with *F. pratensis* (Thomas and Thomas, 1971); and various complex hybrids and backcrosses involving *Lolium* and hexaploid tall fescue (Lewis, 1966, Webster and Buckner, 1971). Synthetic colchiploids of this type represent a substantial evolutionary change and their agricultural potential must, again, be considerable.

6 References

Beddows, A. R. (1953). The ryegrasses in British agriculture, a survey. *W.P.B.S. Bull. Series H*, **17**, pp. 81. Univ. Coll. Wales, Aberystwyth.

Borrill, M. *et al.* (1972). The evaluation and development of cocksfoot introductions. *Ann. Rep. Welsh Pl. Br. Sta.*, 37–42.

Butzer, K. W. (1971). The significance of agricultural dispersal into Europe and northern Africa. In S. Streuver (ed.), *Prehistoric agriculture*, American Museum of Natural History, New York, 313–34.

Carroll, C. P. and **Borrill, M.** (1965). Tetraploid hybrids from crosses between diploid and tetraploid *Dactylis* and their significance. *Genetica*, **36**, 65–82.

Cooper, J. P. (1970). Potential production and energy conversion in temperate and tropical grasses. *Herb. Abstr.*, **40**, 1–15.

Elliott, F. C. (1949). *Bromus inermis* and *B. pumpellianus* in North America. *Evolution*, **3**, 142–9.

Godwin, H. (1965). The beginnings of agriculture in north west Europe. In J. B. Hutchinson (ed.), *Essays on crop plant evolution*, Cambridge, 1–22.

Hanna, M. R. (1961). Cytological studies in *Bromus* species section Bromopsis. *Canad. J. Bot.*, **39**, 757–73.

141

Hanson, A. A. and Carnahan, H. L. (1956). Breeding perennial forage grasses. *U.S.D.A. tech. Bull.*, 1145, pp. 116.

Hartley, W. and Williams, G. L. (1956). Centres of distribution of cultivated pasture grasses. *Proc. 7th intnl Grassld Congr.*, New Zealand, 190–9.

Jenkin, T. J. (1931). The method and technique of selection, breeding and strain-building in grasses. *Imp. Bur. Plant Genet., Herbage Plants Bull.*, 3, 5–34.

Lewis, E. J. (1966). The production and manipulation of new breeding material in *Lolium–Festuca. Proc. 10th intnl Grassld Congr.*, Helsinki, 688–93.

Nath, J. (1967). Cytogenetical and related studies in the genus *Phleum. Euphytica*, 16, 267–82.

Scholz, H. (1975). Grassland evolution in Europe. *Taxon*, 24, 81–90.

Terrell, E. E. (1966). Taxonomic implications of genetics in ryegrasses (*Lolium*). *Bot. Rev.*, 32, 138–64.

Thomas, P. T. and Thomas, H. (1971). Problems of chromosome manipulation in plant breeding. *Ann. Rep. Welsh Pl. Br. Sta.*, 117–22.

Webster, G. T. and Buckner, R. C. (1971). Cytology and agronomic performance of *Lolium–Festuca* hybrid derivatives. *Crop Sci.*, 11, 109–12.

39

Tropical and sub-tropical grasses

Various genera (Gramineae)

J. R. Harlan
University of Illinois Urbana Ill USA

1 Introduction

The establishment of artificial stands for grazing, hay or cut-grass is not generally practised in tropical countries. More often, livestock are grazed on open native or naturalized range or herded along roadways, ditchbanks and waste places. Interest in improved, high-producing pastures is rising, however, as demand for meat and milk continues to grow and prices become more attractive to producers. The most important tropical grasses for these purposes are of African origin and they are most used in tropical America.

2 Cytotaxonomic background

Table 39.1 Tropical and sub-tropical grasses

Common name	Species	x	2n	Origin
Guinea grass	*Panicum maximum*	8	32 16, 40, 48)	East Africa
Elephant grass	*Pennisetum purpureum*	7	28	Wet tropical lowlands of Africa
Kikuyu grass	*P. clandestinum*	9	36	East Africa, highlands
Star grass	*Cynodon nlemfuensis*	9	18, 36	East Africa, savanna
Star grass	*C. aethiopicus*	9	18, 36	East Africa, savanna
Star grass	*C. plectostachyus*	9	18	East Africa, savanna
Bermuda grass	*C. dactylon*	9	18, 36	Asia, Africa, Europe

Most races of guinea grass are tetraploid ($2n = 32$), reproducing by facultative apomixis. Sexual diploid races are known in East Africa (Kenya, Tanzania) and both pentaploid and hexaploid colonies have been reported in Ivory Coast (Combes and Pernes, 1970). Guinea grass is probably introduced to West Africa, however, and the species has not been much studied on a global scale. Improvement by breeding is possible despite apomixis, but little work has been done to date. Unselected clones are most commonly used at present.

Elephant grass is a sexual tetraploid ($2n = 28$) related to pearl millet (*P. americanum*, $2n = 14$). The two can be readily crossed; the hybrid is a sterile triploid with considerable potential as a fodder grass. Relatively little effort has gone into improvement of elephant grass by breeding, but several selected clones are available. Merker grass and Napier grass are names given to certain races or cultivars.

Kikuyu grass is an apomictic tetraploid; the flowers are hidden by sheaths and exert long thread-like stigmas. It flowers very erratically and not at all in some regions. There has been little effort to improve it.

Of the star grasses, *C. plectostachyus* is a sexual diploid ($2n = 18$), genetically well isolated from other species in the genus and easily identified by the very short glumes. The other two species have both diploid and tetraploid races. *Cynodon aethiopicus* can be crossed with *C. nlemfuensis* and *C. dactylon* but the crosses are very difficult to make and the hybrids are sterile. *Cynodon nlemfuensis* is more nearly related to *C. dactylon* and crosses between the two are easy to produce. One such hybrid, 'Coastcross 1', has given a good performance in the southeastern USA (Burton, 1973; Harlan, 1970).

C. dactylon has both diploid and tetraploid races. The widespread weedy races are all tetraploid. Diploid races are found from South Africa to Sudan, Egypt and Palestine, eastward to Afghanistan and southward to southern India. Diploids have been introduced to Hawaii and Arizona, at least, where they have become naturalized, otherwise they appear to be endemic to the region indicated. The species of *Cynodon* are all sexual.

3 Early history

C. dactylon is the only species that appears in early history. It was, and is, a sacred grass to the Hindu because of its support of cattle and appears in Graeco-Roman pharmacopoeias. The rhizomes were considered to have medicinal value. Expressed juice is used as an astringent to stop bleeding and as a diuretic.

4 Recent history

Guinea grass is a tufted perennial adopted to fairly high rainfall (1,400 mm or more). It is now widely naturalized throughout the warm, wet tropics, and artificial stands are used for pasture especially in Latin America. Despite its weedy nature and currently wide distribution, it is a shy seeder and stands are usually established clonally. Quality is considered good for a tropical grass and it is relatively easy to manage. Guinea grass appears to have originated in East Africa but we know little of its subsequent diffusion through the tropics.

Elephant grass is a native of Africa where it grows wild along streambanks and in low spots of the moist savanna and forest margins. It is weedy and now widely distributed throughout wet tropical lowlands. Seeds are contained in fluffy fascicles, easily distributed by wind. Seed production, however, is low and stands are usually established by cuttings. Elephant grass is a tall perennial with vigorous, creeping stolons adapted to open sites in the wet tropics. It can be grown with only 1,000 mm rainfall, but does much better with more. It has a very high yield potential but, if cutting regimes are adjusted to maximum yield, the forage becomes very coarse. It is difficult to manage as a pasture, although it is possible to do so. More often, it is harvested as a cut-grass for feeding green or for silage. This grass is likely to play an important role in any appreciable livestock expansion in the wet tropics.

Kikuyu grass is a low, sod-forming perennial, creeping by both rhizomes and stolons. It is native to the cool East African highlands and has been introduced to South Africa, Sri Lanka (Ceylon) and tropical America. It has made an important contribution in helping to stabilize eroded overgrazed ranges in the Andes and other mountainous areas.

The star grasses are creeping perennials, spreading by seed and stolons but lacking rhizomes. They are adapted to savanna conditions with fair rainfall (800–1,200 mm) but with a pronounced long dry season. They are native to eastern Africa from Ethiopia to South Africa. The three species have often been confounded and it is difficult or impossible to sort out the species actually involved in the older literature.

C. plectostachyus has a restricted distribution associated with the Rift Valleys from Ethiopia to Tanzania. It has not given a good performance outside this region, primarily because it is slow to recover from grazing or cutting. *Cynodon aethiopicum* is native from Ethiopia to Transvaal and is one of the most abundant of the Cynodons of East Africa. It is tough, hard-stemmed and not noted for quality, but much used as a native grass and has given good performance in West Africa. *Cynodon nlemfuensis* ranges from Ethiopia to Zambia and westward to Angola. It is softer, more palatable and has higher digestibility than other species of *Cynodon*. An improved selection '1B–8' has been released in Nigeria.

C. dactylon, called variously Bermuda grass, couch grass, dog's tooth, devil grass and other less complimentary names, is now one of the most widespread plants in the world. Unlike the previous species, it has vigorous rhizomes (one race from Afghanistan and one race from Malagasy excepted) and can become a serious pest of cultivated fields. It is so common and widespread that it is difficult to trace its ultimate origin. Variation patterns suggest that the most vigorous and aggressive of the hardy races can

143

be traced to southwestern Asia, from Turkey to Afghanistan. These are sufficiently winter hardy to penetrate to the latitude of Berlin and are widely naturalized in North America south of latitude 40°N. The tropical weedy races probably spread from both India and Africa.

Bermuda grass is easy to manage and can tolerate enormous abuse. It is the basis of village grazing grounds from the Mediterranean to eastern India. It has played an important role in reducing erosion on overgrazed ranges in the Andes and has rendered enormous service around the world in stabilizing roadsides, ditches, tanks and levée embankments. It is commonly used as a turf grass in tropical and sub-tropical climates. Productive, high-yielding, selected cultivars are, at present, largely confined to the USA, but these have been enormously important in the agricultural economy of the southeastern states (Burton, 1973).

Other warm-country grasses of some importance and their origins are indicated below:

Table 39.2 Minor tropical and sub-tropical grasses.

Common name	Species	x	2n	Origin
Pangola grass	Digitaria decumbens	15	30	Southeast Africa
Weeping love grass	Eragrostis curvula	10	40, 50	Southeast Africa
Rhoades grass	Chloris gayana	10	20, 40	East Africa
Buffel grass	Cenchrus ciliare	?	32, 34, 36, 40, 54	South Africa
Molasses grass	Melinis minutiflora	9	36	Tropical Africa
Bahia grass	Paspalum notatum	10	20, 40	Eastern South America
Dallis grass	Papalum dilatatum	10	40	Eastern South America

5 Prospects

The success of Bermuda grass as a cultivated plant in the southeastern USA has encouraged its use elsewhere. It not only has a high carrying capacity and can tolerate abuse, but is capable of improvement in both yield and quality. Cynodon nlemfuensis is the main source of better quality and crosses between races produce heterosis and higher yield. The plant holds great potential for warm-temperate or subtropical zones.

The quality of elephant grass can probably be improved by crossing with pearl millet and there is no doubt that guinea grass can be improved by breeding. Improvement programmes have not been under way long enough to have had much impact as yet, but a strong trend towards more intensive use of pastures in the tropics is well started and the grasses described here will all play a part.

6 References

Burton, G. W. (1973). Bermuda grass. In M. E. Heath, D. S. Metcalfe and R. E. Barnes (eds), Forages, Ames, Iowa, USA, 3d edn, pp. 755.

Combes, D. and Pernes, J. (1970). Variation dans les nombres chromosomiques du Panicum maximum en relation avec le mode de réproduction. C. R. Acad. Sci., Paris (Série D), 270, 782–5.

Harlan, J. R. (1970). Cynodon species and their value for grazing and hay. Herbage Abstr., 40, 233–8.

McIlroy, R. J. (1972). An introduction to tropical grassland husbandry. London, 2nd edn, pp. 160.

Narayanan, T. R. and Dabadgho, P. M. (1972). Forage crops of India. Indian Council Agric. Res., New Delhi, pp. 373.

Semple, A. T. (1970). Grassland improvement. Cleveland, Ohio, USA, pp. 400.

Vicente-Chandler, J., Caro-Costas, R., Pearson, R. W., Abruna, F., Figarella, J. and Silva, S. (1964). The intensive management of tropical forages in Puerto Rico. Univ. Puerto Rico agric. Exp. Sta. Bull., 187, pp. 152.

Whyte, R. O. (1964). The grassland and fodder resources of India. Indian Council Agric. Res. Dalip Singh, New Delhi, pp. 553.

Currants

Ribes spp. (Grossulariaceae)

Elizabeth Keep
East Malling Research Station
Maidstone Kent England

1 Introduction

Black and red currants are vegetatively propagated shrubs grown on a limited scale in most cool temperate countries. The major producers are the USSR and eastern and northwestern Europe, particularly Britain, West Germany and Poland. Much of the east European crop is exported to western Europe. Currants are little grown in North America since, as alternate hosts of white pine blister rust (*Cronartium ribicola*), they are banned wherever five-needled pines are an important crop.

The bulk of the currant crop is processed, particularly for juice, but also for canning, jamming, liqueurs, pie fillings and pastilles. The black currant is especially valued for the high vitamin C content of its products.

2 Cytotaxonomic background

The genus *Ribes* comprises some 150 species, all diploid ($2n = 2x = 16$), distributed mainly in the temperate regions of Europe, Asia and North and South America. Classification based on morphology into subgenera and sections or series has been remarkably successful in delineating the possibilities of interspecific hybridization. Most intra-sectional (and some intra-subgeneric) combinations produce vigorous, fertile F_1's, while intersubgeneric and many intersectional combinations either fail or produce sterile hybrids which invariably show meiotic irregularities. Black currants are in-included by Janczewski (1907) in the Eucoreosma section (10 spp.) of the subgenus Coreosma, and red currants in the subgenus Ribesia (15 spp.). Four species of Eucoreosma and four of Ribesia have been of importance in the evolution of black and red currant cultivars, respectively.

Black currants

R. nigrum. The European black currant. The sub-species *europaeum*, *scandinavicum* and *sibiricum* are recognized in Russian literature. Northern Europe, central and northern Asia to the Himalayas.

R. dikuscha. A very winter-hardy species with bloomed fruits. Eastern Siberia, northern Manchuria.

R. ussuriense. Resembles *R. nigrum* but suckers. Manchuria to Korea.

R. bracteosum. The Californian black currant. Erect-growing with bloomed fruits borne on very long racemes. Alaska to northern California.

Red currants

R. sativum. Under cultivation has given rise to the large-fruited 'cherry' currants (*R.s. macrocarpum*) and white-fruited forms. Western Europe.

R. petraeum. A variable species, donor of resistance to leaf spot, with fruits ranging from red to almost black. Mountainous regions of Europe and North Africa, Siberia.

R. rubrum. Pink and white forms are known. Central and northern Europe, northern Asia.

R. multiflorum. Remarkable for its dense inflorescences with over 50 flowers. Mountains of southern Europe.

3 Early history

Currants have been domesticated in northern Europe within the last 400–500 years, the first published description probably being that of a red currant in a German manuscript of the early fifteenth century. The black currant (*R. nigrum*) was first recorded in Britain in seventeenth century herbals, which drew attention to the medicinal properties of the fruits and leaves. Of the three species contributing to the early evolution of red currants, *R. sativum* was the only one grown in 1542, *R. petraeum* being introduced into cultivation shortly after this and *R. rubrum* very much later. European currants were taken to North America, probably in the seventeenth century, by early English settlers.

As long-lived perennials of minor dietary importance, early selection appears to have been haphazard and sporadic. Colour forms (white and pink red currants and green black currants) were listed in eighteenth and nineteenth century catalogues but, even today, fruit size and fruiting habit of the older cultivars differ little from those of wild types. There was probably unconscious selection for self-fertility, since selfed progenies of most cultivars show marked inbreeding depression. This suggests that the wild progenitors of the crop plants, like the majority of *Ribes* species investigated (including the subspecies *R. nigrum sibiricum*), were obligate outbreeders.

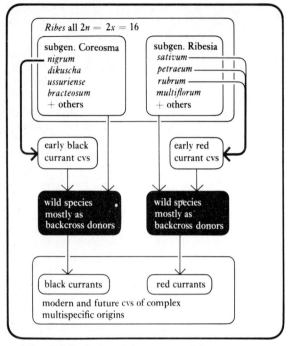

Ribes all $2n = 2x = 16$

subgen. Coreosma
nigrum
dikuscha
ussuriense
bracteosum
+ others

subgen. Ribesia
sativum
petraeum
rubrum
multiflorum
+ others

early black currant cvs

early red currant cvs

wild species mostly as backcross donors

wild species mostly as backcross donors

black currants

red currants

modern and future cvs of complex multispecific origins

Fig. 40.1 Evolution of the currants, *Ribes*.

By the nineteenth century, black currant varieties available in Europe included Baldwin, Black Naples and Black Grape, all still extant, Baldwin being the leading British cultivar today. During the nineteenth century, the numbers of named varieties increased considerably, probably largely through the introduction of plants raised by private individuals or nurserymen from open-pollinated seed of the narrow range of the then existing cultivars. In 1920, Hatton was able to classify 26 varieties, all of pure *R. nigrum* descent, into four main groups of similar (or synonymous) varieties, the types being French Black, Boskoop Giant, Goliath and Baldwin.

In North America, black currant growers apparently relied on European cultivars although, towards the end of the nineteenth century, a number of Canadian-bred varieties derived from earlier importations were released. About this time, the arrival of white pine blister rust on pine seedlings from Europe resulted in severe epiphytotics on pine, leading to the banning of *Ribes* crops (particularly the highly susceptible European black currant) in large areas of the continent.

Like black currants, red currant varieties are very long-lived; Red and White Dutch, still grown today, were described in Mawe's (1778) dictionary of gardening and botany, together with four other red currants. Further varieties were raised thereafter, both in

Europe and North America until, in 1921, Bunyard was able to describe 31 distinct types. He classified these into five groups according to the species they most closely resembled, although recognizing that some included two species in their ancestries. The Raby Castle group, derived from *R. rubrum pubescens*, includes Raby Castle (raised in Britain before 1860) and Houghton Castle (*rubrum* × *sativum*), introduced about 1820. The Versailles group comprises the large-fruited descendants of *R. sativum macrocarpum*, including Versailles (raised about 1835 in France), Cherry (raised before 1840 in Italy) and Fay's Prolific and a number of other American cultivars introduced in the late nineteenth century. The Gondouin group comprises derivatives of *R. petraeum* such as Gondouin (*petraeum* × *sativum*), Prince Albert and Seedless Red. The Scotch group are descendants of *R. rubrum*, while the Dutch group have flowers of the *savtium* type. Further details of early history are given by Knight and Keep (1958) and by Keep (1975).

4 Recent history

By the early twentieth century, a fairly wide range of red currants and a more restricted range of black currants were available in western Europe as basic material for the professional plant breeders who were just beginning to be employed at state institutes and by commercial firms.

At first, breeding was largely confined to intercrossing existing varieties, the main selection criteria being yield, fruit size and quality, and season. In Britain, work of this nature on black currants resulted in a series of cultivars, usually hybrids between representatives of two of Hatton's four groups, released between 1927 and 1962. Wellington XXX, the most widely grown cultivar in Britain today after Baldwin, was introduced in 1927. In general, these cultivars (and others of similar origin from the Continent) did not represent a marked advance on their parents, being equally unreliable in cropping following low temperatures during the flowering season and showing no marked improvement in yield or response to pests and diseases.

However, this work demonstrated the importance of hybrid vigour for commercial performance and the necessity for a broader genetic base if major advances were to be made. Subsequent breeding programmes were markedly influenced by these findings.

The main centres of black currant breeding over the past 30 years have been western Europe, Scandinavia and Russia. In Europe, particularly, the high cost and increasing scarcity of hand labour for this labour-intensive crop, coupled with comparatively low returns, have put a premium on characters that reduce the costs of production. There has been added emphasis on breeding for regular, heavy cropping and for pest and disease resistance. In Russia, breeding for hardiness, with the particular aim of extending the northern limits of black currant growing, is an additional major objective. In some programmes, high vitamin C content is an added selection criterion. Latterly, the advent of mechanical harvesters has had considerable influence on breeding programmes and some recent varieties (e.g. the Dutch Black Reward and the German Invigo) have the erect habit and firm fruit desirable for machine harvesting by shaking.

European breeders have made extensive use of wild or near wild *R. nigrum* selections of Scandinavian origin (e.g. Ojebyn, Brödtorp) as donors of high yield, hardiness and disease resistance. The recently named, heavy cropping British varieties Blackdown, Ben Nevis and Ben Lomond, all at least moderately resistant to American gooseberry mildew (*Sphaerotheca mors-uvae*), include Scandinavian types in their ancestry. Russian cultivars deriving resistance to reversion virus and to its vector, the gall mite (*Cecidophyopsis ribis*), from *R. nigrum sibiricum* are being used in Britain as donors of these characters, although no selections of this origin have yet been released.

In Russia, western European cultivars have been used as donors of self-fertility in crosses with self-sterile local forms of *R. nigrum sibiricum*, which contribute hardiness, fruit size and resistance to leaf spot (*Pseudopeziza ribis*), reversion and gall mite. Several named cultivars of this origin have been released.

Related wild species have been used primarily as donors of pest and disease resistances not found or poorly developed in *R. nigrum* itself. In Russia (and latterly in Britain), *R. dikuscha* has provided strong leaf spot resistance plus hardiness, Primorskij Čempion, much used in later breeding, being one of the first commercial derivatives. Consort (*R. nigrum × R. ussuriense*), bred in Canada and deriving resistance to white pine blister rust from *R. ussuriense*, has been widely used in European black currant breeding. Exceptionally, *R. bracteosum*, of which Jet (released in Britain in 1974) is a first backcross derivative, is contributing agronomic characters, including increased yield potential through many-flowered strigs, rapid and easy handpicking, an erect habit suited to mechanical harvesting by shaking, and late flowering for spring frost avoidance.

Other donor species, the derivatives of which are not yet in commerce, include: the North American *R. sanguineum*, *R. glutinosum*, *R. cereum*, supplying re-sistance to mildew, leaf spot, leaf-curling midge (*Dasyneura tetensi*) and aphids; and the European gooseberry, *R. grossularia*, donor of strong major gene resistance to the gall mite.

Red currant breeding in the twentieth century has been on a relatively small scale. Varieties released since Bunyard's (1921) review include the American Red Lake, now a leading cultivar in Europe, and the Dutch Jonkheer van Tets and Maarse's Prominent. These are in direct line of descent from those raised in the previous century. Other introductions (the Dutch Rondom and the German Heinemanns Rote Spätlese and Mulka) are backcrosses to red currant cultivars of *R. multiflorum* and they derive their heavy cropping, late flowering and ripening and, probably, leaf spot resistance from this species. In Britain, the potential of the far-eastern species, *R. longeracemosum*, as donor of inflorescence characters and disease resistance is under investigation.

In broad outline, modern currant breeding programmes have mostly involved intercrossing locally adapted with genetically dissimilar imported cultivars of complementary phenotypes. Early products of such programmes have already been released both in Europe and Russia. Additional variability has been supplied by related species which are being increasingly used as donors in backcrossing programmes.

It is noteworthy that progress in resistance breeding has been most rapid when resistance has proved to be under major dominant gene control (as in: leaf spot resistance from *R. dikuscha*; rust resistance from *R. ussuriense*; mildew resistance from *R. nigrum scandinavicum*; and gall mite resistance from *R. nigrum sibiricum* and *R. grossularia*). Dominant genes are obviously more rapidly and easily transferred in crops prone to inbreeding depression; and vegetative propagation of selections ensures that the genotype is 'fixed'. The use of major genes has permitted the employment of quick and easy screening methods and makes practicable the combination of several resistances without recourse to the very large progenies that would be necessary for combinations of polygenically controlled responses.

For further details of recent currant breeding see Knight, Parker and Keep (1972), Jennings, Gooding and Anderson (1973) and Keep (1975).

5 Prospects

Contemporary currant breeding is carried out mainly at state-aided institutes and future progress is dependent on the continued willingness of governments to support work on crops of limited acreage.

The future of commercial currant growing will undoubtedly depend ultimately on the use of mechanical harvesters, so that breeding is likely to be aimed increasingly at the upright growth habit and firm, even-ripening fruit needed for machine harvesting by shaking. In red currants, these and the additional requirement for a more regular distribution of the crop over the bush are likely to favour *R. multiflorum* derivatives.

Programmes involving cultivars of widely differing provenance and wild species are just beginning to produce commercial black currant selections so that, in western Europe over the next decade or two, a range of new cultivars outyielding the older types and carrying resistance to major pests and diseases (notably American gooseberry mildew, leaf spot and gall mite) are likely to be introduced. *Ribes bracteosum*, with its combined effect on fruiting, growth habit and yield potential, will almost certainly influence radically the future evolution of black currants. Other species seem unlikely to have a fundamental effect, except on disease and pest response.

The introduction of reliably heavy-cropping varieties which can be machine-harvested will increase the profitability of the crop and may well lead to some increase in acreage, particularly in areas where hand labour is scarce and expensive. The increasing demand for fruit-flavoured, ready-to-eat products such as yoghourt and other dessert confections should ensure an expanding market for some years to come.

6 References

Bunyard, E. A. (1921). A revision of the red currants. *J. Pomol.*, 2, 38–55.

Hatton, R. G. (1920). Black currant varieties. A method of classification. *J. Pomol.*, 1, 65–80, 145–54.

Janczewski, E. de (1907). [Monograph of the currants, *Ribes*]. *Mem. Soc. Phys. Hist. nat. Genève*, 35, 199–517.

Jennings, D. L., Gooding, H. J. and Anderson, M. M. (1973). Recent developments in soft fruit breeding. In *Fruit, present and future*, Royal Horticultural Society, London, 2, 121–35.

Keep, E. (1975). Currants and gooseberries. In J. Janick and J. N. Moore (eds), *Advances in fruit breeding*, Purdue Univ. Press, 197–268.

Knight, R. L. and Keep, E. (1958). Abstract bibliography of fruit breeding and genetics to 1955. *Rubus* and *Ribes* – a survey. *Tech. Commun. Commonw. Bur. Hort. Plantn Crops*, 25, pp. 254.

Knight, R. L., Parker, J. H. and Keep, E. (1972). Abstract bibliography of fruit breeding and genetics 1956–1969. *Rubus* and *Ribes*. *Tech. Commun. Commonw. Bur. Hort. Plantn Crops*, 32, pp. 449.

Avocado
Persea americana (Lauraceae)

B. O. Bergh
Plant Sciences Department University of California Riverside USA

1 Introduction

The avocado has been a major food for the people of Central America for, apparently, several thousand years. It is now grown throughout most of the tropics and subtropics. Major cultivations are established in Israel, South Africa, Chile, Mexico and the United States (California, Florida and Hawaii). California, with usually over 50 kt, is the leading single producer. Only Israel and South Africa have important exports, primarily to western Europe. In these countries it is chiefly a luxury, as is true of the United States and other developed nations; but, in parts of Mexico and more tropical areas in both the New World and the Old, it remains 'the butter of the poor'.

The less tropical avocados have a calorific value of about 124 cal/100 g which is nearly twice that of the lower-oil tropical types. The fruit has up to 4 per cent protein. It supplies a number of vitamins and is exceptionally rich in minerals. The fat is largely unsaturated. In addition, it has special digestive tract benefits, it bears heavily in more tropical areas and its fruits mature all the year round. Certainly it is a valuable food plant and Hume, in 1951, even suggested, perhaps somewhat extravagantly, that the crop had been the most important single contribution made by the New World to the human diet.

Consumers expect fruits to be either sweet or acid (or both) and more or less juicy. The avocado is none of these; its flavour is unique. Eventually, most consumers come to prize its delicate, somewhat nut-like flavour, but adults usually find it bland and disappointing on first trial and have to develop a taste for it.

2 Cytotaxonomic background

For details see Bergh (1969, 1975) and Bergh *et al.* (1973). The 80-odd *Persea* species (all but two or three from the New World) are placed in two graft- and hybridization-incompatible subgenera: Eriodaphne and Persea. All taxa that have been examined in both subgenera have $2n = 2x = 24$.

Persea americana falls in subgenus Persea. It contains a number of varieties, previously classified as separate species (see Bergh *et al.*, 1973), including the three that make up the avocado of commerce: vars. *americana*, *guatemalensis* and *drymifolia*. These are equivalent to the horticultural races known, respectively, as 'West Indian', 'Guatemalan' and 'Mexican'; they are sometimes described as tropical, subtropical and semitropical, on the basis of increasing cold hardiness and general climatic adaptation. There are many differences between them in other botanical traits, including some of great horticultural significance. Thus *guatemalensis* fruits take about a year to mature or about twice as long as those of the other two forms; it has a much thicker and rougher skin; its seed is usually smaller and tighter. The *drymifolia* form is the most salt-sensitive and it has the smallest fruits and the thinnest skin.

The avocado has been crossed with *P. floccosa* but with none of the other wild species that have been tried.

Wild forms are known that may represent ancestors of each of the three varieties (Fig. 41.1). A very primitive form limited to small areas in Honduras and Costa

Fig. 41.1 Occurrence of primitive forms of avocado, *Persea americana*.

Rica could be close to the progenitor of all cultivated types. Primitive *guatemalensis* types are known from isolated spots in Mexico, Guatemala and Honduras. Similarly, primitive *drymifolia* (southern Mexico) and *americana* (Colombia) forms have also been found. The implication is that the three horticultural races were domesticated independently from wild forms in the region Colombia to southern Mexico. The wild forms all have small fruits with large seeds and little pulp so the trend of early selection is evident.

3 Early history

Fossil studies indicate that species of *Persea* grew abundantly in what is now California about 50 million years ago (Schroeder, 1968). The crop apparently originated in Central America. Smith (1969, and earlier) found avocado seeds in connection with human camp sites in southern Mexico as early as perhaps 7000 B.C., with cultivation indicated by about 5000 B.C. He suggested that there was human selection for larger fruit size beginning about 4000 B.C. and that the inhabitants of a second Mexican valley had made much less selective improvement. Both conclusions are questionable but the fact of long-continued and effective prehistoric selection is beyond doubt. Indeed, the avocado cultivars now being grown around the world reflect no significant improvement in horticultural quality over that achieved by early Amerindians. Selection was apparently most effective in *guatemalensis*, judging by its larger fruits and smaller seed, with separation of seed coat from flesh.

The importance of the fruit is reflected in the development of local Indian names for it: the Incan 'palta', the Mayan 'on' and, especially, the Aztec 'ahuacatl' (whence the Spanish 'ahuacate' or 'aguacate' which was corrupted into the English 'avocado'). By the time of the Conquest, Indian people had brought the fruit as far south as Perú on the western side of the continent. Soon afterwards, the Spaniards introduced it to Chile, Venezuela, the West Indies, the Madeira and Canary Islands and elsewhere. Eventually, it became generally established wherever its varieties were climatically adapted but its spread into the Old World tropics was comparatively late (mainly nineteenth century).

The avocado is an outbreeder. Floral dichogamy, with synchronous complementarity of the two flowering types (Bergh, 1969, 1975), maintains outbreeding and heterozygosity. An example of the resulting possibilities for selection is provided by the groves near Atlixco, Mexico. These are usually regarded as hybrids of *guatemalensis* and *drymifolia* but they may simply be the result of evolutionary adaptation to an intermediate environment. In any case, a number of superior commercial forms were selected here, including 'Fuerte' (which is probably still the world's leading cultivar). The Rodiles family owned these groves for many generations and kept planting seeds from the finest of the trees, eventually obtaining thousands of seedlings of unusually high average quality. Nevertheless, the immense variability in quality even amongst such seedlings made asexual propagation highly desirable.

4 Recent history

Asexual propagation was first recorded in Florida, near the end of the nineteenth century (Ruehle, 1963). In this century, clonal cultivars were selected in many regions. Examples include, in California, 'Bacon', 'Zutano' and 'Hass' (rapidly becoming the leading cultivar there and a probable successor to 'Fuerte'); the 'Booth' selections (especially 7 and 8) in Florida; the 'Sharwil' in Australia; the 'Ettinger' in Israel; and, hundreds of others none of which has yet gained much importance.

There are no sterility barriers between the three varieties and recent breeding has primarily involved inter-varietal hybridization. In tropical and near-tropical areas, only *americana* is well adapted but hybrids of it with *guatemalensis* (e.g. the Booth selections) are performing well and are valuable for extending the harvest season. In less tropical regions, hybrids of *guatemalensis* with *drymifolia* predominate. They combine the cold hardiness of the latter with the superior horticultural traits of the former and they bridge the two seasons of maturity. Genetical constitutions vary from largely *drymifolia* ('Ettinger', 'Zutano') through approximately equal proportions ('Fuerte', 'Sharwil') to largely *guatemalensis* ('Hass'). Trihybrids of all three varieties have been developed in Hawaii and in the University of California breeding programme. The result of all this varietal hybridization is that botanical distinctions are breaking down and the avocado is evolving into a polytypic species with all gradations of traits.

All the leading cultivars, including those named above, developed as chance seedlings. Rarely has even the female parent been known. Moreover, all were developed by private individuals. In recent years, the University of California has undertaken a large breeding programme. It has emphasized self-pollination, with the dual aims of direct selection and more uniform (and so more predictable) breeding lines for hybridization. Several dozen selections from it are now under test. Selection programmes have also been undertaken, on a small scale, by institutions in Florida and elsewhere. A major programme is now beginning in Israel.

5 Prospects

Public and private selections from available seedlings (of which there are, of course, vast numbers) will probably continue in avocado-growing regions around the world. There is no substitute for local selection for local adaptation. But the fact that cultivars superior in one country have generally proved to be also superior in distant countries with similar climates suggests continued dominance of the cultivar picture by the major programmes in California, Florida and (in future) probably also Israel. There will probably be continued emphasis on inter-varietal hybrids, (especially because of marketing-period considerations) which will further blur varietal distinctions.

Current breeding objectives include all those attributes of tree, rootstock and fruit (Bergh, 1975) that result in heavy production, good shipping quality, sales appeal and consumer satisfaction. The major disease problem in most regions is *Phytophthora cinnamomi*, causing root rot. Resistance to the fungus is common in species of *Eriodaphne* but members of the two subgenera are, unfortunately, graft-incompatible. The most promising approach appears to be hybridization and selection among the few avocado lines that show a little root-rot resistance with, thereby, a gradual development of commercially-resistant rootstocks. The University of California is searching throughout Central America for additional resistant lines. The more primitive forms of avocado should probably also be established in germ plasm banks, as an insurance against possible new diseases and changing consumer demands.

Since avocado flowers are usually so numerous that only about $0 \cdot 1$ per cent of them produce mature fruit, hand hybridization has been expensive and unrewarding. Returns can be increased by various means (Bergh, 1975). An alternative is to have bees do the hybridizing, either with isolated pairs of clones or else inside a cage; there may, of course, be some self-pollination. When selfing is desired, fruits from near the centre of a monoclone block or from isolated trees supply seeds inexpensively. If such are not available, individual trees can be caged with bees.

6 References

Bergh, B. O. (1969). Avocado. In F. P. Ferwerda and F. Wit (eds), *Outline of perennial crop breeding in the tropics*. Wageningen, The Netherlands, 23–51.

Bergh, B. O. (1975). Avocados. In J. Janick and J. N. Moore (eds), *Advances in fruit breeding*. Lafayette,

Indiana, USA, 541–67.

Bergh, B. O., Scora, R. W. and Storey, W. B. (1973). A comparison of leaf terpenes in *Persea* sugenus Persea. *Bot. Gaz.*, **134**, 130–4.

Ruehle, G. D. (1963). The Florida avocado industry. *Univ. Florida agric. Exp. Sta. Bull.*, **602**, pp. 102.

Schroeder, C. A. (1968). Prehistoric avocados in California. *Calif. Avocado Soc. Yrbk.*, **52**, 29–34.

Smith, C. E. (1969). Additional notes on pre-Conquest avocados in Mexico. *Econ. Bot.*, **23**, 135–40.

Groundnut

Arachis hypogaea (Leguminosae–Papilionatae)

W. C. Gregory and **M. P. Gregory**
North Carolina Agricultural Experimental
Station Raleigh NC USA

1 Introduction

The seeds of groundnut, or peanut, are rich in both oil (43–55%) and protein (25–28%) and make a substantial, if not major, contribution to human nutrition. The crop is grown chiefly for its vegetable oil but the protein fraction presents a favourable amino acid profile (St. Angelo and Mann, 1973). The crop area (1967–69) was about 18Mha and production (in shell) 16 Mt. The crop is produced all round the world in warm temperate and tropical regions, chiefly in the range 36°N–36°S. India, tropical African countries and China are leading producers (McGill, 1973).

2 Cytotaxonomic background

Arachis is a South American genus of 40–70 species, many of them yet undescribed. Most are diploids with $2n = 2x = 20$. Seven sections are recognized by Gregory *et al.* (1973), six by Krapovickas (1973), of which one, sect. Arachis (Axonomorphae), is of concern here. The genus as a whole has a centre of diversity in the Mato Grosso of Brazil. Species in areas of high rainfall are perennial; those in semi-arid areas, annual. Section Arachis contains annual and perennial diploids and two allotetraploids with $2n = 4x = 40$. Of these, *A. monticola*, from northwestern Argentina, is an annual that crosses freely with cultivated *A. hypogaea*; it is presumed to be the current wild descendant of the amphidiploid species ancestral to the cultigen.

Some suggestions as to diploid ancestors have been made. Smartt (1964) showed that an undescribed diploid annual species, sect. Arachis, possessed a pair of small 'A' chromosomes similar to that reported by Husted (1936) in *A. hypogaea*. This distinctive pair was lacking in the erectoid form *A. paraguariensis* and he suggested that the ancestral form of *A. hypogaea* could have arisen by hybridization between two diploid species having these distinct karyotypes.

However, reference to the figure published in Smartt and Gregory (1967) of the somatic chromosomes of the hybrid *A. hypogaea* × *A. helodes* shows that only a single 'A' chromosome occurs. The implication is that *A. helodes* (perennial, sect. Arachis) does not possess a pair of 'A' chromosomes. Since both *A. helodes* and the undescribed annual species belong to the section Arachis, the erectoid *A. paraguariensis* is not required to account for the 'A' chromosome constitution of *A. hypogaea*. In fact, an interspecific hybrid between such annual and perennial species of sect. Arachis is a distinct possibility and such a hybrid would have a karyotype not unlike that of *A. hypogaea–monticola*. Detailed karyotype studies have not as yet been carried out on species in this section and more certain identification of the most likely parents (e.g. *A. duranensis* (annual) and *A. cardenasii* (perennial)) would await the outcome of such studies and of further experimental interspecific crosses. Nevertheless, an experimental re-creation of *A. monticola–hypogaea* is a very real possibility and would be of considerable interest.

The groundnut is thus an allotetraploid believed to have derived directly from a wild allotetraploid (Krapovickas and Rigoni, 1957). That the wild diploid ancestors of this tetraploid were probably fairly nearly related is suggested by the observation (Husted, 1936; Gregory *et al.*, 1951; Smartt, 1964) of occasional meiotic multivalents in *A. hypogaea*, suggestive of segmental allotetraploidy.

All members of *Arachis* are geocarpic and, presumably, all, like the cultivars, are highly inbred. Wild species can sometimes be intercrossed but the hybrids are usully more or less sterile. As a rule, intrasectional crosses are successful, intersectional ones much less so, though some are possible (Smartt and Gregory, 1967; Gregory and Gregory, 1967).

3 Early history

The earliest archaeological records are from Peru, 2000–3000 B.C. (Krapovickas, 1968; Hammons, 1973). The crop was, however, domesticated well to the southeast and must be presumed to have had a much longer history (see below). It was, in fact, domesticated by the predecessors of modern Arawak-speaking peoples at a time which cannot now be determined.

In comparison with wild *A. monticola*, *A. hypogaea* has more robust, non-fragile, pods and shorter carpophores. It may be inferred that these changes, which would be disadvantageous in the wild, were brought about by early human selection and were, in fact, as fundamental in domestication as non-shattering rachises were in the domestication of cereals. In sub-

sequent spread and diversification, a great deal of variability in plant habit and seed characters accumulated.

Krapovickas (1968; see also Krapovickas, 1973) has distinguished five (six are distinguished here; see Fig. 42.1) major geographical groups of cultivated groundnuts in South America. These seem to represent secondary centres of diversity that have developed from a primary centre of domestication in southern Bolivia and northern Argentina (Fig. 42.1). The long-standing notion of the Brazilian origin of *A. hypogaea* (Chevalier, 1933) is not supported by current data.

The differences between the geographical groups in relation to established intraspecific classification may be briefly (and, inevitably, rather superficially) summarized as follows (Krapovickas, 1968, 1973). *Arachis hypogaea* comprises two subspecies and four varieties, the former differing in branching habit.

Name	Type	Differences
A subsp. *hypogaea*		No floral axes on main axis; alternating pairs of vegetative and floral axes along lateral branches.
1 var. *hypogaea*	Virginia	Less hairy, branches short.
2 var. *hirsuta*	Peruvian Runner	More hairy, branches long.
B subsp. *fastigiata*		Floral axes on the main axis; continuous runs of one-many floral axes along lateral branches.
1 var. *fastigiata*	Valencia	Little branched.
2 var. *vulgaris*	Spanish	More branched.

These four basic varietal types are to some extent characteristic of the secondary centres of diversity shown in Fig. 42.1. Centre I contains mostly B (*fastigiata*) types and is thus the centre of Valencia–Spanish groundnuts. Centre II is also mostly B. Centre III contains a unique form of A1. Centre IV is mostly A1 but also shows some B1 types and gives evidence of inter-subspecific hybridization. Centre V contains A1, A2 and a special fastigiate form not found in Centres I and II. Centre VI may be regarded as tertiary.

There is evidence (Krapovickas, 1973; Wynne, 1974) of some slight genetic isolation between the subspecies, presumably a reflection of long separation in the secondary centres.

By the time of Columbus, groundnuts were gener-

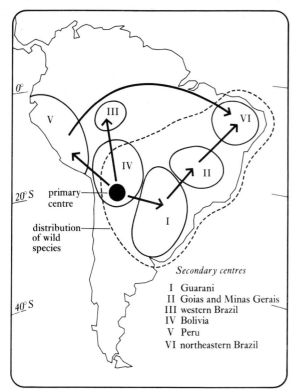

Fig. 42.1 Centres of origin and diversity of the groundnut, *Arachis hypogea*.

Secondary centres

I Guarani
II Goias and Minas Gerais
III western Brazil
IV Bolivia
V Peru
VI northeastern Brazil

primary centre

distribution of wild species

ally distributed through South America, the Caribbean and Mexico and were repeatedly recorded by writers of the time (Krapovickas, 1968; Hammons, 1973). There is no evidence (some earlier writings notwithstanding) that the crop moved out of the western hemisphere before this period. It probably reached eastern Asia, by way of Mexico and the Philippines, in the hands of Spanish travellers; Africa from Brazil by Portuguese contacts about 1500 and, later, India via Africa. Varietal types and diversity present in Asia and Africa broadly agree with these views (Gibbons *et al.*, 1972). First introductions to temperate North America were probably made in the seventeenth century from Africa (Higgins, 1951), although the crop was already established in the Caribbean, Central America and Mexico at that time (Hammons, 1973).

By the middle-to-late nineteenth century, groundnuts had assumed considerable agricultural importance in the USA, India and West Africa. Indeed, Africa was and is now a substantial tertiary centre of diversity (Gibbons *et al.*, 1972).

4 Recent history

Improvement of the crop by individual plant and line

selection began early in the present century (Gregory *et al.*, 1951). The groundnuts of A1 are nearly strict inbreeders; those of B1 and B2 somewhat less so. Experimental crossing on any scale is difficult. Early efforts at improvement by plant and line selection were useful but inevitably limited in scope. Hybridization, followed by pedigree selection of pure lines, was hardly practised until after the late war and was not fully adopted, even in the USA, until very recent years (Gregory *et al.*, 1951; Norden, 1973).

5 Prospects

There can be little doubt that the crop is one with great breeding potential. There is much, largely unexplored, genetic variability, both in cultivars and wild species, awaiting utilization as a new source of adaptation and of resistance to pathogens. Problems in using this variability will certainly be posed by interspecific sterility barriers.

Despite the rather small contribution of wide interspecific hybrids to improvement of modern field crops, *Arachis* may prove an exception. Several hundred interspecific hybrids of *Arachis* and many of their amphidiploid derivatives already exist (Gregory and Gregory, unpublished); and these amphidiploids combined among themselves and with *A. hypogaea* promise the release of a great deal of new variability, indeed, the beginning of a new era of groundnut evolution.

6 References

Chevalier, A. (1933). Histoire de l'arachide. *Rev. Bot. appl. Agr. trop.*, **13**, 146–7, 722–52.

Gibbons, R. W., Bunting, A. H. and Smartt, J. (1972). The classification of varieties of groundnut (*Arachis hypogaea*). *Euphytica*, **21**, 78–85.

Gregory, W. C. and Gregory, M. P. (1967). Induced mutation and species hybridization and de-speciation of *Arachis*. *Ciencia e Cultura*, **19**, 166–74.

Gregory, W. C., Gregory, M. P., Krapovickas, A., Smith, B. W. and Yarbrough, J. A. (1973). Structures and genetic resources of peanuts. In *Peanuts – culture and uses*. Stillwater, Oklahoma, USA, Ch. 3.

Gregory, W. C., Smith, B. W. and Yarbrough, J. A. (1951). Morphology, genetics and breeding. In *The peanut – the unpredictable legume*. Washington, D.C., Ch. 3.

Hammons, R. O. (1973). Early history and origin of the peanut. In *Peanuts – culture and uses*. Stillwater, Oklahoma, USA, Ch. 2.

Higgins, B. B. (1951). Origin and early history of the peanut. In *The peanut – the unpredictable legume*. Washington, D.C., Ch. 2.

Husted, L. (1936). Cytological studies on the peanut, *Arachis*. II. Chromosome number, morphology and behavior and their application to the origin of cultivated forms. *Cytologia*, **7**, 396–423.

Krapovickas, A. (1968). Origen, variabilidad y difusion del mani (*Arachis hypogaea*). *Actas y Memorias XXXVII Congreso International Americanistas*, **2**, 517–34. English translation by J. Smartt in P. J. Ucko and G. W. Dimbleby (eds), *The domestication and exploitation of plants and animals*. London (1969).

Krapovickas, A. (1973). Evolution of the genus *Arachis*. In Rom Moav (ed.), *Agricultural genetics*. New York, 135–51.

Krapovickas, A. and Rigoni, V. A. (1957). Nuevas especies de *Arachis* vinculadas al problema del origen del mani. *Darwiniana*, **11**, 431–55.

McGill, J. F. (1973). Economic importance of peanuts. In *Peanuts – culture and uses*. Stillwater, Oklahoma, USA, Ch. 1.

Norden, A. J. (1973). Breeding of the cultivated peanut (*Arachis hypogaea*). In *Peanuts – culture and uses*. Stillwater, Oklahoma, USA, Ch. 5.

Smartt, J. (1964). Ph.D. thesis, N. C. State University, U.S.A.

Smartt, J. and Gregory, W. C. (1967). Interspecific cross-compatibility between the cultivated peanut *Arachis hypogaea* and other members of the genus *Arachis*. *Oléagineux*, **22**, 455–9.

St Angelo, A. J. and Mann, G. E. (1973). Peanut proteins. In *Peanuts – culture and uses*. Stillwater, Oklahoma, USA, Ch. 17.

Wynne, J. C. (1974). Ph.D. thesis, N. C. State University, U.S.A.

Pigeon pea

Cajanus cajan (Leguminosae–Papilionatae)

W. Vernon Royes
Ministry of Agriculture
Jamaica West Indies

1 Introduction

This hardy pulse is almost entirely grown on poor land by small farmers. It competes successfully in the wild under hard conditions (except water-logging and heavy shade) and is best adapted to tropical grassland and arid regions. It is a shrub which varies in height from $0 \cdot 5$ to 5 m tall and its yield has yet to show a response to added nutrients. Flowering is induced by short days in most types.

World production is currently about 20 kt, with India contributing 95 per cent and the Caribbean, southeast Asia, Pakistan, Malawi and Uganda having measurable production. Indian production may vary from year to year by 20 per cent, but *Cajanus* usually contributes about 4 per cent of world pulse production at an average yield of 600 kg/hectare. It is also grown as a forage crop in the southeastern USA, north-western Australia, Italy and Hawaii.

Some seed is consumed as the popular green (immature) pea but most is processed into 'dahl', the easily stored split pea. The protein contributes 20 per cent of the dry weight and is deficient in the essential sulphur-bearing amino acids and tryptophane.

2 Cytotaxonomic background

The basic chromosome number of *Cajanus* is $x = 11$ and, although more than 30 species have been recognized, the genus is almost certainly one polymorphic species, *Cajanus cajan* (= *Cajanus indicus*). The crop is diploid but some tetraploids and hexaploids have been obtained (and have much reduced yields). *Cajanus* has been crossed with the diploid Indian species, *Atylosia lineata*, *A. sericea* and *A. scarabaeoides* (of which the basic number is also $x = 11$) and a comparison of somatic karyotypes has shown that several chromosomes are similar in the two genera. D. N. De (1974) points to *Atylosia lineata* as being, of the species tested, nearest to *Cajanus* but notes that there are several other *Atylosia* species (e.g. *geminiflora*, *candollei* and *cajanifolia*) which deserve study in this connection. The conclusion is that the crop originated from a yet unidentified species of *Atylosia* by selection for size and vigour of plant, for non-shattering pods and for larger seeds.

3 Early history
The centre of origin of the crop has long been somewhat disputed. D. N. De (1974) states that there is, so far, no archaeological evidence. He cites references to *Cajanus* in northern Indian and Deccan literature of 1600 and 1400 B.P. and concludes that it was a well-known crop in India by the latter date. De favours an origin in peninsular India, recalling the fact that India is the present centre of diversity both of *Cajanus* and of the closely related *Atylosia* and he quotes Engler, Vavilov, Burkill and Murdock in support of this thesis. However, Purseglove (1968) and Herklots (1972) cite an earlier work by Burkill as mentioning the presence of *Cajanus* seed in an Egyptian tomb of the XII Dynasty (earlier than 4000 B.P.). The existence of wild varieties in Africa is not a good reason to accept De Candolle's support of Africa as the centre of origin since 'wild' varieties exist in the Caribbean into which all authors (including De Candolle) agree that the crop was introduced in post-Columbian times. Furthermore, *Atylosia* is a predominantly Indian genus, rare and poorly represented in Africa.

On balance, it looks as though De's argument in favour of an Indian origin has much to commend it and the old dispute can be considered settled. The Egyptian seed, however, still needs interpretation; it may have been misidentified (in which case re-examination by modern methods could be rewarding); or it may indicate that the crop is older than now appears.

Cajanus is referred to in Sanskrit in a lexicon and in a medical text around 1400 B.P. as Tuvari or Tuvarica or Adhaki. It is probable that the two main plant types recognised today, Tur (*flavus*) and Arhar (*bicolor*) have Indian names derived from the Sanskrit words. This suggests that both types were developed at an early stage of evolution of the crop. If one accepts an Indian origin, it is probable (De, 1974) that the crop moved to Malaysia from India around 2000 B.P. on the basis of the Sanskrit names there. Later it was carried to China (about 1500 B.P.) and quite likely reached Australia through the East Indies. The name *Cajanus* is derived from the Malaysian name Katjang. Between 1500 and 1000 B.P. *Cajanus* was carried to East Africa, probably in the region of Zanzibar. From there it moved northwards to the Nile valley and to West Africa. The crop was probably taken to the New World from Zäire or Angola by the Portuguese in early post-Columbian days, prior to the main African slave trade. Menezes has suggested that Guando or Guandu, the Brazilian name for the crop, is a Portuguese transformation of *Cajanus* by Africans in Brazil. This name has spread to Caribbean Latin America as Guandole in the south, Gandole in Puerto Rico, the Dominican Republic and Cuba, and Gungo in Jamaica. The New Plantation colonies of the eastern Caribbean islands refer to *Cajanus* as Pois Angola and Pigeon pea, while Congo pea in Jamaica is a false rationalization of Gungo. It is interesting that the crop is now called Pois de Congo and Ervilha do Congo in French and Portuguese-speaking non-equatorial Africa.

4 Recent history
In India, at the centre of diversity and in the major area of production, the crop exhibits a great range of types showing, collectively, wide adaptation. This has no doubt arisen partly from unsophisticated selection by small farmers of an almost entirely peasant crop and partly from the occurrence of fairly frequent cross-pollination. *Cajanus* is self-compatible and can be self-pollinated but is often crossed by insects. Within India, there appears to be a degree of differentiation between the peninsular Tur varieties (with small, quickly maturing plants, yellow flowers and few light-coloured seeds) and the northern Arhar types (larger, later plants with variegated flowers and more numerous, darker seeds). There is, however, much intergradation, the two are fully interfertile and the validity of the distinction appears to be in some doubt. Conceivably, the two types might represent the relics of two geographically independent domestications (see above).

Sophisticated plant breeding only began about 30 years ago. In India, work at the Indian Agricultural Research Institute (Deshpande *et al.*, 1963) after the late war was centred around combining resistance to *Fusarium* wilt with high seed yields. They carried out hybridization with material from the New World and obtained useful lines. Breeding has continued and has been gathering momentum in the last ten years as attention has gradually been focussed on protein shortage.

Outside India, however, breeders have sought to make *Cajanus* into something other than a small-farmer-crop. Its potential as a forage led Kraus (1926, 1927) to breed it for this purpose in Hawaii as early as 1922. He records the creation of a high yielding variety, New Era. The collection and selection of

155

forage types has also been carried on in northern Queensland more recently. The objectives of such programmes are types which are indeterminate in growth habit, late-maturing and perennial, thus providing browse on arid land all the year round for a number of years.

Selection was first carried out on local and introduced material at the Imperial College of Tropical Agriculture in Trinidad in 1933. The work was sporadic until 1956 when Gooding (1962) initiated a programme of collection, hybridization and selection aimed at increased production of the immature green pea. The objective of this programme was not overall yield but rather a change in the nature of the crop that would allow economic green pea production. This requires varieties which show earliness and determinacy, with concomitant capacity for multiple cropping, dwarfness and suitable pod characteristics such as large seeds, many seeds per pod and good seed/pod weight ratios. This programme has since produced varieties acceptable both for the production of green peas for direct consumption and for canning. Similar work was later carried out in Puerto Rico. In the last ten years, rust (caused by *Uredo cajani*) has posed a problem in the Caribbean and work has started on breeding for resistance.

The usual breeding plan has been hybridization followed by selection under inbreeding (of which *Cajanus* is tolerant). However, crosses between inbred lines show some heterosis and recent work in Uganda has sought to exploit this by avoiding close inbreeding (Khan and Rachie, 1972; Rubaihayo *et al.*, 1972). In this programme, the progeny of high yielding individuals of similar phenotype are bulked into composites and grown under conditions that favour natural crossing. Performance is maintained by recurrent selection. Production problems in East Africa are similar to those of India; breeding aims are wilt (*Vellosiella cajani*) resistance and seed yield.

5 Prospects

The rate of increase in the breeding and other work being carried out on *Cajanus* is striking and augurs well for the massive improvement needed to transform the species into a highly developed crop. There is little doubt that the necessary variability exists and is not in immediate need of conscious preservation. Further variation exists in *Atylosia* and it should be possible to use this by hybridization or even by attempting to recreate *Cajanus de novo*. A much improved knowledge of *Atylosia* is obviously an essential pre-requisite for such a programme.

The future of *Cajanus* probably lies in the breeding of pure lines which will produce grain, green pea and fodder from plants that are amenable to mechanization and will respond to high levels of husbandry (including fertilizing) in large fields. It is doubtful whether the indifferent heterosis shown in controlled crosses would ever make the production of hybrid seed worthwhile even if it were genetically feasible to do so.

However, the immediate reality is the small farmer of India, East Africa and elsewhere who subsists on the dry seed he grows on poor land and with resources that do not include fungicides, insecticides, irrigation or machines. His taxes can contribute to the breeding of improved composites which will not make him rich but which will allow him to eat a little better.

6 References

De, D. N. (1974). Pigeon pea. In J. B. Hutchinson (ed.), *Evolutionary studies in world crops*, Cambridge, 79–87.

Deshpande, R. B., Jeswani, L. M. and Joshi, A. B. (1963). Breeding of wilt resistant varieties of pigeon pea. *Ind. J. Genet.*, **23**, 58–63.

Gooding, H. J. (1962). The agronomic aspects of pigeon peas. *Field Crop Abstr.*, **15**, 1–5.

Herklots, G. A. C. (1972). *Vegetables in Southeast Asia.* London, 232–3.

Khan, T. V. and Rachie, K. O. (1972). Preliminary evaluation and utilization of pigeon pea germ plasm in Uganda. *E. Afr. Agric. For.*, **38**, 78–82.

Kraus, F. G. (1926). Genetic analysis of *Cajanus indicus* and the creation of improved varieties through hybridization and selection. *Proc. Hawaiian Acad. Sci.*, B.P. *Bishop Mus. spec. Publ.*, **11**, 24–5.

Kraus, F. G. (1927). Improvement of pigeon pea. *J. Hered.*, **18**, 227–32.

Purseglove, J. W. (1968). *Cajanus.* In *Tropical crops, Dicotyledons.* London, **1**, 236–41.

Rubaihayo, R. R., Radley, R. W., Khan, T. V., Mukiibi, J., Leakey, C. L. A. and Ashley, J. M. (1972). The Makerere programme. In *Nutritional improvement of food legumes by breeding.* P.A.G. Symposium, F.A.O., Rome.

Chickpea

Cicer arietinum (Leguminosae–
Papilionatae)

S. Ramanujam
Indian Agricultural Research Institute
New Delhi

1 Introduction

Grain legumes (pulses) play an important role in meeting the quantitative and qualitative protein requirements of a large part of humanity, especially in the developing countries of Asia, Africa and Latin America and they also fix large quantities of atmospheric nitrogen. Of the grain legumes, chickpeas stand second in area occupied (10 Mha) and third in quantity (7 Mt). More than 85 per cent of the area, as well of production, is in India, where the crop is mostly grown rainfed in the northern parts during the cool season (October–March). The area sown to the crop and production show considerable fluctuations depending on soil moisture at sowing time. With the development of conditions favouring input-intensive agriculture, chickpea has tended to be displaced by crops of higher potential, especially cereals. Outside India, cultivation is mainly centred around the Mediterranean (Greece, Italy, Portugal, Spain, Morocco, Algiers, Iraq, Iran); Ethiopia and Mexico are the only other countries in which any appreciable areas are grown. There is little international trade.

Chickpea contains 20–22 per cent of crude protein with a digestibility of 80 per cent. The aminoacid profile of the protein is such that it complements that of cereals (especially in lysine). An intake of appropriate proportions of cereals and chickpea can meet fully man's requirement for essential amino acids. The carbohydrate (50–60%) and oil (about 5%) contents can make an appreciable contribution to energy requirements. In India, chickpea is often processed into dhal (split pea) or flour before use, though some is also consumed whole after boiling or roasting.

2 Cytotaxonomic background

Cicer belongs to the tribe Viciae. It is a relatively small genus, with 39 known species distributed mainly in central and western Asia. It has been suggested that *Cicer* exhibits a reticulate combination of the characters of *Vicia* and of the genus *Ononis* of the tribe Trifoliae (Maesen, 1972).

The most common somatic chromosome number reported in the genus so far is $2n = 2x = 16$ (*Cicer arietinum*, *C. chorossanicum*, *C. pinnatifidum*, *C. anatolicum*, *C. bijugum*, *C. cuneatum*, *C. incisum*, *C. judaicum*, *C. microphyllum* and *C. montbretii*). A different number ($2n = 14$) has been reported for *C. pungens* as well as for *C. arietinum* and *C. microphyllum* (as *C. songaricum*). The latter reports are likely to be erroneous, one of the chromosomes in the complement being quite small (about 1μ). The basic number in the genus thus appears to be $x = 8$. This does not agree well with basic numbers (6, 7) of the tribe Viciae (Darlington and Wylie, 1965) but would agree with the tribe Trifoliae, where *Ononis* is placed. However, *Ononis* species counted so far are at a higher ploidy level ($2n = 4x = 32$) than *Cicer* and cannot have contributed to the development of *Cicer*. No information is available regarding the interrelationship between the cultivated *C. arietinum* and the wild species. On morphological grounds, it appears that *C. pinnatifidum*, *C. echinospermum* and *C. bijugum* are closest. Crosses between *C. arietinum* and *C. microphyllum* (as *songaricum*, cultivated in the Himalayan region) as well as the wild *C. pinnatifidum* have not been successful, due to pre-fertilization barriers.

3 Early history

The centre of diversity of the genus lies in western Asia, probably in the Caucasus region and/or Asia Minor (Maesen, 1972). The progenitor of the present-day *C. arietinum* must have been spread by the Aryans both westwards, along the Mediterranean, and eastwards, overland to India. The earliest records of chickpea is from a site in Turkey radiocarbon-dated to 5450 B.C. There is evidence for its cultivation in the Mediterranean area in the third and fourth millenia B.C.

The earliest record of chickpea in India, at Atranji Khera in Uttar Pradesh (Chowdhury *et al.*, 1970), is from 2000 B.C. (thermoluminescent-dating) and presumably represents an overland introduction. Early records from peninsular India are dated A.D. 200–150 B.C., but this may have to be extended, at least to the third to fifth century B.C. (Vishnu-Mittre, 1974), even if an earlier report dating to the second millenium B.C. is taken to be based on a wrong identification, as suggested by Vishnu-Mittre. Etymological evidence suggests that chickpea could have also been introduced into India independently by an oversea route. Thus, the names applied to the crop in the Dravidian language are completely different from the Sanskrit-derived Chana, prevalent in more northerly parts. It is known that there was extensive maritime trade between Egypt, Israel and peninsular India, with flourishing Indian colonies in Memphis.

Ethiopia, the third centre of diversity, could have received the crop from the Mediterranean, contacts with which are known to have existed and where chickpea was grown at least as early as the first millenium B.C.

During the course of its dispersal, chickpea seems to have diverged in two directions. In the western part of its distribution, where it is grown under an equable Mediterranean spring–summer environment, cultivars have been developed with large, owl-head shaped, light-coloured seeds with little wrinkling of the seed coat. The plants are tall and plant parts such as leaflets and flowers are large. In the eastern and southern parts of its distribution, especially in India, where chickpea is raised in the cool dry season, and in Ethiopia, forms with relatively quick maturity and with thin stems having correspondingly small leaflets and flowers have developed. The seeds are small, typically wrinkled, ram-head shaped and dark-coloured. Analysis of the constellation of characters in which the Indian Desi (eastern ecotype) and Kabuli (western) groups differ suggest that the latter has probably been more modified by human selection than the former and can thus be regarded as more advanced (Rohewal *et al.*, 1966). The two types have not, however, become reproductively isolated and, indeed, several varieties occur or have been developed by plant breeders combining characteristics of the two.

4 Recent history

In recent times, the chickpea has been introduced into a few other areas. From the Iberian peninsula, it was carried by sixteenty century Spanish and Portuguese travellers to South and Central America, especially to Mexico. It has also been introduced into East and South Africa and the West Indies by the expatriate Indian population. The large, smooth-seeded types (known as Kabuli) appear to represent a third recent (eighteenth century) wave of introduction into India from the Mediterranean region over the land route.

Considered on a global scale, there has been little variation in the area cropped to chickpea during the last two decades. Productivity has shown some slight improvement, as has total production. The relative stagnation can be traced to the fact that little attention has been paid, until recently, to the improvement of the crop by breeding. It has always been considered as a poor man's food and grown under conditions of limited inputs where other crops cannot be grown. However, substantial breeding efforts have been made in India (Ramanujam, 1972) and in Pakistan in the past two or three decades. Success in improving yielding capacity over the land races has not, however, been very marked. The limited germplasm with which most breeders have worked and the minimum-input conditions bred for must be considered primarily responsible for the small success achieved so far. An important approach in chickpea breeding in India has been to introduce large seed size and responsiveness to improved cultural conditions by crossing indigenous types with the Mediterranean ecotype; several medium-sized types have thus been developed. Breeding for disease resistance or response to fertilizers have not, however, received any attention. Chickpea is predominantly inbred but there is some insect-pollination.

5 Prospects

The world-wide shortage of protein foods and the energy crisis leading to fertilizer shortage has fostered a new interest in grain legumes, including chickpea, and there has been considerable international concern. It is likely, therefore, that there will be greater efforts, both national and international, for the improvement of chickpea than in the past. The major objectives (Swaminathan, 1972) will be to develop a plant type, with high yield potential and responsive to agronomic inputs, which can compete on a more nearly equal footing with high yielding cereals. To achieve this would require a refashioning of the plant type with greater photosynthetic efficiency and reduced photo-respiration loss and more advantageous partitioning of the photosynthate into a larger number of sinks (seeds), rather than into vegetative parts. The development of suitable legume–rhizobium combinations which can fix the large quantities of nitrogen needed for high yields will also be important. Protein content, amino-acid profile, disease resistance and suitability for milling or processing will also be worthwhile objectives.

6 References

Chowdhury, K. A., Sarswat, K. S., Hasan, S. N. and Gaur, R. G. (1970). 4000–3500 years old barley, rye and pulses from Atranjikhera. *Sci, and Cult.*, 37, 531–3.

Darlington, C. D. and Wylie, A. P. (1965). *Chromosome atlas of flowering plants.* London.

Maesen, L. J. G. van der (1972). *Cicer. A monograph on the genus with special reference to the chickpea (Cicer arietinum), its ecology and cultivation.* Wageningen, The Netherlands.

Ramanujam, S. (1972). Some salient results of pulse research. *Indian Fmg*, **21**, 17–18.

Rohewal, S. S., Ramanujam, S. and Mehra, K. L. (1969). Plant type in bengal gram. *Ind. J. Genet.*, **26**, 255–61.

Swaminathan, M. S. (1972). Basic research needed for further improvement of pulse crops in Southeast Asia. In M. Milner (ed.), *Nutritional improvement of food legumes by breeding*. Protein Advisory Group, U.N., New York.

Vishnu-Mittre (1974). The beginnings of agriculture: palaeobotanical evidence in India. In J. B. Hutchinson (ed.), *Evolutionary studies in world crops: diversity and change in the Indian sub-continent*. Cambridge.

Soybeans

Glycine max (Leguminosae-Papilionatae)

T. Hymowitz
University of Illinois Urbana Ill USA

1 Introduction

The soybean is the most important grain legume crop in the world in terms of total production and international trade. In 1973, world production was approximately 58 Mt. The leading producers are the United States, People's Republic of China and Brazil with 74, 12 and 8 per cent of the total production. In addition, Indonesia, Mexico, USSR, Canada, South Korea, Romania, Columbia, Paraguay and Japan grow significant amounts.

In the Far East, the soybean is utilized in liquid, powder and curd forms for human food. Immature green beans and sprouts are considered highly nutritious and consumed in great quantities. In the West, the two primary products are oil and meal. Soybean seeds contain from 20 to 23 per cent oil and from 39 to 45 per cent meal. The oil is converted to margarine, shortening, mayonnaise, salad oils and salad dressings. Most of the meal is used as a source of high protein animal feed for the production of eggs, poultry and pork. The use of soybean protein in the form of concentrates, isolates and textured protein for human consumption offers a partial solution to man's protein needs in a world with a rapidly growing population (Smith and Circle, 1972).

2 Cytotaxonomic background

Plant breeders and geneticists attempting to improve the soybean have often been frustrated by the state of confusion concerning its taxonomy. Over 280 species, subspecies and taxonomic varieties have been listed under *Glycine*. Fortunately, recent studies by Hermann (1962) and Verdcourt (1966, 1970) have greatly clarified the taxonomy of the genus *Glycine*. The nine species are listed in Table 45.1.

Little cytogenetic work other than chromosome counting has been done with the species in the subgenus Glycine. All the species in this subgenus are perennials. None of the species appear to be useful

Table 45.1 The soybean and its relatives

Name	Vernacular	2n	Distribution
Subgenus GLYCINE ($x = 20$)			
1 *G. clandestina*	twining Glycine	40	Australia, south Pacific
2 *G. falcata*	—	40	Australia
3 *G. latrobeana*	purple Glycine	—	Australia
4 *G. canescens*	silky Glycine	40	Australia
5 *G. tabacina*	variable Glycine	80	Australia, southeast Asia
6 *G. tomentella*	—	40, 80	Australia, southeast Asia
Subgenus BRACTEATA ($x = 11$)			
7 *G. wightii*	perennial soybean	22, 44	Africa, southeast Asia
Subgenus SOJA ($x = 20$)			
8 *G. soja*	wild soybean	40	East Asia
9 *G. max*	soybean	40	Cultigen

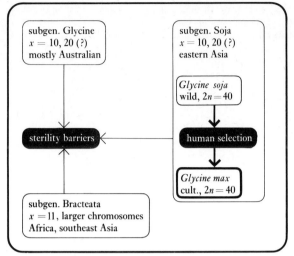

Fig. 45.1 Relationships of soybean, *Glycine max*, and its relatives.

in intensive agriculture. Reciprocal crosses between *G. tabacina* ($2n = 80$) and *G. tomentella* ($2n = ca.$ 80) produce F_1 hybrids that are highly sterile, with from 78 to 81 chromosomes. Attempts to cross the species in the subgenus Glycine with the soybean have not been successful.

Glycine wightii is a perennial climbing vine that has shown promise as a pasture legume in the tropics and subtropics, especially in Brazil and Australia. Verdcourt (1966) divided the species into three subspecies and five taxonomic varieties based on geographical distribution and minor morphological characters. Apparently, *G. wightii* contains both diploid and tetraploid forms. Attempts to cross *G. wightii* with the soybean have not been successful. The chromosomes of *G. wightii* are much larger than those of the species in the subgenera Glycine and Soja. It has been suggested that it might be desirable to treat subgenera Glycine and Soja, with their smaller chromosomes and basic number of $x = 20$ (? $x = 10$) as a distinct genus.

G. soja is an annual twining vine with small, hard, roundish, black, to dark brown seed. It grows wild in the eastern half of the People's Republic of China, adjacent areas of the USSR, Korea, Japan and Taiwan. Hybrids between *G. soja* and the soybean have been made by numerous workers. Current available evidence indicates that there are few, if any, cytogenetic barriers to gene flow between *G. soja* and the soybean.

G. max is an annual, erect, coarse herb that has never been found in the wild. In Chinese and Japanese the cultigen is known respectively, as 'ta tou' and

'daidzu'. Based on chromosome number, chromosome size, morphology, geographical distribution and electrophoretic banding patterns of seed proteins, *G. soja* is most probably the ancestor of the soybean. The major differences between the domesticate, *G. max*, and its putative ancestor, *G. soja*, are as follows: increased seed size, increased oil but decreased protein content in seed, erect habit of growth, larger plant and reduced shattering of mature seed from the pod (Hadley and Hymowitz, 1973).

3 Early history

Current evidence points to the eastern half of northern China, what is essentially today's winter wheat–kaoliang region, as the area in which the soybean first emerged as a domesticate around the eleventh century B.C. Probably, the eleventh century B.C. date will be pushed back in time as additional archaeological evidence is uncovered in the People's Republic of China. The migration of the soybean from the eastern half of northern China to southern China, Korea, Japan and southeast Asia probably took place during the expansion of the Chou dynasty (*ca.* eleventh to third century B.C.) (Hymowitz, 1970).

By 1712, Europeans had become aware of the potential of soybeans through the publications of Engelbert Kaempfer who lived in Japan in 1691 and 1692. From 1875, and for several years afterward, Prof. Frederick Haberlandt of Vienna attempted to expand soybean production in Europe but was not very successful.

The soybean was first mentioned in US literature in 1804. However, the expansion of soybean production in the USA did not come about until the third decade of the twentieth century. The earliest known date for introduction into Brazil is 1882. Soybean production currently is undergoing a rapid expansion in Brazil. For a general review of the development of the soybean industry in the United States and elsewhere, see Probst and Judd (1973).

4 Recent history

During the first three decades of the twentieth century, production was largely confined to the Orient. China, Indonesia, Japan and Korea were the major producers. However, starting in 1924, soybeans began their almost incredible rise to prominence in the United States. Prior to 1924, most US farmers considered the soybean as a hay crop. Between 1924 and 1973, the soybean area in the USA increased from 700,000 ha to 23 million ha. Total production increased from 134,000 t to almost 43 Mt. The total farm value of soybeans produced in the United States in 1973 was nearly 9 billion dollars. In 1973, soybeans became the most important cash crop in the United States, overtaking wheat and maize (Koch, 1974).

The state of Illinois has held the lead in soybean production since 1924. Production was reported for 30 states in 1973. The northern central states of Illinois, Iowa, Missouri, Indiana, Minnesota and Ohio produced 66 per cent of the US total and the Mississippi river delta states of Arkansas, Mississippi, Louisiana and Tennessee produced 16 per cent. In 1973, Illinois alone produced more soybeans than the People's Republic of China, the second leading soybean producer. The most popular cultivars in Illinois are Wayne, Amsoy, Beeson, Corsoy, Cutler and Clark 63.

Starting in the 1940s, the US Department of Agriculture initiated a programme to maintain germplasm and to screen the germplasm for certain economic traits such as chemical components of seed and sources for resistance to specific pathogens. Today, the germplasm collection is located at the University of Illinois, Urbana, and at the Delta Branch Experiment Station, Stoneville, Mississippi. The collection contains about 4,500 entries. It includes plant introductions, cultivars released in the USA, genetic stocks and wild species.

The major emphasis in contemporary breeding programmes is on the development of high-yielding varieties that are high in oil and protein content. Resistance to pathogens and pests, lodging, seed quality, shatter susceptibility of mature seed and height of first pods are important traits considered in breeding programmes. All soybeans destined for commerce must have yellow seedcoats and cotyledons.

Soybeans are responsive to day length. Photoperiodic sensitivity limits adaptation of a cultivar to a narrow belt of latitude. In the Americas, 12 maturity groups have been established for identifying regions of adaptation for soybeans cultivars. Those placed in Group 00, 0 and I are adapted to the longer days in Canada and the northern states of the USA. Proceeding southward are Groups II to VIII. Cultivars placed in Group VIII are adapted to production areas of the Gulf Coast. Soybeans in Groups IX and X are grown south of the continental USA in semitropical and tropical countries of Central America.

Soybeans are completely self-fertile, and the amount of outcrossing under natural conditions is about 1 per cent. However, in certain genotypes outcrossing can occur up to 10 per cent. Current breeding methods are dominated by the self-fertilizing property of the soybean. For an excellent discussion on soybean breeding methods see Brim (1973).

5 Prospects

In the next decade, the continued need for soybean protein to meet the demand for more animal products and especially for use in human foods will continue unabated. In the past, much of the increase in production has come about from an absolute increase in total area planted to soybeans. However, there is a finite amount of land that can be used to grow the crop. Therefore, there will need to be a major breakthrough in yield and this must come about from the development of new, higher-yielding cultivars.

A recent study revealed that for soybean cultivars currently grown in the USA, genetic uniformity is pronounced. Most of the currently grown cultivars are related to 11 lines: Mandarin, Richland, AK, Manchu, Clemson, Tokyo, Mukden, Dunfield, Arksoy, Roanoke and PI 54610. The cultivar Mandarin alone can be traced back in the pedigree of 84 per cent of the cultivars grown in the northern central states. Breeders are repeatedly crossing cultivars with similar genetic backgrounds to produce new cultivars. Occasionally, new germplasm is introduced, e.g. as a source of resistance to a particular pathogen.

The soybean industry is based entirely on germplasm introduced from the Orient. Unfortunately, the germplasm base needed for the improvement of the crop is rapidly being destroyed. The dwindling of these genetic resources is due to the impact of modern technology and the population explosion. The habitats of the wild and weedy relatives of the soybean are being covered by roads, houses, schools and factories. Traditional farmers in Asia, seeking ways to raise their standard of living, are rapidly converting from planting low-yielding 'land races' to superior higher-yielding cultivars developed by modern plant breeding techniques. Soybean 'land races' have originated in traditional farming areas in a number of ways. Individual farm families have grown different lines for hundreds of years and soybeans containing specific traits have been grown out year after year for religious, ceremonial or medicinal value.

The last major international collecting effort was made by W. J. Morse and P. H. Dorsett over 40 years ago. The two US Department of Agriculture seed explorers spent three years (1929 to 1931) collecting over 4,000 soybean seed samples from northeastern China, Japan and Korea. Unfortunately, less than one-third of their collection still survives. The other seed was either thrown out or lost. Obviously, a new effort must be undertaken to collect soybean seed samples from the Orient to provide new germplasm for contemporary plant breeders.

An innovative approach to the development of new varieties is the current effort to use male-sterile lines to promote natural hybridization. Brim and Young (1971) reported that 99 per cent of seed set on male-sterile plants was the result of natural crossing. The method is greatly superior to the hand emasculation method and should be extremely useful in recurrent selection programmes.

In the past, cytological, cytogenetic, and chromosome mapping research has been minimal. The trend in the next decade will be for research priorities to be assigned to fundamental genetic research. To date, only seven linkage groups of two or three loci each have been established, yet about 80 specific gene loci have been identified in the soybean. Little is known about the potential genetic reach of the soybean. Little work has been conducted on the species in the subgenus Glycine or for that matter on the species in genera closely allied to Glycine. If these species were found to be within genetic reach of the soybean, it would open up a new area of potentially useful germplasm for use in breeding programmes.

Soybean breeding, thus far, has been directed towards the needs of temperate countries. The trend in the next decade will be toward breeding the crop for subtropical and tropical countries. Investigations have begun in India, Brazil, Colombia, Nigeria and Uganda. The greatest difficulties in growing soybeans in the lowland subtropics and tropics will be to overcome the day length responsiveness of the species and to develop indigenous cropping patterns distinct from those utilized in the temperate zones.

6 References

Brim, C. A. (1973). Quantitative genetics and breeding. In *Soybeans: improvement, production and uses*, ed. Caldwell, Madison, Wisconsin, pp. 155–86.

Brim, C. A. and Young, M. F. (1971). Inheritance of a male-sterile character in soybeans. *Crop Sci.*, 11, 564–6.

Hadley, H. H. and Hymowitz, T. (1973). Speciation and cytogenetics. In *Soybeans: improvement, production and uses*, ed. Caldwell, Madison, Wisconsin, pp. 97–116.

Hermann, F. J. (1962). A revision of the genus *Glycine* and its immediate allies. *U.S.D.A. Agric. Res. Serv. tech. Bull.*, 1268, pp. 82.

Hymowitz, T. (1970). On the domestication of the soybean. *Econ. Bot.*, 24, 408–21.

Koch, C. (1974). *Soybean digest blue book*, Hudson, Iowa.

Probst, A. H. and Judd, R. W. (1973). Origin, U.S. history and development, and world distribution. In *Soybeans: improvement, production and uses*, ed. Caldwell, Madison, Wisconsin, pp. 1–15.

Smith, A. K. and Circle, S. J. (1972). *Soybeans: Chemistry and technology*. Vol. 1 *Proteins*. Westport, Connecticut.

Verdcourt, B. (1966). A proposal concerning *Glycine*. *Taxon*, 15, 34–6.

Verdcourt, B. (1970). Studies in the Leguminosae-Papilionoideae for the 'Flora of Tropical East Africa': II. *Kew Bull.*, 24, 235–307.

Lentil

Lens culinaris (Leguminosae–Papilionatae)

D. Zohary
The Hebrew University of Jerusalem Israel

1 Introduction

Lentil ranks among the oldest and the most appreciated grain legumes of the Old World, cultivated from the Atlantic coast of Spain and Morocco in the west to India in the east. It is a characteristic companion of wheat and barley cultivation throughout the belt of Mediterranean agriculture. Yields are relatively low (about 50–150 kg/ha), yet lentil stands out as one of the most tasty and nutritious pulses. The seed protein content is high (about 25%) and lentil constitutes an important meat substitute in many peasant communities. Large quantities of lentils are produced and consumed in India, Pakistan, Ethiopia, the Near East and the Mediterranean area. World production in 1971 was estimated at 1·07 Mt. The crop has been successfully introduced to the New World where leading producers are Argentina and Chile, as well as Washington State in the USA.

2 Cytotaxonomic background

Lens is a relatively small genus restricted (in the wild) to the Mediterranean basin and southwestern Asia. In addition to the crop (*L. culinaris* = *L. esculenta*), the genus includes only four wild species (Barulina, 1930; Zohary, 1972). Taxonomically, *Lens* holds an intermediate position between *Vicia* and *Lathyrus*. Its closest affinities seem to be with *Vicia* section Ervum. The four wild species of *Lens* are all slender, ephemeral, small-flowered annuals with characteristic broadly rhomboid, strongly compressed pods carrying 1–2(3) flattened seeds. All members of the genus are apparently predominantly self-pollinated.

We still lack experimental evidence on the genetic affinities between the wild lentils and the crop. But comparative morphology and observations on natural hybridization implicate one wild species, namely *L. orientalis*, in the ancestry of the cultigen (Zohary, 1972). Both in its vegetative and reproductive parts, *L. orientalis* appears as a miniature *L. culinaris*. The

wild plant differs from the cultigen in the characteristic seed dispersal device: the small pods burst at maturity and effectively release the characteristic lenticular seed. The chromosomes of *L. orientalis* and *L. culinaris* appear to be very similar. Both lentils are diploid, with $2n = 2x = 14$.

As its name implies the wild progenitor of the cultivated lentil is a Near Eastern plant. *Lens orientalis* is distributed mainly in Turkey, Syria, Israel, northern Iraq and western and northern Iran. Significantly, this is also the area from which comes the earliest archaeological evidence of lentil cultivation (Fig. 46.1).

Lens culinaris manifests a wide range of morphological variation, both in vegetative and reproductive parts. Scores of distinct varieties and lines have been described by various workers (Barulina, 1930). Conventionally, these cultivars are grouped in two intergrading clusters: (*a*) small-seeded lentils (subsp. *microsperma*), with small pods and small (3–6 mm) seed; this group overlaps in seed size with wild *L. orientalis*; (*b*) large-seeded lentils (subsp. *macrosperma*), with larger pods and with seed attaining 6–9 mm. As in other pulses, increase in seed size is one of the most conspicuous trends under domestication; *macrosperma* forms are to be regarded as the more advanced.

3 Early history

Lentils are definitely associated with the start of the 'agricultural revolution' in the Old World which was initiated by the domestication of einkorn and emmer wheats, barley, pea, flax and lentil. For detailed review see Zohary (1972, 1973) and Zohary and Hopf (1973). Carbonized lentil seeds have been identified from early farming villages in the Near East (dated seventh and sixth millenia B.C.). They are small (2·5–3·0 mm in diameter) and do not occur in large quantities. Larger amounts of seed were discovered in somewhat later phases of the Neolithic settlement of the Near East. Significantly, some of these fifth millenium B.C. remains are already somewhat larger than the wild form and attain 4·2 mm in diameter. This is an obvious development under domestication.

Lentils were closely associated with the spread of Neolithic agriculture to Greece and adjacent Bulgaria. Subsequently, they accompanied the expansion of wheat and barley in the Neolithic and Bronze-Age agricultures of the Near East, the Mediterranean basin and central Europe. Lentil was apparently also an important element in the assemblage of Near Eastern grain crops which were introduced to Ethiopia by the Hamitic invaders.

From the Bronze Age onwards, lentil maintained

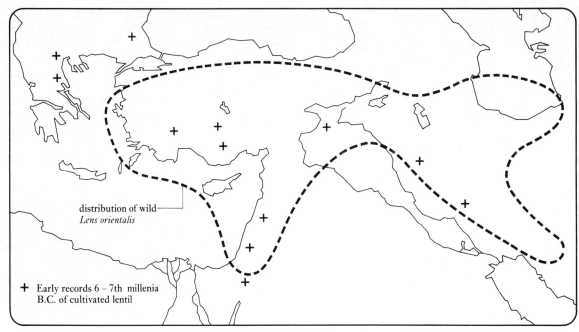

Fig. 46.1 Distribution of wild *Lens orientalis* and early records of cultivated lentils.

itself as an important companion of wheats and barley throughout the expanding realm of the Mediterranean type agriculture, excluding only the cold and humid northern fringe. In classical times, the large seeds which characterize *macrosperma* forms had already been attained and, since then, the crop has changed but little. It has provided a major source of protein to the entire belt of Mediterranean agriculture, from northwestern India to Spain and North Africa. Several centres of genetic diversity have been recognized, in particular, Afghanistan, Turkey and Ethiopia (Barulina, 1930). The crop is still widely grown and highly appreciated where the old traditions are maintained.

4 Recent history

Lentils have been but little altered by modern plant breeding. As in numerous other self-pollinated grain crops, variation has been (and still is) structured in the form of numerous true-breeding lines aggregated in land-races endemic to restricted geographical areas. Since the 1920s, work in lentils has centred on extensive collection and subsequent evaluation of existing varieties (with stress on yield, seed-size and resistance to diseases). In several breeding stations, this introduction-and-evaluation work was supplemented by intervarietal crosses and selection and release of new varieties.

5 Prospects

Tastiness and high quality protein make the lentil an attractive candidate for future development but success will depend primarily on the attainment of higher yields and on the development of varieties better adapted to mechanical harvest. Reduction of flatulence factors in the seed is another goal. The crop offers a considerable amount of variation yet unexploited by the breeder. Development of pure-line cultivars remains the safest (and most promising) breeding strategy. Utilization of hybrid varieties would depend on achieving drastic changes in floral biology, in the lentil indeed a long-range and difficult goal to attain. Moreover it seems unnecessary at this stage.

6 References

Barulina, E. I. (1930). [Lentils of USSR and other countries; a botanico-agronomical monograph.] *Trudy prikl. Bot. Genet. Selek. Supplem.*, **40**, 265–319.

Slinkard, A. E. (1974) (ed.). *The lentil letter*, 1. Saskatoon, Canada.

Zohary, D. (1972). The wild progenitor and the place of origin of the cultivated lentil: *Lens culinaris. Econ. Bot.*, **26**, 326–32.

Zohary, D. and Hopf, M. (1973). Domestication of pulses in the Old World. *Science, N.Y.*, **182**, 887–94.

Zohary, D. (1973). The origin of cultivated cereals and pulses in the Near East. In J. Wahrman and K. Lewis (eds), *Chromosomes today*, **4**, 307–20, New York.

Alfalfa, lucerne

Medicago sativa (Leguminosae–
Papilionatae)

K. Lesins
Department of Genetics University of
Alberta Edmonton Canada

1 Introduction

The world's most important forage crop is called
alfalfa in North and South America and in Iberia. Its
name in the rest of Europe, Oceania and in South
Africa is lucerne. The total area under alfalfa is
approximately 33 Mha (Bolton, 1962; Bolton *et al.*,
1972). In Europe (including European USSR) the
area is close to $9\cdot5$, in North America 13 and in South
America $7\cdot8$ Mha.

The plant is a perennial. Its strong root system
may penetrate to a depth of up to 6 m. From the crown
many over-wintering buds develop into numerous
leafy stems up to $1\cdot2$ m long. One to twelve cuts are
taken per season depending on length of growing
season, availability of nutrients and soil moisture.
Though alfalfa is a heavy user of water it does not
tolerate a high water table; nor does it tolerate acid
soils.

Alfalfa has a papilionaceous corolla with a peculiar
'tripping' mechanism for pollination. Because of this,
only specialized bees are efficient pollinators. Although this mechanism indicates that alfalfa is adapted
to cross-fertilization, some self-fertilization occurs in
cultivated as well as in wild plants.

2 Cytotaxonomic background (Fig. 47.1)

Cultivated *M. sativa* is an autotetraploid ($2n = 32$).
Its closest relative is the wild diploid ($2n = 16$),
taxonomically often known as *M. coerulea*. *Medicago
falcata*, the yellow-flowered alfalfa, is a little less
closely related. This species is known at diploid and
tetraploid levels. *Medicago falcata* is endemic to
the more northern regions, 42 to 62°N latitude,
whereas diploid *M. sativa* is found to the south of
42°N lat. Hybridization between the two taxa may
have been the factor which enabled cultivation of the
crop to spread all over the temperate zone. *Medicago
sativa* has violet flowers and the fruits, commonly
called pods, are spiralled tightly in 3–5 coils; *M.
falcata* has sickle shaped pods and yellow flowers.
The hybrids between *M. sativa* and *M. falcata* are
midway between the parental taxa; flowers in the F_1 are
greenish and the pods are loosely spiralled in 1–3 coils.
In later generations, segregation of both traits takes
place. The taxonomic names applied to hybrid populations (originally not suspected to be of hybrid
nature) are *M. media* or *M. varia*. Another species,
M. glutinosa, hybridizes freely with alfalfa. It is a
tetraploid ($2n = 32$), with loosely coiled pods covered
with glandular hairs. It is endemic to a part of the
Caucasus. A species not well delimited, often confused
with *M. glutinosa*, is *M. glomerata*. In collections from
the Maritime Alps it is a diploid ($2n = 16$). The flowers
are yellow, pods tightly coiled and covered with
glandular hairs. There are indications that *M.
glomerata* also occurs in North Africa (*M. sativa*
ssp. *faurei*), but whether as a diploid or a tetraploid is
not known. Since it readily hybridizes with *M. sativa*
and *M. falcata*, it may be that such forms as *M.
gaetula* and *M. tunetana* are of hybrid nature. *Medicago
glutinosa* and *M. glomerata* may have contributed to
formation of some strains of cultivated alfalfa.

Further away in relationship stands *M. prostrata*.
It is known at diploid and tetraploid levels. It may be
hybridized with *M. sativa* though some interbreeding
barrier is present. Hybridization, with some difficulty,
is possible with the two hexaploid ($2n = 48$) species,
M. cancellata and *M. saxatilis*; also with diploid
M. rhodopea. While hybridization generally is more
successful if partners are at the same chromosome
level, an exception is *M. sativa* ($2n = 16$) $\times$ *M. dzhawak-
hetica* (*M. papillosa*) ($2n = 32$), where hybrids are
much more easily produced at unequal levels. Finally,
trispecies hybrids, *M. sativa* $\times$ (*M. daghestanica* $\times$
M. pironae), have been obtained.

From the above it is obvious that germplasm donors
are available if it is found that they harbour traits
required for improvement of alfalfa.

The chromosomes of *M. sativa* and related species
are small (3 to $4\cdot2\mu$ in length in preparations of un-
pretreated somatic tissues). From pachytene studies,
an ideogram and identification key has been prepared
for the $x = 8$ set of *M. sativa* (Lesins and Gillies, 1972).

3 Early history

M. sativa in its primitive, diploid, state grows in Iran,
eastern Anatolia and, under semi-desert conditions, in
places around the Caspian Sea. The origin of its tetraploid, cultivated, form is probably closely tied with
the growing importance of horses in the ancient world.

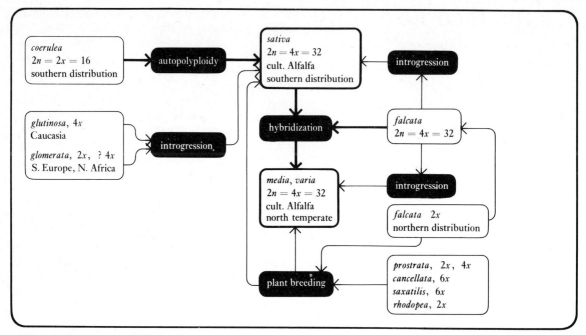

Fig 47.1 Evolution of alfalfa, *Medicago sativa*.

The history of the horse may elucidate how alfalfa came into cultivation: Indo-European nomadic horsemen from the plains of Eurasia moved south and east at the beginning of the second millennium B.C. Some of them crossed the Caucasus. A branch of these warrior horsemen moved along the folds of the central Zagros Mountains, which form the northwest–southeast boundary of the Iranian Plateau. They were assimilated by the native Kassite people. The district later became known as a centre of horse-breeding. In the first part of the second millennium B.C., a new weapon, the light horse-drawn chariot, created a revolution in the nature of warfare. In the middle of the eighteenth century B.C., Kassites seized Babylonia and ruled it for five centuries. To them the horse was a divine symbol and their pottery often shows winged horses. From the kingdom of Mitanni in the fourteenth century B.C., treatises deal with breeding and training of chariot horses and with their diet and veterinary care (Drower, 1969). It is understandable that horse breeders turned their attention to the native alfalfa. According to Hendry (1923), the name itself is traceable to Iranian 'aspasti' and further to old Iranian 'aspo-asti' meaning horse fodder. Arabicized it became alfalfa. Very probably, the more vigorous tetraploid form of alfalfa was noticed and evaluated by the early agriculturists and horse breeders. Mountain-protected fertile valleys in Zagros and the Elburz Mountains and irrigated oases facing the central Iranian Plateau may

well be considered places where alfalfa was first taken into cultivation.

The importance of horses remained great for more than two millennia. Assyrians, in the first part of the first millennium B.C., sent expeditionary forces to Iran, looting metal and horses. People of Ururtu, around Lake Van in eastern Anatolia, were famous horse breeders (Ghirshman, 1954). In the seventh century B.C., Scythians sacked and pillaged a number of countries including Palestine. The prophet Jeremiah foretold: '. . . his chariots shall be as whirlwind: his horses are swifter than eagles'.

Alfalfa, aspasti, is mentioned in a Babylonian text from about the seventh century B.C. The crop came to Europe in the fifth century B.C. during Persian wars with Greece. The Greeks called it Medicai, implying that it came from Media, a part of Iran; *Medica* and *Medicago* are derivatives of that name. In the second century B.C., the Romans acquired the plant from the Greeks. With the Romans the crop was introduced wherever their armies and colonists went. After the fall of the Roman Empire in the fifth century A.D., it is assumed that alfalfa totally disappeared from cultivation in Europe for about one thousand years, with the exception of Spain, where it reappeared with the Moslems in the eighth century. However, where the wild growing *M. falcata* has come into contact with *M. sativa*, some introgression has taken place. Hybrids of *M. media* (*M. varia*) type can be found in areas where as far as memory goes, no *M. sativa* has been planted. In the sixteenth century A.D., alfalfa was introduced from Spain to Italy and from there to the rest of Europe during the next two hundred years. The sixteenth century was also the time when Spaniards came to Peru and Mexico and introduced alfalfa to the New World. From South America it came to California in about 1850. It proved to be very successful there and spread to other southern districts of North America.

It is probable that the spread of alfalfa to the East from Iran also followed the horse. It is recorded that, in 126 B.C., the Emperor of China sent an expedition to Iran to secure specimens of the much admired Iranian horses. While acquiring horses, the expedition also acquired seed of alfalfa which was native to that region and which constituted their principal fodder.

In the eighteenth century A.D., settlers from Europe brought the plant to Australia and New Zealand. In mid nineteenth century it reached South Africa.

4 Recent history

In 1901 field trials were initiated in Minnesota to evaluate the performance of a certain alfalfa strain introduced into the USA by a German immigrant, Wendelin Grimm. The results of the tests were so convincing that Grimm alfalfa became famous. It (and other strains with some *M. falcata* introgression) were winterhardy. The consequent increase in acreage was spectacular. The area of *M. media* type cultivated in the north temperate zone may well be comparable with that of typical *M. sativa* in the south.

In the mid 1920s, bacterial wilt disease became a serious threat to the crop in the USA. Dr Jones, at the University of Wisconsin, identified the organism causing the disease and developed techniques for measuring wilt resistance in alfalfa. This may be considered the starting point of modern alfalfa variety breeding. In 1942, two wilt-resistant varieties were released. In 1954 the spotted alfalfa aphid made its first appearance and quickly threatened alfalfa production in the entire southwest of the USA. Only three years later, two resistant varieties were released. During the period 1957–71, 25 more resistant varieties were put on the market. Recurrent phenotypic mass selection thus proved to be very successful as a breeding method against insect pests and diseases.

In breeding for yield, potential parents are usually field-grown according to polycross, topcross or open-pollination methods. Their progenies are assumed to disclose parental ability for high yield in general and also for high yield in specific combinations. The best combiners are included in synthetic varieties. These breeding methods would provide the desired informa-

tion if the field plots were saturated with cross pollinating agents as is the case with wind pollination. Such saturation, however, is rarely the case in alfalfa. Alfalfa plants vary widely in self-fertility and in yield reduction following inbreeding. In most progenies from tests of two-clone combinations under cages with honey bees, 50 per cent (even 90%) of the progenies originate from selfing; different degrees of selfing have been found in other cases (Busbice *et al.*, 1972). Data obtained for general and specific combining abilities may be misleading, depending on the true amount of cross-pollination taking place in the test plots. Thus, the preoccupation of breeders with averting threats to alfalfa cultivation by diseases and insects and lack of adequate breeding methods have caused yield gains to be less than spectacular; in the USA, top yielding varieties average about 10 per cent above Ranger which was released more than 30 years ago. In several other crops, yield gains have been 40–100 per cent or more during the same period.

5 Prospects

Alfalfa, in spreading from its area of origin to moister and more northern zones, has not been followed by its natural pollinators, the ground-nesting bees. Honey bees are poor pollinators especially in northern areas. Consequently, poor and unpredictable alfalfa seed yields have plagued growers continuously. Domestication of a leafcutter bee (*Megachile rotundata*) may be considered a major breakthrough in alfalfa seed production (Pedersen *et al.*, 1972; Hobbs, 1973). The bee, however, requires high temperatures (above 21°C) and sunny weather to be an efficient pollinator. This restricts alfalfa seed production to areas where other cash crops often give much higher returns. Attempts are therefore being made to develop a self-fertilizing strain which would be tolerant of inbreeding (present author, unpubl.).

Very recently, using cytoplasmic male sterility, breeders have been selecting lines of high specific combining ability with which to make 'hybrids'. Seed production, however, has often been disappointing because of lack of pollinators.

6 References

Bolton, J. L. (1962). *Alfalfa*. London, New York.

Bolton, J. L., Goplen, B. P. and **Baenziger, H.** (1972). In Hanson (ed.), *Alfalfa science and technology*, Madison, 1–34.

Busbice, T. H., Hill, R. R. Jr. and **Carnahan, H. L.** (1972). In Hanson (ed.), *Alfalfa science and technology*, Madison, 283–318.

Drower, M. S. (1969). The domestication of the horse. In

Ucko and Dimbleby (eds.), *The Domestication and exploitation of plants and animals*, London, 471–8.

Ghirshman, R. (1954). *Iran*. Bungay, Suffolk, England (translation).

Hendry, G. W. (1923). Alfalfa in history. *J. Amer. Soc. Agr.*, 15, 171–6.

Hobbs, G. A. (1973). Alfalfa leafcutter bees. *Can. Dept. Agr. Publ.*, 1495, pp. 30.

Lesins, K. and Gillies, C. B. (1972). In Hanson (ed.), *Alfalfa science and technology*, Madison, 53, 86.

Pederson, M. W., Bohart, G. E., Marble, V. L. and Klostermeyer, E. C. (1972). In Hanson (ed.), *Alfalfa science and technology*, Madison, 689–720.

Beans

Phaseolus spp. (Leguminosae-Papilionatae)

Alice M. Evans
Department of Applied Biology University of Cambridge Cambridge England

1 Introduction

The genus *Phaseolus* has contributed crop plants to agriculture in both the Old and the New Worlds. *Phaseolus* beans are valued grain legumes or pulse crops in many tropical countries and are usually consumed as dry beans whereas, in temperate countries, although some of the species are grown as dry beans (notably *P. vulgaris*), varieties have also been developed for fresh pod consumption and for processing as frozen vegetables. They are important sources of protein and calories in human diets in tropical Africa and America where they supplement the carbohydrate staple foods like maize, plantains and cassava. With the increased cost of animal proteins in Europe and North America some species may become important protein-producing crops in these regions too.

2 Cytotaxonomic background

It is now generally agreed that *Phaseolus* originated in the New World and that Old World species previously classified in *Phaseolus* should be assigned to *Vigna*. Westphal (1974) reviews the evidence for the revision of these two genera. There are four cultivated species in the New World:

P. vulgaris: the common, haricot, navy, French or snap bean.

P. coccineus: the runner or scarlet runner bean.

P. lunatus: the Lima (large seeded), Sieva (small seeded), butter or Madagascar bean.

P. acutifolius var. *latifolius*: the Tepary bean.

All the species are diploid with $2n = 2x = 22$. Interspecific hybridization in the genus *Phaseolus* has been carried out by many workers as reported by Smartt (1970), who showed that viable hybrids of varying fertility can be produced beteween *P. vulgaris* and *P. coccineus*. There is some affinity between these two species and *P. acutifolius*, as shown by the formation of viable but sterile hybrids, but less affinity between

P. lunatus and the other three species. In general, hybrids of *P. vulgaris* ♀ × *P. coccineus* ♂ can be produced easily while the reciprocal is produced only with difficulty. Miranda and Evans (1973) have shown that the reciprocal differences in crossability between these two species do not occur with wild populations, suggesting that the unilateral incompatibility has developed under cultivation. *Phaseolus coccineus* is generally cross-pollinated by bees but there appears to be no genetic barrier to self-pollination in this species, as shown by investigations on pollen tube growth (Hawkins and Evans, 1973). The other three species are self-pollinating with only a small amount of out-pollination.

3 Early history

Smartt (1969) has reviewed the evolution of American *Phaseolus* beans under domestication and has shown that the archaeological record in Middle and South America is impressive compared with that of some other crop plants. Radiocarbon datings of 7680 B.P. have been reported for *P. vulgaris* from Callejon de Huaylas, Peru (Kaplan *et al.*, 1973) and of 7000 B.P. from the Tehuacan Valley in Mexico. There is evidence that domestication of *P. vulgaris* also occurred in Brazil and northern Argentina from a wild form of which *P. aborigineus* is a modern survivor. The large limas of *P. lunatus* were domesticated by 4,500 years ago in Huaca Prieta, Peru, and the small-seeded limas by 1,400 to 1,800 years ago in Mexico. *Phaseolus acutifolius* domesticates date to 5,000 and *P. coccineus* to 2,200 years ago in the Tehuacan Valley in Mexico

(Kaplan, 1965).

The four species occupy a range of habitats in the wild, from the cool humid uplands of Guatemala at altitudes of about 1,800 m, where *P. coccineus* grows, to the arid semi-desert conditions in which *P. acutifolius* is found. Warm, temperate conditions are best for *P. vulgaris* which occurs in an ecological transition belt at 500–1,800 m altitude, with the highest frequency of wild types at *ca.* 1,200 m in the Mexico–Guatemala area (Miranda, 1974). *Phaseolus lunatus*, on the other hand, is found in the tropics and sub-tropics (Mackie, 1943).

For both *P. vulgaris* and *P. lunatus* archaeological findings suggest separate domestications in Central and South America from conspecific geographic races, the small seeded types being native to Central America and the large seeded types to South America. Whether the Central American is the primary centre and the South American a secondary one, with migration from the primary to the secondary, or whether domestication was polycentric is an unsolved problem. If the former, then the question arises as to how migration took place in prehistoric times and how the large seeded forms of South America arose from the small seeded Middle American types.

All four species are treated as annuals in cultivation but, in the wild, *P. coccineus* and *P. lunatus* are perennial. Annual and perennial forms of wild *P. vulgaris* are found but *P. acutifolius* is strictly annual.

In post-Columbian times the two most important species spread widely. There is evidence that Spanish galleons took lima beans across the Pacific to the

Fig. 48.1 Relationships of the four cultivated species of *Phaseolus*.

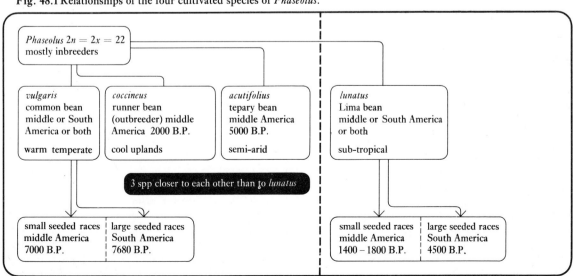

Philippines and thence to Asia, also from Peru to Madagascar (Westphal, 1974). The slave trade took the lima and the common bean from Brazil to Africa and subsequently inland to the trade routes. The latter reached Europe in the early sixteenth century. In Europe the French bean spread rapidly in the sixteenth and seventeenth centuries and reached England by 1594 (Purseglove, 1968). Beans also spread eastwards through the Mediterranean basin and into middle Europe. Beans were cultivated all over Italy in the seventeenth century and in Greece, Turkey and Iran. In North America these beans spread through California (5,000 to 2,000 B.P.) and into what are now the western areas of the USA. In the eastern part of the USA there were numerous introductions from Europe in the late nineteenth century.

The dispersal of the two most widely grown species is thus fairly well documented and both species are suitable models for the study of crop plant evolution. Wild populations are still available, primitive cultivars are found in Middle and South America, land races in the African tropics, and a range of more advanced cultivars in Europe and North America.

4 Recent history

The most obvious effect of domestication and of subsequent plant breeding in *Phaseolus* is the modification of growth habit. To take *P. vulgaris* as an example, in the germplasm collection of 5,000 accessions of wild material and cultivars examined at Cambridge, five races have been described under field conditions. The pattern of evolution appears to be as follows (Evans, 1973):

Middle America

Small seeded, indeterminate climbers	*Race 1(a)*
Indeterminate semi-climbers	*Race 2*
Indeterminate bush types	*Race 3*
Determinate, multi-noded	*Race 4*

South America

Large seeded, indeterminate climbers	*Race 1(b)*
Determinate bush types	*Race 5*

In the runner and lima beans no indeterminate bush types occur but the latter has many determinate bush varieties while the former has only one (Hammond's Dwarf). In the tepary, all cultivars are indeterminate bush types (Smartt, 1969). There have also been reductions in the number of branches and leaves as a consequence of domestication and an increase in leaf size and stem diameter in all species except *P. acutifolius*.

The most important changes which have taken place in the fruiting characters of these species are in pod and seed size, although there has also been an appreciable increase in flower size in *P. vulgaris* and *P. coccineus*. Seed numbers per pod have changed in *P. vulgaris* and *P. acutifolius*, where up to nine seeds are found in wild forms but rarely more than five in most cultivars. There has been no significant change in *P. lunatus* and *P. coccineus*. The latter, however, shows a tremendous increase in pod and seed size from 6 to 30 cm in pod length and from 100 to 1,000 mg for individual seed weight. In all species, an increase in permeability of the seeds to water has taken place under domestication which is an important property for uniform germination and for ease of cooking (Miranda, 1974).

A remarkable range of testa colours occurs in the seeds of *Phaseolus* species. In *P. vulgaris*, five main colour groups have been identified, namely: white, black, red, ochre and brown. Superimposed on these ground colours may be various patterns of spots, flecks and stripes. The genetics of seed coat and flower colour in *P. vulgaris* has been studied by Prakken (1970).

Pod structure has also been altered in cultivars, with a reduction in dehiscence and in fibre content. Three distinct pod textures are found in the common bean: the parchmented, which are very fibrous and dehisce strongly at maturity; the leathery types which are less dehiscent but split readily along the sutures; and the fleshy or stringless pods which are indehiscent and do not split readily. Varieties with parchmented pods are used only for dry seed production; leathery podded varieties can be used for green pod production when young, or haricot production when fully mature. Fleshy podded or stringless types are used entirely for the production of green pods (Smartt, 1969).

A further change which has taken place with the dispersal of *Phaseolus* beans to temperate regions of the world is in response to photoperiod. Since the genus is native to the tropics many cultivars from the region are photoperiod sensitive and responsive to short days. There are also day-neutral types available and all the temperate cultivars have been selected for day-neutrality or tolerance of long days.

5 Prospects

Most breeding work has been carried out on *P. vulgaris* and *P. lunatus*, primarily in the USA but also in Europe for the former species. There is currently considerable interest in the breeding of these pulse crops in Africa and Latin America, particularly with the development of national programmes and the efforts of the International Crop Research Institutes in

Colombia for the common bean and in Nigeria for the lima bean. The world production of *P. vulgaris* approaches 11 Mt of dry beans and under 2 Mt of green beans.

These pulse crops are notoriously low in yield when compared with the cereals, partly no doubt because more breeding effort has been devoted to the cereals. The first aim in breeding must be to breed for high yield potential. Breeders divide over the issue of whether to breed for the strict ideotype as defined by the physiologist, where there is generally specific adaptation to a given environment, or whether to breed for adaptability to a range of environments and hence to select for low genotype-environment interaction. Adams (1974) has made a case for a bean ideotype in *P. vulgaris* and Evans (1974) for breeding for adaptability. There is general agreement that the determinate bush varieties with pods held high off the ground are the most desirable for mechanical harvesting. There may, however, be a place for the indeterminate bush types in the tropics, since they provide an assurance of continued growth after temporary environmental stress (Leakey, 1971).

High yield potential cannot be considered in isolation from disease and pest resistance. Zaumeyer and Thomas (1957) have reviewed the status of the main diseases of *P. vulgaris*. These include: anthracnose (*Colletotrichum lindemuthianum*) which is one of the most destructive diseases and is now almost worldwide; root rots (caused by *Fusarium oxysporum f. phaseoli*); bacterial blights (caused by *Xanthomonas* spp.); rust (caused by *Uromyces appendiculatus*); and three types of viruses causing mosaics. In the lima bean, the main diseases include downy mildew (*Phytophthora phaseoli*) and pod blight (*Diaporthe phaseolorum*). Pest resistance is also important in the tropics, particularly to the pod borer and to the weevils which attack bean seed in storage. Recently, there has been concern in the USA over the genetic vulnerability to disease of the advanced cultivars, particularly the Michigan navy beans, which have a narrow genetic base. Attempts are being made to overcome this potential hazard and one way of doing this is to employ the recurrent selection procedures which ensure maximum use of genetic resources.

In the last decade some emphasis has been placed on protein content and quality. The beans are rich in protein (about 22%) but, as in most legumes, the proteins are deficient in the sulphur amino acids. Unlike the situation in cereals, there is no unequivocal evidence that there is a significant negative association between protein percentage and yield. Although correlations between yield and protein are negative, the coefficients are generally low and there is every hope that high protein and high yield can be combined in these crops. The heritability of protein content is nevertheless low because of the strong influence of the environment.

Methionine and cystine are the first limiting amino acids, followed by tryptophan, but these crops are high in lysine and so can supplement cereal proteins, which are generally low in lysine.

The presence of anti-nutritional factors such as trypsin inhibitors, haemagglutinins and flatus factors presents another problem and work is in progress to try to reduce these components by genetic means or to ensure adequate cooking to destroy the toxic factors. Hutchinson (1970) points out the need for the breeder to work closely with the biochemist and the nutritionist.

The breeder also has a further task particularly in breeding grain legumes for developing countries. The symbiosis whereby legumes gain N compounds from the rhizobia in the root nodules is not always efficient. Hence the difficult breeding project of selecting an efficient combination of legume host and rhizobial symbiont for the production of effective root nodules is one of importance, particularly in the tropical world. However, unequivocal responses in seed yield to inoculation with *Rhizobium* in beans have so far rarely been reported. This may, at least in part, be attributed to the N reserves of these large seeded legumes, in contrast to the lack of reserves in the small seeded leguminous crops; or it may be that response is shown in forage legumes grown for their vegetative parts but not in grain legumes grown for seed.

Finally, reference has already been made to the narrow genetic base of some of the advanced cultivars. There is now a greater awareness of the need for the conservation of the genetic variation available in these crops. Germ plasm seed banks are being developed and utilized and the International Institutes will help considerably in this venture to ensure that breeding programmes do not reach an evolutionary endpoint.

6 References

Adams, M. W. (1974). Plant architecture and physiological efficiency in the field bean. *In Potentials of field beans and other food legumes in Latin America*, CIAT, Cali, Colombia (1973) Series, Seminar 2E, 266–78.

Evans, A. M. (1973). Genetic improvement of *Phaseolus vulgaris*. In *Nutritional improvement of food legumes by breeding*. P.A.G. Symposium, F.A.O. Rome (1972) 107–15.

Evans, A. M. (1974). Exploitation of the variability in plant architecture in *Phaseolus vulgaris*. In *Potentials of field beans and other food legumes in Latin America*. CIAT, Cali, Colombia (1973) Series, Seminar 2E, 279–86.

Hawkins, C. F. and Evans, A. M. (1973). Elucidating the behaviour of pollen tubes in intra- and interspecific pollinations of *Phaseolus vulgaris* and *P. coccineus*. *Euphytica*, **22**, 378–85.

Hutchinson, J. B. (1970). The evolutionary diversity of the pulses. *Proc. Nutr. Soc.*, **29**, 49–55.

Kaplan, L. (1965). Archaeology and domestication in American *Phaseolus* beans. *Econ. Bot.*, **19**, 358–68.

Kaplan, L., Lynch, T. F. and Smith, C. E. (1973). Early cultivated beans (*Phaseolus vulgaris*) from an intermontane Peruvian valley. *Science*, N.Y., **179**, 76–7.

Leakey, C. L. A. (1971). The improvement of beans (*Phaseolus vulgaris*) in East Africa. In *Crop improvement in East Africa*. C. A. B. Farnham Royal, 99–128.

Mackie, W. W. (1943). Origin, dispersal and variability of the Lima bean, *Phaseolus lunatus*. *Hilgardia*, **15**, 1–29.

Miranda, C. S. (1974). Evolutionary genetics of wild and cultivated *Phaseolus vulgaris* and *P. coccineus*. *Ph.D. thesis, University of Cambridge*.

Miranda, C. S. and Evans, A. M. (1973). Exploring the genetical isolating mechanisms between *Phaseolus vulgaris* and *P. coccineus*. *Ann. Rep. Bean Improv. Coop.*, **16**, 39–42.

Prakken, R. (1970). Inheritance of colour in *Phaseolus vulgaris*. A critical review. *Meded. Landbouwhogeschool, Wageningen*, **70**(23), 1–38.

Purseglove, J. W. (1968). *Tropical Crops: Dicotyledons* I. London, 284–310.

Smartt, J. (1969). Evolution of American *Phaseolus* beans under domestication. In P. J. Ucko and G. W. Dimbleby (eds.), *The domestication and exploration of plants and animals*. London, 451–62.

Smartt, J. (1970). Interspecific hybridization between cultivated American species of the genus *Phaseolus*. *Euphytica*, **19**, 480–89.

Westphal, E. (1974). *Pulses in Ethiopia, their taxonomy and agricultural significance. Phaseolus*. Agric. Res. Rep. Wageningen, **815**, 129–76.

Zaumeyer, W. J. and Thomas, H. R. (1957). *A Monograph study of bean diseases and methods for their control. USDA tech. Bull.*, **868**, pp. 255.

Peas

Pisum sativum (Leguminosae-Papilionatae)

D. Roy Davies
John Innes Institute Norwich England

1 Introduction

The pea crop constitutes one of the four most important seed legumes, with an estimated total world production of approximately 10 Mt. It is grown most extensively in cool countries, flourishing in northern Europe, parts of Russia and China and the northwestern USA, though also having an important role at high altitudes in the tropics and as a winter crop in some hotter regions. As such, the crop constitutes an important source of protein for human consumption. Traditionally, the crop has been grown for harvesting as fresh peas or as a dry mature product but, in many areas of the world, the use of the crop has changed materially in recent years. A substantial proportion of the crop grown in northern Europe and North America is now harvested as immature peas for freezing; this product has become one of the most important 'convenience foods' demanded by the twentieth-century housewife and a new industry has developed to support the large-scale production and processing of immature peas.

2 Cytotaxonomic background

Taxonomists have classified the members of the genus *Pisum* into a number of species but more recent analyses have questioned the validity of separating many of the groups. Thus, Ben-Ze'ev and Zohary (1973) intercrossed the cultivated *P. sativum* with *P. elatius*, *P. humile* and *P. fulvum* and confirmed earlier reports that there was no cytogenetic basis for considering the first three as anything but members of a single species, namely *P. sativum*. The genus is now considered to have only two species, namely *P. fulvum* and *P. sativum* (Davis, 1970). All members of the genus are self-pollinating diploids ($2n = 14$) and all intercross freely, although *P. fulvum* is more readily crossed with *P. sativum* when the latter is used as the maternal parent. The F_1 seeds of the cross *P. fulvum* × *P. sativum* are shrunken whereas, in the reciprocal cross,

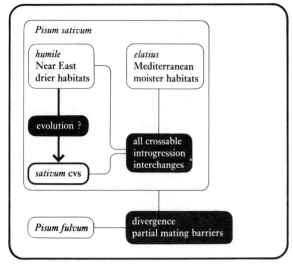

Fig. 49.1 Relationships of cultivated *Pisum sativum* and its wild relatives.

although the F_1 seeds are normal, the seedlings show a stunting of growth and other developmental abnormalities. Meiotic analyses of such F_1 hybrid plants indicate that the two species have diverged chromosomally, since quadrivalents, trivalents and univalents are observed at metaphase I. *P. elatius* (and certain sub-populations of *P. humile*) differ from *P. sativum* by a chromosomal translocation (Ben-Ze'ev and Zohary, 1973). Whenever such a translocation is present in a sub-population, all members are homozygous for that chromosome change. The so-called perennial pea, *P. formosum*, is now ascribed to a separate genus, *Vavilovia* (Davis, 1971). Where cultivated peas are found in close proximity to *P. humile* and *P. elatius*, hybrids are detected and it is assumed that such introgression has helped to generate the variation now seen in the cultivated crop.

3 Early history

Neither the wild progenitor nor the early history of the pea crop is known. Excavations of Neolithic settlements (*ca.* 7000 B.C.) in the Near East and in Europe have revealed carbonized pea seeds. These had a smooth surface (like the present-day cultivated varieties), indicating that the cultivation of the pea is as old as that of wheat and barley (Zohary and Hopf, 1973). Vavilov (1949) considered the probable centres of origin to be Ethiopia, the Mediterranean and Central Asia, with a secondary centre of diversity in the Near East; there is no definitive knowledge as to which of these is the primary source and which are merely centres of diversity. *Pisum elatius* is a tall

climber with long pods and small seeds (though little smaller than those found in some cultivars of *P. sativum*). It is found in the more humid regions of the Mediterranean. *Pisum humile* has a smaller habit, resembling *P. sativum* in this respect, and is widely distributed in the Near East; it occupies drier habitats than *P. elatius*. Nevertheless, intergrading forms, both between these two groups, and between them and the cultivated forms, are found. Modern cultivars differ from the wild forms in generally having larger seed and a shorter compact habit, but it must be stressed that there are also both tall and small-seeded cultivars. Some of the primitive forms have slightly bitter seed and, in general, have a tough seed coat which allows them to remain dormant in the soil for long periods of time. On the basis of morphological and cytological evidence, Zohary and Hopf (1973) suggest that those sub-populations of *P. humile* which show no chromosomal divergence from *P. sativum* could be the ancestral form from which the cultivars were derived. This conclusion must, however, be considered tentative.

4 Recent history

A detailed account of the recent history of peas is given by Berger (1928). Almost certainly, the crop has been grown by man for several thousand years. Greek and Roman writers mentioned peas, but not until the sixteenth century were any varieties described; first, a distinction was made between the field peas with their coloured flowers, small pods and long vines, and the garden peas, usually with white flowers and large seed. Later, the edible podded peas (those lacking a parchment layer in the pod) and tufted or Scottish (fasciated) peas were mentioned, followed rapidly by more detailed descriptions of a large number of varieties. In 1787 Knight began his controlled intercrossing of varieties, peas being the first crop in which controlled breeding was undertaken for the production of new varieties. In the nineteenth century, English breeders produced some of the classic pea varieties which persisted in horticulture for so long; in their experiments they observed, but failed to interpret, some of the phenomena later analysed by Mendel.

Until the last thirty years, breeding objectives had remained constant; particular varieties were selected for the fresh market, others for harvesting as dried peas and yet others for canning. For the last of these three uses, either dried peas are soaked or immature seeds are processed. For dried and canned peas, the colour and quality of the product are important, good quality being readily recognized by agronomists but difficult to define in biochemical terms. Traditionally, round seeded forms have been used for dried peas; for canning or for seeds harvested in the immature state, either round or wrinkled forms are used; wrinkled seeds have higher sugar and lower starch contents.

During the last two decades, a new industry has developed for the production of immature (wrinkled) peas for freezing and this has imposed new demands on breeders. Hitherto, this kind of crop has been grown on a horticultural scale but production has now reached agricultural dimensions. The primary requirement is for peas of a given maturity stage (those either too young or too old being rejected), and it is this that has imposed new objectives in breeding. The criterion of maturity stage is strictly defined, harvesting being controlled to within a matter of hours. The demand is, therefore, for a crop in which as many peas as possible are at the correct developmental stage at the time of harvest. The simplest way of achieving this is by increasing the number of pods per node, as all the pods at a node mature more or less simultaneously; some varieties have only one, most current varieties two or three, but genotypes with four pods per node are available. An increased simultaneity of development of pods at successive nodes can also be achieved by the introduction of certain genes regulating flower development.

5 Prospects

There must be considerable scope for yield improvement since advances comparable to those achieved in the cereal crops have clearly not been attained. Improved resistance to diseases caused by *Fusarium* species, *Ascochyta pisi* (pod spot) and *Peronospora viciae* (downy mildew), as well as to those caused by viruses, is of importance, particularly in more humid areas. An interesting and possibly more speculative development is the production of genotypes with markedly different foliage characteristics. Several single gene mutations are available in peas which affect the development of the leaf or stipule; for example, one of these (*af*), when homozygous, converts the leaflets to tendrils; such a phenotype could have agronomic advantages (Snoad and Davies, 1973). The 'sward' characteristics of such plants are entirely different from those of the normal type; the reduced amounts of foliage could lead to a more rapid and easier processing of the crop for freezing and, particularly in wetter regions, to easier drying of the crop grown for seed. The more open and drier 'sward' could also lessen the development of pathogens.

With the increasing cost and demand for protein, there is scope for re-examining the potential role of the pea crop as stock feed; this will involve developing new criteria in breeding and selecting forms with high protein and improved contents of methionine and cystine. In this context, there may well be a need, particularly in northern Europe and North America, to broaden the fairly narrow genetic base on which the breeding of cultivated peas has previously been founded. Unfortunately, efforts to conserve genetic variability have been left rather late and, though collections of old cultivars and breeding material exist in many countries, few thorough collections of germ plasm have been assembled in centres of origin or diversity (Gentry, 1971).

6 References

Berger, A. (1928). Peas. In Hendrick, Hall, Hawthorn and Berger (eds.), *Vegetables of New York* 1, 1–132.

Ben-ze'ev, N. and Zohary, D. (1973). Species relationships in the genus *Pisum O. Israel J. Bot.* 22, 73–91.

Davis, P. H. (1970). *Flora of Turkey*, 3, 370–3. Edinburgh.

Gentry, H. S. (1971). Pisum resources, a preliminary survey. *Plant Genet. Resources Newsl.* 25, 3–13.

Snoad, B. and Davies, D. R. (1972). Breeding peas without leaves. *Span* 15, 87–9.

Vavilov, N. I. (1949). The origin, variation, immunity and breeding of cultivated plants. *Chron. Bot.* 13, 1–54.

Zohary, D. and Hopf, M. (1973). Domestication of pulses in the Old World. *Science*, N.Y., 182, 887–94.

Clovers

Trifolium spp. (Leguminosae–Papilionatae)

Alice M. Evans
Department of Applied Biology University of Cambridge Cambridge England

1 Introduction

The genus *Trifolium* includes approximately 250 annual and perennial species, commonly called clovers, which are native to the humid, temperate regions of the world. About 25 species are of significance as food for grazing animals and, of these, about 10 are agriculturally important. Clovers are components of natural grasslands and are also cultivated in association with companion grass species, in simple or in complex seeds mixtures. Their main value lies in their ability to fix atmospheric nitrogen through the bacteria of the genus *Rhizobium* which inhabit their root nodules, but they are also rich in minerals and trace elements.

True clovers are trifoliate and carry the flowers in head-like inflorescences which may have many florets per head (as in red clover) or only few (as in subterranean clover). The number of seeds per pod varies according to species from one to eight (Whyte *et al.*, 1953).

2 Cytotaxonomic background

The ten species which have found places in agriculture are as follows:

Table 50.1 The cultivated clovers

Section	Species (annual, perennial)	Common name	Chromosome number ($2n =$)
Lotoidea	*repens* (per.)	White	32
	hybridum (per.)	Alsike	16
	ambiguum (per.)	Kura, Caucasian	16, 32, 48
Vesicrastrum	*resupinatum* (ann.)	Persian	16
	fragiferum (per.)	Strawberry	16
Trifolium	*pratense* (per.)	Red	14
	incarnatum (ann.)	Crimson	14
	alexandrinum (ann.)	Egyptian, Berseem	16
Trichocephalum	*subterraneum* (ann.)	Subterranean	16
Chronosemium	*dubium* (ann.)	Yellow suckling	14

Red and white clovers are the two most widely known perennial species; the former is used mainly as a hay legume while the latter, because of its creeping nature, is a long-lived pasture legume. Alsike can replace red clover on more acid soils in northern Europe and in Canada; strawberry clover can replace white clover in wet conditions and has been used in the USA, Australia and New Zealand. The remaining perennial species is Kura or Caucasian clover which is of limited importance but has been used in apiaries in the USA because of its high nectar yield. The annual species include: Persian, Egyptian, crimson and subterranean clovers, all used as summer annuals in the Mediterranean and the Near Eastern regions; the last named is also grown extensively in Australia. Yellow suckling clover is the only other annual and is sometimes included in seeds mixtures in Britain for short leys, where it may persist by self seeding.

The annual species are generally self-fertile although crimson and Egyptian clovers require insect visitation to trip the flowers. The perennial species are self-incompatible, with the partial exception of strawberry clover which has self-fertile forms. Self-incompatibility in red clover is of the gametophytic (oppositional allele) type, under the control of S-genes.

Polyploidy is not common in *Trifolium* and most species are diploid with $2n = 14$ or 16 ($x = 7$, 8). White clover is, however, a tetraploid with $2n = 4x = 32$ but it shows regular bivalent formation at meiosis. There is some evidence that it is an amphidiploid and its related diploid species are thought to be *T. nigrescens*, a self-incompatible Mediterranean annual (Evans, 1962) and *T. occidentale*, a self-compatible perennial, indigenous to southern England, south-western France and Spain (Gibson and Beinhart, 1969). However, Chen and Gibson (1970a) have suggested that, since meiosis of hybrids between *T. repens* and autotetraploid *T. occidentale* ($4 \cdot 5^{IV}$) is similar to that of autotetraploid *T. occidentale* ($4 \cdot 64^{IV}$), *T. repens* may well be an autotetraploid of *T. occidentale* which has undergone diploidization. No conclusive evidence is yet available on the origin of *T. repens* but studies of meiosis of *T. repens*, *T. nigrescens*, *T. occidentale* and their hybrids indicate that these three species have high pairing affinity (Chen and Gibson, 1970b). *Trifolium uniflorum*, a self-incompatible autotetraploid perennial, is also thought to share the same genome as *T. occidentale* and *T. repens* but it is in some way differentiated (Chen and Gibson, 1972). *Trifolium ambiguum*, with diploid, tetraploid and hexaploid races, has some affinity with *T. hybridum* (Evans, 1962).

Several workers have attempted to produce interspecific hybrids in *Trifolium*, particularly to introduce new genetic variation into red and white clovers, but there has been no great success. Evans (1962) has

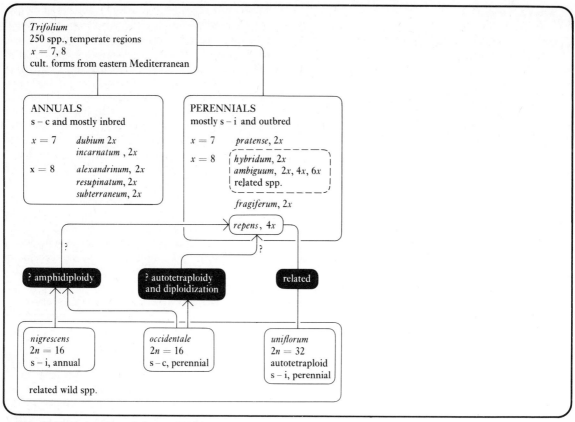

Fig. 50.1 Relationships of clovers, *Trifolium*.

suggested methods of overcoming interspecific incompatibility by grafting techniques and embryo culture, the former to screen for graft-compatible genotypes that might also be sexually compatible and the latter to salvage hybrid embryos which would otherwise abort. Neither interspecific hybridization nor polyploidy has been important in the evolution of the genus *Trifolium*; the rarity of wide hybridization in the forage legumes may well be associated with the predominant adaptation to insect pollination.

3 Early history

The centre of diversity of the clovers is in the eastern Mediterranean region. It is not always easy to distinguish between the distributions of wild clovers (which are determined primarily by climatic and natural biotic factors) and those of cultivated species (which have been deliberately or unconsciously selected by man and his grazing animals). However, the deliberate cultivation of pastures is so recent that documentary evidence is available on the spread of forage legumes. Red clover was cultivated in southern

Europe in the third and fourth centuries A.D., was introduced into Spain in the sixteenth century and, through Holland and Germany, to England in 1650. Wild red clover plants are stemmy and have few leaves while the cultivated types are more leafy and generally more luxuriant.

The settlement of North America, Australia and New Zealand resulted in a great increase in the distribution of forage legumes and their adaptation to new environments. For example, subterranean clover, a Mediterranean annual, has found an important place in the winter rainfall areas of Australia. White clover has spread to most temperate regions and, although there are no marked differences between wild and cultivated populations, there is a decrease in leaf size with increasing altitude and latitude in the northern hemisphere (Davies and Young, 1967).

Since the deliberate cultivation of forage crops has developed in comparatively recent times, the clovers present a group which is in active process of domestication (Cooper, 1965). The success of perennial forage legumes depends on successful co-operation with two unrelated organisms, namely rhizobial bacteria for nitrogen fixation and insects (chiefly bumble and honey bees) for pollination, the important species being out-pollinating. Evolution of all three organisms must occur in parallel if a successful relationship is to be maintained (Davies, 1971). White clover has a simple pollination mechanism and presents no problems; red clover, on the other hand, has evolved a longer corolla tube and can only be efficiently pollinated by long-tongued bumble bees (*Bombus* spp.). In strawberry clover, self-fertility has evolved from ancestral self-incompatibility, possibly as a result of the absence of suitable pollinating insects at the northern limits of distribution (Davies, 1971). As to the nodule organism, there is evidence that at high altitudes and latitudes in Europe, rhizobia ineffective for white clover are frequent.

With the dispersal of white clover, changes have occurred by natural selection and there is evidence of adaptation to local environments, for example, in the distribution of wild and cultivated populations having genetic polymorphisms controlling cyanogenesis. Leaves of some cyanogenic plants, when removed from the plant, release easily detectable amounts of hydrogen cyanide. The production of HCN depends upon the complementary action of one dominant gene (*Ac*) for glucoside and another (*Li*) for the enzyme. *Trifolium repens* plants can thus be of four phenotypes: glucoside plus enzyme, (G+E); glucoside and no enzyme (G); no glucoside but with enzyme (E); and neither (0). Daday (1965) has reported on how these polymorphic populations are maintained. He found that the distribution of the frequencies of the *Ac* and *Li* alleles showed a regular cline with altitude and latitude and depended upon the winter temperature in the northern hemisphere. A fall of 1° in the January mean temperature corresponds with a fall of 4·23 per cent in the frequency of *Ac* and 3·16 per cent in the frequency of *Li*. The frequencies of glucoside and enzyme alleles are also highly correlated. Opposing selective forces have been described. At higher temperatures, selective predation by rabbits and slugs of acyanogenic plants and general superiority of fitness of the cyanogenic plants give the latter a selective advantage. At low temperatures, because cold activates the enzyme, with consequent release of HCN which inhibits respiration, cyanogenic plants are at a disadvantage while acyanogenic plants also show superior fitness. These selection pressures are thought to be sufficient to maintain north–south clines in Europe and North America.

Another polymorphism in white clover is that of the leaf mark, presence of mark being dominant to absence and the various marks showing a multiple allelic series. Each genotype of amphidiploid white clover has two loci. The selective advantage associated with leaf marks is not understood, but it is possible that it is associated with the attraction of pollinating insects. Daday has reported that almost all the plants in his Mediterranean samples had an intense leaf mark and the proportion of plants without marks increased towards the more northerly latitudes of Europe. Leaf marks are absent from the self-pollinating species, crimson, Egyptian and Yellow suckling clovers.

4 Recent history

Purposeful breeding of clovers is very recent. The perennial forage legumes present particularly complex breeding problems. They are used under a variety of conditions; for example, for hay, silage or grazing. Furthermore, they are seldom grown as pure stands but usually with a companion grass, so management procedures become important. All these considerations have to be taken into account in designing a breeding programme. Breeding clovers is not only a genetic exercise in handling populations but it is also an ecological exercise in understanding adaptation and an agronomic exercise in devising the appropriate managements so that assessment techniques are as near to agricultural practice as possible.

The breeder can base his improvement on locally adapted forage varieties which have developed under the selective action of appropriate agronomic environments. Alternatively he can introduce plant material from homologous climatic areas elsewhere. Cooper (1965) has stressed the importance of understanding the nature of adaptation to particular environments. For example, in a Mediterranean environment the main limiting factor is summer drought and most species are either winter annuals, surviving the dry season as seed, or, if perennial, show summer dormancy. Most adapted populations from this region show the ability to grow at low temperatures and low light intensities. White clover populations from Spain introduced into New Zealand give considerably better winter production than do indigenous populations, but they are often more susceptible to frost damage.

In contrast, local varieties from northern Europe show a different seasonal pattern of growth. Winter cold is the limiting factor and most forage legumes are perennial, having considerable frost resistance and, often, winter dormancy. Leaf production at low temperature and low light intensities is poor.

These primary climatic effects can also be influenced by different agronomic systems; for example, in red clover. The early flowering broad red clover has evolved under a system of one-year leys, with two or more cuts per year while the late flowering single-cut types have developed under a system of grazing followed by late cutting. A similar situation in relation to management is seen between the cultivated Dutch white clover which has been grown for seed for many years and the wild white clover indigenous to old grazed pastures.

In his choice of material the plant breeder must be aware of these features. However, most adaptive characters are polygenically controlled and, if an adapted population is introduced into a new environment or subjected to a new management, considerable genetic variation between genotypes may be revealed. The plant breeder may utilize 'ready made' populations introduced from other regions or he may hybridize these with his local indigenous material (Davies, 1969). Although the unit of study for the plant breeder is the single plant, a number of genotypes with good combining ability have to be built up into synthetic or hybrid varieties in red clover and white clover. Such varieties must also show adequate ecological combining ability to grow satisfactorily with companion grass species under prescribed systems of management. In the self-pollinated clovers often grown as pure stands many of these problems do not exist and these types are bred as pure line varieties.

One approach to clover breeding in the diploid species, alsike and red clover, has been by way of induced polyploids. Tetraploid varieties of these species have been marketed, particularly in Sweden (Bingefors and Ellerström, 1964) but they have not yet had much agricultural impact, largely due to reduced seed fertility. However, the persistency, increased resistance to diseases, such as clover rot, (*Sclerotinia trifoliorum*) and eelworm (*Ditylenchus dipsaci*) in tetraploid red clover suggest that polyploids will have to place in future practice.

5 Prospects

The breeding of forage legumes has not received as much attention as grass breeding. In Britain this has largely been due to the fact that fertilizer nitrogen has been subsidized by the state so that grass production *per se* has been successful without too much consideration being given to the clover component. The emphasis in clover breeding has been to select genotypes which (paradoxically, perhaps) tolerate high levels of nitrogenous fertilizer and to select rhizobial strains which will operate at high N levels. In addition, in white clover, there have been attempts to select taller growing plants with longer petioles which would

be better able to compete for light with grass species under high fertilizer conditions. The recent energy crisis must tend to change this situation. With the increasing cost and lower availability of nitrogenous fertilizers, the clover contribution to grasslands is likely to become very important. The importance of legumes as soilage crops may also tend to increase, since nitrogen inputs can be reduced by planting legumes in rotation with cereals, or in alternate rows with other crops, followed by ploughing-in as green manure (Pimentel *et al.*, 1973). This will give impetus to the study of clovers, not only as soil enrichers but also as agents of N transfer in the grass sward. The latter is thought to occur by the excretion of N compounds from the nodules of the clover into the soil, by release of N through breakdown of clover nodules and roots and by transfer of N through the grazing animal.

The conservation of genetic resources for clover improvement is likely to be encouraged by the recent emphasis on plant introduction. The utilization of populations of clovers from other regions and the hybridization of these types with indigenous material, should, in principle, lead to release of new genetic variability which should in turn ensure flexible evolutionary response to future needs.

6 References

Bingefors, S. and **Ellerström, S.** (1964). Polyploidy breeding in red clover. *Zeit. Pflzücht*, **51**, 315–34.

Chen, Chi-Cheng and **Gibson, Pryce B.** (1970*a*). Meiosis in two species of *Trifolium* and their hybrids. *Crop. Sci.*, **10**, 188–9.

Chen, Chi-Cheng and **Gibson, Pryce B.** (1970*b*). Chromosome pairing in two interspecific hybrids of *Trifolium*. *Can. J. Genet. Cytol.*, **12**, 790–4.

Chen, Chi-Ching and **Gibson, Pryce B.** (1972). Barriers to hybridization of *Trifolium repens* with related species. *Can. J. Genet. Cytol.*, **14**, 381–9, 591–5.

Cooper, J. P. (1965). The evolution of forage grasses and legumes. In J. B. Hutchinson (ed.), *Crop plant evolution*, Cambridge, 142–65.

Daday, H. (1965). Gene frequencies in wild populations of *Trifolium repens*. IV. Mechanisms of natural selection. *Heredity*, **20**, 355–65.

Davies, W. Ellis and **Young, N. R.** (1967). The characteristics of European, Mediterranean and other populations of white clover (*T. repens*). *Euphytica*, **16**, 330–40.

Davies, W. Ellis (1969). Herbage legumes – special considerations. *Occ. Symp. Br. Grassld. Soc.*, **5**, 21–7.

Davies, W. Ellis (1971). Host–pollinator relationships in the evolution of herbage legumes in Britain. *Sci. Progr. Oxf.*, **59**, 573–89.

Evans, A. M. (1962). Species hybridization in the genus *Trifolium* 1. Methods of overcoming the barriers to species incompatibility. *Euphytica*, 11, 164–76.

Evans, A. M. (1962). Species hybridization in the genus *Trifolium* 2. Investigating the pre-fertilization barriers to compatibility *Euphytica*, 11, 256–62.

Gibson, Pryce B. and Beinhart, G. (1969). Hybridization of *T. occidentale* with two other species of clover. *J. Hered.*, **60**, 93–8.

Pimentel, D., Hurd, L. E., Bellotti, A. C., Forster, M. J., Oka, I. N., Scholes, O. D. and Whitman, R. J. (1973). Food production and the energy crisis. *Science*, N.Y., 182, 443–9.

Whyte, R. O., Nilsson-Leissner, G. and Trumble H. C. (1953). *Legumes in Agriculture*. FAO Agric. Studies, 21, pp. 367.

Field bean

Vicia faba (Leguminosae–Papilionatae)

D. A. Bond
Plant Breeding Institute Cambridge
England

1 Introduction

Vicia faba is an important legume grain in much of the north temperate zone and at higher altitudes in the cool season of some sub-tropical regions. It is also occasionally used in mixtures with other crops for silage or for green manure.

Its usefulness to man almost certainly derives from it being an erect plant with easily threshed pods containing large seeds of high protein content. FAO statistics show China to have had, in 1971, 3 Mha out of the world's estimated 4·7 Mha; other important bean-producing countries are Italy (though the area is decreasing due to *Orobanche*), Spain, the UK, Egypt, Ethiopia, Morocco, USSR, Mexico, Brazil and Peru. Until recently, there was very little cultivation of *Vicia faba* in North America or Australasia.

In some Asian, African and Mediterranean countries the green or ripe seeds provide a substantial part of the protein in human diet. In western Europe the use of fresh or preserved *V. faba* is confined to restricted areas and to the large-seeded varieties, smaller-seeded varieties being cultivated on a wider scale for animal feed and, on a limited scale, for racing pigeons. Few countries supply all their animal protein feed as field beans however, and the crop also functions as a beneficial break from cereals. There is some world trade in *V. faba* (about half the UK crop being exported) but its price depends on the world price of other legume grains, especially soya.

2 Cytotaxonomic background

Vicia faba is a diploid with $2n = 2x = 12$ and no polyploids are known. The chromosomes are characterized by their large size, high DNA content and the one long chromosome; there is evidence, however, of variability in karyotype, including translocations. Other *Vicia* species have $2n = 10, 12, 14$ and none can be hybridized with *V. faba* (Schäfer, 1973). No wild ancestor of *V. faba* is known; it has been suggested (Zohary and Hopf, 1973) that some other species of *Vicia* (e.g. *V. narbonensis*, *V. galilaea*) had a common ancestor with *V. faba* but there is no direct evidence of this and chromosome numbers, size and DNA content suggest that any evolutionary relationship must have been in the very remote past. Future studies of DNA type and repeated sequences may, however, provide data which would further the understanding of the origin of *V. faba*.

Muratova's (1931) intraspecific classification, based mainly on seed size, has been widely used. She recognized subsp. *faba* (vars *faba*, *equina*, *minor*) and subsp. *paucijuga*. However, Hanelt (1972*b*) considered *paucijuga* to be only a geographical race of subsp. *minor* and recognized subsp. *faba* (vars *faba*, *equina*) and subsp. *minor* var. *minor*. Cubero (1974) distinguished only the four varieties (*minor*, *equina*, *faba* and *paucijuga*). Sub-varieties have been named but fine systematic subdivision serves little purpose in this partially allogamous species in which there is considerable variation within populations and no sterility barriers between sub-species.

3 Early history

The bean is unlikely to have been among the first crops to be cultivated and Schultze-Motel (1972) concluded from archaeological evidence that *V. faba* was introduced to agriculture in the late neolithic period. There have been no prehistoric finds east of a line running close to the coast from Israel to Turkey and Greece (Hanelt *et al.*, 1972); Cubero (1974) therefore supposed the centre of origin to be in the Near East, with the species radiating out in four directions: (*a*) to Europe, (*b*) along the North African coast to Spain, (*c*) along the Nile to Ethiopia, and (*d*) from Mesopotamia to India. Secondary centres of diversity may have later become established in Afghanistan and Ethiopia.

By the Iron Age, the culture of *V. faba* was fairly well established in Europe, including Britain; there have been finds from 1800 B.C. in Egypt and there are numerous references from classical Greek and Roman literature. But there is no evidence that the species was grown to any extent in China before A.D. 1200; Hanelt (1972*a*) thinks it was then brought there at the beginning of the silk trade and, by the sixteenth

Fig. 51.1 Distribution of indehiscent pods in *Vicia faba* (after Hanelt, 1972*b*).

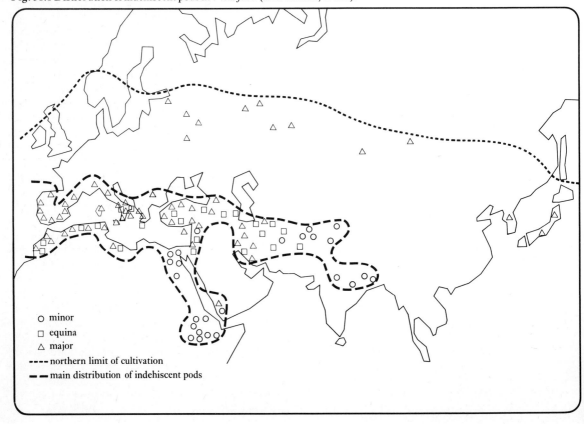

○ minor
□ equina
△ major
---- northern limit of cultivation
▬ ▬ main distribution of indehiscent pods

:entury, considerable cultivation had developed there. Almost all Chinese *V. faba* are of the large-seeded *major* type (Muratova, 1931) and large-seeded types are not known to have existed anywhere until about A.D. 500. Thus the *minor* subspecies is presumably the more primitive but, according to Cubero (1973), it now has the least genetic diversity, so that the greater evolutionary potential now lies in the *major* and *equina* groups. Figure 51.1 is based on Hanelt's (1972*b*) map of the distribution of types with indehiscent pods. This character is recessive to shattering; the latter is therefore thought to be the more primitive condition but non-shattering types have been favoured by selection in the more arid regions. Within this zone, *minor* can be seen to have accumulated at the eastern ends (Ethiopia and India) and *major* at the western end. *Major* varieties have moved to Russia, China and Japan in recent times. In central and northern Europe all three (*major*, *equina* and *minor*) are now cultivated but most of them have pods capable of shattering.

Vicia faba is generally thought to have reached Mexico and South America in the hands of the Spaniards, there being no convincing evidence of Amerindian cultivation in pre-Columbian times. Most of the central and South American *V. faba* are of the *major* type.

Homer refers to *V. faba* as black-seeded and it is fairly certain that the light-buff testa (which is recessive to black) appeared quite recently.

The bean is partially allogamous and differences in degree of outcrossing have been demonstrated. Normally, the flowers must be tripped and pollination is dependent upon bee activity. However, the ability to self in the absence of tripping (autofertility) is present to a high degree in subsp. *paucijuga*, in some populations from India and Africa and, to some extent, in Mediterranean populations. It is associated with short-season adaptation, few flowers per node, short or medium plant height and strong tillering capacity. The northward migration of these types (mainly into Europe from the early Iron Age onwards), followed in subsequent migrations by crossing with local strains which by then had become established, must have resulted in a greater yield potential through increased plant height, lengthened vegetative and flowering period and more flowers per node, though with fewer stems per plant (Paul, 1974). The need for setting of the first flowers became less and it now appears that only the heterozygosity-dependent form of autofertility is expressed in central European populations, particularly in *equina* types. After some loss of heterozygosity, mechanisms favouring cross-pollination (an increased need for tripping and weak self-incompatibility in some inbreds) begins to operate. Hence, there has developed an adaptive adjustment of the breeding system such that the species maintains a degree of heterozygosity and gene flow without being wholly dependent upon insect pollination.

In summary, it appears that the main features of the early history of the bean were adjustment of length of life-cycle, plant habit and pod dehiscence to habitat; these changes were accompanied by local increase in seed size, by local adjustment of the breeding system towards more or less inbreeding and by the development of varied seed colours.

4 Recent history

Much selection, conscious but unrecorded, must have taken place in recent times, for example, for large seededness; and winter hardiness was probably selected independently in *equina* and *minor* types. But recent written records of attempts to study the plant with a view to breeding probably began with Darwin (1858) who noted reduced pod set in a cage from which bees were excluded. More recently, Sirks (1931) analysed genetically a large number of variants (which unfortunately were later lost). Muratova (1931) carried out systematic and geographical studies. In the 1950s, work at Cambridge, England, and Dijon, France, brought understanding of the variable amounts of natural outcrossing and the regulation of the breeding system through hybridity.

Since 1960, greater variability has been induced, assembled and investigated through mutation and cytological work at Svalöf, Sweden; through the study of collections and classification at Gatersleben, Germany; and through the study of genotype relationships at Cordoba, Spain. Some of the variability has been exploited in achieving small increases in yield, earlier maturity, resistance to a few diseases (some viruses) and in the breeding of seed types suited to special purposes (such as the preserving of broad beans in the UK and Holland).

Work at the Welsh Plant Breeding Station has recently suggested that, given sufficient breeding effort, the autofertility of Asian and African populations could perhaps be transferred to beans adapted to British conditions. Exploiting the autofertility of hybrids and the considerable general heterosis that goes with hybridity has been attempted at the Plant Breeding Institute, Cambridge, and at the Institut National de Recherche Agronomique, France, but sufficient control over the cytoplasm-restorer male-sterility system for large-scale multiplication has not yet been achieved.

181

Other breeding methods which have been used include mass selection and progeny-test selection under open pollination but control of pollination in enclosures is being increasingly employed, particularly for: (*a*) the production of inbreds (especially for the evaluation of lines in regard to characters of low heritability and the subsequent composition of synthetics); and (*b*) control over recurrent selection in populations.

In summary, current varieties are all either mass-selected populations or synthetics, attempts so far to produce either 'hybrid' varieties or pure lines with high performance having failed.

5 Prospects

The F_1 hybrid method of breeding will no doubt be further investigated but the outcome cannot be predicted with certainty because the economics of seed costs for the grower have still to be ascertained, even if the male sterility can be controlled. It is perhaps more probable that breeding will push the species towards complete self-fertility. If this situation were reached there would still be some insect visitation and natural crossing but, in view of the fact that varieties appear and disappear from recommended lists very quickly, a greater responsibility would then rest with breeders to preserve genetic variability. A move towards inbreeding could be, in the longer term, disgenic.

Objectives in temperate regions seem to be changing towards stabilizing yield rather than improving maximum yield. Uncertainty of pollination, which may be overcome by autofertility, is only one of the limiting factors. The need to reduce flower loss may require a trend towards a more determinate plant structure, such as that which exists in certain mutants and in some of the Asian, African and Mediterranean populations; determinate growth may provide a more stable basis of yield though it would be less responsive to good seasons than the tall, indeterminate habit which has evolved in the long seasons of central and north-western Europe.

There is likely to be more emphasis on breeding for adaptation to new areas of cultivation (such as Canada) and for improving intrinsic characters (such as protein content and quality), for the reduction of toxic factors, and for suitability for industrial processing (such as textured vegetable protein). The white-flowered mutant, with its associated low tannin content and improved digestibility, is likely to be increasingly employed in this sort of context.

There is some variability to exploit in the above factors and in resistance to pests, disease and other hazards but the variation is not inexhaustible. The prospects for bean improvement would be greatly enhanced by successful hybridization with other species of *Vicia* with consequent release of valuable variability in such characters as self fertility and resistance to pathogens.

6 References

Cubero, J. I. (1973). Evolutionary trends in *Vicia faba*. *Theoret. appl. Genet.*, **43**, 59–65.

Cubero, J. I. (1974). On the evolution of *Vicia faba*. *Theoret. appl. Genet.*, **45**, 47–51.

Darwin, C. (1858). *Gdnrs. Chron.*, 828.

FAO (1971). *Production Yearbook*, **25**, 164–6.

Hanelt, P. (1972*a*). Zur Geschichte des Anbaues von *Vicia faba* und ihre Gliederung. *Kulturpfl*, **20** 209–23.

Hanelt, P. (1972b). Die infraspezifische Variabilität von *Vicia faba* und ihre Gliederung. *Kulturpfl.*, **20**, 75–128.

Hanelt, P., Schäfer, H. and **Schultze-Motel, J. von** (1972). Die Stellung von *Vicia faba* in der Gattung *Vicia* und Betrachtungen zur Entstehung dieser Kulturart. *Kulturpfl.*, **20**, 263–75.

Muratova, V. (1931). Common beans (*Vicia faba*). *Bull. appl. Bot. Genet. Plant Breed.*, *Suppl.*, **50**, pp. 285.

Paul, C. (1974). Herkunft und evolutionistische Variabilität. *Göttinger Pflzücht Seminar* 2, 5–10.

Schäfer, H. I. (1973). Zur Taxonomie der *Vicia narbonensis* – Gruppe. *Kulturpfl.*, **21**, 211–73.

Schultze-Motel, J. von (1972). Die archäologischen Reste der Ackerbohne *Vicia faba* und die Genase der Art. *Kulturpfl.*, **19**, 321–58.

Sirks, M. J. (1931). Beitrage zu einer Genotypischen Analyse der Ackerbohne *Vicia faba*. *Genetica*, **13**, 209–631.

Zohary, D. and **Hopf, M.** (1973). Domestication of pulses in the Old World. *Science*, N.Y., **182**, 887–94.

Cowpeas

Vigna unguiculata (Leguminosae–
Papilionatae)

W. M. Steele
Department of Agricultural Botany
University of Reading England
formerly Department of Botany
University of Nairobi Kenya

1 Introduction

Cowpeas are an ancient crop now grown as a pulse, as a vegetable or for fodder throughout the tropics and subtropics. They are chiefly important as a source of protein (and especially of lysine) in the staple cereal diets of subsistence and peasant farming communities of semi-arid Africa and Asia, who eat the seeds, young leaves and pods and feed the haulms to livestock. When disease-resistant cultivars shall have become available, cowpeas will perhaps become even more important in the predominantly starchy diets of the humid tropics. In advanced tropical agriculture, cowpeas are sometimes grown as a pulse, but mostly for forage or fodder.

Cultivated cowpeas are annual herbs with a great range of growth habits and response to photoperiod and great variation also in seed characters. More or less erect, determinate, day-neutral types are commonly grown as sole crops for seed or forage, whereas prostrate, indeterminate, short-day types are interplanted with cereals in peasant agricultures.

Cowpeas are not important in world trade, so production data are scarce; but seed production for human use is probably greatest in India and in West Africa, northern Nigerian annual production being estimated as 1 Mt from 5–6 Mha.

The literature of cowpeas has recently been reviewed by Summerfield *et al.* (1974).

2 Cytotaxonomic background

Vigna is a pantropical genus of about 170 species, 120 in Africa (66 endemic), 22 in India and southeast Asia (16 endemic), and a few in America and Australia (Faris, 1965). Its affinities with *Phaseolus* and *Dolichos* have led to a confused classical taxonomy, recently clarified by Verdcourt (1970) who recognized five subspecies of *V. unguiculata*. Two subspecies are wild: subsp. *dekindtiana* in the African savanna zone and Ethiopia; and subsp. *mensensis* in forests, with scabrous, dehiscent pods and seed dormancy not found in the cultivars. The common cultivated cowpea everywhere is subsp. *unguiculata*; the other cultivated subspp., *cylindrica* and *sesquipedalis*, are widespread in India and the Far East but, though they have been introduced to Africa, are not found in traditional African farming systems.

The cytotaxonomy of *Vigna* is relatively simple, being uncomplicated by polyploidy ($2n = 2x = 22$) and with apparently little genetic and no chromosomal divergence of the cultivars from their putative ancestor. The five subspecies of *V. unguiculata* are interfertile, but all attempts to hybridize cultivars with other *Vigna* species, notably *V. luteola*, *V. marina* and *V. nilotica*, proposed as wild progenitors have failed (see Faris, 1965).

No theory for the origin of the cultivars in Asia can explain the distribution and truly wild status of subsp. *dekindtiana*; on the contrary, the evidence all suggests that subsp. *unguiculata* was domesticated from it in Africa.

3 Early history

West Africa and India are both modern centres of diversity of cultivars; we must therefore account for domestication in Africa, dispersal within Africa and eastwards to Asia, and diversity in two widely separate areas.

Circumstantial evidence arising from the study of cowpeas in the ancient West African cereal farming system strongly suggests that their early history in Africa was closely associated with that of sorghum (*Sorghum bicolor*) and pearl millet (*Pennisetum typhoides*). Doggett (1970) believes that these cereals were domesticated in Ethiopia 5–6 thousand years ago, subsequently spreading throughout the African savanna zone and to Asia. Archaeological evidence from West Africa, though scarce, indicates that sorghum was grown there in neolithic times (third millenium B.C.).

Though it has yet to be studied in living collections, the diversity of wild cowpeas in Ethiopia is impressive, and contrasts with their homogeneity in West Africa. Faris (1965) concluded that the cultivars were domesticated from subsp. *dekindtiana* in West Africa because this is their only centre of diversity within its range. It is however, equally probable that the West African centre of diversity is secondary and that cowpeas were domesticated with (or soon after)

183

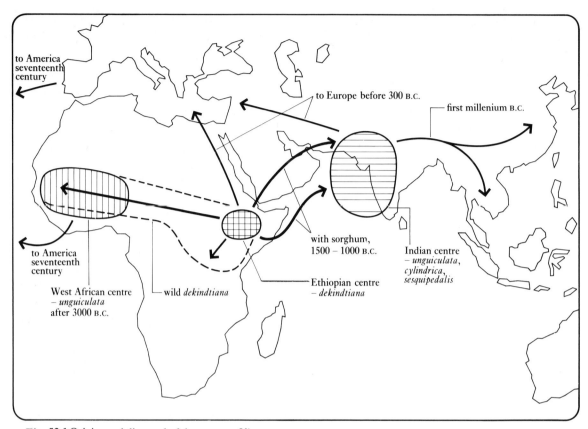

Fig. 52.1 Origins and dispersal of the cowpea, *Vigna*.

sorghum and pearl millet in Ethiopia; or that they had a 'diffuse' origin in the savanna zone after dispersal of the cereals. Only archaeology and palaeoethnobotanical studies in Ethiopia and the southern Sahara can tell.

Once established in West Africa, diversity probably accumulated and cultivars were adapted to the cereal farming system evolved; the expansion of the crop would have provided opportunities for the survival of chance mutants, for natural hybridization of inbred cultivars and for introgression from sympatric populations of subsp. *dekindtiana*. Though cowpeas are usually regarded as more or less cleistogamous inbreeders, up to 2 per cent outcrossing has been reported from Senegal.

Alexander and Coursey (1969) suggest that early West African cereal farmers reached the fringe of the rain forest zone in the third millenium B.C., where their contact with hunting/gathering communities who ate wild yams led to the development of yam cultivation in West Africa. Neither the husbandry of yams, nor the prolonged, bimodally distributed rainfall in the transition zone between forest and savanna, would have favoured the cultivation of spreading, short-day cowpeas which are adapted to the cereal farming system. On the other hand, day-neutral cultivars of subsp. *unguiculata* are common there now, grown alone or interplanted with yams; it seems likely that they evolved from short-day types and that, subsequently, the demand for protein favoured their spread in the yam zone.

The logical sequel to this theory of the origin and spread of cowpeas in Africa is that subsp. *unguiculata* arrived in Asia with sorghum and pearl millet which, Doggett (1970) suggests, occurred more recently than 1500 B.C. Once in Asia, there could be no further gene exchange with wild cowpeas, but the expansion of the crop in India would have provided opportunities for the accumulation of genetic diversity, from which conscious selection probably established subspecies *cylindrica* (for forage) and *sesquipedalis* (for its long, immature pods eaten as a vegetable). From India, the cultivars spread into southeast Asia and the Far East and ultimately reached Europe from Asia or Egypt (see Chevalier, 1944); they were referred to in Greek texts of 300 B.C. and, by the end of the Roman era, were established as a minor food legume in southern Europe.

The morphological and physiological changes attendant upon domestication may be summarized as follows. The great diversity of West African subsp. *unguiculata* contrasts with the relative homogeneity of modern populations of wild subsp. *dekindtiana*. These consist of prostrate or twining, short-day plants with short (4·5–10 cm) scabrous dehiscent pods and small, smooth, dark speckled seeds (3–6 mm long; 100 weigh less than 4 g) with testas sufficiently impermeable to water to confer a degree of dormancy. Thus the loss of pod dehiscence and seed dormancy must have accompanied domestication.

The subsequent change of greatest agricultural significance in West African cultivars was the divergence, under the influence of human selection, of upright, day-neutral types, adapted to the climate and agronomy of the root and tuber farming system, from spreading, short-day cultivars adapted to the cereal farming system of the semi-arid savanna zone. A similar but more extreme change (not associated with root crops) in Asia gave rise to the distinct subsp. *cylindrica* consisting of strictly erect, day-neutral plants.

Variation in West African cultivars of subsp. *unguiculata* includes all of the flower, pod and testa colours described by Saunders (1960), but most day-neutral cultivars have relatively small, smooth, cream or brown seeds (shorter than 7 mm; 100 weight less than 10 g) and pods shorter than 13 cm. Most short-day cultivars have large, white seeds with rough testas (7–12 mm long; 100 weigh up to 32 g) and pods up to 22 cm long.

Though the proportions of essential amino acids in cowpea seeds vary enough to warrant an extensive survey, there are no major differences in this respect between the seeds of wild cowpeas and cultivars.

4 Recent history

The Spanish introduced cowpeas to the New World tropics in the seventeenth century and day-neutral cultivars of subsp. *unguiculata* must have been taken there with the slave trade from West Africa. The crop was first grown in the southern United States in the early eighteenth century; recently, diverse cultivars have been bred there both for seed (often for canning) and forage production.

Breeding has yet had no impact in the West African and Indian centres of diversity, which still provide opportunities for sampling a wide range of genotypes; large collections are maintained in West Africa, India and the United States (see Summerfield *et al.*,

1974). Thus genetic erosion does not seem to have been serious.

5 Prospects

The objective of current research in Africa and Asia is to increase the utilization of cowpeas in human diets; traditional farming systems will have to be altered to achieve this aim, and many 'land races' of cowpeas will cease to be grown. Breeders hope to produce determinate cultivars which are highly productive as densely sown sole crops, utilizing determinate growth from subsp. *cylindrica* and desirable seed types from subsp. *unguiculata*; attention is also being given to resistance to locally important pathogens and to the very destructive insect pests which are the chief factor limiting the yield of sole crops, at least in Africa. The protein content of seeds can be increased (the range reported is 22–35%) and its nutritive value will no doubt be improved by increasing the proportion of sulphur-containing amino acids.

New breeding methods to achieve these objectives include modified recurrent selection applied to inclusive germplasm pools. Laborious intercrossing programmes are required to establish the genetic base for the application of this technique to an inbreeding crop; consequently, the potential of male sterility to effect large savings of time and labour is being investigated. The search for new sources of germplasm includes attempts to obtain interspecific hybrids (especially with the Asiatic 'mung' bean group, previously referred to *Phaseolus*), and the inclusion of subsp. *dekindtiana* in breeding for resistance to pathogens and insect pests (by incorporating scabrous, siliceous or hairy pods).

6 References

Alexander, J. and **Coursey, D. G.** (1969). The origins of yam cultivation. In Ucko, P. J. and Dimbleby, G. W. (eds.), *The domestication and exploitation of plants and animals*. London, pp. 405–25.

Chevalier, A. (1944). Le dolique de Chine en Afrique. *Rev. Bot. appl. Agric. trop.*, **24**, 128–52.

Doggett, H. (1970). *Sorghum*. London.

Faris, D. G. (1965). The origin and evolution of the cultivated forms of *Vigna sinensis*. *Can. J. Genet. Cytol.*, 7, 433–52.

Saunders, A. (1960). Inheritance in the cowpea (*Vigna sinensis*) II. *S. Afr. J. agric. Sci.*, 3, 141–62.

Summerfield, R. J., Huxley, P. A. and **Steele, W. M.** (1974). Cowpeas, *Vigna unguiculata*. A Review. *Field Crop Abstr.*, 27, 301–12.

Verdcourt, B. (1970). Studies on the Leguminosae–Papilionoideae for the Flora of East Tropical Africa IV. *Kew Bull.*, **24**, 507–69.

Onion and allies

Allium (Liliaceae)

G. D. McCollum

Agricultural Research Service
US Department of Agriculture
Beltsville Maryland USA

1 Introduction

Many aspects of the food alliums have been discussed by Jones and Mann (1963). Among the most popular of vegetables, onions and their relatives (garlic, leek, Japanese bunching onion, chives, Chinese chives, rakkyo) with the oniony or garlicky flavours and odours of alkyl sulphides, are used as food or condiment in most countries of the world and grown in nearly all cool areas. In 1970, world production of onions, principally *Allium cepa*, was an estimated 11·7 Mt, led by the USA, Japan, Spain, Turkey, Italy and Egypt. World production of garlic, led by Spain, Argentina and Italy, probably exceeds 500 kt. Onions are an important commodity in world trade. Egypt and The Netherlands export the most; the United Kingdom imports the most.

2 Cytotaxonomic background

Allium is a diverse genus of the family Liliaceae, subfam. Allioideae, with over 600 species in the north temperate zone. Most of the North American species have a basic $x = 7$, whereas the Old World Alliums are predominantly $x = 8$ (Traub, 1968). The domesticated Alliums, except chives, come from the Near East or from central or eastern Asia; they all have $x = 8$. The common onion, *A. cepa* (including also shallot, multiplier and potato onion) has $2n = 16$ and is known only in cultivation, with its primary gene centre thought to be Afghanistan. The nearest related species (*A. vavilovii*, *A. oschaninii*, *A. pskemense* and *A. galanthum*) are wild diploids ($2n = 16$) of central Asia (Fig. 53.1). *Allium fistulosum*, $2n = 16$, the Japanese bunching or Welsh onion, is unknown in the wild but the related *A. altaicum* ($2n = 16$) is a wild plant of northern Mongolia and adjacent Siberia. The chive, *A. schoenoprasum*, with $2n = 16, 24, 32$, is a variable and widespread species ranging from the Arctic to far southwards in the mountains of North America, Europe and Asia. Garlic, *A. sativum*, known only in cultivation, is diploid, $2n = 16$. Some authors consider the diploid *A. longicuspis*, endemic to central Asia, to be its wild ancestor. The extremely variable *A. ampeloprasum* ($2n = 16, 24, 32, 40, 48$) grows as a wild, often weedy, plant from southern Europe and north Africa through the Middle East to western and southern USSR. Two cultivated forms of *A. ampeloprasum*, garden leek (syn. *A. porrum*) and the similar kurrat (syn. *A. kurrat*), are tetraploid with $2n = 32$, while a third form, great-headed garlic, is hexaploid. Rakkyo, *A. chinense*, a native of central and eastern China and cultivated throughout eastern Asia, is $2x$, $3x$ and $4x$. The Chinese chive, *A. tuberosum*, a wild and cultivated plant throughout eastern Asia, is $2x$ and $4x$.

In Traub's (1968) classification, common onion, Japanese bunching onion, chives and rakkyo occupy four separate subsections of Section Cepa; leek and garlic are put together in Section Allium; Chinese chives is in Section Rhizirideum.

3 Early history

Jones and Mann (1963) and Helm (1956) have reviewed the history of the food species of *Allium*. Supposedly wild *A. cepa* collected by nineteenth century botanists in western Asia was more likely one of the related species that range from northern Iran and Afghanistan, northwards into the central Asiatic region of the USSR (see Wendelbo, 1971). According to Vavilov, the primary centre of origin of onion and garlic is central Asia and of the leek the Near East and Mediterranean. The Near East is a secondary centre for onion and the Mediterranean a secondary centre for garlic and the large-size types of onion.

References to onions, garlic and leek as food, medicine or religious objects can be traced back to the 1st Egyptian Dynasty (3200 B.C.) and to the biblical account of the exodus of the Israelites from Egypt (1500 B.C.). Introduction into Egypt must have occurred much earlier. The use of onions and garlic in medicine in India in the sixth century B.C. can be inferred from later Indian writings. By the time of the Greek and Roman authors, from Hippocrates (430 B.C.) and Theophrastus (322 B.C.) to Pliny (A.D. 79), several onion cultivars were named and described as long or round; white, yellow or red; mild or pungent.

As onion culture spread, cultivars evolved more and more diversity in shape, colour, flavour, keeping quality and critical adaptations to cultivation in new climates and environments. The most important adaptive traits involved bulbing response to day-

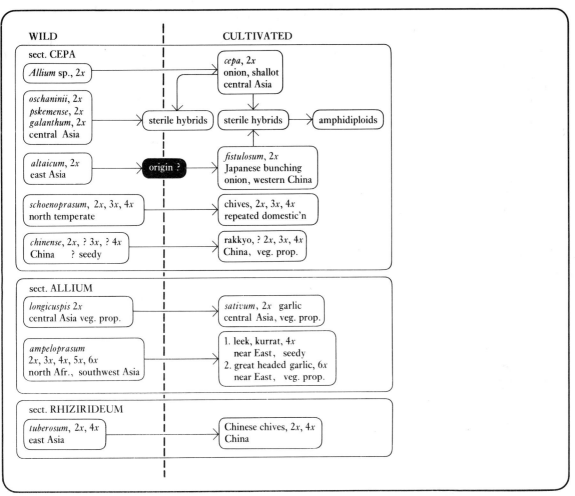

Fig. 53.1 Relationships of the cultivated species of *Allium* and their wild relatives.

length and high temperatures and bolting response to low temperatures, characteristics which onion breeders must still deal with today. The modern onion is self-compatible but is an outbreeder that is subject to inbreeding depression. We can assume that it has retained the breeding system of its early ancestors and has evolved independently of contacts with other species for several thousand years. Experimental hybrids with related wild species (*A. galanthum*, *A. oschaninii*, *A. pskemense*) are highly sterile.

Since prehistoric times, *A. fistulosum* has been the main garden onion of China and Japan. Grown for its edible tops or long leaf bases (it is not bulbous), it developed an array of cultivars perhaps as diverse as that of the common onion in southwestern Asia and Europe. The centre of origin of *A. fistulosum* is regarded as central and western China; how it is related to

the similar wild *A. altaicum* is in dispute. *A. fistulosum* is an outcrosser and is self-compatible. It will cross with *A. cepa* but the hybrids are highly sterile.

Leek and kurrat, traceable, like onion and garlic, to ancient civilizations of the Near East, closely resemble wild *A. ampeloprasum*, differing mainly in a lesser tendency to form bulbs. Leek was known in Europe in the Middle Ages and is still popular there, many varieties having been selected for long, white, edible leaf-bases and green tops, winter hardiness, and resistance to bolting. Only the green leaves of kurrat, grown in Egypt and other countries of the Near East, are eaten. Leek and kurrat are interfertile and it may be supposed that they would cross readily with wild *A. ampeloprasum* of the same (4*x*) ploidy level. The species, at the 4*x* level at least, is known to be a self-compatible outbreeder (Ved Brat, 1965) that shows inbreeding depression. The hexaploid *A. ampelo-prasum*, great-headed garlic, is vegetatively propagated by its large cloves or smaller ground bulblets. It flowers profusely but sets few or no seeds.

Garlic is an ancient crop of central Asian origin, known only in cultivation and mentioned in ancient Chinese, Egyptian and Greek writings. The early domestication of garlic took quite a different turn from that of the largely seed-propagated leek and onion: it became exclusively vegetatively propagated by cloves and, in those cultivars that still bolt, by inflorescence bulbils. Some modern cultivars may produce flowers mixed with the bulbils but the flowers never set seeds. Garlic thus presents an interesting problem as to the origin of the many cultivars, differing in maturity, bulb size, clove size and number, scale colour, bolting, scape height, number and size of inflorescence bulbils and presence or absence of flowers. We do not know how much variation was selected while *A. sativum* or its ancestors were still sexual and how much has arisen by bud mutation after garlic became vegetatively propagated, viviparous and sterile. The presumed wild ancestor, *A. longicuspis*, is itself viviparous and seedless.

Cultivated chives, said to have a Mediterranean origin, are not recorded earlier than sixteenth century Europe. They have not changed much from the polymorphic, wide-ranging wild *A. schoenoprasum* and may have been brought into cultivation many times from wild populations. Both outbreeders and inbreeders are known (Ved Brat, 1965). Chinese chives, *A. tuberosum*, and rakkyo, *A. chinense*, have been domesticated since ancient times in the Orient. Chinese chives is probably little changed from the wild; it is an abundant seed producer. Rakkyo is vegetatively propagated by bulb multiplication; cultivated forms are nearly all tetraploid, sparsely-flowering and seed-sterile. Flower structure indicates that both rakkyo and Chinese chives are outbreeders, but information about their breeding systems is lacking.

4 Recent history

Onions, garlic and leek, all common vegetables in Europe in the Middle Ages, were mentioned by Chaucer (about 1340) and written about by many of the herbalists. *Allium cepa* was likely to have been planted in the West Indies by Columbus (1494) and reintroduced to the New World many times. It was cultivated in Massachusetts in 1629. Vilmorin listed 60 varieties in 1883. Development has proceeded by mass or individual selection at the diploid level. Involvement with other species has been rare, though the relatively unimportant top onion, *A. cepa* var. *viviparum*, may

have originated through some past hybridization between *A. cepa* and *A. fistulosum*. An amphidiploid *A. cepa × fistulosum* ('Beltsville Bunching') was introduced by H. A. Jones in 1950 as a green bunching onion superior to *A. fistulosum*. 'Delta Giant' shallot is a backcross of shallot (*A. cepa*) with a *fistulosum*-shallot amphidiploid.

Modern methods of onion breeding are discussed by Jones and Mann (1963) and Kuckuck and Kobabe (1962). In the 1920s and 1930s, many programmes were started, not only to improve horticultural qualities but also to develop cultivars resistant to pests and diseases. As knowledge of onion diseases increased, many sources of resistance were reported. *Allium fistulosum*, especially, was found to be resistant or immune to pink root, downy mildew, smut, yellow dwarf and thrips. However, in spite of early optimism, these resistances remain unexploited in onion breeding because of hybrid sterility and limited genetic recombination. Efforts to use *A. galanthum*, *A. pskemense* and *A. oschaninii* have met similar difficulties (McCollum, 1974). Progress, instead, has come slowly by field and greenhouse selection for resistance within the *A. cepa* material itself. The approach has been largely empirical. One can infer from breeding results that polygenic rather than major-gene inheritance of resistance is usual in onions. The best example of a success is the series of pink root-resistant cultivars field-selected by H. A. Jones and associates. These cultivars make up a large part of the short-day onion crop of the southwestern USA.

The major event of modern onion breeding was the discovery by H. A. Jones in 1925 of a male-sterile *A. cepa* plant, Italian Red 13-53, at the University of California, Davis. Jones and his co-workers showed that the sterility resulted from an interaction of a recessive nuclear gene locus with a specific cytoplasm. They showed how to transfer the cytoplasm to inbred parent lines and proposed a system for commercial production of hybrid seed. From this first proposal (1943) and the first commercial hybrid release (1952), up to the present, public and private onion breeders in the USA have been occupied with developing hybrid cultivars that are uniform, high yielding, long keeping and disease resistant. Breeding of hybrids spread to many other countries. It was based initially on the US cytosterile material but some countries are now using male steriles found locally. Total seed production for all types of onions reported in the USA was less than 30 per cent hybrid in 1973, although the proportion of main crop bulb onions grown from hybrid cultivars is much higher. The use of hybrids has not increased as rapidly as expected because of unsolved production problems that reduce seed yields.

No event comparable with the change to hybrid onions has occurred to interrupt the breeding history of the other Alliums, the modern cultivars of which are generally the result of mass or individual plant selection over long periods of time.

5 Prospects

The current emphasis on developing hybrid onion cultivars is not likely to change, given the superiority of hybrids in seedling vigour and in uniformity of appearance, size and maturity. The methods of simultaneously transferring cytoplasmic male sterility and developing inbred parent lines by backcrossing, progeny testing and selfing will continue, as will the largely empirical random testing to identify superior hybrid combinations. The efficiency of this process may be improved as more biometrical genetic studies of economic characters and disease resistances are undertaken. Three-way and double-cross hybrids may come into use, partly to improve seed yields. Selection will put more emphasis on internal quality as efforts increase to develop cultivars for specific purposes: mild flavour for the fresh market; pungency and high solids content for the dehydration industry; and single centre bulbs (i.e. suppression of doubling) for processing as french-fried onion rings. Knowledge of the volatile flavour, odour and lachrymatory components and of other biochemical constituents (such as those related to dormancy and long storage life) can be expected to contribute increasingly to quality breeding.

Restrictions on the use of chemicals to control pests and diseases will continue the pressure on breeders for cultivars resistant to: *Alternaria* blight, *Botrytis* neck rot and leaf blight, *Fusarium* basal rot, pink root (*Pyrenochaeta*), *Sclerotium* white rot, smut (*Urocystis*), bacterial rots (*Pseudomonas* and *Erwinia*), yellow dwarf virus, and to nematodes (*Ditylenchus*), thrips and maggots (*Hylemya*). Resistance to some of them has been found within locally adapted *A. cepa* material but, where it is lacking, breeders may have little choice but to introduce unadapted foreign germplasm into their programmes.

However, the resultant broadening of the genetic base could be beneficial because many US programmes have become rather narrowly based, especially since the beginnings of hybrid onion development. All US male-sterile lines used in commercial seed production derive their sterile cytoplasm from Jones's original plant and many of the parental lines now used in hybrid production today are closely related. The US collection of *A. cepa* germplasm, including so-called primitive cultivars from Turkey, Iran, Afghanistan and Pakistan, is somewhat limited and has had relatively little impact on US onion breeding. Even less use is being made of several potentially valuable but difficult-to-use related species. *Allium fistulosum*, *A. galanthum*, *A. oschaninii* and *A. pskemense* cross with *A. cepa* but the hybrids are highly sterile. Accessions in living collections in the USA under the name *A. vavilovii* are interfertile with *A. cepa* but their identities and origins need checking. *Allium farctum* ($2n = 16$), recently described by Wendelbo from West Pakistan and Afghanistan, is a new species for Section Cepa. Its crossability with *A. cepa* has not been determined. There is one report that *A. roylei* ($2n = 16$) from the western Himalayas can be crossed with *A. galanthum*.

Resistance to some diseases has been found only in Alliums even less closely related to onion than the ones just mentioned, for example, the white rot resistance of *A. tuberosum* and the *Botrytis* neck rot resistance of *A. schoenoprasum*. Newly developing techniques of genetic transfer (e.g. in cell culture) may be required if the interspecific barriers are to be overcome.

Onion seed fields often produce uneconomically low yields or fail entirely. Considerable public and private cooperative effort is under way in the US to study and remedy this situation. Attention is being given to: cultural requirements of the seed plants; disease, insect, and weed control; bee pollinator activity (Shasha'a *et al.*, 1973); nectar attractiveness; availability of pollen; and environmental hazards to pollination, fertilization, and seed development. The problem may be, in part, that the reproductive processes of modern onion lines are not yet adapted to tolerate the frequent periods of hot, dry weather that occur in those parts of the USA in which onion seed is now produced.

Little change is foreseen in the breeding of the other Alliums. Genetic male sterility discovered in Japan could be used for hybrid seed production of *A. fistulosum* by rogueing of fertiles and vegetative propagation of the sterile clones. Such use of genetic male sterile leek has been proposed by Schweisguth (1970). There is considerable interest in improving garlic cultivars but, until the causes of sterility are determined and overcome (Koul and Gohil, 1970), the only feasible breeding method is the slow one of selecting spontaneous or induced somatic mutations.

6 References

Helm, J. (1956). Die zu Würz- und Speisezwecken kultivierten Arten der Gattung *Allium. Kulturpflanze*, **4**, 130–80.

Jones, H. A. and Mann, L. K. (1963). *Onions and their allies*. London and New York, pp. 286.

Koul, A. K. and Gohil, R. N. (1970). Causes averting sexual reproduction in *Allium sativum*. *Cytologia*, **35**, 197–202.

Kuckuck, H. and Kobabe, G. (1962). Küchenzwiebel, *Allium cepa. Handbuch der Pflanzenzüchtung*, 2nd edn, 6, 270–312.

McCollum, G. D. (1974). Chromosome behavior and sterility of hybrids between the common onion, *Allium cepa*, and the related wild *A. oschaninii*. *Euphytica*, **23**, 699–709.

Schweisguth, B. (1970). Etudes préliminaires à l'amélioration du poireau *A. porrum*. Proposition d'une méthode d'amélioration. *Ann. Amélior. Plantes*, **20**, 215–31.

Shasha'a, N. S., Nye, W. P. and Campbell, W. F. (1973). Path coefficient analysis of correlation between honey bee activity and seed yield in *Allium cepa. J. Amer. Soc. hort. Sci.*, **98**, 341–7.

Traub, H. P. (1968). The subgenera, sections and subsections of *Allium. Plant Life*, **24**, 147–63.

Ved Brat, S. (1965). Genetic systems in *Allium*. III. Meiosis and breeding systems. *Heredity*, **20**, 325–39.

Wendelbo, P. (1971). Alliaceae. In K. H. Rechinger (ed.), *Flora Iranica*, **76**, pp. 100. Graz, Austria.

Flax and linseed

Linum usitatissimum (Linaceae)

A. Durrant
Department of Agricultural Botany
University College of Wales
Aberystwyth Wales

1 Introduction

Linum usitatissimum is an annual, self-fertilizing plant grown for its fibre (fibre flax), or for its seed (oil flax, seed flax, linseed), or for both (dual purpose flax). Fibre flax varieties are up to 120 cm tall, are scarcely branched (at the customary high seed rates) and are grown in the more temperate regions throughout the northern hemisphere, especially in the USSR. Linseed varieties are shorter and more branched and are grown in warmer regions, in Argentina, Uruguay, India, USA, Canada and the USSR.

The fibre of commerce is obtained from fibrous (essentially cellulose) bundles running the length of the stem and forming a ring in the cortex, the bundles individually consisting of overlapping strands, averaging 4 cm in length. Flax fibre is stronger than cotton or wool and linen fabrics have special qualities of strength, durability, gloss, water absorption and drying. Linen thread is used in gloves, footwear, netting and sports gear. The seed contains 40 per cent linseed oil, a drying oil used in varnishes, paints and linoleum. The residue after oil extraction is a valuable high protein cattle food called linseed cake or, when ground, linseed cake meal. Linseed meal is the crushed, unextracted seed.

2 Cytotaxonomic background

Linum is a genus of nearly 200 species spread over the temperate and warm temperate zones of the northern hemisphere, most abundantly in Europe and Asia but with about 50 species in America. Species are inbreeding or outbreeding (some heterostylous) annual or perennial herbs or half-shrubs, many of which have been brought into the garden.

Basic chromosome numbers in the genus are remarkably variable ($x = 8, 9, 10, 12, 14, 15, 16$) but there are evident nodes at $2n = 2x = 18$ and $2n =$

$2x = 30$; flax belongs to the second, $2n = 30$, group (but see also below). Bivalents have been found in at least one haploid ($n = 15$) plant. Hybrids have been obtained between several species of the $2n = 18$ group and between several of the 30 group, many showing one or more translocations (Gill and Yermanos, 1967). The *L. pratense* group ($2n = 18$) in North America is close to the *L. perenne* group ($2n = 18$) of Eurasia and is thought to have entered via Siberia and Alaska. The higher numbered groups ($2n = 30, 36$) were probably separate introductions to America. Many species have been assigned more than one chromosome number (e.g. *L. usitatissimum*, $2n = 30, 32$), and only recently has the necessary task of reclassifying the many species, subspecies and forms begun (Mosquin, 1971; Ockendon, 1971; Robertson, 1971).

Although the cytotaxonomy of the genus is confused, it is clear that cultivated flax is one of a well-defined group of North African and Eurasian species with $2n = 30$ chromosomes: *L. africanum, L. angustifolium, L. corymbiferum, L. decumbens, L. nervosum* and *L. pallescens*. They produce fertile hybrids (Gill and Yermanos, 1967) and are possibly variants of a single species. *Linum usitatissimum* is nearest to the highly variable *L. angustifolium* ($2n = 30, 32$), a strongly branching and tillering perennial or biennial with dehiscent capsules, common in southern Europe and western Asia, less frequent in western Europe, with which it is easily crossed and differing from it cytologically by one translocation. The flowers of cultivated flax are adapted for self-fertilization, although cases of substantial cross-fertilization have been recorded; *L. angustifolium* cross-fertilizes more readily and is said to outcross naturally with *L. usitatissimum*.

3 Early history

Flax, possibly *L. angustifolium*, is associated with the earliest records of civilization in the deposits of the Swiss Lake Dwellers, and *L. usitatissimum* was cultivated for a thousand or more years B.C. in Egypt and the Middle East. Linen was worn in ancient Egypt and the Egyptians used it to wrap their mummies, the quality sometimes equalling the finest produced today. Linseed oil was used in embalming. The early Greeks and Romans also cultivated flax for fibre and seed, the Romans probably stimulating its culture in Britain where they later established a linen industry. The plant (or its products) is mentioned in the Bible, and by Virgil, Ovid, Cicero and Pliny. References reveal the associated vocabulary of the time, e.g. *linteum domi retum* (homespun linen), *negotiatio lintea* (linen trade). 'Flax' and 'flaxen' are from the Anglo-Saxon *fleax*. Elsewhere, fibre flax has been grown from the earliest times in European Russia and Central Asia. In India linseed oil was used in ancient rituals.

Fig. 54.1 Relationships and evolution of flax and linseed, *Linum usitatissimum*.

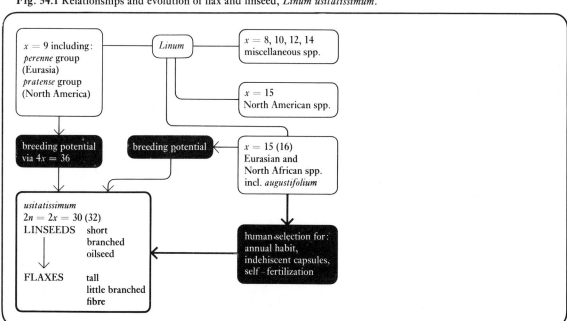

From a study of more than a thousand indigenous types in Europe and Asia, Vavilov (1951) defined two geographical areas, one in southwestern Asia (India, Afghanistan, Turkestan) containing seed types, and the other with fibre types in the Mediterranean countries (from Spain and Algeria to Greece and Egypt), with a transitional region (Asia Minor, Caucasus, Black Sea and Caspian Sea area) containing dual purpose forms. The great diversity of forms found as India is approached suggests that flax originated here and spread northwards and westwards. The tall unbranched fibre types have a shorter vegetative period than the shorter, more branched seed types, and the indigenous fibre types of the USSR could have arisen by natural selection from the southwest Asian seed types as they spread northwards. The Mediterranean fibres may have come from the dual purpose forms in the transitional area or, possibly, from another centre of diversity in Abyssinia; or they could have spread from Russia via Finland and eastern European countries. Others have suggested that fibre types originated in the Baltic region and became shorter, more branched and with larger seeds as they spread southwards and that Indian linseeds may have been derived from Central Asian, Egyptian and Abyssinian forms. More probably, any southerly migrations that occurred were subsequent to the spread northwards at an earlier period.

To sum up, the origin of cultivated flax is not known for certain. The most likely progenitor is the widely distributed, biennial or perennial, *L. angustifolium* (Fig. 54.1) but other species of the group could have contributed. Domestication involved selection for annual habit and non-shattering capsules (dehiscent and semi-dehiscent forms still exist) and more efficient self-fertilization. The place of origin is also unknown; there could have been several centres now obscured by subsequent migrations but procumbent linseeds probably appeared first, then the erect, followed by dual-purpose and finally fibre types.

4 Recent history

Tithes were levied on fibre flax crops in England in the twelfth century and, later, a proportion of all cultivated land had to be sown with it. Seed was taken to the New World in the seventeenth century by the early Colonists. Cultivation migrated westwards during the next 200 years as new settlements arose because, apart from supplying fibre for hand spinning, it acted as a good first crop on newly broken ground and wilt disease reduced the acreage sown on old cultivated land. Fibre flax cultivation gradually declined to a negligible level in the United States, to be replaced by linseed which is now an important crop. Large quantities of linseed are grown in Argentina, and in Canada where fibre flax cultivation also continues.

Among western nations, flax remained the most important vegetable fibre almost up to the beginning of the present century when cotton took its place. In Britain, fibre flax acreage declined from a peak in the 1860s, when large quantities were grown in Scotland and Northern Ireland, to a few thousand acres at the present time. Both linseed and fibre crops showed temporary increases in acreage under the stimulus of the Second World War and a successful programme of fibre flax cultivation was speedily initiated in New Zealand (Hadfield, 1953).

Several cultivar groups are recognised in the USSR, from the flaxes of the Baltic region (Riga, Pskov flax), tall, unbranched, fine quality fibre flaxes which have contributed to the development of many of the world's best fibre flaxes, to the drought and frost-resistant linseeds of Central Asia. At one time Russia was practically the sole exporter of fibre flax seed.

In India, where flax is grown for its seed, there are two main linseed types, both markedly different from the Russian, the one grown in the river alluvium of the Ganges and the Indus with reduced taproot and numerous secondary roots near the soil surface, and the other (Peninsular type) grown further south, with deeply penetrating taproot and laterals at a greater depth. Genetically, the difference could be small, for an Indo-Ganges type has been obtained by irradiating Peninsular type plants.

Flax breeding began to develop after 1900, slowly at first but, by 1930, several countries had produced improved lines. Among them were the Stormont (Northern Ireland) varieties, and the Liral series bred by the Linen Industry Research Association at Lamberg, near Belfast. Flax is an inbreeder and, naturally, pedigree pure line breeding methods were adopted. Early studies of the inheritance of wilt disease reaction in the USA resulted in the cultivation of wilt resistant linseeds. Forty years ago, Flor, in Canada, obtained evidence for a gene-for-gene relationship between virulence of flax rust and resistance in the host plant and he later developed flax varieties for identifying rust genotypes (Flor, 1955).

Besides disease resistance, other characters which have received attention are: seed size, yield, oil content, iodine number, height, uniform ripening and drought, cold and herbicide resistance. In fibre flax, fine quality and length of fibre have to be balanced against lodging resistance. The twentieth century has thus seen, to some extent, the replacement of local populations by improved cultivars, but at a time of substantial, if local, decline of the crop.

5 Prospects

There is sufficient reserve of intraspecific genetic variation among the innumerable forms of *L. usitatissimum* to continue to supply breeders with material for crop improvement. Irradiation has also provided mutations of potential economic value such as larger seeds, longer straw, increased fibre content, disease resistance and high seed yield (Bari, 1971). Many autotetraploids have been produced but they are practically sterile and have other undesirable qualities. *Linum usitatissimum* can be crossed easily with the ($2n = 30$) species listed above. These fertile hybrids could be an important source of genetic material, including genes for disease resistance, and they are also the most promising for further studies on the origin of cultivated flax. Other crop improvement possibilities exist among the derivatives of crosses with autotetraploids ($2n = 4x = 36$) of species with $2n = 18$ chromosomes.

Another source of genetic variation is the variation directly induced in flax by environmental effects (Durrant, 1972). When plants of some fibre flax varieties (for example, Stormont Cirrus and Liral Prince) are grown in soils of different mineral content and at a higher temperature than normal (in a cool greenhouse instead of out-of-doors), large differences appear in their progeny. Parents grown with an additional nitrogen fertilizer at these higher temperatures give progeny which are larger than their parents and up to five times larger than the progeny of parents treated with a phosphatic fertiliser. These large and small induced types (genotrophs) appear because the environment induces changes in the chromosomes so that they are genetically different and they breed true. Neither of them are as tall as their parents and, despite the large difference in size, they look more like linseeds than the parental, fibre type plants. Further genetic changes occur when the plants carrying different induced changes are crossed, or are outcrossed to other varieties not grown in the inducing environments.

Other changes that can be environmentally induced include: anther filament length, presence of hairs in the capsules, peroxidase isoenzymes and the amount of nuclear DNA. Induced changes in amount of nuclear DNA can be reverted to the same amount as in the original fibre flax parents by growing the plants for a period of three years at a lower temperature. This unconventional breeding behaviour must have played some part in the evolution of flax. For example, as the linseeds spread northwards, lower temperatures could

have contributed directly to the appearance of northern forms by inducing a tall, fibre flax, aided by natural selection favouring those plants that were genetically able to respond to the environment in this way. The changes obtained experimentally under the warmer conditions may be a partial reversal of this process. More information is required, however, on the nature of the genetic changes before their full evolutionary significance can be assessed.

6 References

Bari, G. (1971). Effects of chronic and acute irradiation on morphological characters and seed yield in flax. *Radiation Bot.*, **11**, 293–302.

Durrant, A. (1972). Studies on reversion of induced plant weight changes in flax by outcrossing. *Heredity*, **29**, 71–81.

Flor, H. H. (1955). Host–parasite interaction in flax rust–its genetics and other implications. *Phytopath.*, **45**, 680–85.

Gill, K. S. and **Yermanos, D. M.** (1967). Cytogenetic studies on the genus *Linum*. I. Hybrids among taxa with 15 as the haploid chromosome number, *Crop. Sci.*, **7**, 623–27.

Gill, K. S. and **Yermanos, D. M.** (1967). Cytogenetic studies on the genus *Linum*. II. Hybrids among taxa with nine as the haploid chromosome number. *Crop Sci.*, **7**, 627–31.

Hadfield, J. W. (1953). *Linen flax fibre production in New Zealand*. NZ Linen Flax Corp. and NZ Deps. Agric., Indust. and Comm., and Sci. indust. Res.

Mosquin, T. (1971). Biosystematic studies in the North American species of *Linum*, section Adenolinum (Linaceae). *Canad. J. Bot.*, **49**, 1379–88.

Ockendon, D. J. (1971). Taxonomy of the *Linum perenne* group in Europe. *Watsonia*, **8**, 205–35.

Robertson, K. R. (1971). The Linaceae in the southeastern United States. *J. Arnold Arboretum*, **52**, 649–65.

Vavilov, N. I. (1951). The origin, variation, immunity and breeding of cultivated plants. *Chronica Bot.*, **13**, 1–366.

55

Okra

Abelmoschus esculentus (Malvaceae)

A. B. Joshi and M. W. Hardas
Indian Agricultural Research Institute
New Delhi India

1 Introduction

Okra is an annual vegetable crop, grown from seed, in tropical and sub-tropical parts of the world, in the latter only in the hotter parts of the year. Its tender green fruits are used as a vegetable and are generally marketed in the fresh state, but sometimes in canned or dehydrated form. The economic botany and uses of the crop have been reviewed by Schery (1954) and by Singh *et al.*, (1975).

2 Cytotaxonomic background

Okra is known by many local names in different parts of the world. The botanical name generally associated with it is *Abelmoschus esculentus*, earlier designated *Hibiscus esculentus* under the section Abelmoschus of *Hibiscus*, established by Linnaeus in 1737. The generic name, *Abelmoschus*, proposed by Medikus, was upheld by Hochreutiner (see Joshi *et al.*, 1974) on the basis of the constant distinguishing feature of caducous calyx. Index Kewensis lists over 30 species of *Abelmoschus* in the Old World, 4 in the New World and 4 in Australia.

Varying chromosome numbers ($n = 36$; about 33; 59 to 72) have been reported for *A. esculentus* (Joshi *et al.*, 1974). The same authors reported on the cytotaxonomy of *A. esculentus* in India in which they repeatedly found $n = 65$. They also reported on its hybrids with *A. tuberculatus* ($n = 29$), a new species recently described from India. According to them, *A. esculentus* is an amphidiploid comprising two genomes: one with 29 chromosomes (T'), homologous with that of the related wild species, *A. tuberculatus* (T), and the other with 36 chromosomes (Y). The latter genome of *A. esculentus* exhibited very little homology with the M genome of *A. moschatus* ($n = 36$) and greater, but still very incomplete, homology with the F genome of *A. ficulneus* (also $n = 36$). In order to obtain a fuller picture of the evolutionary origin of *A. esculentus*, they have pointed out the need to determine the present-day geographical distribution of *Abelmoschus* taxa that might have contributed the 36-chromosome genome to it. Related to this is also the need to determine the relationships of material (of Ceylon and Japanese origins) for which low chromosome numbers ($n = 33, 36$) have been recorded.

Kuwada (review in Joshi *et al.*, 1974) also studied the cytology of the hybrid between *A. esculentus* and *A. tuberculatus* and its amphidiploid, but reported different chromosome numbers for the two species.

Gadwal (1966, see Joshi *et al.*, 1974) reported on meiosis in the sterile hybrids of *A. esculentus* ($n = 65$, T'$+$Y) with *A. pungens* ($n = 69$) and *A. tetraphyllus* ($n = 69$). Kuwada (1957) reported on the cytology of the sterile hybrid, *A. esculentus* ($n = 62$)$\times A. manihot$ ($n = 34$), and its amphidiploid.

In his search for a donor of immunity to yellow-vein mosaic, the late Dr H. B. Singh, at the Indian Agricultural Research Institute, New Delhi, studied a number of cultivars from West Africa (where the crop is known as gumbo). He identified one accession from Ghana as being immune; its cytotaxonomy is under study. Hybrids of this cultivar with Indian okra ($n = 65$) were only partially fertile; those between the Ghanaian accession and *A. tetraphyllus* ($n = 69$) were completely sterile.

From all this one can only conclude that okra is probably not a single species but a polytypic complex which exhibits both high polyploidy and hybridity and of which the parental wild species are yet undetermined. Even basic chromosome numbers are unknown: thus the haploid numbers $n = 29, 36$ must themselves be polyploid so Indian okra, with $n = 65$, though proximately tetraploid, could well be basically 16-ploid.

3 Early history

The cultivated okra is surely of Old World origin. According to Vavilov, it was probably domesticated in the Ethiopian region; according to Murdock in West Africa (see Joshi *et al.*, 1974). There is a late nineteenth century record of its occurrence in the wild state on the White Nile in the Sudan (Purewal and Randhawa, 1947, see Singh *et al.*, 1975). No dates can be given for the migration of cultivated okra from Africa to the Mediterranean region and to Asia. It was apparently introduced to tropical and subtropical America during the early post-Columbian period.

The accepted view that okra is African should not, however, be taken as established. The cytotaxonomy of *Abelmoschus* is yet so confused that an Asian origin of whole or part of the cultigen does not seem impossible. It will be recalled that Indian okra bears genomes (T', Y) that show a varying degree of homology with the genomes of two Indian wild species (T, F) and that African–Indian hybrids show some sterility (see above). The crop could perhaps yet turn out to be polyphyletic.

4 Recent history

Most breeding research on okra is reported from India (Singh *et al.*, 1975) and the USA (Boswell and Reed, 1962). Smooth-fruited varieties are preferred over varieties whose fruits are prickly even in the green tender stage. The latter are rendered smooth and eatable on cooking. Local preferences for length of marketable fruits vary. Processing industries generally prefer shorter fruits of manageable size. Local preferences vary also for five-angled, many-angled or round-fruited types. Young green fruits are rich in vitamin C but content of this vitamin falls as the fruits grow in age and length, though still green and tender. Pusa Sawani variety of okra in India gives fruits up to 18 cm long, tender and stringless. Until recently, this variety was reputed for its freedom from the symptoms of the yellow–vein–mosaic virus disease in the field ('symptomless carrier' condition). A dual-purpose, ornamental-vegetable type, with red fruits, has been reported from southern India (Anon, 1966).

5 Prospects

Okra breeders seem to be currently paying attention to breeding for resistance to diseases and pests and to early fruiting. Several new sources of resistance to yellow–vein–mosaic have been located by Prem Nath (1970). Sources of tolerance or resistance to root-knot nematodes have been located (review in Singh *et al.*, 1975); to *Fusarium* wilt by Grover and Singh (1970) and Corley (1965); and to *Verticillium* wilt by Silveira *et al.* (1970). Mutation experiments for induction of new variability have been reported by Nandpuri *et al.* (1971) and Kuwada (1970). Mutagens used were gamma and X-rays, respectively.

6 References

Anon. (1966). Red *bhendi* – a promising ornamental-cum-economical type for kitchen garden. *Madras agric. J.*, **53**, 297.

Boswell, V. R. and Reed, L. B. (1962). Okra Culture. *Leafl. US Dep. Agric.*, **449**, pp. 8.

Corley, W. L. (1965). Some preliminary evaluations of okra plant introductions. *Bull. Ga. Exp. Sta.*, **145**, pp. 16.

Grover, R. K. and Singh, G. (1970). Pathology of okra (*Abelmoschus esculentus*) caused by *Fusarium oxysporum* f. *vasinfectum*, its host range and histopathology. *Ind. J. agric. Sci.*, **40**, 989–96.

Joshi, A. B., Gadwal, V. R. and Hardas, M. W. (1974). Okra. In Hutchinson, J. B. (ed.), *Evolutionary studies in world crops. Diversity and change in the Indian subcontinent.* Cambridge, 99–105.

Kuwada, H. (1957). Cross compatibility in the reciprocal crosses between amphiploid and its parents (*Abelmoschus esculentus* and *A. manihot*) and the characters and meiotic division in their hybrids obtained among them. *Jap. J. Breed.*, **7**, 103–11.

Kuwada, H. (1970). [X-ray induced mutations in okra (*Abelmoschus esculentus*)]. *Tech. Bull. Fac. Agric. Kagawa Univ.*, **21**, 2–8.

Nandpuri, K. S., Sandhu, K. S. and Randhawa, K. S. (1971). Effect of irradiation on variability in okra (*Abelmoschus esculentus*) *J. Res., Punjab Agric. Univ.*, **8**, 183–8.

Prem Nath (1970). Problem oriented breeding projects in vegetable crops. *SABRAO Newsl. Mishima*, **2**, 125–34.

Schery, R. W. (1954). *Plants for man.* London.

Silveira, A. P. Da, Cruz, B. P. B., Bernardi, J. B. and Silveira, S. G. P. Da. (1970). [Response of some varieties of *Hibiscus esculentus* to *Verticillium* wilt]. *Biologico*, **36**, 63–8.

Singh, H. B., Swarup, V. and Singh, B. (1975). *Three decades of vegetable research in India.* ICAR, Tech. Bull., New Delhi.

56

Cotton

Gossypium (Malvaceae)

L. L. Phillips
North Carolina State University
Raleigh NC USA

1 Introduction

World cotton production is based on the seed fibres of four species of *Gossypium*. By far the most important of these is *G. hirsutum*, from which nearly 95 per cent of the 60 million bale (13 Mt) world crop was harvested in 1972–73. Lint length of *G. hirsutum* ranges from slightly less than an inch to $1\frac{3}{8}$ inches (classed as medium to long staple). Extra long staple ($1\frac{3}{8}$ inches and over) cotton from *G. barbadense* comprises about 5 per cent of world cotton production and short staple lint (less than $\frac{7}{8}$ inch) from *G. herbaceum* and *G. arboreum* contributes less than 1 per cent of the world total. Though cotton species are basically tropical perennials (and early cultivations were in the tropics and maintained as ratoon perennials) cotton today is an annual crop of temperate regions as well as the tropics. The major producing countries of *hirsutum* cotton are the USA and the USSR. Extra long staple production is dominated by Egypt, Sudan and the USSR, while India and Pakistan produce virtually all the short staple cotton.

2 Cytotaxonomic background (Fig. 56.1)

The genus contains 30 diploid species ($2n = 26$) and four tetraploids ($2n = 4x = 52$), the latter group including the commercially important *G. hirsutum* and *G. barbadense*. The diploids are conventionally divided into six genome groups based on cytological affinities. The A genome includes *G. herbaceum* and *G. arboreum*, the only diploids with true (spinnable) lint. Wild *G. herbaceum* is distributed in the savanna vegetation of southern Africa; *G. arboreum* is known only from cultivation. The three species of the B genome have allopatric distributions in southern and northern Africa and in the Cape Verde Islands. The C genome group is tentatively assigned to seven Australian species but there are probably three distinct cytotypes represented in the assemblage. The eleven taxa of the D genome are basically Mexican, but there are distributional range extremes in Arizona, USA and in Peru and the Galapágos Islands. The four E genome species occupy a range from eastern Africa, through the tip of the Arabian peninsula, to Pakistan. The single species of the F genome is found in eastern Africa. A rather extensive literature on the comparative cytology of interspecific hybrids of *Gossypium* diploids has been summarized by Phillips (1966).

The New World tetraploids are natural amphidiploids containing A and D genomes (Beasley, 1940; Skovsted, 1937). Gerstel (1953) has identified the A genome donor as *G. herbaceum* and Phillips (1962) has indicated that *G. raimondii*, a Peruvian species, is the most closely related known D species to the contributor of the New World D genome.

The time and place of the origin of the ancestral amphidiploid have been the subject of much conjecture, largely because the A genome contributor is not extant in the New World. Ancient origin theories have included proposals: first, that the A and D species were sympatric on the supercontinent of Pangaea and hybridized before the Jurassic–Triassic separation of South America and Africa; second, that the origin occurred on a Cretaceous trans-Pacific land bridge; or, third, in Eocene North America following A species migration via Beringia. At the other extreme, there have been proposals that the amphidiploid arose after the introduction of *G. herbaceum* to South America by man. Variations of both ancient and recent origin theories include proposals that the progenitors of *G. hirsutum* and *G. barbadense* involved different D genome species.

Most of these proposals were made in the first decade following the demonstration of the amphidiploid nature of tetraploid cotton and before the cytogenetics of triploid and synthetic hexaploid hybrids involving the tetraploid species and those of the A and D genome groups had received appreciable attention. These studies (summarized by Phillips, 1962, 1964, 1966) have indicated a very close homology between the amphidiploid A and D genomes and the chromosomes of *G. herbaceum* and *G. raimondii*, respectively; an ancient origin for the amphidiploid therefore assumes little structural change of its chromosomes, or those of its donor species, in 90–120 million years. Yet, it is within this same time span that a major part of *Gossypium* evolution has taken place and there exists today considerable structural chromosome diversity between the species of different major evolutionary lines (i.e., the genome groupings). On the other hand, there has been sufficient time since the origin of the amphidiploid and the earliest archaeological date for

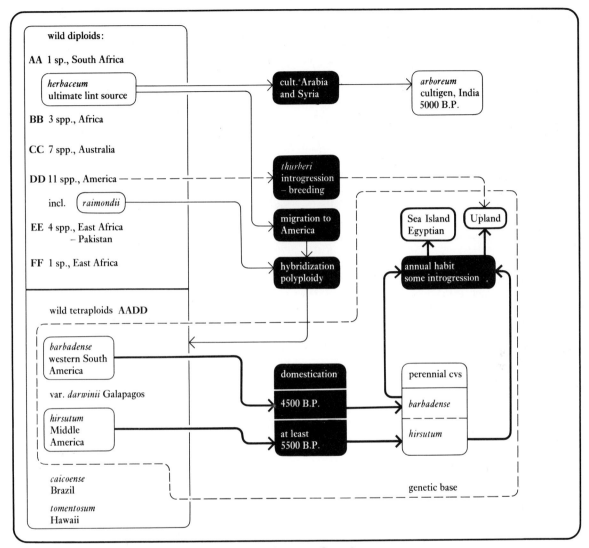

Fig. 56.1 Evolution and relationships of the cultivated cottons, *Gossypium* spp.

tetraploid cotton (*ca.* 5500 B.P.) for the evolution of at least four species, each exhibiting considerable genetical and cytological diploidization, and the dispersal of one of them to the Hawaiian Islands. It would seem that man could hardly have been involved in the origin of the amphidiploid.

In summary, the origin of the New World amphidiploid probably took place since the start of the Pleistocene in northern South America following oceanic drift of an A genome propagule from Africa. (Seeds of some Gossypiums remain viable after 2–9 months' immersion in sea water.) Four species (and several biotypes that deserve infraspecific recognition) have evolved concomitantly with migration and/or dispersal to

northeastern and western South America (*G. caicoense* and *G. barbadense*, respectively), throughout Mesoamerica and the Caribbean (*G. hirsutum*), and to the Galápagos and Hawaiian Archipelagoes (*G. barbadense* var. *darwinii* and *G. tomentosum*, respectively). Studies of the comparative cytology of the amphidiploid taxa (Gerstel, 1953) indicate that they are monophyletic; that is, they evolved from a single interspecific hybrid combination.

3 Early history

The diploid species, *G. herbaceum* and *G. arboreum*, have a long history of cultivation in the Old World. Cloth fragments, dated to 5000 B.P., have been found in the Indus Valley of Pakistan (Gulati and Turner, 1928). Gerstel (1953) has shown that *G. herbaceum* and *G. arboreum* differ cytologically by a major reciprocal chromosome translocation and that the *G. herbaceum* chromosome-end arrangement is primitive. *Gossypium herbaceum* was probably first cultivated in Arabia and Syria before finding its way to the Indian subcontinent where *G. arboreum* arose under cultivation. Concomitantly with its differentiation into several races, *G. arboreum* became the dominant species throughout Africa and Asia. During the last 100 years the New World tetraploids have supplanted the diploids in all of this area except for peninsular India.

Archaeological sites dated to *ca.* 4500 B.P. in central-coastal Peru have yielded cotton boll and fibre remains with characteristics intermediate between those of wild *G. barbadense* populations extant in the Galápagos Islands and near the Guayas Estuary in Peru and Ecuador, and primitive *G. barbadense* lines from the same region (Stephens and Mosely, 1974). Thus these archaeological cottons represent an early stage in the domestication of *G. barbadense*. There is a distinct biotype of *G. barbadense* (var. *brasiliense*) distributed throughout Amazonia and in the Antilles, but it is unclear whether this type represents an

Fig. 56.2 Distribution and centres of origin of the New World tetraploid cottons, *Gossypium hirsutum* and *G. barbadense*.

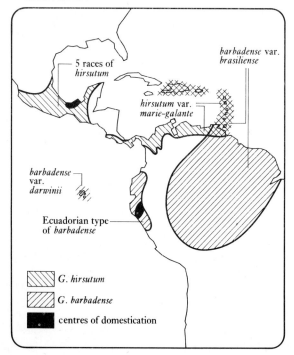

barbadense var. brasiliense

5 races of hirsutum

hirsutum var. marie-galante

barbadense var. darwinii

Ecuadorian type of barbadense

▨ G. hirsutum

▨ G. barbadense

■ centres of domestication

evolutionary line originating from the Peru–Ecuador type (mixtures of both types occur along the inter-Andean valleys of the Amazon River system) or from an independent centre of *G. barbadense* domestication in eastern South America.

The oldest cotton remains in the Americas are from south-central Mexico and are dated about 5500 B.P. (Smith, 1968). The remains are very likely those of *G. hirsutum* but, since they have the characteristics of fully domesticated types, this archaeological cotton does not represent the earliest domestication of *G. hirsutum*. The wild forms of *G. hirsutum* are so widely dispersed (Gulf coast of Mexico, Caribbean coast of South America, Caribbean Antilles, several Pacific islands) that they offer little help in locating possible centres of domestication. Stephens (1973) recognized two probable centres of *G. hirsutum* domestication, namely the Gulf coast of Mexico and the Caribbean coast of northwestern Columbia.

Mention should here be made of the other two species of the New World tetraploid group, neither of which is cultivated. One of these, *G. caicoense*, is now known only from small populations in the states of Rio Grande do Norte and Bahia, Brazil. The taxon has been in experimental culture only for a short time (1965) and is accorded specific status because *G. hirsutum* × *G. caicoense* and *G. barbadense* × *G. caicoense* F_2 progenies segregate for chlorophyll deficiencies and other aberrant types. This behaviour is typical of *G. hirsutum* × *G. barbadense* and other interspecific F_2s but is unknown in intraspecific crosses. *Gossypium tomentosum* is unique among the New World tetraploids in having seed fibres that are unconvoluted (not spinnable) and strongly adherent to the seed coat. This species is endemic to several islands of the Hawaiian archipelago; no cotton resembling *G. tomentosum* in fibre characteristics has been collected in the Americas.

4 Recent history

Cotton was grown and used extensively in pre-Columbian Mesoamerica and South America and, during the first 300 years of the Colonial period, perennial *G. hirsutum* and *G. barbadense* were disseminated to other areas by two main routes: first, a Spanish route from Cuba (which served as a marshalling point for all trade from Spanish America) to Spain and the Spanish Mediterranean (Sicily, Malta, Cyprus); and, second, a Portuguese route from Brazil to Africa and India. The cottons introduced by these routes were greatly superior to the Old World diploids and gradually supplanted *G. arboreum* and *G. herbaceum* as sources of raw cotton. These perennial introductions have, in turn, been replaced by three independently derived annual types of New World cotton, namely, Upland (*G. hirsutum*), Sea Island (*G. barbadense*) and Egyptian (*G. barbadense*).

Upland, which made its appearance in the mid-eighteenth century in the southeastern United States, probably originated from Mexican introductions but some of the early varietal names are the same as those used in eighteenth century West Indian islands and some exchange of germ plasm between the Caribbean and southeastern USA is likely. Upland was grown primarily for home use until the advent of Whitney's saw gin in 1793 but, between this time and the start of the Civil War, Upland production increased 1,000-fold while its culture spread through the mid-south to Texas. Cotton production in the USA was drastically reduced during the Civil War, and what had been a trickle of Upland types to other cotton growing areas became a deluge as Upland-based cotton agriculture, in traditional cotton-growing regions as well as in new ones (e.g. South America), proliferated throughout the tropics and subtropics to offset the American deficit.

Two other developments have affected Upland cotton in the USA and, indirectly, all other *hirsutum*-growing regions. In the early years of this century the boll weevil moved into the US cotton belt from Mexico and, within two decades, the belt, from the Carolinas to Texas, was facing a crisis. Since only early maturing varieties could produce satisfactorily in coexistence with the boll weevil, there was a very considerable narrowing of the genetic base in Upland cotton. Also, in the early years of this century, a search for germ plasm in southern Mexico resulted in the introduction of the first Acala material to the USA. The original Acala accessions, and subsequent collections from Guatemala, have provided the genetic base for Upland production in the irrigated lands of the southwestern USA. Acala varieties have been widely introduced into the major cotton-producing regions of the world during the last two decades.

Sea Island cotton is a *G. barbadense* type that arose in the coastal plains and offshore islands of the Carolinas and Georgia in the late eighteenth century; its origin is obscure but it does carry a genetic marker (Ck^y; Stephens, 1974) that is largely restricted to *G. barbadense* types from eastern South America and the West Indies. Sea Island is distinguished by its very high quality lint (very long and fine) and annual habit, characteristics not found in wild forms of either *G. hirsutum* or *G. barbadense*. However, Stephens (1974) has experimentally derived annual types with transgressive segregation for lint length from hybrids

between primitive forms of *G. barbadense* and *G. hirsutum* and this seems a reasonable explanation for the origin of Sea Island cotton. Sea Island strains were rather late maturing and therefore proved vulnerable to the boll weevil; the production of Sea Island cotton was abandoned in the USA about 1920, though some attempts were made to revive its cultivation in the 1930s. The production of Sea Island cotton in the West Indies (where it had been introduced from the USA) persisted until about 1960.

Egyptian cotton is also based on a form of *G. barbadense* and its origin is traceable to a single plant in a Cairo garden in the year 1820. Traditionally, this, Jumel's, cotton is assumed to have come from Nigeria via the Portuguese South America-Africa trade route (Hutchinson, 1949). However, Stephens (1974) indicated that the Egyptian cottons may have more phylogenetic affinity with Peruvian than with Amazonian *G. barbadense* and he propose that Jumel's cotton was of Mediterranean origin, by way of the Spanish trade route from western South America.

Egyptian cotton has probably derived much of its fibre quality from Sea Island lines which were introduced to Egypt from the southeastern USA in the latter part of the eighteenth century and contributed to the breeding of improved Egyptian varieties. Some of this material, in turn, was introduced to the southwestern USA and figured prominently in the parentage of the Pima varieties of *G. barbadense* grown in that area.

Though naturally partly cross-pollinated in the field, the cottons are tolerant of close inbreeding and are bred by pedigree methods as inbred pure-lines. Quantitative genetic studies have shown that the traits considered to be the components of yield (boll size, boll number, lint percentage, etc.) are controlled by dominance and/or epistatic gene action, whereas most fibre characteristics are influenced by additive gene action. Heterosis for yield is 10–25 per cent for intraspecific *G. hirsutum* F_1 hybrids, and somewhat higher for intraspecific *G. barbadense* F_1s; yield component heterosis for interspecific crosses ranges from 25 to 50 per cent. The possible commercial-scale exploitation of first generation heterosis has been researched for 20 years but has been frustrated by the unavailability of satisfactory genetic and/or cytoplasmic male sterility and the fact that the relatively heavy pollen requires insects as vectors. The 'male-sterile' half of this problem is nearing solution but, at least in the USA, where, on most cotton acreage, natural outcrossing is less than 15 per cent, shortage of insect vectors will continue to be a major obstacle to hybrid cotton production.

5 Prospects

The trend of cotton breeding in the future will be toward a greater exploitation of wild and feral germ plasm resources to broaden the genetic base of the cultivated species. Though most of the diploid species have been in experimental culture for some time and 2,500–3,000 accessions of wild and feral New World tetraploid types are also available to cotton breeders, very little use has been made of this potential source of variability. One notable exception is the introgression of fibre strength from *G. thurberi* (a D genome diploid) into several varieties of Upland cotton by way of the *G. arboreum* × *G. thurberi* synthetic amphidiploid. The explanation for this lack of interest in exotic germ plasm is that, historically, selection pressure in cotton has been almost exclusively on earliness and yield. Over the past 40 years, with improvements in breeding, planting and harvesting techniques and the massive use of insecticides, cotton yields have been increased several-fold; moreover, the perennial, small-bolled and sparsely-linted wild and feral accessions are unlikely sources of earliness or yield potential.

Two relatively recent trends are changing cotton breeding objectives and expanding the utilization by cotton breeders of hitherto untapped germ plasm resources. First, advances in fibre technology research have defined and described the components of fibre quality (i.e. length, strength, elasticity, fineness, relative wall and lumen diameter) and have made available to the breeder the testing procedures necessary for measuring the efficiency of cotton quality breeding programmes. Second, concern over environmental pollution related to cotton production, historically a very heavy user of chlorinated hydrocarbon and organophosphate insecticides, has resulted in government-directed curtailments of cotton insecticide use and has given impetus to a search for non-chemical methods of controlling insect pests. Collections of wild diploids, of wild and feral tetraploids and of obsolete varieties are being screened for fibre quality and insect resistance in all major cotton-producing areas. Most of this renewed interest is concentrated on tetraploid germ plasm because the chromosome engineering problems are much more intractable in transferring desirable variability from the diploids to *G. hirsutum* or *G. barbadense* than within, or between, these two species.

The development of glandless-seeded cotton has been in progress for twenty years but, because of its great economic potential, research in this problem-area continues. The high-protein seed of normally glanded cotton contains a polyphenol (gossypol) that is toxic to non-ruminant animals; although gossypol can be chemically removed from cottonseed oil and meal, the process is costly and reduces protein quality. With extant breeding stocks and procedures, the glands can be genetically removed from the entire plant (including the seeds) but glandless plants are considerably more vulnerable to insect pests than are glanded individuals. The economic potential of glandless cottonseed is considerable (annual world cottonseed production is about 25 Mt) but even more important is the fact that gossypol-free cottonseed would become an available dietary source of protein to a human population that consumes almost no cottonseed protein at the present time. These considerations ensure that the glandless seed-glanded plant phenotype will be an important breeding objective in the coming years.

6 References

Beasley, J. O. (1940). The origin of the American tetraploid *Gossypium* species. *Amer. Nat.*, **74**, 285–6.

Gerstel, D. U. (1953). Chromosomal translocations in interspecific hybrids of the genus *Gossypium*. *Evolution*, **7**, 234–44.

Gulati, A. N. and Turner, A. J. (1928). A note on the early history of cotton. *Ind Cent. Cotton Cmttee., tech. Lab. Bull.*, **17**, 1–10.

Hutchinson, J. B. (1949). The dissemination of cotton in Africa. *Empire Cott. Growing Rev.*, **26**, 1–15.

Phillips, L. L. (1962). Segregation in new allopolyploids of *Gossypium*. IV. Segregation in New World × Asiatic and New World × wild American hexaploids. *Amer. J. Bot.*, **49**, 51–7.

Phillips, L. L. (1964). Segregation in new allopolyploids of *Gossypium*. V. Multivalent formation in New World × Asiatic and New World × wild American hexaploids. *Amer. J. Bot.*, **51**, 324–9.

Phillips, L. L. (1966). The cytology and phylogenetics of the diploid species of *Gossypium*. *Amer. J. Bot.*, **53**, 328–35.

Smith, C. E., Jr. (1968). Plant remains. In D. S. Byers (ed.), *Prehistory of the Tehuacan Valley*, **1**. Austin, Texas, U.S.A., 220–55.

Stephens, S. G. (1973). Geographical distribution of cultivated cottons relative to probable centers of domestication in the New World. In A. M. Srb (ed.), *Genes, enzymes and populations*. New York, 239–54.

Stephens, S. G. (1974). The use of two polymorphic systems, nectary fringe hairs and corky alleles, as indicators of phylogenetic relationships in New World cottons. *Biotropica*, **6**, 194–201.

Stephens, S. G. and Mosely, M. E. (1974). Early domesticated cottons from archaeological sites in central coastal Peru. *Amer. Antiquity*, **39**, 109–22.

Skovsted, A. (1937). Cytological studies in cotton. IV. Chromosome conjugation in interspecific hybrids. *J. Genet.*, **34**, 97–134.

Breadfruit and relatives

Artocarpus spp. (Moraceae)

Jacques Barrau
Muséum National d'Histoire Naturelle
Paris France

1 Introduction

In Malaysia and Oceania several species of *Artocarpus* are ancient domesticates grown for their fruits, sometimes also for their fibrous bark, latex and timber. Chief among them is *A. altilis* (= *A. communis*), the breadfruit, a staple food plant in several Pacific islands. The starchy fruits are eaten roasted, steamed, baked or sometimes made into dough; they may also be preserved by desiccation or as fermented paste. When present, the seeds are also eaten, as are those of the two following species. The champedak (*A. integer*) and the jackfruit (*A. heterophyllus*) have a more westerly distribution than the breadfruit and are grown mainly in southeast Asia and Indonesia. They are generally seedy and their fruits and seeds are eaten raw or cooked. In addition, several wild species are sometimes used as food, especially in the Malaysian archipelago, New Guinea and western Micronesia.

Of the three cultigens the breadfruit and the jackfruit are now widely grown in the lowland tropics but only in Oceania does one of them, the breadfruit, attain great importance; it probably deserves greater scientific attention than it has had in view of its potential value for other subsistence economies.

For general reviews of economic botany of these crops, see Jarrett (1959), Fosberg (1960), Ochse *et al.* (1961), Barrau (1962), Purseglove (1968).

2 Cytotaxonomic background

The genus *Artocarpus* (about 50 spp.) is native to southeast Asia. The wild progenitor of the breadfruit has not been identified but it probably belonged to the series Incisifolii from the eastern borders of Indonesia, western Micronesia and New Guinea (Fig. 57.1) (Jarrett, 1959). The other two domesticates, jackfruit and champedak, are closely related and belong to a different series, the Cauliflori. All three are basically diploid with $2n = 2x = 56$ but, in addition, triploidy seems to be common among seedless breadfruit clones.

The main facts are summarized in the following table.

Table 57.1 The breadfruit and its relatives

Name	$2n =$	Pre-colonial distribution
A. altilis (breadfruit) (= *A. communis*)	56 (also 84)	Malaysia to eastern Polynesia
A. integer (champedak) (*A. champeden*)	56 (?)	Malaysia
A. heterophyllus (jackfruit)	56	Malaysia to India

3 Early history

The three species seem to be of ancient domestication in Vavilov's Indomalayan centre but dates are neither known nor can be inferred. The breadfruit, a multi-purpose plant used for food, fibre, latex and timber, must have attracted man's attention very early, just as it struck the first European explorers of the Pacific with wonder. The crop, domesticated in the eastern parts of the archipelago, may be assumed to have accompanied (along with coconuts, bananas and aroids) the human migrations that peopled Oceania 1000–2000 B.P. There it became a staple food plant and local selection isolated a large number of clones, many of them seedless. Parthenocarpy may be inferred but, unfortunately, there appear to be no experimental studies on the point; nor has the relation between the presumed parthenocarpy and triploidy (which would imply a degree of seed sterility) been investigated. Superficially, an interesting analogy with bananas is apparent and would be well worth study. Though the mechanisms are unknown, it seems clear that the crop has been fairly profoundly modified in cultivation.

All three species are monoecious but diclinous, bearing separate male and female inflorescences on specialized laterals. All three are commonly propagated by root-cuttings, breadfruit nearly always so, jackfruit and champedak also often by seed. They are outbreeders and jackfruit seedlings, as would be expected, are very variable (Purseglove, 1968).

One might guess that introgression with local wild species has played a part in the evolution of the three cultigens but there is no direct evidence on the point. *Artocarpus mariannensis*, of western Micronesia, is taxonomically similar to the breadfruit and has sometimes been implicated in the evolution of the crop.

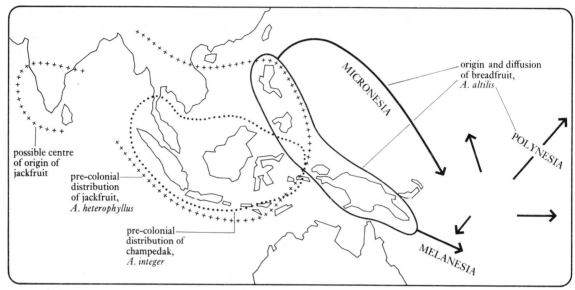

Fig. 57.1 Origins and dispersal of breadfruit, *Artocarpus altilis*, and its relatives.

4 Recent history

Breadfruit and jackfruit were widely spread through the tropics in the eighteenth to nineteenth centuries but attained little importance outside their native areas. The story of the introduction of breadfruit to the West Indies (1793) and the associated mutiny on the *Bounty* has often been told: Purseglove (1968) gives a summary and references.

None of the three crops has been accorded any significant plant breeding effort, or even clonal selection, which seems a pity in view of, at least, the breadfruit's potential for food production. The stimulus may yet be provided by a serious disease, Pingelap, which has appeared in the last 20 years in the islands of Micronesia. Characterized by either a sudden foliage wilt or by die-back, it is of unknown origin (O'Connor, 1969) but it would seem that resistant clones can be found.

Studies of variability in the breadfruit have been reported by Jarrett (1959) and Fosberg (1960). Some ten years ago there was, in Apia, Western Samoa, a collection of Oceanian breadfruit clones established there for comparative study and breeding purposes.

5 Prospects

The three crops are now what they have been for millenia, components, of varying importance, of subsistence economies. That the breadfruit and jackfruit, at least, have considerable potential seems clear but deliberate breeding will be necessary to exploit it.

6 References

Barrau, J. (1962). *Les plantes alimentaires de l'Océanie, origines, distribution et usages.* Marseille, pp. 275.

Fosberg, F. R. (1960). Introgression in *Artocarpus* (Moraceae) in Micronesia. *Brittonia*, **12**, 101–13.

Jarrett, F. M. (1959). Studies in *Artocarpus* and allied genera. III a revision of *Artocarpus* subgenus *Artocarpus*. *J. Arnold Arb.*, **40**, 113–368.

Ochse, J. J., Soule, M. J., Djikman, M. J. and Wehlburg, C. (1961). *Tropical and subtropical agriculture*, New York, **1**, 647–56.

O'Connor, B. A. (1969). *Exotic plant pests and diseases, a handbook of plant pests and diseases to be excluded from spreading within the area of the South Pacific Commission.* S.P.C., Noumea, New Caledonia.

Purseglove, J. W. (1968). *Tropical crops. Dicotyledons.* London, **2**, 377–86.

Hemp

Cannabis sativa (Moraceae)

N. W. Simmonds
Scottish Plant Breeding Station
Edinburgh Scotland

1 Introduction

Hemp, an ancient crop plant, provides a fibre, a psychotomimetic drug and a minor food and a fixed oil from the seeds. The economic botany of the crop has been reviewed by Purseglove (1968), fibre aspects by Kirby (1963).

The plant is an annual herb that commonly attains a height of 6–8 m, but has been known to grow to 12 m. Yields of the bast fibre average $2 \cdot 0$–$2 \cdot 5$ t/ha. The drug is contained in resin glands in all parts of the plant but drug yield is low unless harvest is confined to the upper parts of female plants of special cultivars grown in hot climates, as is the case in the production of Ganja in India.

Though native to central Asia, the species is ecologically very adaptable and now grows as a crop and/or as a weedy escape throughout the world, temperate and tropical. The USSR is the leading fibre producer, India the leading drug producer. Legitimate international trade in the drug is negligible.

2 Cytotaxonomic background

The genus *Cannabis* is usually regarded as monotypic, though a number of species other than *C. sativa* have been named. According to Schultes (1970), in an authoritative review, no clear infraspecific taxa can be recognized, despite the publication of many names. There is no evidence of intraspecific genetic differentiation (Small, 1972).

Hemp is diploid, with $2n = 2x = 20$ (Menzel, 1964; Small, 1972). Like the related *Humulus*, the crop is basically dioecious and is wind-pollinated. The male is heterogametic (XY) and there is variability both in sex ratio (females tend to exceed males) and in sex-expression (females are more variable than males). Genetic studies of monoecious lines have not clearly established whether such lines are all XX or are XX, XY and YY, so the control of sex expression is not yet fully understood. Tetraploid families have shown an excess of females, suggesting that XXXY is female. For a review of the subject, see Westergaard (1958). Earlier cytological evidence (Westergaard, 1958) suggested that the male chromosome complement was heteromorphic and this was fully confirmed by Menzel (1964) on the males of four dioecious stocks; two monoecious lines were homomorphic and presumptively XX.

Sex-expression is affected by many environmental factors including day length (short days favouring monoecism) and various experimental treatments (as by gibberellins). Unless genetically monoecious strains are available (see below) seed-raised populations contain (roughly) equal proportions of male and female plants; for resin production, males are usually rogued out.

Hemp is thus an outbreeder directly derived from an outbred wild ancestor. Whether truly wild populations survive anywhere does not seem to be known; Schultes (1970) thinks probably not. Certainly, escaped, weedy forms are very widespread and reciprocal introgression between escaped or wild forms and cultivated could be expected and, indeed, almost certainly occurs (Schultes, 1970).

3 Early history

Hemp as a crop must have originated somewhere in temperate Asia. Schultes (1970) tends to favour the idea of a diffuse origin somewhere in the huge area from the Caspian and the Himalayas to China and Siberia. Probably, the first use was for fibre and hemp may be one of the oldest non-food crops. Li (1973) states that it was the only fibre available to ancient peoples of northern and north-eastern China and eastern Siberia; the plant has probably been cultivated in China for at least 4,500 years and evidence of the use of the fibre there goes back to Neolithic times.

The historical and other records assembled by Schultes (1970) suggest that the crop reached western Asia and Egypt during the period 1000–2000 B.C. and was perhaps taken to Europe by Scythian people about 1500 B.C. before it became established in the Mediterranean area. There is, however, a record of a hempen fabric from Turkey dated 700–800 B.C. Hemp became a significant fibre crop in Europe from about A.D. 500 onwards and reached the USA in 1632.

Schultes infers, no doubt correctly, that there must have been selection for larger and less dormant seeds, non-shattering inflorescences and perhaps also reduced perianths and thin testas, as well as for fibre characters. Kirby (1963) notes that varieties differ in photoperiodic reponse so hemp resembles maize, rice, soya, potatoes and many other crops in respect of this adaptation.

The narcotic properties of the plant were recognized in India by 1000 B.C. but, perhaps surprisingly, were not known in the Mediterranean before post-classical times and in Europe perhaps much later still (? eighteenth century). The drug was introduced to western medicine about 1840, following the work of O'Shaughnessy in India. It was not recognized as 'official' in the US National Formulary until 1941. For a review of medical aspects, see Mikuriya (1969).

4 Recent history

There has been little breeding effort and what has been done seems to be poorly documented. Fibre varieties are tall, little branched and as little seed fertile as possible; drug and oil varieties are short, branched and seedy. Male plants are inferior to female in fibre, seed and drug production; some effort went into developing monoecious fibre-producing lines in Germany before the late war but the impression is that modern varieties (e.g. in Korea – Ree, 1966) are still predominantly dioecious in constitution and developed by bi-parental mass-selection for fibre content. Ree noted that intervarietal hybrids showed the expected heterosis but that seed production was too expensive to permit practical exploitation of hybrid varieties; extensive hand-rogueing of males would presumably be involved in their production.

As to drug production, it seems to be established that there are differences between varieties in gross contents of resins and in balance of specific constituents (Joyce and Curry, 1970; Turner and Hadley, 1973) but there is no biochemical-genetic information. The old belief that female plants have higher drug contents than male has recently been challenged by Agurell (in Joyce and Curry, 1970). He showed that active resin content on a wt/wt basis was the same for the two sexes but that females were more productive because more leafy.

5 Prospects

If hemp-fibre production were ever to be re-developed, drug-free varieties would, for obvious social reasons, be essential. A first requirement for their production would be biochemical-genetic understanding which is at present entirely wanting. Since the resins are produced in leaf glands, induction and study of glandless mutants (compare glandless mutants of *Gossypium*) could be rewarding. Population structure would pose problems; the choice would lie between (a) mass-selected dioecious lines; (b) monoecious lines or; (c) hybrids. Given that hemp is an outbreeder, the

order of preference would probably be, on *a priori* grounds: (c), (a), (b). This expectation might, however, be modified by differences between sex-types in fibre content.

There is, nowadays, considerable interest in the pharmacology of the drug and its re-admission to medical use is not inconceivable. Biochemical genetic understanding would be no less essential for breeding drug varieties than for breeding drug-free fibre or oil-seed types.

6 References

Joyce, C. R. B. and Curry, S. H. (eds.) (1970). *The botany and chemistry of Cannabis*. London.

Kirby, R. H. (1963). *Vegetable fibres*. London, 46–61.

Li, H. L. (1973). The origin and use of *Cannabis* in eastern Asia, their linguistic–cultural implications. In *Cross-cultural perspectives on Cannabis*. Chicago.

Menzel, M. Y. (1964). Meiotic chromosomes of monoecious Kentucky hemp (*Cannabis sativa*). *Bull. Torrey bot. Club*, **91**, 193–205.

Mikuriya, T. H. (1969). Historical aspects of *Cannabis sativa* in western medicine. *New Physician*, **18**, 902–8.

Purseglove, J. W. (1968). *Tropical crops. Dicotyledons*. London, **1**, 40–4.

Quimby, M. W., Doorenbos, N. J., Turner, C. E. and Masond, A. (1973). Mississippi-grown marihuana, *Cannabis sativa*; cultivation and observed morphological variations. *Econ. Bot.*, **27**, 117–27.

Ree, J. H. (1966). Hemp growing in the republic of Korea. *Econ. Bot.*, **20**, 176–86.

Schultes, R. E. (1970). Random thoughts and queries on the botany of *Cannabis*. In Joyce, C. R. B. and Curry, S. H. (eds.), *The botany and chemistry of Cannabis*, London, 11–38.

Small, E. (1972). Interfertility and chromosomal uniformity in *Cannabis*. *Canad. J. Bot.*, **50**, 1947–9.

Turner, C. E. and Hadley, K. (1973). Constituents of *Cannabis sativa* II. Absence of cannabidiol in an African variant. *J. pharm. Sci.*, **62**, 251–4.

Westergaard, M. (1958). Sex determination in plants. *Adv. Genet.*, **9**, 217–81.

Fig

Ficus carica (Moraceae)

W. B. Storey
University of California Riverside
California USA

1 Introduction

The fig is an esteemed tree fruit in many countries in the tropical, subtropical and moderately temperate regions of the world. In most countries it is grown only for local consumption, either as fresh fruit or as dried fruit and preserve. However, in several Mediterranean countries (Algeria, Greece, Italy, Portugal, Spain and Turkey) large amounts are produced for export, dried and as paste. The long-term average annual production of dried figs and fig paste by all countries including those in the Near and Middle East is about 238 kt.

For comprehensive coverage on many subjects relating to the fig, see Condit (1947, 1955).

2 Cytotaxonomic background

Ficus is a genus of about 2,000 species of tropical and sub-tropical trees, shrubs and vines, worldwide in distribution (Condit, 1969). Besides *F. carica*, the only other species cultivated for fruit is *F. sycomorus*, the sycamore of Scripture. It is grown locally, mostly in Egypt and Israel. Several other species have edible fruits, which, however, lack palatability.

The basic chromosome number of the genus is $x = 13$. Somatic chromosome numbers have been reported for slightly over 100 species (Condit, 1928, 1935, 1964). Excepting five African species which have somatic numbers of 52 ($2n = 4x = 52$) and *F. elastica* var. *decora*, which is triploid ($2n = 3x = 39$), all are diploids with $2n = 2x = 26$.

Cytotaxonomic relationships in *Ficus* are unknown. However, *F. carica* has been hybridized with *F. palmata*, *F. pseudo-carica* and *F. pumila*.

3 Early history

According to Greek mythology, during the War of the Titans, Zeus was pursuing Ge and her son Sykeus when Ge, in order to save Sykeus, metamorphosed him into a fig tree. The city of Sykea in the ancient country of Cilicia derived its name from this myth. An Athenian myth credits the goddess Demeter with having revealed 'the fruit of autumn' (the fig), to human beings. The generic name *Ficus* comes from *ficus Ruminalis*, a sacred fig tree which, according to Roman legend, sheltered the infants Romulus and Remus who were to become the founders of Rome (about 746 B.C.) while they were being cared for by a she-wolf and a woodpecker. The tree was named for Ruminia, a goddess who watched over suckling animals. The specific epithet *carica* refers to Caria, an ancient political division of Asia Minor bordering on the Aegean Sea.

Botanical evidence points to the fertile region of southern Arabia, where wild fig and caprifig trees are still to be seen, as the centre of origin of *F. carica*. No date has been determined for the time of its domestication but it must have been millenia ago. In the course of time, it spread in cultivation through all the countries of southwestern Asia and the Mediterranean basin. It was carried to all parts of the world during the age of exploration and colonization which followed Columbus's discovery of the New World in 1492 and circumnavigation of the world by Magellan's ships in 1519–21.

The fruiting behaviour of the fig is peculiarly complicated. To understand changes under domestication, a pomological digression is necessary. The so-called 'fruit' of all *Ficus* species is the syconium, a complex, enlarged, fleshy, hollow peduncle bearing numerous, minute, pedicelled flowers on the inner wall. The inner chamber is open to the outside through a narrow pore, the ostiole. The true fruits are drupelets that develop from the ovaries of the enclosed flowers. The syconium is a collective or multiple fruit and is, in a sense, spurious, in that it consists of vegetative tissue on which the true fruits are borne.

The flowers of *F. carica* are unisexual. The species is gynodioecious; that is, there are two sex forms: the caprifig, which is monoecious (hermaphrodite); and the fig which is pistillate (female).

Caprifigs bear three crops of syconia annually, the *mamme*, *profichi* and *mammoni*, which mature respectively in spring, summer and autumn in the northern hemisphere. Caprifig syconia tend to be dry, chaffy and unpalatable. However, they do serve as habitation for the fig wasp (*Blastophaga psenes*) which goes through three synchronized life cycles annually. Some caprifig clones are maintained solely as sources of *Blastophaga* for pollinating (caprifying) the flowers of one class of fig cultivar (the Smyrna type); others are maintained for use in breeding.

Depending on cultivar, figs may or may not produce

two crops annually: a first or 'breba' crop which emerges from overwintering latent buds initiated in the previous autumn growth and matures in June–July; and a second or main crop which develops on current growth in June–August and matures in August–November. All cultivars produce the main crop but not all produce the breba.

The caprifig has staminate flowers and pistillate flowers with short styles through which female *Blastophaga* lay eggs in the ovules. The larvae develop in the ovary at the expense of the ovule. The fig has pistillate flowers with long styles, which prevent the wasp from laying eggs in the ovules, and vestigial staminate flowers or none.

The differences between fig and caprifig are determined either by a recessive gene with pleiotropic effects or by two closely linked recessives, probably the latter, as follows:

G, normal allele, short-styled pistils
g, recessive allele, long-styled pistils
A, normal allele, androecium present
a, recessive allele, androecium suppressed

Thus caprifigs are *GA/GA* and *GA/ga* and figs *ga/ga*. The various possible matings segregate as expected. Homozygous caprifigs are rare and have, indeed, only recently been identified. In the wild, in the presence of *Blastophaga*, one would expect caprifigs to set few seeds, so the mating system would approach dioecism (with male heterogamety) rather than gynomonoecism.

There are no sex-linked vegetative characters distinguishing seedling caprifigs from figs. Identifying the sex of an individual growing on its own roots can only be done when the first syconia appear, usually at an age of 3–7 years. The time can be shortened to 1–3 years, however, by grafting scions from 6- to 8-month old seedlings on to the branches of older trees.

Female trees (i.e. figs proper) have, in the course of evolution, become differentiated into three groups distinguished by characters concerned with fruit-setting, as follows:

Smyrna type: The breba crop is lacking or scanty and tends to be caducous. The main crop is usually adequate but drops prematurely if not caprified; the drupelets contain fertile seeds.

San Pedro type: The breba crop is usually good but the main crop syconia drop unless caprified (there are exceptions); caprified figs contain fertile seeds, while the breba crop and uncaprified main crop figs may be 'seedless' or may contain 'seedlike bodies', that is, parthenocarpic drupelets.

Common type: This may or may not produce a breba crop; no caprification is necessary for the main crop; both breba and main crop figs may or may not contain

Fig. 59.1 Evolution of the fig, *Ficus carica*.

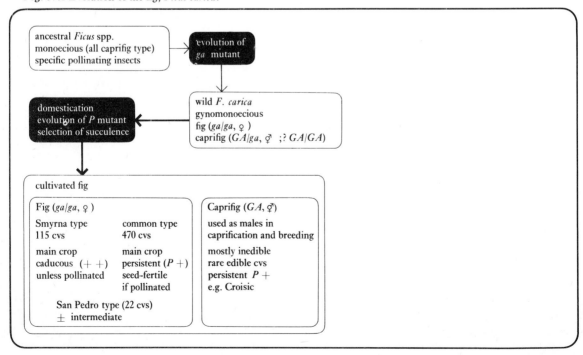

parthenocarpic drupelets.

In summary, the Smyrna type which depends on pollination to set a crop and is seedy must be regarded as basic. The Common type, which sets a crop with or without empty parthenocarpic drupelets, is the most advanced. The San Pedro type is more or less intermediate. The main character distinguishing Common figs from the other two groups is, therefore, whether or not the syconia are caducous in the absence of pollination. The genetic control of this character is as follows:

$+$ wild type recessive, ovules normal, syconia caducous
P dominant mutant, ovule-lethal, syconia persistent

Since P is only male-transmitted, fig $\times$ caprifig matings produce either all caducous or 1 caducous: 1 persistent progeny, according to parental genotype. It will be noted that all persistent plants are heterozygous $(P+)$. This dominant P mutation was therefore a significant feature of fig evolution.

At last count, 605 more or less valid cultivars were recognized even after the reduction of synonyms (Condit, 1955). These were distributed among the horticultural types as follows: Smyrna, 115; San Pedro, 22; Common, 470. Doubtless many obscure cultivars in limited local cultivation in various countries escaped notice.

It is noteworthy that all important cultivars produced on a large scale commercially today, for example: Sari Lop (syns. Lob Injir, Calimyrna), Dottato (syn. Kadota), Verdone (syn. White Adriatic) and Brunswick (syn. Magnolia), were selected centuries ago by persons unknown. Some have been in cultivation for a very long time. Sari Lop for example has been grown in the Meander Valley of Turkey for possibly as long as 2,000 years; Dottato of Italy was praised by Pliny (A.D. 23–79); Verdone has been grown in the Adriatic region for hundreds of years and was introduced into England in 1727 by Richard Bradley. Franciscana has been grown at Estepona, Spain, for centuries. It was introduced into Mexico and Chile by Spanish missionaries soon after the conquest. In 1768, Franciscan missionaries from Mexico established it in California where it has become known as Black Mission.

4 Recent history

In 1895, the United States Department of Agriculture obtained scion-wood of the Chiswick fig collection from the Royal Horticultural Society of England. Scions of every cultivar, 66 in all, were grafted successfully on trees in the orchard of the California Nursery Company at Niles, California. This collection served as foundation stock for early attempts at breeding.

The first attempts to create new cultivars by hybridization were probably those of Burbank, followed by those of Swingle (1908) and Hunt (1911). From 1928 to 1968, Ira J. Condit carried on studies of fig genetics and cytology and engaged in fig breeding at the University of California, Riverside. He developed two new Adriatic-like cultivars, Conadria and DiRedo, for planting commercially as drying and paste figs, and two other cultivars, Excel and Flanders, for home gardens. At present, an especially promising new cultivar, Tena, which was developed by W. B. Storey for drying and paste, is being tested in commercial orchards. The five cultivars named here seem to be the only ones ever to have been introduced into the trade from a planned breeding programme.

Success in breeding is due largely to the use of the caprifig Croisic because: (a) it is the only caprifig known to have the succulence and palatability of the fig proper; and (b) it does not require habitation by the fig wasp to set fruit. It is the progenitor of every persistent $(P+)$ caprifig used for breeding in California today. Its origin is unknown. It was first described by Solms-Laubach (1882) who saw it growing at Croisic, France. It was being cultivated near Cordelia, California by 1893. Its genetic behaviour is like that of other caprifigs, but in its progenies are at least some seedlings that produce succulent syconia. Apparently, succulent caprifigs are not attractive to *Blastophaga*, for the syconia are scarcely inhabited.

5 Prospects

FAO reported that, in 1967, fig breeding was being done only in the USA and the USSR (UN 1961, 1965, 1967). For a review of current fig breeding work, see Storey and Condit (1969) and Storey (1975).

The current objectives of fig breeding in California are to improve tree and fruit characters of both fig and caprifig with special reference to the commercial production of dried figs and fig paste. The main objectives with respect to fig cultivars are: (a) syconia equal to those of Sari Lop in attractiveness, size, shape, and quality but not requiring caprification (an expensive process which tends to spread the disease caused by *Fusarium moniliforme*, var. *fici*); (b) early to midseason maturity and dropping of fruit to avoid the spoilage which occurs after midseason. Desirable fruit characters are: golden or greenish yellow skin like that of Sari Lop which dries to a light-straw colour; white meat; amber pulp; characteristic fig flavour; medium size and weight (about 20 per kg); ovoid or pyriform shape; small ostiole with tightly closed eye to exclude disease-carrying insects.

Fig

60

The main characters to be bred for in caprifigs which are to be used for further fig breeding are succulent syconia like those of Croisic, desired size, shape and colour and persistence on the tree without benefit of *Blastophaga*. For caprifigs to be used for caprifying Smyrna type cultivars, the characters are: caducous syconia (when not inhabited) to protect growers from buying numerous blanks; both early and late maturing for full coverage of the period of anthesis; dry, hollow syconia to encourage hatching and excitation of *Blastophaga*; good pollen production.

6 References

Condit, I. J. (1928, 1933, 1964). Cytological and morphological studies in the genus *Ficus*. I, II, III. *Univ. Calif. Publ. Bot.*, **1**, 233–44; **17**, 61–74.

Condit, I. J. (1947). *The fig*. Waltham, Mass.

Condit, I. J. (1955). Fig varieties: a monograph. *Hilgardia*, **23**, 323–538.

Condit, I. J. (1969). *Ficus: the exotic species*. Berkeley, Calif., USA.

Eisen, G. (1896). Biological studies on figs, caprifigs, and caprification. *Calif. Acad. Sci. Proc.*, Ser. 2, **5**, 897–1001.

Hunt, B. W. (1911, 1912). Fig breeding. *Georgia Univ. Bull.*, **11**, 146–8; **12**, 106–10.

Solms-Laubach, H. G. (1882). Die Herkunft, Domestication, und Verbreitung des gewöhnlichen Feigenbaums. *Abhandl. K. Ges. Wiss. Göttingen*, **28**, 1–106.

Storey, W. B. (1975). Fig. In J. Janick and J. N. Moore (eds.), *Advances in fruit breeding*. Purdue, USA, 568–89.

Storey, W. B. and I. J. Condit. (1969). Fig (*Ficus carica*). In F. P. Ferwerda and F. Wit (eds.), *Outlines of perennial crop breeding in the tropics*. Wageningen, The Netherlands, 259–67.

Swingle, W. T. (1908). Breeding new and superior figs. *Calif. State Fr. Growers Conv. Rep.*, **34**, 187.

United Nations Food and Agriculture Organization (1961, 1965, 1967). *World list of plant breeders*. Rome.

Hops

Humulus lupulus (Moraceae)

R. A. Neve
Wye College Ashford Kent England

1 Introduction

Although first cultivated for reputed herbal properties, hops are now used almost exclusively in brewing. Their value for this purpose lies in the soft resins which, during the brewing process, are converted to isohumulones to provide the characteristic bitterness of beer. Hops also contain essential oils which contribute to beer flavour though opinions differ on the importance of these. Minor uses are in hop pillows (for their alleged soporific value), for bread making (mainly in East Africa and probably to prevent spoilage of yeast cultures), while the young shoots are sometimes eaten like asparagus.

The bacteriostatic action of hop resins on Gram positive organisms was probably the original reason for the use of hops in brewing, the consequent bitterness being a subsidiary effect. Pasteurization and aseptic brewing methods have made the preservative value of hops unimportant and they are now used only for their flavour effects.

The hop plant is a climbing, dioecious perennial that dies back to ground level each autumn. Commercially, it is propagated by cuttings and is grown up permanent wirework systems which may be up to 7·5 m high. Extension growth of hop shoots of 25 cm per day has been recorded. The lupulin glands which contain the resins and essential oils are produced mainly on the bracteoles of the inflorescences (cones) of the female plant although they also occur on leaves and male flowers.

Hops are indigenous only in the northern hemisphere in latitudes above 32°, the limiting factor being daylength. They have been introduced into the southern hemisphere and are grown commercially in Australia, New Zealand, South Africa and South America.

World production of hops totals about 100 kt, worth 200 million dollars annually.

In Europe the traditional problems for hop growers

have been the damson-hop aphid (*Phorodon humuli*) and powdery mildew (*Sphaerotheca humuli*) but new problems arose with the introduction of downy mildew (*Pseudoperonospora humuli*), apparently from Japan in the 1920s, and outbreaks of virulent forms of *Verticillium albo-atrum* in England in the 1930s and later in Germany.

A general account of the botany, cultivation and utilization of the crop is given by Burgess (1964).

2 Cytotaxonomic background

Some authorities have sub-divided the species, on the basis of morphological differences associated with geographical distribution, into *H. americanus* in America and *H. cordifolius* in Japan, but Davis (1957) considered that there was too great a similarity between specimens for these distinctions to be maintained. It has, however, been shown that the Y chromosomes of male hops show morphological differentiation which corresponds with those sub-divisions. Ono (1955) described five sex chromosome types and, of these, only the Winge type has been found in European, and only the New Winge and 'homotype' in indigenous American hops. Sinoto and New Sinoto types, which both have a sex quadrivalent, have only been recorded in Japanese hops. All naturally occurring plants are diploid ($2n = 2x = 20$).

The only other *Humulus* species is the annual *H. japonicus* ($2n \, ♀ = 16, \, ♂ = 17$) from Japan and China, while the closely related *Cannabis* is native to Central Asia. It seems likely that the genus *Humulus* originated in Asia and spread eastwards to America and westwards to Europe.

3 Early history

There is a reference to the use of hops in beer-making in the Finnish saga, *Kalevala*, thought to date back some 3,000 years, but the first reliable written records date from the ninth century in Europe and the cultivation of the crop was well established in central Europe by the thirteenth century. Hop cultivars in this area have, until very recently, been extremely uniform and limited to two basic types, namely the Hallertauer hop in Bavaria and the Saaz type in Czechoslovakia and in parts of Germany (Thompson, 1972). It seems likely that these were selected from within local indigenous populations for their superior preservative and flavouring characteristics. The impossibility of distinguishing between cultivated and wild plants suggests that little, if any, improvement of the indigenous hops was achieved during the primary domestication.

Although hops are indigenous in England (Godwin,

1976), records show that Flemish planters introduced their cultivation to this country in the sixteenth century; it seems likely that they brought planting material with them, although hybridization with indigenous types may well have occurred subsequently. Records suggest that the Golding type has been cultivated in England for at least 250 years but that the Fuggle hop which, until recently, represented 80 per cent of the acreage, was a chance seedling raised in 1861.

There is no doubt that the early American settlers took hop cuttings with them but the characteristics of the present-day American cultivars indicate with some certainty that they, too, developed as a result of hybridization between the introduced and indigenous plants, the selection of new types being a necessary corollary to the movement of the hop-growing industry from its original location in the eastern states to the much more arid conditions in the west.

The principal cultivar in Yugoslavia, the Styrian Golding, is almost certainly Fuggle introduced from England, while the main Japanese cultivar Shinshuwase, reputed to be a hybrid between Saaz and White Vine, is morphologically more like an American than a European hop (Ono, 1959). Thus, although hop growing is widespread throughout the northern hemisphere, the traditional cultivars are very limited in number, there being only about eight of any significance, and most of these are probably inter-related. This remarkable uniformity is a reflection of the very conservative attitude of brewers.

4 Recent history

The work of brewing chemists in the late nineteenth century identified the alpha-acid fraction of the soft resins as the source of most of the 'bittering power' of hops and indicated that the value of hops to the brewer was proportional to their alpha-acid content. The American Cluster hops had higher alpha-acid contents than had any European varieties and Salmon in Britain set out to breed varieties superior in this respect (Thompson, 1955). He achieved great success by introducing wild hops from North America into his breeding programme. Many of his hybrids had greatly enhanced alpha-acid contents but this character was usually associated with an aroma not generally acceptable to brewers. Subsequent breeding improved the aroma and Salmon's variety, Northern Brewer, now grown in many countries, is considered to have a generally acceptable aroma and has nearly twice the alpha-acid content of the old European types. It has also been used extensively as a parent in breeding programmes, particularly in Europe.

The occurrence of a virulent strain of *Verticillium*   209 *albo-atrum* in England led Keyworth to screen a wide selection of Salmon's breeding material for resistance to the pathogen, which causes wilt disease. All the resistant lines were descended from wild American parents and, though the genetics of the resistance mechanism have not been elucidated, there is some evidence of complementary gene action between American and European genotypes. Although Northern Brewer was highly susceptible to wilt in England it has proved to be highly resistant in Germany, where the disease is also serious.

Resistance to downy mildew was the aim of a breeding programme, begun in Germany in 1922, which has developed slowly but successfully by, apparently, accumulating polygenic resistance solely from the mid-European population. Resistant varieties have recently been released both in Germany and Britain, the resistance in the latter being based on German selections introduced into the breeding programme.

Although Salmon reported resistance to powdery mildew in 1917, it was only in 1972 that a commercially acceptable, resistant variety became available in England. Three major genes for resistance have been identified, two characterized by typical hypersensitive reactions, the third developing resistance at a later stage of invasion by the pathogen (Liyanage *et al.*, 1973).

In most countries hops are grown seedless by the elimination of all male plants from the hop-growing districts. Only a few countries, including England, have traditionally grown seeded hops but market pressures may make it necessary for them, too, to change to seedless production. To do this successfully, males would have to be eliminated from very large areas whereas the cultivation of sterile varieties would make this difficult task unnecessary. Partly for this reason, triploid breeding programmes were begun in the 1950s and triploids are now grown commercially in New Zealand. In England, however, the excessive vigour of triploids and their later maturity have been serious problems, while their seed contents, though low, are generally above the 2 per cent permitted in Europe for seedless hops.

Breeding of hops is complicated by their dioecious character. Not only are they cross-breeding, but the inflorescences of male plants are of no commercial value so the genotype of males in relation to brewing characteristics cannot be assessed phenotypically. One attraction of triploid breeding is that the use of tetraploid females and diploid males ensures that the parent with known commercial characteristics contributes more to the genotype of the progeny.

The breeding of polyploids led to the discovery that tetraploids with XXXY sex chromosomes produce fertile male and female inflorescences in roughly equal numbers. The brewing characteristics of such plants can be determined from their female cones and those selected can then be used as male parents.

Nevertheless, the main emphasis is still on diploid breeding; sterile diploids have been found with seed contents much lower than those of triploids and these are being further developed.

5 Prospects

The advantages of cultivars with high alpha-acid contents are now generally recognized by brewers and there is no evidence that the limits of improvement have been reached. Resistance to the major diseases is available and will be increasingly incorporated into new cultivars, but resistance to hop aphids and spider mites is still being sought.

Collections of wild hops have been assembled but they do not appear to have useful characteristics other than those already available; however, an extensive collection being made in Yugoslavia may yield interesting results (Wagner, 1975).

It would be commercially valuable if the habit of the plant could be modified so as to avoid the need for the very expensive wirework up which they are at present grown, while enabling a modified harvester to be used which would 'combine' the crop in the field. No suitable variants have been discovered though a recessive gene carried by many wild American plants does, when homozygous, produce sterile plants which only grow 2–3 cm tall. Mutation breeding might produce the required character but this technique has a limited chance of success in a dioecious plant like the hop. The probability of obtaining any recessive mutation in the homozygous condition is so low that it would be necessary to rely upon a suitable dominant mutant being produced.

6 References

Burgess, A. H. (1964). *Hops*. London.

Davis, E. L. (1957). Morphological complexes in hops (*Humulus lupulus*) with special reference to the American race. *Ann. Mo. bot Gdn.*, **44**, 271–94.

Godwin, H. (1976). *The history of the British Flora*. Cambridge, 2nd edn.

Liyanage, A. de S., Neve, R. A. and Royle, D. J. (1973). Resistance to hop powdery mildew. *Rep. Dept. Hop Res. Wye Coll.*, 1972, 49–50.

Ono, T. (1955). Chromosomes of common hop and its relatives. *Bull. Brew. Sci.*, **2**, 3–65.

Ono, T. (1959). The hop in Japan. *Bull. Brew. Sci.*, **5**, 1–11.

Thompson, F. C. (1955). An appraisal of the new (Wye) varieties of hops and their place and future in brewing.

61

J. Inst. Brew., **61**, 210–6.

Thompson, F. C. (1972). The hop growing districts in Europe. *Brygmesteren*, **2**, 41–51.

Wagner, A. (1975). *Divji hmelj, Humulus lupulus, v Jugoslavia*. Doct. Diss. Univ: Ljubljana.

Bananas

Musa (Musaceae)

N. W. Simmonds
Scottish Plant Breeding Station
Edinburgh Scotland

1 Introduction

The bananas are large, stooling, herbaceous perennial plants which are propagated vegetatively by means of corms. The basal corms are surmounted by 'pseudostems' of leaf sheaths. Inflorescences are terminal and are initiated near ground level, being then thrust up the centre of the pseudostem by elongation of the true stem. Basal flower clusters are female and persist through growth to form the fruit bunch; distal flower clusters are male and the flowers and bracts that subtend them are commonly deciduous.

The first ('plant') crop is taken about a year after planting; it is followed by the first ratoon and, thereafter, with regular pruning of excess suckers, cropping becomes more or less continuous. Fruit bunches are generally harvested unripe and ripened under cover or, in the case of exported fruit, after more or less prolonged refrigerated transport. The bananas are tropical by origin and are intolerant of frost. In cultivation they are distributed throughout the warmer countries and hardly transgress 40°N and S latitudes. In the tropics proper they constitute a major local foodstuff (e.g. in East Africa) as well as providing the basis for an important export trade to temperate countries. World production is estimated to be about 20 Mt of which about 15 per cent enters the export trade. For general treatments of the bananas as a crop (including breeding) and of banana evolution see Simmonds (1966, 1962).

2 Cytotaxonomic background

The wild bananas are all diploid and are distributed in southeast Asia and the Pacific (Table 61.1).

There are also a few (3–5) more species of uncertain affinity of which two have different chromosome numbers ($2n = 2x = 14$, 18). The related genus *Ensete* has $2n = 2x = 18$ and is distributed in tropical Africa and southeast Asia.

Table 61.1 Classification of the bananas

Section	$2n = 2x =$	Remarks
Eumusa	22	13–15 species, southeast Asia, including *Musa acuminata* and *M. balbisiana*
Rhodochlamys	22	5–7 species, southeast Asia
Australimusa	20	5–7 species, New Guinea area
Callimusa	20	6–10 species, southeast Asia

The vast majority of cultivated bananas (the Eumusa cultivars) originated from *M. acuminata* and *M. balbisiana* so have chromosome numbers based on 11 (22, 33, 44). The small and unimportant Fe'i group of cultivars had a quite independent origin from species of Australimusa and are diploid with $2n = 2x = 20$.

Two other members of the group are of some economic importance. First, Manila hemp, long a source of marine cordage, derives from the sheath fibres of ábaca, *Musa textilis*. This species, a member of Australimusa with $2n = 2x = 20$, is native to and cultivated in the Philippine Islands and Borneo. It has been but little bred and there is no evidence that cultivars are significantly removed from their wild progenitors. Second, *Ensete ventricosum*, native to highland east Africa, has long been cultivated in the highlands of Ethiopia for pseudostem starch and fibre. Again, though the agriculture is an ancient one, there is no evidence that the species has evolved in cultivation.

The rest of this chapter is devoted to the fruit-bearing Eumusa cultivars.

3 Early history

The key to understanding of banana evolution lies in the analysis of parthenocarpy and sterility in the edible diploids. Wild banana fruits are trilocular berries in which the seeds are surrounded by a mass of sweetish-acid, starchy, parenchymatous pulp. If unpollinated, ovaries generally swell slightly and remain as persistent empty shells. Fruit growth is proportional to seed content and it is clear that the growth of pulp depends upon a stimulus from the developing seeds. By contrast, in edible bananas, the stimulus to pulp growth is autonomous; seeds are unnecessary (though, if present, they do further stimulate fruit growth). Genetic studies show that the autonomous stimulus that culminates in vegetative parthenocarpy is due to several (at least three) complementary dominant genes which are present (though of unknown distribution) in wild forms of *Musa acuminata*. They also show that it is possible to have thinly parthenocarpic fruits which are (more or less) edible if unpollinated but heavily seeded if pollinated. The typical seedless edible banana is therefore the product of two evolutionary processes: parthenocarpy and sterility, both essential.

Sterility in the edible diploids is basically genetic female sterility, as inferred (though not precisely analysed) from breeding experiments. Thus many female-sterile but male-fertile edible diploids, natural and experimental, are known. In addition, gametic sterility, manifest in both ovules and pollen, is often superimposed; it has been caused by the carrying over from wild ancestors and the accumulation during clonal existence of structural chromosome changes (inversions and translocations).

Edibility evolved first in wild *Musa acuminata*. The species is very variable and taxonomic evidence indicates that the primary centre was the Malay peninsula (possibly including closely neighbouring territories). With high probability, human selection favoured parthenocarpy *plus* seed-sterility, leading to the production of edible seedless fruits, followed by male sterility consequent upon structural heterozygosity. Edible diploids are still widely but thinly distributed in southeast Asia but have largely been supplanted by the polyploids now to be considered; there remains a considerable centre of diploid diversity in the New Guinea area, however.

Most cultivated bananas are triploid ($2n = 3x = 33$) and it has been shown that triploid plants are more vigorous and their fruits grow faster (and therefore larger) than diploids. From taxonomic evidence it is clear that there are three major groups of triploids (A, *acuminata* genome; B, *balbisiana* genome): AAA, AAB, ABB. There are also rare hybrid diploids (AB) and tetraploids (AAAB, AABB, ABBB; see Richardson *et al.*, 1965). There is no evidence that *M. balbisiana* ever evolved edibility on its own (but for a contrary view see Vakili, 1967), so that the hybrid groups of cultivars originated by outward migration of edible diploid, male fertile AA types into the area of *M. balbisiana*, followed by crossing and polyploidy. Probably, *balbisiana* genomes conferred a degree of hardiness and resistance to seasonal drought in the monsoon climates north of the primary centre of origin (Fig. 61.2).

Including somatic mutants (of which there are many) there are several hundred clones (perhaps 500 – K. Shepherd pers. comm.) in existence of which AA, AAA, AAB and ABB are the most numerous; AB, AABB, AAAB and ABBB are much rarer. Curiously, no natural AAAA clones are known, though they have been abundantly produced experimentally.

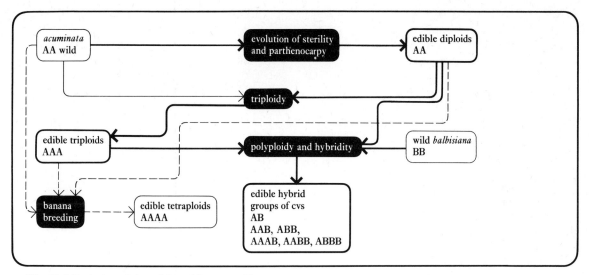

Fig. 61.1 Evolution of the cultivated bananas, *Musa*.

No dates can be assigned to the early evolution of the bananas. With high probability, the essential events occurred millenia rather than centuries ago. The crop reached West Africa before European contact and a few clones were carried thence in the late fifteenth century to the New World. It is likely that bananas entered Africa from Malaysia (rather than from India) by way of Madagascar and/or the east coast in the hands of Indonesia migrants and travelled up the great lakes. More or less contemporaneously (during the first millenium A.D.) the crop moved eastwards across the Pacific to the further islands. There is no good evidence of pre-Columbian presence in the New World.

Little is known about the Fe'i group of cultivars. They occur (but now only rarely) in New Guinea and Pacific islands. They are diploids (with $2n = 2x = 20$) and, phenotypically, clearly belong in Australimusa, with the wild New Guinea species *Musa maclayi* as a very likely progenitor. An independent origin of edibility must be inferred.

Fig. 61.2 Evolutionary geography of the bananas, *Musa*.

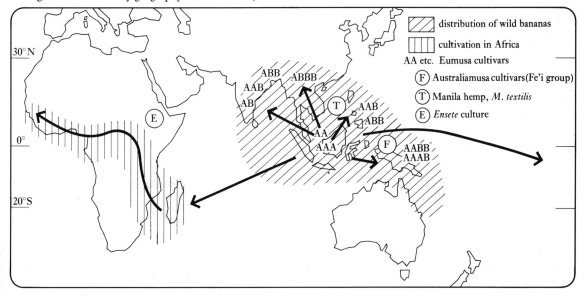

4 Recent history

By the end of the sixteenth century, bananas were widely spread throughout the tropics. Relatively few clones had moved out of the centre of origin in southeast Asia and this pattern persists today; diversity declines in the sequence Asia–Africa–America.

The banana export trades developed mainly in tropical America in the later nineteenth century and have depended for 100 years on remarkably few clones, all triploids (AAA), namely: Gros Michel and various mutant members of the Cavendish group. These are well adapted to large-scale cultivation and ocean transport to distant markets but are susceptible to certain diseases (review in Wardlaw, 1961). Indeed bananas constitute perhaps the best example in the history of agriculture of the pathological perils of monoclone culture.

Gros Michel is susceptible both to Panama disease (banana wilt) caused by *Fusarium oxysporum* f. *cubense* and to leaf spot (Sigatoka) caused by *Mycosphaerella musicola* which appeared in the New World in 1933. It is, however, rather tolerant of (maybe not resistant to) the burrowing nematode, *Radopholus similis*. The Cavendish group of cultivars is resistant to wilt but very susceptible to leaf spot and nematodes. Leaf spot can be controlled by spraying and nematodes by fumigation. Panama disease is uncontrollable in a susceptible variety and this fact provided the stimulus for banana breeding. Nowadays, resistance to all three diseases is the aim; but other diseases, not mentioned here, may yet have to be included.

Banana breeding began in Trinidad and Jamaica in the early 1920s and has been continued in the West Indies ever since. A detailed review is given in Simmonds (1965), updated and amplified by Shepherd (1968) and by Menendez and Shepherd (1975).

Essentially, the approach ultimately developed has been first to breed edible diploids (AA) which were vigorous, productive, female sterile, male fertile and disease resistant. These are crossed as males on to the disease susceptible but otherwise acceptable Gros Michel (AAA). Gros Michel is, if pollinated, very slightly seed fertile and, with much labour and difficulty, progenies can be built up. About one third of the seedlings are tetraploid (AAAA), the products of female restitution $(33+11)$; miscellaneous aneuploids, heptaploids $(33+33+11 = 77)$ and other high polyploids are discarded. In practice, tetraploids from Gros Michel are too large and, since about 1960, its semi-dwarf mutant, Highgate, has been substituted for it. Fortunately, the mutation behaves as a dominant with good penetrance in tetraploid progeny.

Early results (1930s and 1940s) were unpromising in that, though disease-resistant clones with quite good bunches were produced in some numbers, they were not good enough and none was really successful. The move (1940s onwards) to improve the male parents resulted in better bunches but worse disease resistance.

Fig. 61.3 Summary of banana breeding.

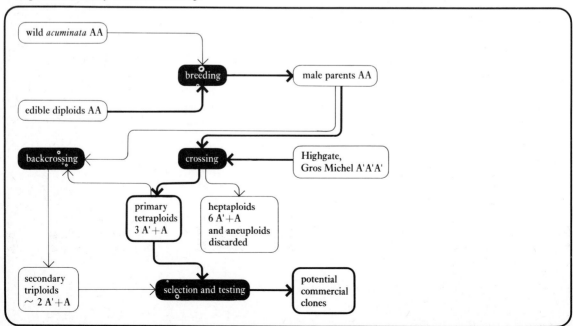

5 Prospects

Currently (Shepherd, 1968), intense selection for disease resistance as well as fruit characters in the male parents is practised and a mood of cautious optimism prevails. There may yet be obstacles (in storage characters for example) but, as Shepherd points out, success could save the West Indian trades some £3 million annually, the estimated cost of disease losses and control measures.

The use of secondary triploids (Fig. 61.3) deserves comment. Early backcrosses to wild AA clones (predictably) yielded only rubbish. But later combinations of improved diploids by primary tetraploids have produced occasional very attractive bunches, though of poor fruit quality (Shepherd, pers. comm.). There are possibilities here, but the prospects do not seem good. Future export bananas are still likely to be tetraploid.

The genetic base of banana breeding is narrow. Given that the object is to breed for export trades, the only available female parent is Gros Michel/Highgate, the Cavendish group being completely seed-sterile. On the male side, the base has been deliberately widened over the years by collecting wild and edible *acuminata* diploids. Rather good collections of these are now available, along with large stocks of other materials both wild and cultivated. Little effort has ever gone into breeding bananas for local use in the tropics but there is no doubt that much could be done and that the materials are available. Most established polyploid clones are rather seed sterile but fertile enough, if pollinated, that useful tetraploid progeny can be obtained.

Looking to the future, it seems likely that the polyclonal small-farmer plantings of Asia and Africa will persist more or less unchanged though they could be enhanced by breeding for local use if the effort were made. The export trades will have to continue to rely basically on the Cavendish group until primary tetraploids (or, less likely, secondary triploids) from Highgate replace or supplement them. Indeed, without clonal diversification, these trades can hardly be expected to survive indefinitely.

6 References

Menendex, T. and Shepherd, K. (1975). Breeding new bananas. *World Crops*, 27, 104–12.
Richardson, D. L., Hamilton, K. S. and Hutchison, D. J. (1965). Notes on bananas 1. Natural edible tetraploids. *Trop Agriculture, Trin.*, 42, 125–37.
Shepherd, K. (1968). Banana breeding in the West Indies. *Pest Arts News Summ.*, 14, 370–9.
Simmonds, N. W. (1962). *The evolution of the bananas.* London.
Simmonds, N. W. (1966). *Bananas.* 2nd edn, London.
Vakili, N. G. (1967). The experimental formation of polyploidy and its effect in the genus *Musa. Amer. J. Bot.*, 54, 24–36.
Wardlaw, C. W. (1961). *Banana diseases.* London.

Clove

Eugenia caryophyllus (Myrtaceae)

F. Wit
Hamelakkerlaan 38 Wageningen The
Netherlands; *formerly* Foundation for
Agricultural Plant Breeding Wageningen
The Netherlands

1 Introduction

The clove is the dried, full grown, aromatic flower bud of the clove tree, a medium-sized evergreen tropical tree. It was one of the spices which attracted European sea-traders to the Far East in the sixteenth century. Modern uses, apart from culinary purposes, include the manufacture of kretek cigarettes and the production of clove oil by distillation. Kretek cigarettes contain a mixture of tobacco and shredded cloves in a ratio of about 2 or 3 to 1. The industry, unique to Indonesia, is an important outlet for cloves. Clove oil, especially clove leaf oil, is the main source of artificial vanillin.

World production of cloves amounts to some 20–30 kt, most of which comes from the islands of Zanzibar and Pemba. Madagascar and Indonesia are the second and the third largest producers. Madagascar is the only source of clove leaf oil.

The cultivation of the clove requires little care and is mainly in the hands of smallholders. The crop is harvested by climbing the trees and picking by hand the clusters of buds growing on the tips of the branches.

The most serious diseases are 'sudden death' in Zanzibar and 'matibudjang' in Indonesia. These diseases may kill trees soon after maturity. They closely resemble each other but are perhaps not identical (Tojib Hadiwidjaja, 1956; Nutman and Roberts, 1971).

2 Cytotaxonomic background

As the correct name for the clove tree *Eugenia caryophyllus* is accepted here, but there are many synonyms. The old genus *Eugenia* comprises a great number of tropical and subtropical trees, all aromatic. Many of them are cultivated as fruit trees.

The nearest relative to the clove is a wild tree which is common in the forests on the lower mountain slopes in many islands of the Moluccas and in New Guinea. These trees, locally known as wild cloves, have larger and less aromatic flower buds and leaves than the cultivated trees. They are more vigorous in the juvenile phase, resistant to the 'matibudjang' disease, and more variable in many characters. Their essential oil content is much lower and the oil has a different composition. Hybrids between wild and cultivated cloves are fertile but have not been studied cytologically. Opinions differ as to whether they should be regarded as conspecific or whether this wild form should be separated as *E. obtusifolia*.

3 Early history

The clove tree must have been in cultivation for more than 2,000 years, for cloves were imported into China during the Han period (Crofton, 1936). Chinese traders re-exported them to neighbouring countries, keeping their origin secret. Cloves were known both in Europe and India towards the close of the second century. The quest for cloves for the markets of the West led to the discovery of the Spice Islands by the Portuguese in the beginning of the sixteenth century. This group of five small volcanic islands in the North Moluccas in Indonesia is considered to be the centre of origin of the cultivated clove. Assuming the wild species of the region to be the ancestor of the cultivars, it may be inferred that early evolution involved the reduction of size of leaf and bud and selection of oil characters.

The growing trade induced the planting of cloves in many Moluccan islands until the Portuguese monopoly of the trade was taken over by the Dutch East India Company early in the seventeenth century. Clove cultivation was then confined to Amboina and a few other small islands in the South Moluccas which were within easy control. Clove growing outside the permitted area was punished by severe penalties and every possible effort was made to destroy the trees there. This action may well have led to a great loss of genetic diversity.

Despite Dutch efforts to stop the spread of the crop, French rivals collected several hundred young plants from one or two remote places in the northern Moluccas in 1770–72. They were taken to l'Ile de France (Mauritius) and Bourbon (la Réunion). Only a few plants survived and one tree in la Réunion is thought to have been the ancestor of the whole clove industry in la Réunion and Madagascar. The introduction to Madagascar is said to date from about 1820. The crop was introduced to Zanzibar, either from la Réunion at the end of the eighteenth century or from Mauritius in 1818.

The present populations in Malaysia and Indonesia (the oldest industry outside the Moluccas) derive from introductions from the Moluccas in about 1800.

It is apparent from the above that the clove industry outside the Moluccas is, at most, two centuries old. Considering the long juvenile phase (six to ten years) and the high average age of the tree under favourable conditions (about 50–60 years), this period spans very few generations. Spontaneous mutation can have done little to increase genetic diversity, so variability in the secondary centres of cultivation will be mainly determined by the diversity of the introduced planting material.

There are no reports on variation present in the Spice Islands in the sixteenth century, only one type of tree being described. The nature of the area of early distribution (many small islands and a few larger ones with isolated habitats), the practice of taking seedlings from the crop found growing under productive trees and the breeding system, predominantly self-pollination tending to homozygosis, would all have favoured the development of numerous local varieties. The general picture in all secondary centres is, however, one of great uniformity. Neither in Zanzibar nor in Madagascar have varietal differences been observed. It is highly probable, therefore, that their populations derived from one or very few trees, representing one rather uniform variety. From Pemba and Sumatra alone have two types of tree been mentioned, differing mainly in the number of buds per cluster and in the length of the peduncles.

The lack of diversity in the clove populations of Amboina was noted by Rumphius (1741) in the middle of the seventeenth century. Only after close observation was he able to distinguish (in addition to the common type) two other types of tree, differing mainly in the size and the colour of the ripe cloves. One of the divergent types was found in a few isolated areas, the other in one place only. During botanical surveys in Indonesia shortly before the late war, the second type could not be found again; apparently it had become extinct. However, so far as can be concluded from Rumphius's concise description, it may have been identical with the Zanzibar variety and with the type of tree mentioned from the Spice Islands in the sixteenth century. As both the kretek industry and the smokers appeared to prefer Zanzibar cloves to the Indonesian product, several hundred seeds were secretly brought by the Dutch from Zanzibar to Indonesia in 1932. So the variation present in Amboina in the seventeenth century may have been restored.

4 Recent history

Clove breeding began in Zanzibar and Indonesia between the two world wars. In Zanzibar, the most important objectives were resistance to the 'sudden death' disease, regular and heavy bearing and a shape adapted to easy harvesting. Seedlings from selected mother trees were supplied to farmers (Tidbury, 1949). No data are available to show whether these seedlings produced better results than those from unselected trees.

The clove breeding programme in Indonesia aimed in the first place at the development of planting material which could produce a substitute for the imported Zanzibar cloves used in the kretek industry. Other aims were the improvement of yield and resistance to 'matibudjang' disease.

From superior-looking trees in various centres, yields were periodically recorded and samples were tested for their kretek qualities. Progenies were raised from superior trees, partly after open pollination, partly after selfing or crossing. Some of the material survived the war and could be studied in maturity.

The most important result of this work was the demonstration of heterosis in hybrids between the Zanzibar and the Amboina strains. These hybrids were much more vigorous than both parental types in the juvenile phase and yielded considerably more when mature. The cloves were of the same kretek quality as those from pure Zanzibar trees. Hybrids between the cultivated and the wild clove tree were very vigorous but inferior in quality. Data on the heritability of important characters and on long-term yield comparisons of progenies of different parents are still lacking. In the Zanzibar introductions, no relation could be found between yield in the first harvest years and in later years, nor between the yield and cluster size of the mother-trees and their selfed progenies. The investigations have, however, been too incomplete to draw the conclusion that selection within the Zanzibar strain will not be effective.

In Madagascar, clove breeding started more recently. The general approach is tree selection, especially for regular yield, testing small progenies of promising mother-trees and renewed attempts at vegetative propagation. Some promising results have been obtained by approach grafting on rootstocks of *Eugenia jambolana* (Dufournet and Rodriguez, 1972). Temporary successes have, however, been gained before (Wit, 1969).

5 Prospects

The Indonesian clove industry has grown greatly in recent years and will be able to supply a large part of the needs of the kretek industry, the largest consumer (Castles, 1965). World consumption is unlikely to increase considerably and a rapid rise in world production would be undesirable. However, breeding for high and regular yield and for resistance to 'sudden death' and 'matibudjang', will remain important, to reduce the costs of production. The general approach will no doubt still be by way of the establishment of selection gardens in each major producing area, where progenies of promising trees can be tested against material from other growing areas and countries. In the course of time it should become known whether useful genetic variation exists within local populations or whether selection should be performed between such populations. It is perhaps time that international competition were replaced by some kind of international collaboration.

In each major producing country, a world-wide collection of clove trees should be set up as a living museum. In such collections studies should be performed on varietal differences, mating system and so forth. Most important from the point of view of the evolution of the clove tree is the realization that the genetic base of the crop in most countries is very narrow and that inter-varietal hybrids tend to be heterotic. Populations in different areas in Indonesia resemble land races. They seem to consist of varying mixtures of few slightly different genotypes. A future development might be the widening of the genetic base by hybridization and the exploitation of hybrid vigour. The utilization of heterosis could be greatly increased by the development of a satisfactory method of vegetative reproduction and by the discovery of self-incompatible mother-trees, for example in Indonesian material (Wit, 1969).

6 References

Castles, L. (1965). Cloves and kretek. *Bull. Indones. econ. Stud.*, **2**, 49–59.

Crofton, R. H. (1936). *A pageant of the Spice Islands.* London.

Dufournet, R. and Rodriguez, H. (1972). Travaux pour l'amélioration de la production du giroflier su la côte orientale de Madagascar. *Agron. trop.*, **27**, 633–8.

Maistre, J. (1964). *Les plantes à épices.* Paris.

Nutman, F. J. and Roberts, F. M. (1971). The clove industry and the diseases of the clove tree. *Pest Arts News Summ. (PANS)*, **17**, 147–65.

Rumphius, G. R. (1971). *Herbarium Amboinense,* **2**, 1–13. Amsterdam.

Tidbury, G. E. (1949). *The clove tree.* London.

Tojib Hadiwidjaja (1956). 'Matibudjang' disease of the clove tree (*Eugenia aromatica*). *Contr. Gen. Agr. Res. Sta. Bogor,* **143**, 1–74.

Wit, F. (1969). The clove tree. In Ferwerda and Wit (eds), *Outlines of perennial crop breeding in the tropics.* Wageningen, 163–74.

Olive

Olea europaea (Oleaceae)

N. W. Simmonds
Scottish Plant Breeding Station Edinburgh
Scotland

Note: This chapter was condensed from a text in French, kindly provided by Monsieur H. Elant of F.A.O., Rome. The editor and publishers make grateful acknowledgement for this text and for the figure.

1 Introduction

The olive is one of the most characteristic plants of the Mediterranean basin. The crop, an ancient one, is of immense historical importance as the principal source, for a period of millenia, of edible oil for the people of the Mediterranean area. Though ancient, it has changed but little and is now in decline, a decline which can be arrested only by adaptation of oleiculture to modern conditions. The leading producers are Spain, Italy and Greece, in that order, with a total tree population of the order of 400–500 million. For a review of the crop, see Morettini (1950).

2 Cytotaxonomic background

The olive is one of the relicts of the tropical mid-Tertiary flora of the Mediterranean, the only member of the genus to have survived in the area. Two botanical varieties are usually recognized: cultivated *sativa* and wild *oleaster*. The latter, in the light of evolutionary views outlined below, must be regarded as escaped rather than truly wild or in any sense ancestral to the cultivars.

Olea contains about 35 species distributed widely from Africa to New Zealand. Three species deserve mention in relation to the evolution of the crop. First, there is *O. laperrinii*, with leaves silvery-white beneath, native to the Saharan mountains (Benichow, 1962); related forms, sometimes included in the species, are found in the Macaronesian archipelago (Canary Islands and Azores) and North Africa. Apparently related forms occur by the Red Sea and Persian Gulf. Second, there is *O. africana* (= *chrysophylla* = *verrucosa*) which is African, from far southern Africa to the Sudan-Egypt border, (Verdoorn 1956); it, like the next, has narrow leaves, golden beneath. The third

species is *O. ferruginea* (= *cuspidata*), native from Iran to the Himalayas.

Cultivated olives are nearly all diploid with $2n = 2x = 46$; occasional triploids and tetraploids have been reported and also one case of polysomaty ($2n = 55$). It is to be noted that the large fruited cultivars that have been examined (e.g. Gordat) are diploid. The chromosomes of related wild species do not seem to have been counted.

3 Early history (Chevalier, 1948; Ciferri, 1950a, b)

The parents of the cultivated olive are not known but are generally inferred to have been two species, one a form with narrow leaves golden beneath and the other (the 'proto-*Olea*' of Ciferri), the contributor of the oily pulp character. The former could have been *O. africana* or *O. ferruginea* mentioned above; 'proto-*Olea*' is thought to have become extinct. Whatever the parents, the crop probably originated as a hybrid swarm in the mountains of the eastern Mediterranean: the Taurus, Amanus and Lebanese mountains, as far as High Galilee.

A hybrid origin is indicated by the great diversity of kernel types found by M. Dunand at Byblos (on the coast of Syria) and dated to the fourth millenium B.C. Some were more or less identical to those of *O. africana*, others with thin shells, elongated and sometimes curved.

From the eastern Mediterranean, the crop moved westwards, generating a secondary centre of diversity in the Aegean area and what might be regarded as a tertiary one further west still, in the area southern Italy–Tunisia (Fig. 63.1). The Aegean centre is characterized by large-fruited types which occur only as 'impurities' in Near Eastern populations. The Tunisian populations in the tertiary centre may have been influenced by local wild forms. The crop was known to the Romans by 600 B.C.

The history of the crop has been dominated by the longevity of the tree, by its capacity for regeneration by suckering from the rootstock and by the low rate of vegetative multiplication that has (until very recently) been possible. No doubt, chance seedlings have generated a supply of new clones over the millenia but there is evidence, nevertheless, that the rate of clonal turnover is very low. Thus, some Tunisian clones are thought to have survived from Roman times. The general pattern is of cultivars composed of mixtures of clones, fairly homogeneous and with recognizable dominant types, sometimes widespread but sometimes broken up into a diverse mosaic in diverse environments (for example, mountainous areas).

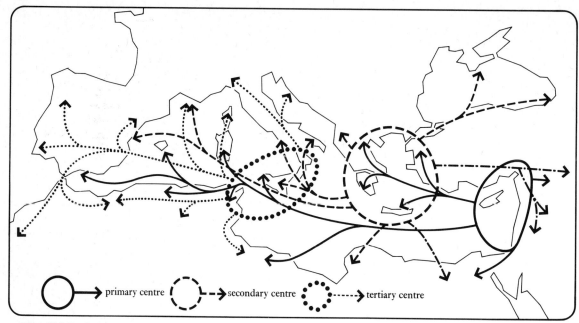

Fig. 63.1 Evolutionary geography of the olive, *Olea europaea*. With acknowledgements to H. Elant.

4 Recent history

This pattern, of clonal mixture, has persisted from ancient times down to the present. Even the extensive plantings of the late nineteenth and early twentieth centuries, lacking a means of rapid vegetative multiplication, merely reduced variability without establishing outstanding clones.

Attempts to classify olive clones have been carried on in a number of places for a good many years but are, even now, far from complete (Cantarelli, 1962; Scaramuzzi, 1950). It is unfortunate that the very promising studies by Ciferri and Brevigleri in Italy, using the concept of 'proles', were never followed up. More recently, Rollo–Romero, in Spain, has started to develop a numerical–taxonomic approach which promises well. We are still far from a comprehensive understanding of the variability of the crop.

There have been many attempts, by horticulturists and nurserymen, to select clones for yield (and regularity of yield), oil content and resistance (to, e.g. *Cycloconium oleaginum* (leaf spot)); selected clones have, indeed, been distributed but only in one trial (in France) have they been field-tested before distribution. It seems very likely, *a priori*, that superior clones exist and could be isolated but the practical problems of propagation and time-scale are severe. However, mist-propagation of mature leafy cuttings is becoming widely used and clonal selection proceeds, though testing is still very rare (Almeida, 1963).

Olive breeding (as distinct from clonal selection) has hardly been touched. The earliest attempt was begun 40 years ago by Ortego-Nieto at Jaën (southern Spain) using open-pollinated seeds. Many seedlings were killed by frost and the survivors showed curious juvenile characters and some high oil contents. They were not superior to local cultivars. Controlled crossing has yet to be exploited. There is some evidence of an erratic self-incompatibility and the crop is wind pollinated.

5 Prospects

Attempts to classify olive clones and to isolate and propagate superior selections continue. There is general awareness of the need to test selections but little action. Meanwhile the needs of mechanical harvesting by shaking are becoming apparent, namely: easily detached fruits of synchronous maturity. The temptation is to try to adapt established plantings for the purpose but, in the long run, specially bred clones will be essential.

6 References

Almeida, F. J. de (1963). Acerca do melhoramento da oliveira. *Bot. Junta nac. Azeite*, **65–66**, 1–18.

Benichow, A. (1962). Recherches critiques sur l'olivier de Laperrine. *Mem. Inst. Rech. Sahar.*, *Alger*, **6**, pp. 55.

Cantarelli, C. (1962). Cultivar e pigmentazione delle olive: essaine analitico dei prolifenoli delle drupe di diverse cultivar dell'Italia centrale. *Atti I Conv. naz. Olivicolo-Olearia, Spoleto*, 283–93.

Chevalier, A. (1948). L'origine de l'olivier cultivée et ses variations. *Rev. internat. Bot. appl. Agric. trop.*, **28**, 1–25.

Ciferri, R. (1950*a*). Dati ed ipotesi sull'origine e l'evoluzione dell'olivo. *Olearia*, 115–22.

Ciferri, R. (1950*b*). Éléments pour l'étude de l'origine et de l'évolution de l'olivier cultivée. *Act. XIII Congr. internat. Oléicult.*, **1**, 189–94.

Farres, R. P. and Riera, F. J. (1950). Contribution à la cytogénétique de l'olivier. *Act. XIII Congr. internat. Oléicult.*, **1**, 428–43.

Morettini, A. (1950). *Olivicoltura*. Rome.

Scaramuzzi, F. (1950). Pour la description et la classification des races d'olivier cultivé. *Act. XIII Congr. internat. oléicult.*, **1**, 68–74.

Verdoorn, I. C. (1956). The Oleaceae of southern Africa. *Bothalia*, **6**, 549–639.

Coconut

Cocos nucifera (Palmae)

R. A. Whitehead

Payhembury Honiton Devon England;
formerly Coconut Industry Board Jamaica West Indies

1 Introduction

The coconut palm is an important crop throughout the tropics at lower altitudes. Its main economic products, derived from the large fruits (colloquially 'seeds' or 'nuts'), are copra (the dried solid endosperm or 'meat' of the nut), the edible oil extracted from this and desiccated coconut for which the endosperm is ground prior to drying. The oil is used for cooking and in the manufacture of margarine and soap. Coir fibre, of lesser importance, is obtained from the fibrous mesocarp (or husk) of the fruit.

The Philippines are by far the largest producers of copra, followed by Indonesia, with India and Sri Lanka (Ceylon) approximately equal third. About two thirds of the crop is consumed locally, the rest being exported, mainly to the United States and western Europe.

2 Cytotaxonomic background

Cocos nucifera has $2n = 2x = 32$. The genus is monospecific, though species now included in several other genera have at various times been included in *Cocos*. No truly wild coconuts are known and nothing is known of cytogenetic relationships with South American relatives.

The species is very variable and occurs in suitable places throughout the inter-tropical zone. Distinctive local populations have often been given local names but such names are merely useful provisional labels until more is known. There are more so-called varieties (and more variability) in southeast Asia than in Africa or South America. Despite uncertainty about the status of varieties, the distinction made between tall (or common) and dwarf types, of which there are several, is a useful one. They are referred to as *typica* and *nana* by some authors. Talls tend to be mainly outbreeding, dwarfs usually inbreeding and smaller fruited. Niu Leka dwarf palms, found in some Pacific islands, resemble talls in characteristics other than height. Dwarfness is (probably, not certainly) due to a single gene of heterozygous penetrance, so that heterozygotes are semi-tall.

3 Early history

The place of origin of the coconut has been disputed. Some writers have argued in favour of a South American origin, others for an origin in Polynesia or Asia. Frémond *et al.* (1966) have summarized the main reasons for considering a southeast Asian origin the most probable; variability and the numbers of local names and uses are greater here than elsewhere (Wester, 1920; Tammes and Whitehead, 1969).

Sufficient facts are now available for some inferences to be drawn on the spread of coconuts from the main centre of diversity in southeast Asia to other areas. Places remote from the main dispersal route would tend to have few seeds arriving and less variable resulting populations than places near to the main route. Variability would also be low in areas in which initial establishment or subsequent growth was difficult, due to unfavourable environments. Large land masses and high elevations would form barriers to dispersal.

A good case can be made for Central and South America as being at the ends of both eastward and westward migrations from southeast Asia. The main points are: (*a*) palms on the Pacific coasts of tropical America are distinct from those in the Caribbean and on the Caribbean/Atlantic coasts (Whitehead, 1965, 1966*a*, 1968*a*; Harries, 1971); (*b*) it seems likely that spread across the Pacific was from west to east; (*c*) spread across the Atlantic was from east to west.

On the first point, the Caribbean islands and the Atlantic/Caribbean coasts have Talls similar to the Jamaica Tall whereas, on the Pacific coasts, the Panama Tall (previously called San Blas) occurs. Exceptions to the general pattern occur (e.g. in Nicaragua, where Jamaica Tall type palms are found on both coasts and in Venezuela and Surinam where both types are found (Harries, 1971)) but are not difficult to explain.

On the second point, there is little barrier to spread through the Pacific from New Guinea to French Polynesia and thence to western tropical America, though the distance between the last two is considerable. This factor would reduce the number of seed-nuts likely to be transported. Spread might have been by drifting or it might have been assisted by man. With few exceptions, coconuts on Pacific islands more resemble the Panama Tall than the Jamaica Tall in major characteristics (Whitehead, 1966*a*). Populations in Fiji, Tonga, Western Samoa and New Guinea are more variable than those in French Polynesia, suggesting that there was more frequent spread to these areas or that the environment was more suitable for the crop.

There is fairly good evidence for point (*c*). Coconuts

Fig. 64.1 Dispersal of the coconut, *Cocos nucifera*.

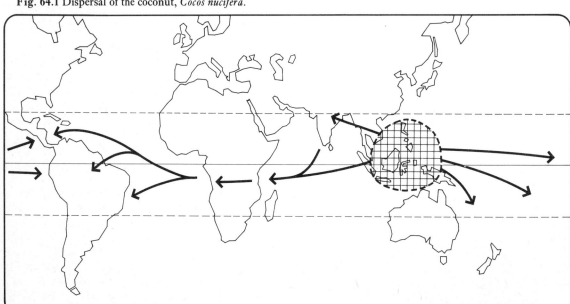

in the Caribbean islands and on the eastern coasts of tropical America broadly resemble those of West Africa (Frémond *et al.*, 1966; Whitehead, 1965, 1966*a*). They are essentially similar to those of East Africa, the Seychelles, Mozambique, Sri Lanka (where active planting has been done since at least the twelfth century) and to some varieties in India. Palms bordering the Atlantic are of relatively recent introduction. Although the Spaniards found coconuts on the western coast of Panama, there were none on either side of the Atlantic at the time of Columbus (Purseglove, 1963; but see also Cook, 1910). It therefore seems reasonable to assume that Atlantic–American palms arrived relatively recently from southeast Asia via Africa; Portuguese shipments to Brazil and Spanish introductions to Puerto Rico would have been possible in the fifteenth century.

The spread of coconuts to East Africa from southeast Asia, perhaps via the Seychelles, presents no serious problems but the details are not known. The number of seeds initially introduced into West Africa was probably small; the most likely route would have been overland. Frémond *et al.* (1966) have commented on the genetic poverty of West African populations.

It thus seems clear that the Jamaica Tall and the Panama Tall coconuts are distinct geographical races at opposite ends of the range, the tropical American mainland forming a barrier that has, until recently, prevented the merging of the two populations. The relatively slow germination of the Jamaica Tall supports the broad pattern of dispersal described (Whitehead, 1965).

Diseases such as Lethal Yellowing and Cadang-Cadang and pests (for example, Rhinoceros beetle) have exercised significant selection in some countries, as have also peculiar local environments such as coralline soils, drought and high rainfall (Frémond *et al.*, 1966). Hurricanes have influenced coconut evolution, initially reducing variability (50 per cent or more of a stand may be destroyed) but sometimes necessitating seed introduction from neighbouring areas.

Some relatively isolated islands in the South Pacific, notably Rennell, Rotuma and the Wallis Islands, have a high proportion of large-fruited palms. These probably originated from fruits selected by fishermen for large nut size (and hence larger 'water' content). In the New Hebrides and the Solomons, fruit size is small but nut numbers are high. This sort of situation might be the result of initial selection for high nut numbers when plantations were established.

Dwarf palms, of various kinds, have been preserved and propagated in many areas because they are ornamental, early bearing and easily harvested. They are often extremely uniform, having originated from very few seeds. In Jamaica, two palms from seeds introduced in 1940 had over 15,000 descendants by 1966 (Whitehead, 1966*b*). Sometimes variable hybrid populations are found.

4 Recent history

Though the use of chosen parents and nursery selection have, for many years, been recommended plantation practice, the large number of seedlings required has often meant that only weak selection was possible. Outstanding parents, high yielding and of good combining ability, have been identified but most of the seed released to growers has simply been taken from healthy palms which appeared to be high yielding. Systematic breeding started only recently, though some basic research was done, notably in Indonesia in 1930–42 and in India and Sri Lanka. Attempts were made to assemble collections but numbers of seeds successfully introduced were usually few and fairly unrepresentative.

Reasons for the delay in starting breeding are not hard to find. They include: dependence on seed production; small number of nuts per bunch and per pollination; long generation time and the large amount of land needed; shortage of information on coconut botany (including variability) and breeding techniques; cost of prospecting and collecting on any worthwhile scale.

A programme for the genetic improvement of the crop and some early results of work started in West Africa in 1949 were outlined by Frémond at a symposium on Tropical Crop Improvement held during the 10th Pacific Science Congress in Hawaii in 1961. Accounts by workers from other areas indicated that breeding work was then contemplated or had started at a number of centres. The demand for improved planting material at about this time was stimulated by the urgent need to replace old plantations and to provide materials resistant to several serious diseases. At the Hawaii Symposium, FAO was asked to initiate a world taxonomic survey and to maintain a list of genetic stocks. Further resolutions, to facilitate exchange of germ plasm and information, were made at an FAO Conference later in the same year, when the difficulties and high cost of surveys were emphasized. Since that time, surveys and the collection and exchange of germ plasm have developed under the encouragement of FAO (Fig. 64.2). In addition, information on variability in the major producing areas has accumulated and a Variety Consideration Panel has

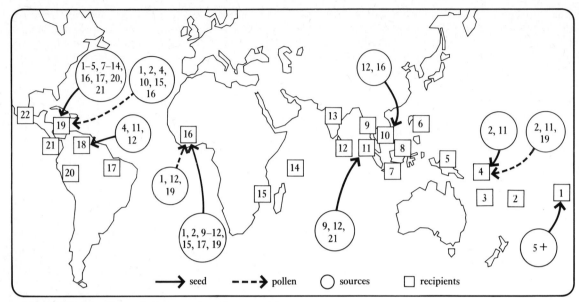

Fig. 64.2 Recent dispersal of coconut breeding material.

recently been formed to study problems of classification.

The main objectives in coconut breeding can be summarized as: (*a*) resistance to diseases and, where possible, pests; (*b*) early bearing; (*c*) high yield of copra; (*d*) varietal adaptability. The emphasis depends on the local conditions. In West Africa, increased yield of copra or oil per hectare is the main aim (Frémond *et al.*, 1966) whereas in Jamaica, resistance to Lethal Yellowing disease is critical (Whitehead, 1968*b*).

In general, though a start has been made, breeding has yet had no impact on coconut evolution. Producing populations are still overwhelmingly the locally adapted ecotypes generated by past evolutionary processes.

5 Prospects

Coconuts are essentially outbreeders and heterosis has been abundantly demonstrated in a wide range of inter-origin crosses. This heterosis is, no doubt, in part a reflection of the narrow genetic base of most local strains (or ecotypes). Since clonal propagation on a commercial scale is inconceivable with present technology (though perhaps a distant future possibility using cell or tissue culture techniques), improvement is likely to depend upon the systematic identification of parental lines of good combining ability. The product may be either improved populations or 'hybrids'. Coconuts lend themselves fairly well to hybrid seed production, using hand emasculation,

and some such populations (e.g. of economically acceptable dwarf×tall crosses) are already in use. Mass pollination by mechanical means is a possibility. It may well be that fairly rough control of hybridity would be economically acceptable, especially if hybrids could be identified in the nursery. In the longer term, the outcome is likely to be extensive recombination between local ecotypes that were previously isolated and themselves too narrowly based genetically to provide an effective source of improvement.

Meanwhile, the local acceptance of the Malayan Dwarf has been significant in providing a valuable supplement to traditional stocks. Not only has it performed well on its own and yielded productive hybrids, but it has also proved a source of resistance to Lethal Yellowing disease in Jamaica where most of its crosses show resistance (Whitehead, 1966*b*, 1968*b*). Some F_1 hybrid seed is now being released (Malayan Dwarf×Panama Tall, known as Maypan).

If coconut breeding is to maintain the impetus it now promises, several conditions must be fulfilled. There will be continuing need for exploration, collection, maintenance and botanical analysis of local populations, for genetic analysis of economic characters and disease resistance and for further search for disease resistance coupled with the development of rapid screening methods; above all, there will be a need for continuity, lack of which has been almost a traditional hazard in coconut research.

6 References

Cook, O. F. (1910). History of the coconut palm in America. *Contrib. U.S. Nat. Herb.*, **14**, 271–342.

Frémond, Y., Ziller, R. and de Nucé de Lamothe, M. (1966). *Le cocotier.* Paris.

Harries, H. C. (1971). Coconut varieties in America. *Oléagineux, Ann. 26*, **4**, 235–42.

Purseglove, J. W. (1963). Some problems of the origin and distribution of tropical crops. *Genet. agraria*, **17**, 104–22.

Tammes, P. L. M. and Whitehead, R. A. (1969). Coconut. In F. P. Ferwerda and F. Wit (eds), *Outlines of perennial crop breeding in the tropics.* Wageningen, The Netherlands, 175–88.

Wester, P. J. (1920). The coconut palm; its culture and uses. *Philipp. agric. Bur. Bull.*, **35**, 18–21.

Whitehead, R. A. (1965). Speed of germination, a characteristic of possible taxonomic significance in *Cocos nucifera. Trop. Agriculture Trin.*, **42**, 369–72.

Whitehead, R. A. (1966a). *Sample survey and collection of coconut germ plasm in the Pacific Islands.* H.M.S.O., London.

Whitehead, R. A. (1966b). Some notes on dwarf coconut palms in Jamaica. *Trop. Agriculture Trin.*, **43**, 277–94.

Whitehead, R. A. (1968a). *Collection of coconut germ plasm from the Indian/Malaysian region, Peru and the Seychelle Islands (1966–1967) and testing for resistance to Lethal Yellowing disease in Jamaica.* F.A.O., Rome.

Whitehead, R. A. (1968b). Selecting and breeding coconut palms (*Cocos nucifera*) resistant to lethal yellowing disease. A review of recent work in Jamaica. *Euphytica*, **17**, 81–101.

Oil palm

Elaeis guineensis (Palmae)

J. J. Hardon

Plant Breeding Station Van der Have BV
Rilland The Netherlands

1 Introduction

The oil palm is the most rapidly expanding plantation crop in the tropics. World exports of palm oil and kernel oil increased from 0·86 Mt in 1962 to 1·45 Mt in 1972. The increase continues and exports are expected to reach 3 Mt in 1982. Palm oil at present follows soja oil as the second most important oil crop (Hartley, 1972). In the past decade, Malaysia has established itself as a main producer, accounting for over half of the total world exports. Indonesia comes second, followed by Zaïre and the Ivory Coast. In Nigeria, a traditional oil palm country, trade has dropped sharply over the past years due to economic restrictions but should regain its former importance in coming years. The oil palm is found in most West African countries, while new developments are taking place in several other places with suitable climates (e.g. Brazil, Colombia, Ecuador, Panama, Costa Rica, the Solomon Islands, Papua, Thailand and the Philippines).

The oil palm has its main distribution within 10s of the equator. The palm has a single growing point from which fronds emerge in a regular sequence at the rate of 20–26 fronds per annum. It is monoecious, producing separate male and female inflorescences. Both are panicles born on woody stalks. Fruits (drupes) are situated on secondary branches and consist of an orange coloured mesocarp which contains 'palm oil', a hard lignified shell (endocarp) and a white kernel containing 'kernel oil'. Fruits are ovoid or elongated, 2–5 cm in length and weighing 5–20 g. Fruits are tightly packed on large ovoid fruit bunches. The number of fruits per bunch varies from 50 to 100 in young palms, up to 1,000–3,000 in older ones.

Individual bunches weigh from less than 1 kg to 20–50 kg. After harvesting, bunches are immediately brought to factories where fruits are stripped off and mesocarp oil extracted in mechanical presses. From the press-cake, kernels are removed and the shell cracked and separated from the kernel. Kernels are usually sold as such and the oil extracted by more specialized machinery. Palm oil and kernel oil have different properties and are generally used separately. Kernel cake forms a valuable animal feed.

2 Cytotaxonomic background

Elaeis has been classified with the genus *Cocos* in the tribe Cocoidea. Three species are distinguished thus:

(1) *E. guineensis*, endemic to West Africa, the major economic species and generally referred to simply as 'Oil Palm'.

(2) *E. oleifera* (formerly *Corozo oleifera* or *E. melanococca*), endemic to the northern Amazon basin, extending into Central America as far north as Costa Rica.

(3) The relatively little known *E. odora* (formerly *Barcella odora*) reported to occur in various places in the Amazon region.

In spite of the fact that the centres of origin are on different continents, close relationship between *E. guineensis* and *E. oleifera* is suggested by the fact that the species hybridize readily and produce fertile offspring (Hardon and Tan, 1969).

Systematic relationships suggest that the genus must have originated in the New World; but how and when it became established in West Africa is open to speculation. *Elaeis guineensis* and *E. oleifera* differ in several morphological characteristics and are readily separable, which suggests that isolation must have occurred some considerable time ago. This is supported by the presence of fossil pollen similar to that of *Elaeis* in Miocene strata in the Niger Delta (Zeven, 1967).

The chromosome number is $2n = 2x = 32$. As with most palms, the cytology has not been studied in any detail. The basic number, $x = 16$, is common in related genera.

3 Early history

Early Portuguese, Dutch and English seafarers from the fifteenth century onwards mention palm oil (and palm wine) from the coast of West Africa. Numerous vernacular names for the oil palm in various West African languages adds linguistic support to the

Fig. 65.1 Distribution of the oil palm, *Elaeis guineensis*, and its relatives.

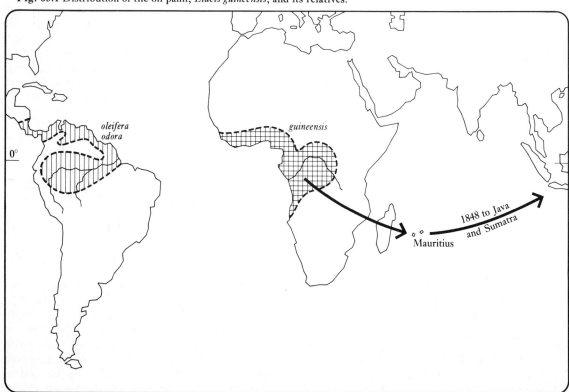

important role the oil has played in the village economy for many centuries (Zeven, 1967, 1972). The natural habitat is banks of rivers, lakes, swamps and other places too wet for rain forest. Most authors consider it to be originally a species of the transition zone between savanna and rain forest, but now more widely spread. The palm cannot maintain itself in competition with rain forest species. Its present wide distribution throughout the rain forest belt of West Africa, from Cape Verde down to Angola, is likely to be due to human disturbance, providing suitable habitats in forest clearings and around temporary settlements. Most oil palm populations fall into this semi-wild or 'camp-follower' category and they form the main source of palm oil and kernels in West Africa (Zeven, 1972). Principal centres of such populations are found in the Ivory Coast, eastern nigeria and in the Congo basin extending into Angola. The palms so grown are little, if at all, removed from the wild type.

In South and Central America, *E. oleifera* has not been used to the same extent. Low human population density and a more migratory pattern of existence of the South American Indians in the rain forest zone did not provide the same conditions as in West Africa for the palms to become established in sizeable groves. There are reports that the oil has minor uses for lamps and as an edible oil in the Amazon basin and in Colombia.

4 Recent history

The world market for vegetable oils grew fast during the later part of the nineteenth century; palm and kernel oils became widely used in stearic candles, margarine and soap. Industrial use was found in the tin-plate industry and palm oil was even used as a lubricant. To secure a reliable supply of these oils, Sir William Lever obtained extensive concessions, first in Sierra Leone and later (1911) in the Congo.

Oil mills were erected and this stimulated more efficient exploitation of existing palm groves and the planting of selected material. During the same period, plantation culture was started in Sumatra, followed a few years later by Malaysia, and a new industrial crop was born. The evolution of the crop really started from this point.

Until recently, the whole plantation industry in southeast Asia was based on material originating from four specimens in the Botanical Garden of Bogor, Java. Two of these four palms still survive in 1974. They were obtained as seedlings in 1848 from Mauritius where oil palms must have been introduced from Africa as ornamental plants at an earlier date.

Descendants of the original Bogor palms were

distributed throughout the Indonesian archipelago in small trial plantings and as ornamentals. Extensive rows of avenue palms, lining roads on tobacco estates in Deli on Sumatra, furnished the seeds for the first estates in that area around 1911. Material of this origin is generally referred to as the Deli Dura. It immediately proved to be high yielding and with good fruit characters. Such a narrow and, in a sense haphazard, genetic origin is rather typical of a number of other plantation crops, notably rubber and coffee. It is a curious fact that, in all these cases, the first choice of material appears in retrospect to have been extremely fortunate, indeed superior to most later introductions from the centres of origins of the crops concerned.

Selection programmes started in the early 1920s, employing at first simple mass selection, followed later by controlled crosses to produce bi-parental families. Both total yield of fruit bunches and fruit characters were taken into account in selection. The efficiency of this selection cannot be judged. There are no comparative trials incorporating material from various generations of selection and at least part of the overall improvement in yield observed must be ascribed to improved standards of agriculture and management.

During the same period (from 1922) oil palm breeding was started in the Congo (now Zaïre) by INEAC (Institut National pour l'Étude Agronomique du Congo Belge) at Yangambi (Pichel, 1956). Whereas the starting material in Sumatra and Malaysia consisted of a type with thick-shelled fruit (*E. guineensis* var. *dura*), in Zaïre the programme started from open-pollinated bunches of palms with thin-shelled fruits (*E. guineensis* var. *tenera*) collected at Eala and Yawenda. The Tenera form of the palm was considered superior because fruits with a thin endocarp (shell) had correspondingly more oil bearing mesocarp per fruit (Table 65.1).

Table 65.1 Composition of Dura and Tenera fruits

	Mesocarp % of fruit	Shell % of fruit	Shell thickness (mm)
Dura	20–65	25–45	2–8
Tenera	60–90	4–20	0·5–4

Tenera × Tenera crosses were found to segregate in 1:2:1 ratio of Dura, Tenera and a third type (Pisifera), characterized by the absence of a shell in the fruits, which however generally abort at an early stage of development (Beinaert and Vanderweijen, 1941). This discovery of single-gene inheritance of shell thickness allowed the production of pure Tenera (sh+, sh−) families by crossing Dura (sh+, sh+)

(female) with Pisifera (sh—, sh—) (male). Thus the heterozygote is the agriculturally desirable phenotype.

Table 65.2 Components of yield of Dura and Tenera full sibs in a breeding experiment in Malaysia

	Bunch yield (t/ha)	Fruit per bunch, %	Mesocarp per fruit %	Oil per Mesocarp %	Oil per bunch %	Yield of oil (t/ha)
Dura	27·0	65	61	50	19·8	5·2
Tenera	27·0	62	78	50	24·1	6·6

The superiority of Tenera over Dura (20–25%) in the production of mesocarp oil is shown in Table 65.2. Breeding programmes were, accordingly, changed after the Second World War from family and individual palm selection to selection of Dura and Pisifera lines which, in combination, would give high-yielding Tenera families.

The programme of INEAC in Zaïre, was, in principle, based on a diallel between six Teneras whereby high-yielding combinations were reproduced by crossing Duras derived from selfing one Tenera parent with Pisiferas of a selfing of the other Tenera parent. In Malaysia and Sumatra, Pisiferas were obtained from Tenera crosses of new introductions and by introgressing Teneras into Deli Dura and backcrossing these to Teneras and Pisiferas of unrelated new introductions. In the Ivory Coast, IRHO (Institut de Recherche des Huiles et Oléagineux) expanded its breeding programme in the 1950s based on a large exchange programme involving crosses between materials of various origins: Deli Dura (Malaysia and Sumatra), Yangambi Tenera (Zaïre), La Mé Tenera (Ivory Coast) and Pobé Tenera (Dahomey). Results suggested that highest yields were obtained by crossing Deli Dura with Pisifera of Yangambi and La Mé origin (Meunier and Gascon, 1972). This led to a breeding programme based on reciprocal recurrent selection. The same system was adopted by the Nigerian Institute for Oil Palm Research (NIFOR) but using material of various Nigerian origins (Sparnaaij, 1969).

5 Prospects

The strategy to be employed in oil palm breeding has been the subject of recent discussions (Hardon et al., 1972) which have been essentially concerned with the relative importance of general combining ability (g.c.a.) and specific combining ability (s.c.a.). In Malaysia, emphasis is placed on the creation of new, genetically more variable, populations to allow efficient utilization of g.c.a. In the Ivory Coast, mild inbreeding within existing populations is tolerated and commercial seed production is based on inter-population ('inter-origin') crosses. The expectation is that this will allow exploitation of both g.c.a. and s.c.a. There is a little biometrical information on genetic variances and realized progress under different systems of selection, so there is no immediate way of resolving these questions.

In practice, however, differences are not quite so evident. In Malaysia and Indonesia, as well as in the Ivory Coast, most of the current commercial planting material is based on crosses between Deli Dura and Pisifera mainly of Yangambi origin. It is a sobering thought that, in this cross-pollinating species, a considerable part of the commercial planting material in southeast Asia, parts of West Africa and in South America, still, in 1974, goes back to four botanical specimens in Java and only a few palms in Zaïre. However, the use of Nigerian and La Mé Pisifera is increasing but, due to the long generation time, changes in breeding methods and planting material will necessarily be slow.

The narrowness of the genetic base of present breeding populations is generally realized and there is increasing interest in the collection of wild material. Collecting expeditions have been carried out by INEAC in Zaïre and by NIFOR in Nigeria (Hartley, 1967). The IRHO (Meunier, 1969) and MARDI (Malaysian Research and Development Institute) are presently involved in extensive collecting programmes. In Malaysia, considerable attention is given to crop-physiological analysis. Results suggest (Hardon et al., 1972) that a 'model palm' has a high harvest index (ratio of harvested dry matter to total dry matter) and a high tolerance of competition, whereby competition for light through sizes and structure of the canopy plays an important role.

An obvious short cut in oil palm breeding would be vegetative reproduction. So far, clonal multiplication has not been possible, but work on tissue culture is in progress in various research centres.

Interspecific hybrids have been produced between *E. guineensis* and *E. oleifera* (Hardon and Tan, 1969) and F$_1$, F$_2$ and various backcross generations are under observation. Of prime interest is the low height increment of the hybrids, a character inherited from the *oleifera* parent. Hybrid bunch yields are promising but oil content of the mesocarp is lower than in *E. guineensis* though oil quality is good.

In West Africa, some work is being done on breeding for resistance against *Fusarium* wilt. Seedling tests have been developed and repeatable results obtained. The question is whether resistance or tolerance in the seedling stage offers any useful protection at maturity.

66

6 References

Beinaert, A. and **Vanderweijen, R.** (1941). Contribution a l'étude génétique et biométrique des variétés d'*Elaeis guineensis*. *Bull. INEAC., Série sci.*, **27**, pp. 101.

Hardon, J. J. and **Tan, G. Y.** (1969). Interspecific hybrids in the genus *Elaeis*. I. Crossability, cytogenetics and fertility of F1 hybrids of *E. guineensis × E. oleifera*. *Euphytica*, **18**, 372–9.

Hardon, J. J., Hashim, Mokhtar and **Ooi, S. C.** (1972). Oil palm breeding: a review. In *Advances in oil palm cultivation*. I.S.P., Kuala Lumpur, Malaysia.

Hartley, C. W. S. (1967). *The oil palm*. London.

Hartley, C. W. S. (1972). The expansion of oil palm planting. In *Advances in oil palm cultivation*, I.S.P., Kuala Lumpur, Malaysia.

Meunier, J. (1969). Étude des populations naturelles d'*Elaeis guineensis* en Côte d'Ivoire. *Oléagineux*, **24**, 195–201.

Meunier, J. and **Gascon, J. P.** (1972). General schema for oil palm improvement at the IRHO. *Oléagineux*, **27**, 1–12.

Pichel, R. (1956). L'amélioration du palmier au Congo Belge. *Bull. Agric. Congo Belge*, **48**, pp. 67.

Sparnaaij, L. D. (1969). Oil palm. In Ferwerda and Wit (eds), *Outlines of perennial crop breeding in the tropics*. Wageningen, 339–87.

Zeven, A. C. (1967). *The semi-wild oil palm and its industry in Africa*. Wageningen.

Zeven, A. C. (1972). The partial and complete domestication of the oil palm (*Elaeis guineensis*). *Econ. Bot.*, **26**, 274–9.

Date palm
Phoenix dactylifera (Palmae)

J. H. M. Oudejans Baarn The Netherlands; *formerly* CIBA-GEIGY Ltd. Basle Switzerland

1 Introduction

The date palm is of ancient cultivation in the desert belt of North Africa and the Middle East; the commercial groves of southern California are a very recent development. The world population of date palms is estimated at 93 million trees, yielding about 1·9 Mt of fruit annually, with Iraq, Iran and Egypt as the chief producers. For many Arab peoples the date is the staple carbohydrate food. According to moisture content of the fruit, varieties are classified as soft, semi-dry or dry, the former types being mainly exported and the latter consumed locally.

The date is a robust feather-leaved palm with basal suckers. It is dioecious and usually wind- (sometimes insect-) pollinated. The fruits, drupes with fleshy pericarps, are borne in large bunches and yields may be as high as 30–100 kg per tree. Palms bear in 5–7 years.

2 Cytotaxonomic background

Chevalier (1952) recognized 12 species of *Phoenix*. The six species examined cytologically by Beal (1937) were all diploid with $2n = 2x = 36$ and of rather constant karyotype. Beal could detect no heteromorphic pair that might have represented sex chromosomes. Interspecific hybrids, both natural and experimental, are numerous and fertile. Natural hybrids of *P. dactylifera* with *P. reclinata*, the only two *Phoenix* species producing basal suckers, reportedly produce fruits of good quality (Munier, 1974). *Phoenix atlantica*, the so-called 'spurious date', could be a possible ancestor or an intermediate form. Its variety *maroccana* produces edible fruits.

Date seedlings, as would be expected of an outbreeder, display much variability, as also do cultivars of different regions. This variability led Chevalier to postulate that, if one original progenitor of *P. dactylifera* ever existed, it must have been modified by hybridization with other species in different parts of the range, for example, with: *P. sylvestris* in Indo-Pakistan, *P. reclinata* and variant types in northeastern Africa and *P. canariensis* in northwestern Africa and Spain.

All species of *Phoenix* are dioecious but, as noted above, there is no evidence of a heteromorphic sex chromosome pair. The sex ratio in the date is about 1:1 and occasional female fertility on male palms has been noted. It is not known which sex is heterogametic; in the absence of cytological differentiation, genetic analysis of hermaphrodites would be necessary to determine this.

3 Early history

The genus *Phoenix* was widely distributed in Europe in Tertiary times but retreated southwards as the climate changed, remaining abundant in the North African region until desiccation there restricted its distribution.

The cultivation of the date possibly dates back to Neolithic times. The earliest record is a sample of stones from Egypt dated about 4500 B.C. Traces from the third millenium B.C. from western India, records of the Sumerian civilization and the etymology of the name (from the Hebrew, 'dacheb') jointly support Chevalier's (1952) suggestion that domestication occurred simultaneously in different places in the area between the Indus and the Atlantic.

Since the wild ancestor (or ancestors) of the palm is/are unknown, the changes that occurred during and after domestication cannot be inferred. It seems safe to assume, however, that selection for succulence of fruit must have been a major feature. Vegetative propagation by suckers was no doubt practised from earliest times.

The interfertility of *Phoenix* species and the occurrence of natural hybrids (noted above) suggests, perhaps, that the whole, or a large part of, the genus could be regarded as the 'primary gene pool'.

Another ancient cultural practice is artificial pollination; it is mentioned in the cuneiform texts of Ur (about 2300 B.C.). In seedling stands with approximately equal numbers of male and female trees, natural pollination would secure adequate fruit setting. Since, however, male trees produce very abundant pollen, they are retained in a proportion of one male to 25–50 females and pollination was traditionally (and still is locally) carried out by hand. The process, involving much climbing, is laborious and is being displaced by machinery that blows the pollen through long tubes (el Baradi, 1968).

4 Recent history (Oudejans, 1969)

Originally, populations consisted of chance seedlings

from which outstanding individuals were selected and vegetatively propagated. Such clones are not only the numerous local varieties but also the important and widely distributed ones. This completely empirical practice of selection is certainly still going on in many primitive agricultural communities.

Only comparatively recently has deliberate breeding been started. Nixon, in California about 1930, made a detailed study of metaxenia, the phenomenon by which the pollen exercises a direct influence on the maternal tissues of the fruit, inducing differences in time of ripening or size and colour. Since the pollen of certain males can thus have a favourable effect on maturity and fruit quality, it proved advantageous to propagate these males into clones. Nixon (1966) defined the promising male tree as one which would: flower synchronously with prospective female clones; produce abundant viable pollen; and show favourable metaxenia effects.

Male clones for use in commercial plantings are thus one objective of date breeding; their breeding value, the performance of their progeny, is immaterial. Commercial clones, however, are all female and, if it is desired to make intervarietal crosses, it is necessary first to obtain the male counterparts of some, at least, of the selected females. This has been attempted in California by back-crossing repeatedly to the chosen female (Nixon and Furr, 1965). The programme, up to 1965, had produced 37 back-cross families, some at the BC 4 stage, but, because of the long life cycle (6·5 years from seed to first flowering), the success or otherwise of the work is yet unknown. Inbreeding depression might be predicted to follow from backcrossing an outbreeder such as the date. In Arizona, inbreeding experiments with the progeny of the variety Deglet Noor were abandoned after three generations because of disappointing results (Nixon and Furr, 1965).

In Morocco and Algeria, work has been done on a devastating disease called 'Bayoud', caused by *Fusarium oxysporum* f. sp. *albedinis* which is transmitted by contaminated soil and plant parts. The disease threatens the existence of excellent varieties such as Deglet Noor but resistant clones (e.g. Tazerzait, Takerboucht) are known. Field testing for resistance is possible and clones and seedlings are being screened in Morocco (Louvet and Toutain, 1973; Toutain, 1973).

5 Prospects

The date palm, with its long life cycle, dioecious habit and low rate of vegetative propagation, poses peculiar problems for the breeder. Little has been achieved to date and rapid progress in the future is not to be expected. That there is genetic potential for improvement is, however, not in doubt: the numerous clones and seedlings that exist in all date-growing countries have yet hardly been exploited by breeding.

Accelerated vegetative propagation of young suckers under mist irrigation in glasshouses is being developed in Algeria. In the longer run, *in vitro* propagation, now being studied by Beauchesne and Poulain in Antwerp, offers exciting possibilities.

Morphological characters of the fruit bunch, physiological characters of the palm (for example, salt tolerance) and disease resistance will no doubt claim the breeders' attention. Besides resistance to Bayoud, resistances also to Graphiola leaf disease and to mites have been reported.

6 References

Baradi, T. A. el (1968). Date growing, a review. *Trop. Abstr.*, **23**, 473–9.

Beal, J. M. (1937). Cytological studies in the genus *Phoenix*. *Bot. Gaz.*, **99**, 400–7.

Chevalier, A. (1952). Recherches sur les *Phoenix* africaines. *Rev. int. Bot. appl. Agric. trop.*, **32**, 205–33.

Louvet, J. and Toutain, G. (1973). Recherches sur les Fusarioses, VIII. *Ann. Phytopathol.*, **5**, 35–52.

Munier, P. (1974). Le problème de l'origine du palmier dattier et l'atlantide. *Fruits d'Outre-Mer*, **29**, 235–40.

Nixon, R. W. (1966). Growing dates in the United States. *U.S.D.A. agric. Inform. Bull.*, **207**, pp. 50 (revised).

Nixon, R. W. and Furr, J. R. (1965). Problems and progress in date breeding. *Rep. Date Growers' Inst.*, **42**, 2–5.

Oudejans, J. H. (1969). Date palm. In F. P. Ferwerda and F. Wit (eds). *Outlines of perennial crop breeding in the tropics*. Wageningen, The Netherlands, 243–57.

Sesame

Sesamum indicum (Pedaliaceae)

N. M. Nayar
Central Plantation Crops Research Institute
Vittal Karnataka State India

1 Introduction

Sesame has been claimed to be the most ancient oil seed known to man though, perhaps, according to such archaeological records as are available, linseed should be accorded this position. Certainly, the crop has been widely esteemed for a very long time. Sesame is an annual that matures in 70–140 days, but mostly in 105 days or less. The small, ovate seeds are contained in capsules, borne on short peduncles, in the axils of the upper leaves of the main stem and branches. The oil content of the seed is high (45–60%). The oil, valued for its high quality and stability, is used mostly for culinary purposes. The meal contains more than 20 per cent protein which is of high quality (except for low lysine content). The oil, the seed and even the leaves have been ascribed several medicinal and other desirable properties.

Sesame is grown throughout the tropics and subtropics but, possibly because of its low productivity (332 kg/ha seed in 1972), ranks only ninth among the top 13 vegetable oil crops which make up 90 per cent of world production. In 1972, it was grown over 5·8 Mha and produced 1·9 Mt of seed (FAO data). India, China, Sudan, Mexico, Venezuela and Burma are the world's leading producers.

For general treatments of the crop, see Joshi (1961), Purseglove (1968) and Weiss (1971). Origins and cytogenetics have been discussed by Joshi (1961) and by Nayar and Mehra (1970).

2 Cytotaxonomic background

Pedaliaceae is a small family consisting of about 16 genera and 60 species. About 37 species have been described in *Sesamum*. Besides *S. indicum* (the predominant cultivated species), *S. angustifolium*, *S. radiatum* and *Ceratotheca sesamoides* are also cultivated to a small extent in Africa. Most species occur in three regions: tropical Africa, India, including Sri Lanka

(Ceylon), and the Far East. About 25 taxa are distributed in tropical Africa, nine in the Indian region, four in the East Indies; and one each has been described from Crete and Brazil. A few taxa occur in more than one region and two, *capense* and *schenckii*, occur in all the three major regions of distribution.

The cytogenetics of the group is little understood and the chromosome numbers of only three genera are known. Somatic chromosome numbers of nine *Sesamum* taxa have been determined, as follows: $2n = 2x = 26$ in 5 cases; $2n = 4x = 32$ in 3 cases; and $2n = 8x = 64$ in 2 cases. The basic numbers are thus $x = 8, 13$. The predominant cultivated species, *indicum*, and the two widely distributed forms, *capense* and *schenckii*, are diploids sharing the basic $x = 13$. Interspecific crosses have been attempted in about 24 combinations, involving *indicum*, *schenckii*, *grandiflorum*, *alatum*, *capense* (all $2n = 26$), *prostratum*, *laciniatum*, *angolense* (all $2n = 32$), *radiatum* and *occidentale* (both $2n = 64$) and only about two-thirds have succeeded. The results from these studies and from distributional and morphological data suggest close affinities between *radiatum* and *occidentale* (which may even be conspecific), between *prostratum* and *laciniatum* and between *indicum* and its ssp. *malabaricum*. They suggest also that *grandiflorum* may be synonymous with *schenckii*, and also *ekambaramii* and *alatum* with *capense*.

3 Early history

There are numerous archaeological, prehistorical and literary references to sesame from the Middle East, Egypt, Iran and India dating from about 4300 years B.P. The evidences from Malaya and China are also ancient. All of them testify to the high esteem in which both the seed and oil were held by the peoples of these regions and the many uses in cooking, medicine, toiletry and ritual. There are records of the crop in the Assyrian tablets, in the writings of Herodotus, Xenophon, Theophrastus, Pliny, Marco Polo and of other ancient travellers and chroniclers. Archaeological evidence indicates that sesame was being cultivated in Palestine and Syria during the chalcolithic period, about 3000 B.C.

Many earlier writers (for example, Hiltebrandt, Burkill, Dalziel and Portères) considered Africa to be the ultimate source of the crop. This view was based on the presence of nearly two-thirds of *Sesamum* species in tropical Africa and the dominant position of the crop in the economies of (and the presence of variability in) several African countries. Even if sesame were known in the Middle East from prehistoric times, this region does not possess any wild

Sesamum species. We can only speculate as to whence in Africa the Middle East could have received the crop.

The largest sesame acreage in the world is in India and it is also an important crop in Burma. The plant shows considerable variability in both these countries. There is no doubt about its antiquity in the Indian subcontinent. Charred sesame seeds have been recorded from Harappa (*ca.* 3500–1700 B.C.), the oldest archaeological site in the region. Etymology supports the idea of antiquity; thus the Sanskrit word for oil, *taila*, is almost synonymous with that for sesame, *til*.

De Candolle suggested that India might have received sesame from the Far East but this is now known to be wrong. Both Watt (in his great *Dictionary*) and Vavilov considered that it had polytopic origins and the regions other than India in which they considered sesame to have originated were Ethiopia (including Somalia and Eritrea) and central Asia. There is no reason to assume that the crop could have originated in central Asia or the Middle East except that its use in these areas since prehistoric times has been well chronicled. No wild taxa occur there, the variability present in the cultivated *indicum* is no greater than that found in several other regions and, today, the crop is of only local importance in the region.

Recently, Nayar and Mehra (1970) proposed that sesame could have originated in either the Ethiopian region or in peninsular India or even in both independently. The two regions could be considered botanically contiguous as they share several species and genera, have long had cultural and commercial contacts over both land and sea, and have at least two wild species (*capense* and *schenckii*) in common. In addition, a wild taxon described as *S. indicum* ssp. *malabaricum*, which is completely interfertile with the cultivated *indicum*, occurs on the Malabar coast. It may be either an escape from cultivation or a companion weed species of the cultivated *indicum*.

The progenitor species of the cultivated *indicum* is unknown as no wild sesame is known except the ssp. *malabaricum* just mentioned. Its potential as an ancestral species to the cultivated *indicum* has yet to be studied. Among the truly wild forms, the two species *capense* (including *alatum* and *ekambaramii*) and *schenckii* (including *grandiflorum*) occur in Africa, India and the Far East and thus have the widest distributions in the genus. Both possess the same chromosome number as the cultivated *indicum* ($2n = 2x = 26$) and both have produced viable F_1 hybrids with *indicum*. Both the F_1 hybrids are sterile but *schenckii* (as *grandiflorum*)-*indicum* shows some end-season fertility. Taking into account morphological differences between the seeds of *capense* and *indicum*, Nayar and Mehra (1970) advanced *schenckii* as the more likely ancestral species but there is yet little experimental evidence to support this proposition. Earlier, Portères had suggested that sesame arose through hybridization between *alatum* (= *capense*) and *radiatum* and even that the related *Ceratotheca sesamoides* might have contributed to some of the cultivated races. Zukovsky, however, felt that *radiatum* could not have been a progenitor of *indicum* because it possesses a narcotic odour which the cultivated species does not have and it has quite a different chromosome number.

4 Recent history

Sesame exhibits a great range of variation, as has been amply documented by the world-wide surveys of Vavilov and Hiltebrandt. While tropical Africa, the Middle East, the Indian subcontinent and China have been growing sesame from ancient times, the crop has been introduced to East Africa (south of Ethiopia, presumably, in fact, a reintroduction) and America only in the last 200 years.

The crop is predominantly self-fertilized, though some cross-fertilization by insects also takes place. Wind pollination is not known to occur.

Yield and, to a lesser extent, disease resistance (particularly to phyllody virus) have long been the major breeding objectives. The oil content of sesame is relatively high but intervarietal variation is known to occur for this character and breeding for higher oil content has been taken up. Lately, attention has been paid in several countries to adaptation to mechanized harvesting which requires near-synchronous flowering and indehiscent capsules; some success has already been achieved.

5 Prospects

Weiss (1971) has produced statistics to show that world production is at best static, in contrast to the great increases in production being recorded by other sources of edible oils such as soybeans, coconut, oil palm and olive. Consequently, sesame's share of world trade (previously about 10%) has been declining. This is true despite the facts that sesame oil is one of the best qualitatively, the oil content of the seed is comparatively high, the crop is well adapted to a wide range of conditions and the growth period is relatively short. No doubt, the present average oil yield (150 kg/ha) is the lowest of any major oil crop (1,000 kg/ha for oil palm, 270 kg for ground nut, 225 kg for sunflower,

200 kg each for olive and soya) but, according to FAO records, oil yields above 500 kg/ha are obtained from sesame in several countries such as Egypt, El Salvador, Guatemala, Honduras and Afghanistan. Clearly, the crop has better potential than is now realized, provided attention is paid to improvement and management. Given the sort of international efforts that have been launched for the improvement of cereals and legumes, there is no need to assume that sesame's yield potential would be less than those of other oil-seed crops when it is recalled that they have been receiving a much greater research effort. And, with its intrinsic qualities of good palatability, high nutritive value and stability of both oil and meal, sesame will always be more acceptable to the discriminating consumer than most of the other edible vegetable oils.

6 References

Joshi, A. B. (1961). *Sesamum, a monograph*. Indian Central Oil Seeds Committee, Hyderabad, India.

Nayar, N. M. and Mehra, K. L. (1970). Sesame: its uses, botany, cytogenetics and origin. *Econ. Bot.*, **24**, 20–31.

Purseglove, J. W. (1968). *Tropical crops. Dicotyledons*. London.

Weiss, E. A. (1971). *Castor, sesame and safflower*. London.

Black pepper

Piper nigrum (Piperaceae)

A. C. Zeven
Institute of Plant Breeding (I.v.P.)
Agricultural University Wageningen The
Netherlands

1 Introduction

Pepper is one of the oldest spice crops, the dried fruits constituting the pepper of commerce. Access to supplies from southeast Asia was a major stimulus to the great navigations of the fifteenth and sixteenth centuries. The crop is grown by smallholders in tropical countries such as India, Sarawak, Indonesia and Brazil. Only one or a few clones are grown in any one country. This entails the danger of disastrous epidemic disease, such as has occurred in Indonesia and Sarawak in recent years. World production is about 80 kt.

Other *Piper* species are cultivated. They are *P. aduncum*, a soil-conserving plant; *P. angustifolium*, which yields Folio Matica; *P. betel*, the betel vine, for its leaves which are chewed with betel nut (*Areca*); *P. cubeba*, the cubebe or tailed pepper, for it aromatic 'tail'; *P. guineense*, the guinea pepper; *P. longum*, for its unripe spikes; *P. methysticum*, for its roots which provide a toxic, soporific beverage (kava); and *P. ornatum*, an ornamental.

For general treatments of the crop see Purseglove (1968) and Waard and Zeven (1969).

2 Cytotaxonomic background

The somatic chromosome number varies between $2n = 48$ (Sharma and Bhattacharyya, 1959), $2n = 52$ (Mathew, 1958; Martin and Gregory, 1962), $2n = 104$ (Mathew, 1958) and $2n = ca.$ 128 (Darlington and Wylie, 1955). Sharma and Bhattacharyya suggested that the original basic chromosome number (x) was 12 and that x now varies from 12 to 16. The cultivars studied by Mathew and by Martin and Gregory had $x = 13$. It is suggested that one of the causes of weak F_1 plants is a difference in chromosome number of their parents.

3 Early history

Wild pepper plants grow in the Western Ghats of Malabar, southwestern India (Gentry, 1955). This region must be presumed to be the centre of origin of the crop. In early times, pepper was spread to southeast Asia by means of cuttings. Seeds could probably not have been used for dissemination because of their short longevity (7 days). Some clones have reached a wide area of distribution. Thus cv. Banka in Indonesia resembles cv. Kamchay in Indochina and also gave rise to cv. Kuching in Sarawak.

The pepper vine is a perennial, developing, as aerial parts, the terminal stems, stolons or runners, and lateral branches (see Purseglove, 1968; Waard and Zeven, 1969). Wild *Piper* species (including *P. nigrum*) are dioecious but human selection has resulted in monoecious clones. Thus cvs. Banka and Kalluvalli have perfect flowers, while cv. Kuderavalli has perfect female and male flowers. By contrast, cv. Uthirancotta appears to possess female flowers only. Male plants are easily recognized by their vigorous vegetative growth. Wild *Piper nigrum* must be an outbreeder. The clonal cultivars must be predominantly self pollinated so incompatibility does not seem to be present. The mode of pollination is uncertain.

4 Recent history

Pepper, historically a south Asian crop, was only relatively recently introduced into Africa, the Pacific Islands and Central and South America.

5 Prospects

Formerly, clonal evaluation was the main type of selection work. Some of these clones may have developed from volunteer seedling plants. More recently (since 1952), deliberate hybridization programmes have been developed in several places, notably India, Indonesia, Sarawak and Puerto Rico. The main object is resistance to *Phytophthora* foot-rot and the nematode, *Meloidogyne*. Apparently, no adequate source of resistance to *Phytophthora* is available and grafting on resistant rootstocks has proved impossible. Hence the collection of parents should be augmented with wild material of *P. nigrum* and related species. The genetic base of the crop is undoubtedly very narrow.

Other breeding objectives are perfect flowers (hermaphroditism) and high yield (many long spikes with many rows of flowers and full setting of big seeds having high alkaloid contents).

6 References

Darlington, C. D. and Wylie, A. P. (1955). *Chromosome atlas of flowering plants*. London.

Gentry, H. S. (1955). Introducing black-pepper into America. *Econ. Bot.*, **9**, 256–68.

Martin, F. W. and Gregory, L. E. (1962). Mode of pollination and factors affecting fruit set in *Piper nigrum*. *Crop Sci.*, **2**, 295–9.

Mathew, J. (1958). Studies on Piperaceae. *J. Ind. bot. Soc.*, **37**, 155–71.

Purseglove, J. W. (1968). *Tropical Crops. Dicotyledons*, **2**, London, 441–50.

Sharma, A. K. and Bhattacharyya, N. K. (1959). Chromosome studies on two genera of the family Piperaceae. *Genetica*, **29**, 256–89.

Waard, P. F. W. de and Zeven, A. C. (1969). Pepper, *Piper nigrum*. In F. P. Ferwerda and F. Wit (eds), *Outlines of perennial crop breeding in the tropics*. Wageningen, 409–26.

Buckwheat

Fagopyrum (Polygonaceae)

C. G. Campbell
Canada Department of Agriculture
Research Branch Morden Manitoba
Canada

1 Introduction

Buckwheat is a crop of secondary importance yet it has persisted through centuries of civilization and enters into the agriculture of nearly every country in which grain crops are cultivated. The plants are not grasses but the seeds (strictly, achenes) are usually classified among the cereal grains because of similar usage. The grain as such is generally used as animal or poultry feed, and dehulled groats are cooked as porridge and the flour is used in pancakes, biscuits, noodles, cereals, etc. The grain, however, contain one or more dyes which, as a result of fluorescence, are photodynamically active. They can produce an irritating skin disorder, on white or light-coloured areas of skin or hide, under conditions of heavy consumption of buckwheat and exposure to sunlight (De Jong, 1972).

The crop is usually consumed more or less locally though international trade is increasing. The protein of buckwheat is of excellent quality (Coe, 1931) and this, coupled with the plant's ability to do well on the poorer soils, perhaps accounts for its widespread usage.

2 Cytotaxonomic background

Three species, all diploid with $2n = 2x = 16$, are generally recognized: *cymosum*, *esculentum* and *tataricum*, perennial, common and Tartary buckwheat, respectively. Notch-seeded buckwheat (*F. emarginatum*) is cultivated in northeastern India and China. Its grains differ from Tartary and common by having the angles or edges of the hull extended into wide, rounded margins or wings. As the plant resembles common buckwheat in all other respects and is completely compatible with it, most workers consider it a form of *F. esculentum* (Bailey, 1917). The perennial species, *F. cymosum*, has rhizomes and differs from common and Tartary buckwheat in its shoots, branching and racemes. The cytotaxonomy of buckwheat has been little studied but it is generally believed that *F. cymosum* was the original parent of both common and Tartary buckwheat (Krotov, 1963).

3 Early history

The buckwheats originated in temperate eastern Asia. The perennial species, *F. cymosum*, the probable ultimate source of both common and Tartary buckwheat, is native to northern India and China. Wild forms of *F. esculentum* are found in China and Siberia. Tartary buckwheat is reported as growing in the Himalaya of northeastern India and China under cooler and harsher climatic conditions, to which it is better adapted than common buckwheat.

Comparison of *F. cymosum*, the putative common ancestor, with the two cultivated forms suggests that selection for annual habit, non-shattering inflorescence and low seed dormancy was operative. Tartary buckwheat, in having more fragile heads and more dormant seed than the common form, seems to be the less advanced of the two cultigens. The breeding system of *F. cymosum* is unknown but it may reasonably be assumed to share with *F. esculentum* a heteromorphic incompatibility system; *F. tataricum* is self-compatible and homomorphic (see below). It looks, therefore, as though one cultigen, *tataricum*, has moved towards inbreeding under domestication, though less evolved morphologically than the other.

Although buckwheat is known to have been cultivated in China for 1,000 years, it is not believed to be very ancient (Hunt, 1910). The first written records are Chinese scripts of the fifth and sixth centuries. It was introduced into Europe in the Middle Ages, reaching Germany early in the fifteenth century (Hughes and Henson, 1934). From there it spread and was cultivated for several centuries in England, France, Spain, Italy, Germany and Russia. It has also found a place in the agriculture of Africa and Brazil. It was introduced early into the American Colonies, having been relatively much more important then than now.

4 Recent history

The recent history of both common and Tartary buckwheat has generally been one of decline. The USSR is by far the biggest producer but, even there, the crop has gone down, as it has also in France, once a major producer, and the USA. Canada's acreage declined steadily until the early 1960s but has increased since then as a result of demand for export.

Common buckwheat is a self-incompatible, sexually-propagated crop and, as such, does not lend itself

readily to improvement by breeding. Sharma and Boyes (1961) have studied the heteromorphic sporophytic type of incompatibility involved and have presented evidence that the inheritance of style length is controlled by a single gene. However, short-styled flowers also differ from long-styled ones in having longer stamens, larger pollen grains, the thrum stylar incompatibility reaction and the thrum pollen incompatibility reaction. Sharma and Boyes (1961) postulated an *S* supergene for this complex of characters, similar to that proposed for *Primula*. They further suggested that the supergene could be broken down into sub-units by crossing-over. Although the mechanism is not known, there is evidence of genetic change at the *S* locus. Thus, homomorphic, highly self-compatible diploid lines have been isolated; their value in breeding programmes has yet to be determined as severe inbreeding depression occurs.

Fagopyrum tataricum is self-compatible but less work has been done on it than on *F. esculentum*. Recently, however, workers in the USSR have begun to tackle this crop. Attempts to transfer the self-compatibility of *F. tataricum* to *F. esculentum* have proved unsuccessful. A study of pollen tube growth in the interspecific cross showed that pollen tubes of *F. esculentum* reached the base of *F. tataricum* styles, whereas those of *F. tataricum* were inhibited in *F. esculentum*. No hybrid seed was obtained (Morris, 1951).

Since seed size can generally be increased by raising the ploidy level to tetraploid and as seed size and yield are important factors considered in most breeding programmes, a considerable number of autotetraploid lines have been produced. Such lines generally have much lower fertility than their diploid counterparts and the considerable effort put into breeding at the tetraploid level has resulted in few new varieties.

Buckwheat has been relatively free from damage by insects or diseases and no major programmes of resistance breeding have been carried out. As quality parameters have been only vaguely defined, the major emphasis in most breeding programmes has been yield. The use of heterosis has been investigated but the incompatibility system, coupled with the very heterogeneous nature of the crop, has led to little achievement in this field.

5 Prospects

The present trend of breeding by mass selection or maternal selection still seems to be fairly effective and will probably continue. The severe inbreeding depression which occurs in the first few generations of selfing makes it highly unlikely that the self-compatible lines now available will be utilized to any large extent in utilizing the heterotic capabilities of the crop. Emphasis can be expected to continue on increasing grain yield, as there has been little improvement on this respect from most breeding programmes. Any improvement in disease or insect resistance will probably be accomplished by introgression from desirable lines.

One factor that will probably affect breeding programmes is the photoperiodic response of the crop. Buckwheat has been considered indeterminate since it will produce flowers over a wide range of photoperiods. However, recent work indicates that the photoperiod influences the time and rate of flowering as well as plant habit (Shustova, 1965). Varieties responding to different photoperiods have been developed and grown commercially in Japan and are being utilized in other programmes in adapting lines to various regions.

One aspect that must be considered in the future is the collection and maintenance of genetic material. Few expeditions have taken place but there is good international collaboration on maintenance and distribution of seed stocks. Further collecting, to obtain a more representative sampling of native genetic resources, should result in a fairly satisfactory situation.

6 References

Bailey, L. H. (1917). *Cyclopedia of American agriculture*, 5th edn, **2**. New York and London.

Coe, W. R. (1931). Buckwheat milling and its by-products. *U.S. Dep. Agric. Circ.*, **190**, pp. 11.

De Jong, H. (1972). Buckwheat. *Field Crop Abstr.*, **25**, 389–396.

Hughes, H. D. and **Henson, E. R.** (1934). *Crop production principles and practices*. New York.

Hunt, T. K. (1910). *The cereals in America*. New York and London.

Krotov, A. S. (1963). *Grechikha*. Moscow and Leningrad.

Morris, R. M. (1951). Cytogenetic studies on buckwheat. *J. Hered.*, **42**, 85–9.

Sharma, K. D. and **Boyes, J. W.** (1961). Modified incompatibility of buckwheat following irradiation. *Can. J. Bot.*, **39**, 1241–6.

Shustova, A. P. (1965). [Effect of light conditions on the growth, development and yield of buckwheat]. *Fiziologiya Rast.*, **12**, 782–8.

Strawberry

Fragaria ananassa (Rosaceae)

J. K. Jones
Department of Agricultural Botany,
University of Reading England

1 General introduction

The cultivated strawberry, *Fragaria ananassa*, is grown extensively in most temperate and in some sub-tropical countries. It is one of the most important soft fruits and, in 1970, world production was more than $1 \cdot 3$ Mt. The main production is in Europe which, together with the USSR and the USA, produced more than 1 Mt. Yields of 10–13 t/ha are now obtained from seasonal fruiting varieties outdoors and higher yields still are attained in glasshouses. In favourable environments in which ever-bearing varieties can be cropped for most of the year, yields in the range 50–100 t/ha have been reported.

Strawberry plants are perennial herbs with short, woody stems or stocks and rosettes of leaves. All species and most cultivated varieties are seasonal and produce a sequence of inflorescences and the stolons or runners. Plants formed on runners are used in propagation and each variety is a clone. Plants are cropped for a few years only and then replaced by maiden runner plants. The berry is a false fruit, an enlarged fleshy receptacle, growth of which is stimulated by the development of the true fruits (achenes). Much fruit is eaten fresh and the remainder is processed for canning, for jams and conserves, for freezing and for flavouring drinks and confectionery.

2 Cytotaxonomic background

At least 46 species of *Fragaria* have been described, but many may not be truly distinct. The species form a polyploid series, from diploid to octoploid, with a basic chromosome number of $x = 7$. *Fragaria vesca*, the commonest wild diploid, is distributed throughout north temperate Europe, Asia and America and occurs also in North Africa and South America. Other diploids have more restricted distributions in Asia, although *F. viridis* occurs widely in Europe and Asia. The three known tetraploids occur in eastern Asia only, and the only hexaploid species, *F. moschata*, is distributed in western Asia and Europe.

Fragaria ananassa is octoploid ($2n = 8x = 56$) and is derived from crosses between the American octoploids *F. chiloensis* and *F. virginiana*. *Fragaria chiloensis* occurs discontinuously along the Pacific coast, commonly on sand dunes, from the Aleutian Islands to California, and then in Chile where it also grows inland at altitudes up to 1,600 m. A subspecies grows in Hawaii. *Fragaria virginiana* grows in open woodland and hill meadows in North America, from the east coast to the Rocky Mountains and from New Mexico to Alaska. It is rare near the Pacific coast but some natural hybrids with *F. chiloensis* have been

reported. Natural introgression may have contributed to the diversity of both species, especially in *F. virginiana* of which some variants have been described as separate species. A third octoploid species, *F. iturupensis*, has recently been reported from the Kuril Islands, northeast of Japan (Staudt, 1973).

Cytological studies of interspecific hybrids and of natural and induced polyploids indicate that there has been little differentiation in the homology of chromosome sets in any of the species. Each of the natural polyploids appears to be an autopolyploid (though possibly with some regulation of chromosome pairing) and this prevents any simple analysis of evolutionary relationships. The separate distributions of the tetra-

Fig. 70.1 Evolution of the strawberries, *Fragaria*.

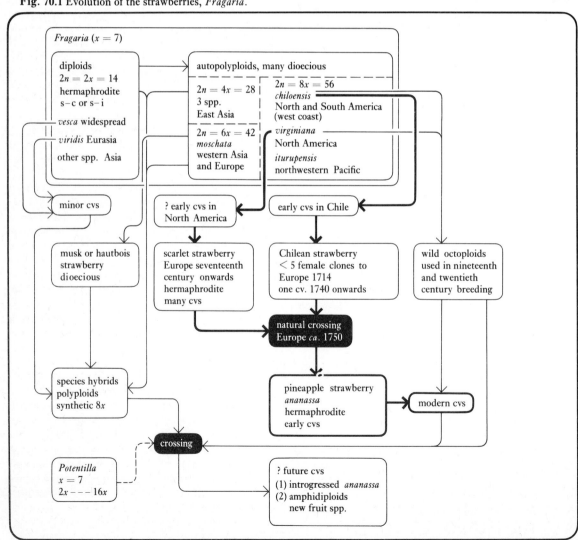

ploid, hexaploid and octoploid species suggest that each ploidy group originated independently and this has stimulated interest in the ways in which such changes in chromosome number may have occurred. Two possibilities are by unreduced and by doubled unreduced gametes, both of which have been shown to occur (Ellis, 1962; Bringhurst and Senanayake, 1966). However, the finding of *F. iturupensis* (Staudt, 1973) extends the range of the octoploids towards that of tetraploid *F. orientalis* on mainland Asia. Further studies of the Asian species are needed.

Several diploids, including *F. vesca*, are monoecious, self-compatible and mostly inbreeding, but three diploid species are self-incompatible. Most of the polyploids are entirely or predominantly dioecious, although hermaphrodite forms occur and have been selected in cultivation; these are self-compatible, but cross pollination is either essential or advantageous for full fruit set.

3 Early history (Darrow, 1966; Wilhelm and Sagen, 1975)

The fruits of all *Fragaria* species are palatable and it is probable that they were collected in the wild though of this there is no actual record. Strawberries (probably *F. vesca*) were tended in gardens by the Romans in 200 B.C. and possibly earlier; this species was planted in England and France during the fourteenth and, more extensively, in the fifteenth and sixteenth centuries, when some ever-bearing forms were used. Herbalists record that these strawberries were grown for ornament and medicine as well as for food. Many improved alpine varieties of *F. vesca* were selected in the nineteenth century, especially in France, and some are still grown on a small scale.

The other European species were also used, *F. viridis* on a small scale and *F. moschata* quite extensively. The latter, the musk or hautbois strawberry, was established in cultivation by the sixteenth century and, in 1765, Duchesne in France demonstrated that it is dioecious and showed why it was necessary to retain the apparently sterile male plants as well as the fruitful female ones. Distinct varieties were selected in England and France and the names Black, Apricot and Raspberry suggest that considerable variation existed. Selection did not markedly increase the size of the fruit but this species is still grown on a small scale for its unique flavour and aroma.

American octoploids were used before Europeans settled in America. North American Indians used fruit of *F. virginiana* to flavour bread and beverages and there are indications that they planted as well as gathered. *Fragaria chiloensis* was cultivated in Chile

before the Spaniards arrived and Chilean Indians had different names for wild and cultivated forms; they propagated the latter from runners. The fruits were used fresh and in wine and also as raisins after drying in the sun. The Spaniards, impressed by the large firm fruits of three colours (red, yellow and white), introduced the cultivated *F. chiloensis*, which they called *frutilla*, into Peru and Ecuador, where some varieties are still grown.

Fruits of *F. virginiana* are not much larger than those of *F. vesca* but they have different colour and flavour. Seeds were taken to Europe by 1556, but the first certain record of *F. virginiana* in cultivation dates from 1624. Some of the first introductions did not grow well but further introductions were made and self-compatible hermaphrodite forms selected. Varieties (known as the Scarlet strawberry) became popular for fresh fruit and conserves during the eighteenth century and at least 30 (and probably many more) varieties were used by 1820. Most of these were later replaced by *F. ananassa*, quite quickly in Europe and more slowly in America. A few varieties are still grown for the special flavour, but the disadvantage of small fruit size has not been removed by selection.

Although the large fruits of the cultivated Chilean strawberry had been reported during the sixteenth century, the first introduction to Europe was in 1714. Possibly influenced by the vegetative propagation used in Chile, Frézier brought plants rather than seeds and five survived the journey to France. Some (or all) of these could have been of the same clone. The plants were given to different gardens and it seems likely that only one or two of them formed the basis of cultivated *F. chiloensis* in Europe and later became the source of *F. ananassa*. All five plants were female and they mostly remained a sterile novelty, but fruits were obtained locally in Brittany. By 1740, growers had realized that satisfactory fruiting was obtained if *F. moschata* and/or *F. virginiana* were grown near to female *F. chiloensis* plants. A system of planting in alternate rows developed and the strawberry of Plougastel, named after the village in which it was grown extensively by 1750, became popular in Paris and London. This Chilean clone was grown until the end of the nineteenth century but attempts to grow it elsewhere in France were unsuccessful and no other variants of *F. chiloensis* were available.

The two American species stimulated interest in new strawberries but the crossing that produced the first plants of the new cultivated strawberry, *F. ananassa*, was accidental. Neither the date nor the location of origin is certain but the documentation suggests that the first plants were obtained and selected in a botanic garden rather than by the farmers growing the successful variety of *F. chiloensis*. It seems likely that interest in finding a new strawberry was maintained where the female *F. chiloensis* fruited rarely and never produced large fruits like those reported from Chile.

The new Pineapple or Pine strawberry was first described by Miller in London in 1759; Miller obtained the plant from a garden near Amsterdam where it had first appeared about 1750. He considered it to be a variant of the Chilean strawberry but noted that it differed in several ways, including having hermaphrodite flowers and a large pink fruit with a flavour reminiscent of pineapple. Duchesne obtained plants from London and observed that they had characters of both *F. chiloensis* and *F. virginiana*. It is probably significant that Duchesne was aware of dioecy in *Fragaria* and of the successful cross-pollination of *F. chiloensis* in Brittany. He knew that the two species were grown together in gardens and so suggested, in 1766, that *F. ananassa* was a hybrid of *F. chiloensis* and *F. virginiana*; he then showed that the cross produced viable seeds and progeny.

There are reports that similar new strawberries were already being grown in other gardens about the same time; these may have originated separately from other crosses between one of the few clones of *F. chiloensis* and different forms of *F. virginiana*. Since *F. moschata* was probably growing in the same gardens, and *F. chiloensis* × *F. moschata* hybrids can be vigorous and partly fertile, it (*F. moschata*) could have been the pollen parent of some of the new hybrids. However, since Duchesne (who had obtained progeny from this cross) did not notice any characteristic of the musk strawberry in his *F. ananassa* material, it seems unlikely that *F. moschata* was a parent of these first plants of *F. ananassa*. It has since been repeatedly confirmed that *F. chiloensis* and *F. virginiana* cross readily and that some of the progeny are vigorous, fertile and similar to *F. ananassa*. This origin of a new crop species from a cross between two species that were already cultivated separately, and without change in chromosome number, is unusual among crop plants.

4 Recent history

Early introductions of *F. chiloensis* and *F. virginiana* must have been highly heterozygous and several varieties of the Pineapple strawberry (i.e. *F. ananassa*) were selected before 1800. These were grown along with varieties of *F. virginiana* and their success stimulated further breeding especially in England and, rather later, in France and America. Many of the early selections were from progenies of *F. ananassa*, but the collections of *F. virginiana*, the original Chilean *F.*

chiloensis and some wild *F. chiloensis* from California were also used. No special breeding methods were necessary, since any useful segregant could be maintained vegetatively.

The genetic base of early breeding programmes was narrow, although not as narrow as is sometimes thought, since some American breeders used the native octoploids, especially *F. virginiana*. An early stimulus to extend the genetic base was the occurrence of plants which fruited for a longer time, similar to the ever-bearing forms that were known in *F. vesca*. A perpetual flowering segregant was selected in France in 1866 and a number of ever-bearing varieties was produced. At about the same time, American breeders were impressed by the availability of fruit of *F. chiloensis* in California throughout most of the year. More recently, other collections of both progenitor species have been used in breeding; as a result, it has been realized that both species have extensive distributions and corresponding ranges of genetic variation.

Useful genetic variation in other characters has been obtained from the wild octoploids. For example, *F. ovalis* (*F. virginiana*) has been used in the breeding of American varieties which are resistant to drought and low temperatures and both *F. chiloensis* and *F. virginiana* have provided resistance to pests and diseases such as *Verticillium* wilt, red stele (*Phytophthora fragariae*) and aphids. In some cases it has been shown that such useful characters can be separated from unwanted characters by three or four backcrosses (Scott *et al.*, 1972). Several characteristics make *F. ananassa* suitable for this sort of interspecific transfer; the recent origin from two species and the occasional hybridization with each progenitor, the apparent lack of differentiation of the chromosome sets in the autopolyploids and the clonal reproduction of heterozygous genotypes. A recurrent problem in strawberry breeding from the early years onwards has been the occurrence of partial sterility. Sometimes it appears to be associated with a recurrence of dioecy but it may also be a consequence of interspecific hybrid origin.

5 Prospects

The main objectives in breeding are still yield, fruit characters and resistance to pests and diseases but there is increasing interest in suitability for new methods of propagation, cultivation, harvesting and processing. Characters such as high yield from first year plants, the restricted production of runners and uniformity of fruit size and quality are thus of increasing importance. Similarly, the extension of fruiting time (either by using remontant or ever-bearing varieties, or by growing plants in protected environments (such as glasshouses and polyethylene tunnels) generates problems of pollination and fruit set. The need for enhanced self-pollination or for flowers adapted to different pollinators is indicated.

It seems unlikely that there will be substantial changes in breeding methods in the near future, although several possibilities are being investigated. For example, the production of seed-propagated F_1 hybrids would have economic attractions but would present formidable difficulties. The production of inbred parental lines, whether by repeated selfing or from polyhaploid tetraploids, would certainly not be easy. Controlled crossing would require effective gametocides or a return to dioecy. Seed propagation would facilitate control of viruses (although established methods of meristem propagation are effective enough). Another possibility lies in the use of apomictic seeds induced by alien pollination.

There is still much unused genetic variation in the natural octoploids, both in the wild species and in *F. ananassa*. Some genetic variation has probably been lost from the cultivated strawberries during the last hundred years, presumably because the requirements of commercial production necessitated other priorities. For example, American varieties in 1870 were described as having flavours comparable with those of apple, apricot, cherry, grape, mulberry, raspberry and pineapple. Even allowing for some exaggeration, this indicates a range of variability which is now simply not being used. This (and similar variation in other characters) is more likely to be available now in the wild octoploids than in *F. ananassa* itself.

The genetic base of strawberry breeding is being further extended by the use of other species. The variation in these is mostly unknown but preliminary evidence suggests that useful variation is present in, for example, diploid *F. nilgerrensis* which occurs throughout the humid sub-tropics of Asia. Hexaploid *F. moschata* has been used as pollen parent for the production of new varieties of *F. ananassa* but cytological details have not been reported. Diploid *F. vesca* has been combined with *F. ananassa* in decaploids ($10x = 70$) which are fertile but not yet exploited. Other higher polyploids have been produced but present evidence suggests that octoploidy is optimal. A series of synthetic octoploids of diverse origin has been produced and crossed with natural octoploids; many of the hybrids are sufficiently fertile to be used in breeding (Jones, 1966). Since most species combinations can be obtained, either directly or indirectly, it now seems probable that any useful genetic variation in the genus can be made available

for breeding; so the entire genus potentially becomes the genetic base.

Possibly the acutest economic need is for characters that facilitate mechanical harvest. There are reports that at least 60 per cent of the crop can already be collected from some varieties by one particular machine and improvements in both variety and machinery should surely be possible. Some of the essential characters, such as uniformity of ripening and firmness of fruit, are available in *Fragaria* species but the taller, stronger, erect inflorescences that would make harvesting less damaging to the plants are not available in the genus. These characters (and others) are present in related genera, the most likely source being the closely related *Potentilla* ($x = 7$, $2n = 2x - 16x$).

Many *Fragaria* × *Potentilla* crosses stimulate seed development, but viable hybrids have been obtained with only four *Potentilla* species. However, since much of the work has been on a small scale, without the use of advanced techniques for overcoming crossing barriers, and since few species of *Potentilla* have been tried, the potential may be considerable. All the hybrids are sterile, but two amphidiploids are fertile, *F. ananassa* × *P. palustris* ($14x = 98$) and *F. chiloensis* × *P. glandulosa* ($10x = 70$) (Ellis, 1962; Asker, 1971; Bringhurst and Barrientos, 1973). These cross with octoploid strawberries and such *Potentilla* crosses may supply useful characters. However, since all the hybrids and their progeny have the *Fragaria* habit, the present prospects for constructing new species with radically different morphology do not seem good.

6 References

Anderson, W. (1969). *The strawberry: a world bibliography.* Scarecrow Press, New Jersey, USA.

Asker, S. (1971). Some viewpoints on *Fragaria* × *Potentilla* intergeneric hybridisation. *Hereditas*, **67**, 181–90.

Bringhurst, R. S. and **Barrientos, F.** (1973). Fertile *Fragaria chilovensis* × *Potentilla glandulosa* amphiploids. *Genetics*, **74**, 530.

Bringhurst, R. S. and **Senanayake, Y. D. A.** (1966). The evolutionary significance of natural *Fragaria chiloensis* × *F. vesca* hybrids resulting from unreduced gametes. *Amer. J. Bot.*, **53**, 1000–6.

Darrow, G. M. (1966). *The strawberry. History, breeding and physiology.* New York.

Ellis, J. R. (1962). *Fragaria–Potentilla* intergeneric hybridisation and evolution in *Fragaria*. *Proc. Linn. Soc. Lond.*, **173**, 99–106.

Jones, J. K. (1966). Evolution and breeding potential in strawberries. *Sci. Hort.*, **18**, 121–30.

Scott, D. H., Draper, A. D. and **Greeley, L. W.** (1972). Interspecific hybridisation in octoploid strawberries. *Hort. Sci.*, **7**, 382–4.

Staudt, G. (1961). Die Entstehung und Geschichte der grossfruchtigen Gartenerdbeeren, *Fragaria × ananassa. Züchter*, **31**, 212–18.

Staudt, G. (1973). *Fragaria iturupensis*, eine neue Erdbeerart aus Ostasien. *Willdenowia*, **7**, 101–4.

Wilhelm, S. (1974). The garden strawberry: a study of its origin. *Amer. Sci.*, **62**, 264–71.

Wilhelm, S. and Sagen, J. (1975). *History of the strawberry from ancient gardens to modern markets.* (In press.)

Cherry, plum, peach, apricot and almond

Prunus spp. (Rosaceae)

Ray Watkins
East Malling Research Station
Maidstone Kent England

1 Introduction

Cherries and plums (after apples and pears) are two of the most important fruit crops grown in the cooler temperate regions of the world. The similarly important peaches and apricots overlap the warmer portion of the plum and cherry growing areas but, like the cherries and plums, also require adequate winter chilling. Almonds, an increasingly important nut crop, are grown in regions with a Mediterranean climate but even so they also must have at least a short period of winter chilling.

The world crop is estimated to be $5\cdot5$ Mt for peaches, 4 Mt for plums, $2\cdot5$ Mt for cherries, $1\cdot25$ Mt for apricots and $0\cdot75$ Mt for almonds. All are major trading commodities either fresh or, to a steadily increasing extent, processed (Childers, 1973).

2 Cytotaxonomic background

The genus *Prunus* is part of the sub-family Prunoideae of the family Rosaceae and all have a basic chromosome number of $x = 8$. Rehder (1954) has classified *Prunus* into 77 species. The sub-genera, sections and major crop plants are shown, with chromosome numbers (Knight, 1969), in the following classification:

Subgen. PRUNOPHORA
Sect. Euprunus (6 spp.; $2x - 6x$; Europe to China)
  P. domestica, European plum ($6x$); *P. insititia*, Damson plum ($6x$); *P. cerasifera*, Cherry plum ($2x, 3x, 4x, 6x$); *P. salicina*, Japanese plum ($2x, 4x$).
Sect. Prunocerasus (13 spp; $2x$; North America).
  P. americana, American plum ($2x$).
Sect. Armeniaca (6 spp.; $2x, 3x$; Europe to Korea).
  P. armeniaca, Apricot ($2x$).
Subgen. AMYGDALUS
Sect. Euamygdalus (6 spp.; $2x, 8x$; western Asia to China).
  P. persica, Peach ($2x$); *P. amygdalus*, Almond ($2x$).

Sect. Chamaeamygdalus (1 sp.; 2*x*, 3*x*; southeastern Europe to eastern Siberia).
Subgen. CERASUS
Sect. Microcerasus (8 spp.; 2*x*; north temperate).
Sect. Pseudocerasus (14 spp.; 2*x*, 3*x*, 4*x*; central Asia to Japan).
Sect. Lobopetalum (2 spp.; 4*x*; China).
Sect. Eucerasus (3 spp.; 2*x*, 3*x*, 4*x*, 5*x*; Europe to central Asia).
   P. avium, Sweet cherry (2*x*); *P. cerasus*, Sour cherry (4*x*).
Sect. Mahaleb (3 spp.; 2*x*; Europe, western Asia, North America).
Sect. Phyllocerasus (1 sp.; China).
Sect. Phyllomahaleb (1 sp.; 2*x*; Manchuria, Korea, Japan).
Subgen. PADUS (11 sp.; 4*x*; Europe, Asia, North America).
Subgen. LAUROCERASUS (2 spp.; 8*x*, *ca* 22*x*; Europe to western Asia).

The most frequently achieved hybridization of *Prunus* crop species (Kester, 1969; Knight, 1969; Rehder, 1954; Zylka, 1970, 1971) with other species has involved the subgenera and sections listed in Table 71.1.

Table 71.1 The subgenera and sections most frequently involved in interspecific hybridization with *Prunus* crop species

Crop plant	Interspecific hybridization	
	Major importance	Minor importance
European plum (*P. domestica*)	PRUNOPHORA Euprunus Prunocerasus	CERASUS Microcerasus AMYGDALUS Euamygdalus
Damson plum (*P. insititia*)	PRUNOPHORA Euprunus	
Cherry plum (*P. cerasifera*) and Japanese plum (*P. salicina*)	PRUNOPHORA Euprunus Prunocerasus Armeniaca	CERASUS Microcerasus AMYGDALUS Euamygdalus
American plum (*P. americana*)	PRUNOPHORA Euprunus Prunocerasus	CERASUS Microcerasus AMYGDALUS
Apricot (*P. armeniaca*)	PRUNOPHORA Euprunus Prunocerasus Armeniaca	CERASUS Microcerasus AMYGDALUS Euamygdalus
Peach (*P. persica*)	AMYGDALUS Euamygdalus Chamaeamygdalus PRUNOPHORA Euprunus	CERASUS Microcerasus PRUNOPHORA Prunocerasus Armeniaca
Almond (*P. amygdalus*)	AMYGDALUS Euamygdalus Chamaeamygdalus	PRUNOPHORA Euprunus Armeniaca
Sweet Cherry (*P. avium*) and Sour cherry (*P. cerasus*)	CERASUS Microcerasus Pseudocerasus Eucerasus	CERASUS Lobopetalum Mahaleb

For purposes of genetic transfer AMYGDALUS and PRUNOPHORA form one group but, within this group, almonds and damson plums are somewhat isolated from the mainstream of genetic transfer, although this may merely reflect less attention by fruit breeders rather than intrinsic isolation. Transfer between the AMYGDALUS–PRUNOPHORA group and the sweet and sour cherry group is only rarely direct, being usually via the Microcerasus section of CERASUS. The AMYGDALUS part of the group seems to be more loosely connected to the Microcerasus bridge than the PRUNOPHORA portion of the group.

The subgenus PADUS is only very weakly linked to other *Prunus* (via CERASUS and AMYGDALUS) if the number of hybrids between PADUS and other species is considered as a guide. Using this criterion, the subgenus LAUROCERASUS appears to be even more isolated from other subgenera.

3 Early history (Fig. 71.1)

It seems probable that the first diploid *Prunus* species arose in central Asia, and that the Eucerasus section, containing the sweet and sour cherries, were early derivatives of such an ancestral *Prunus* species. The present day CERASUS species in the section Microcerasus are probably closer to this ancestral *Prunus* species than the commercial cherries in the Eucerasus section, since the Microcerasus form a genetic bridge for hybridization purposes between the Eucerasus and the AMYGDALUS and PRUNOPHORA subgenera.

The North American cherry species *P. besseyi* and *P. pumila* must have been derived from the central Asian *Prunus* centre of origin relatively recently since they also possess, to a high degree, the Microcerasus bridging role. The Chinese *P. tomentosa*, also an important present-day element of the Microcerasus

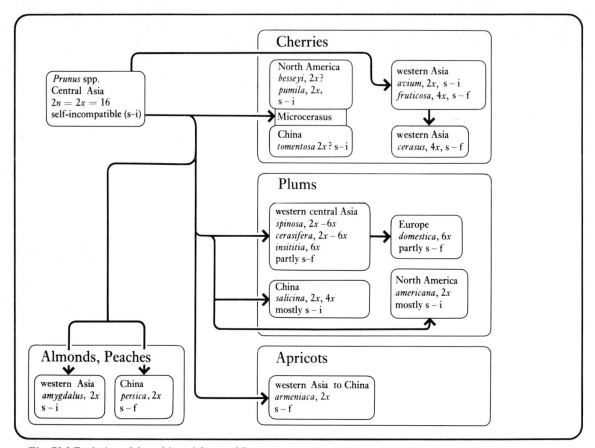

Fig. 71.1 Evolution of the cultivated forms of *Prunus*.

bridge, is probably a branch from the link between the North American Microcerasus and the *Prunus* centre of origin.

The tetraploid sour cherry, *P. cerasus*, probably originated either from diploid *P. avium* or from *P. avium* and tetraploid *P. fruticosa*. If *P. fruticosa* was involved in the ancestry of *P. cerasus*, then the sour cherry must have evolved before the spread from western Asia of both the sweet and sour cherry to their secondary centre of origin in Europe, since the centre of origin of *P. fruticosa* remains in western and central Asia.

Genetically speaking, the plums appear to hold the centre of the *Prunus* stage, since they have the greatest genetic diversity of any subgenus and are a link between the other major subgenera. The centres of origin for plums include: Europe for *P. domestica* (European plum); western Asia for *P. insititia* (Damson plum); western and central Asia for *P. cerasifera* (Cherry plum); China for *P. salicina* (Japanese plum); and North America for *P. americana* (American plum).

Selection by man in early historic times in the various centres of origin and also in the many areas to which they were distributed in early and more recent historic times has ensured the retention of more genetic diversity than if selection had been practised at only one centre.

Prunus americana and the other plum species of the section Prunocerasus, all with a North American centre of origin, appear to be closely related to the Asian and European plums of the section Euprunus. Indeed, the separation of plums into Prunocerasus and Euprunus seems to have a geographical rather than a genetical basis. It seems likely that Prunocerasus plums were established in North America relatively recently (possibly at the same time as the wild cherries *P. besseyi* and *P. pumila*) since it is difficult to imagine that they could have moved eastwards from central Asia to North America at a very early stage and still remain relatively unchanged. If Prunocerasus and Euprunus are part of a single section, then *P. salicina* of China may be a branch from the link between the North American plums and the plums from central Asia (in the same way as the Chinese *P. tomentosa* is associated with the link for the wild Microcerasus cherries).

Prunus domestica, the European plum, is the most recent crop species to have arisen within PRUNO-PHORA. *Prunus instititia*, *P. cerasifera* and *P. spinosa* were probably involved in its ancestry.

The apricot (*P. armeniaca*), unlike the plums which are also part of the subgenus PRUNOPHORA, is normally self-pollinated (like the peach) and, also like the peach, has a primary centre of origin in western China. Unlike the peach, however, the apricot has a secondary centre of origin in western Asia. *Prunus armeniaca* appears to be further from the centre of the genus *Prunus* than the plums. The factors which contributed to selection for self-fertility must have also thereby contributed indirectly to isolating the species.

Support for the hypothesis that the peach (*P. persica*) originated in western China comes from a consideration of the present distribution of wild species and a study of early Chinese, Indian and Fertile Crescent writings which show that it was known and cultivated in China before it was known further west.

The almond (*P. amygdalus*), which is closely related to the peach, became established in a separate centre of origin in an area extending from central to western Asia. The peach and the apricot both developed self-fertilizing breeding systems in the harsh environment of western China while the almond, in the slightly more moderate environment of central and western Asia, remained, like most diploid *Prunus* species, self-incompatible (Vavilov, 1951).

The ancestral species from which both peach and almond arose probably split at an early stage somewhere in central Asia, with the peach evolving to the east in western China and the almond evolving and moving westward while still maintaining a foothold in the ancestral homeland of central Asia.

The peach, with its Chinese centre of origin close to where most of the present-day Microcerasus species occur (except those which evolved further eastwards in North America), appears to be closer to the genetic centre of *Prunus*, and hence closer to the Microcerasus bridge to the cherry group, than the almond, which has a centre of origin to the west of all the Microcerasus species (except the relatively uncommon *P. incana* and the related *P. prostrata*).

4 Recent history

In early historic times all fruit crops were grown from seed; subsequently, especially for the most variable crops and particularly in areas with the most advanced agriculture, the best selections were propagated vegetatively on seedling rootstocks, usually of the same or closely related species. Quite recently there has been a trend towards selection and breeding of uniform vegetatively propagated rootstocks as a replacement for variable seedling types (Tydeman, 1962).

Theophrastus, in his history of plants (300 B.C.), mentioned the cherry and it is probable that it was domesticated several centuries before this time in the region of Asia Minor or Greece.

The cultivated forms of the cherry spread throughout the Roman Empire to include the area already covered by the wild forms of *P. avium* and *P. cerasus*. However, there was a decline in interest in cherries after the fall of the Roman Empire and, as late as 1491, an important German book, the *Herbarius*, listed only two types of cherry – Sweet and Sour – even though the Romans had recognized numerous types many centuries earlier. Following the decline during the Middle Ages, cherries were again extensively planted in Europe, especially in Germany, from the sixteenth century onwards.

Cherries were taken to North America by the early settlers from Europe and spread with them across the continent. The sweet cherries were more favoured in the west and the sour cherries in the east.

Most sweet cherry cultivars are self-incompatible and therefore care is needed in the selection of pollinators. In contrast, sour cherries are usually self-compatible and so a single variety can usually be safely planted in a solid stand.

The cherry reproduces fairly true from seed, particularly the tetraploid sour cherry. This is partly due to the natural isolation of Eucerasus cherries from other CERASUS species. They developed to the west of the central Asian CERASUS centre of origin whilst most other CERASUS species evolved to the east. With so many cherry varieties so similar to each other, it is common to find cultivars grouped by type rather than by individual name: Morello-type, Napoleon-type, Montmorency-type, Duke-type (hybrids between sweet and sour cherry).

Breeding and selection in cherries has been largely confined to commercial *P. avium* or *P. cerasus* varieties, particularly the former, and, consequently, cherries have remained until very recently (Zylka, 1970) more isolated from the *Prunus* gene pool than other *Prunus* fruit crop species.

The five main groups of plum evolved in three distinctly different areas of the world until about 200 years ago; since then they have been used increasingly as a common gene pool in breeding. All five plum groups, and many wild plum species, can be readily hybridized and together form an outstanding pool of genetic variability. In contrast to cherries, many wild species have been successfully used in breeding programmes, especially in North America. The dominant commercial group is the hexaploid European plum, *P. domestica*, but it has gained very significant contributions from both wild and cultivated diploid species.

European breeding programmes have mostly involved crosses between *P. domestica* varieties and are aimed at improving fruit characters while North American breeding has been more concerned to add useful adaptation-to-environment characters to *P. salicina* and *P. domestica* from a range of species.

Apricots, which were probably first cultivated in western Asia only a few centuries B.C., were introduced into Europe during the Roman era. The production of late-flowering types (to avoid the hazards of spring frosts) and of types adapted to the environments of new areas of production have been major breeding objectives. Firmer flesh, higher fruit quality, disease resistance and hardiness have also been important objectives.

The apricot hybridizes with the plum and so-called 'plumcot' varieties have been selected. Yellow fruited types which resemble the purple apricot (*P. dasycarpa*) can be produced by crossing apricot with the Myrobalan plum (*P. cerasifera*).

The peach spread from China to Europe during the last 2,000 years and was subsequently taken to North America and other parts of the world, such as South Africa and Australia, where it is now of major importance. Breeding is aimed at producing better quality fruit, although selection for adaptation to new environments has been especially important in North America, where the short life of the trees is of considerable concern.

Almonds (*P. amygdalus*) spread at an early stage from Asia to the Mediterranean and subsequently, with European settlement, to North America and also became important there. There has been a very considerable increase in production in North America, especially in California, during this century. In Mediterranean and Asian countries such as Turkey, almonds are almost exclusively grown from seed whereas in North America they are mostly propagated vegetatively. Breeding, in this self-incompatible crop, is mainly concerned with kernel quality although selection for late-flowering, self-fertility and early harvest are also of major importance (Kester, 1969).

5 Prospects

The most significant current trend in *Prunus* breeding is related to studies of the inheritance of economic characters (of both varieties and wild species) and the subsequent transfer of such characters to new commercial varieties. An integral part of this trend will be the increased use of wild species, particularly as contributors to the breeding of scion varieties where adaptation to new environments (including ones with new or greater risks from pests and diseases) is important. In addition, wild *Prunus* species will provide valuable characters needed in the breeding of new vegetatively propagated rootstocks. Clonal rootstocks (including triploids and pentaploids) will be especially valuable for those crops and in those areas in which dependence at the moment is on very variable seed-propagated rootstocks. Many species will be tested and little known types such as *P. jacquimontii* may conceivably provide valuable rootstock parents for both the cherry and plum sides of the Microcerasus bridge.

The cultivated plums and related wild species (and to a lesser extent the apricot and its relatives) of PRUNOPHORA appear to be in a particularly favourable position for the exchange of genetic material both within the subgenus and also with AMYGDALUS and CERASUS and are likely to be increasingly used for this purpose.

In contrast, commercial cherries are the most isolated group. If the potential for future improvement of the cherry crop is to be raised to a satisfactory level, it will be necessary to intensify the effort devoted to hybridizing with species outside the Eucerasus section of CERASUS (Zylka, 1970). The prospects for producing new rootstocks (including triploids as well as diploids and tetraploids) from such crosses are very good. Less well known species such as *P. rufa*, *P. cerasoides* (related to *P. campanula*), *P. mugus* (related to *P. incisa*), *P. dawykiensis* (possibly *P. canescens* × *P. dielsiana*), *P. schmittii* (*P. canescens* × *P. avium*) and *P. wadei* (*P. pseudocerasus* × *P. subhirtella*) may prove valuable, in addition to species which have already been widely tested as rootstocks. The well-known Japanese ornamental cherries (*P. serrulata*) are a possible source of resistance to the aphid, *Myzus cerasi*.

It is unlikely that the subgenus LAUROCERASUS will have much to contribute in the immediate future to the production of better varieties. The possibilities for using the subgenus PADUS in the near future are only slightly better and it is unlikely that they will contribute very much, except in the breeding of new rootstocks, where species such as *P. nepalensis* and *P.*

cornuta have shown some promise for use with cherry (Singh and Gupta, 1971).

Looking somewhat further into the future, the potential contribution of wild *Prunus* species to the production of new varieties, suited to an even greater range of environments than those available to our present *Prunus* fruit crops, is considerable, providing there is adequate long-term planning. To be successful such planning must be initiated before the advancing tide of 'modern' agriculture destroys the remaining islands of natural genetic variability. In the attempts of developing countries to mimic the advances made in developed countries by replacing indigenous types and wild stands with western varieties, there is a real danger that a vital portion of the natural genetic variability will be permanently lost, to the detriment of future genetic advance. The pending switch from seed to vegetatively propagated almonds in Turkey, part of the centre of origin for *P. amygdalus*, is an example of such a hazard.

Cell fusion techniques (followed where necessary by chromosome doubling) hold some promise for reducing the number of steps necessary to effect transfers between more distantly related species within the AMYGDALUS–PRUNOPHORA–CERASUS portion of *Prunus*. Using these and related techniques, it will probably become practicable to consider using PADUS, LAUROCERASUS and other Rosaceous genera, and even families outside of the Rosaceae, in the creation of new fruit crops or for making dramatic changes to our present varieties.

6 References

Childers, N. F. (1973). *Modern fruit science*. Rutgers University, New Jersey, 5th edn.

Kester, D. E. (1969). In R. A. Jaynes (ed.), *Handbook of North American nut trees*. Geneva, New York, 302–14.

Knight, R. L. (1969). Abstract bibliography of fruit breeding and genetics to 1965; *Prunus. Tech. Commun. Commonw. Bur. Hort. Plantn Crops*, 31, pp. 649.

Rehder, A. (1954). *Manual of cultivated trees and shrubs*. New York, 2nd edn.

Singh, R. N. and Gupta, P. N. (1972). Rootstock problem in stone fruits and potentialities of wild species of *Prunus* found in India. *Punjab hort. J.*, 157–75.

Tydeman, H. M. (1962). Rootstocks. In *Handbuch der Pflanzenzüchtung*, 6, Berlin, 2nd ed., 547–72.

Vavilov, N. I. (1951). The origin, variation, immunity and breeding of cultivated plants. *Chronica bot.*, 13, pp. 364.

Zylka, D. (1970). *Die Verwendung von Wildarten der Gattung Prunus in der Sortenzüchtung und als Unterlage*. Wiesbaden.

Zylka, D. (1971). Die Verwendung von wilden Kirscharten in der Sortenzüchtung und als Unterlage. *Gartenbauwissenschaft*, 36, 261–91, 417–44, 557–72.

Apple and Pear

Malus and *Pyrus* spp. (Rosaceae)

Ray Watkins
East Malling Research Station
Maidstone Kent England

1 Introduction

Apples and pears are the most important fruit crops of the cooler temperate regions. The world crop is estimated to be 20 Mt for apple and 6 Mt for pear.

There is a considerable international trade in apples and pears both within major regions such as Europe and also between continents in both hemispheres. For example, there is a very significant sale of fruit from South Africa, Australia and New Zealand to Europe, particularly to the United Kingdom.

During the last 20 years a very significant increase in apple and, to a lesser extent, pear production has occurred in those portions of the mountainous regions of countries such as India where the necessary degree of winter chilling (mostly at elevations above 1,000 m) is combined with satisfactory light and temperature conditions during the summer (mostly at elevations below 3,000 m).

2 Cytotaxonomic background

Malus and *Pyrus* genera belong to the sub-family Pomoideae of the family Rosaceae and all have a basic chromosome number of $x = 17$. Rehder (1954) has classified the genus *Malus* into 25 species (distributed in five sections) and the genus *Pyrus* into 15 species (in a single section).

Most plants in most species in both genera are diploid, although tetraploids and/or triploids have been frequently, or even usually, found in several species. The *Malus* species in the section CHLOROMELES (all indigenous in North America) are the major exception being mostly either triploid or tetraploid (Wilcox, 1962), although *M. ioensis*, possibly the stem species of the section, is diploid. Despite the fact that all species which have contributed to cultivated varieties are diploid, a relatively high proportion of scion varieties, especially of apple, are triploid. Selection by man has been strongly in favour of plants which resulted from the fertilization of unreduced ovules by normal haploid pollen.

Cytologically and genetically, $2n = 2x = 34$ plants usually behave as normal diploids. Nevertheless, over the years, there have been a number of suggestions, often conflicting (Knight, 1963), regarding the possible origin of *Malus* and *Pyrus* by secondary polyploidy from species with lower basic chromosome numbers. However, the origin from more primitive types was so remote in time that clarification of the issue is likely to be extremely difficult.

Hybridization between the two genera is difficult and derivatives rarely survive. There is no evidence to suggest that recent hybridization between the genera has contributed to the evolution of cultivated varieties of either crop. The two genera are sufficiently distinct to suggest that they arose separately from more primitive species. Intergeneric hybrids between other genera in the Pomoideae (*Sorbus* × *Malus*, *Sorbus* × *Pyrus* and *Cydonia* × *Pyrus*) appear to be more readily achieved than *Pyrus* × *Malus* hybrids (Knight, 1963). Even the relatively rare genus *Docynia* from the Himalayan region is probably more closely related to *Malus* than *Pyrus* is to *Malus*.

Hybridization between most wild species within each of the two genera occurs readily. The resulting hybrids have been of value as ornamental plants, as rootstocks and, to an increasing extent in recent years, as contributors of disease and pest resistance in fruit breeding programmes.

It has been traditional to classify the cultivated apple as *M. pumila* (formerly *M. communis* or *Pyrus malus*), although some workers prefer to split this species into *M. domestica* (the cultivated apple), *M. pumila* (including var. *paradisiaca*, the paradise apple) and *M. sylvestris* (the wild crab apple). However, since the apple is self-incompatible and can be hybridized readily with most other *Malus* species it is probable that most European and Asian species have contributed to the evolution of presentday cultivars.

The cultivated pear is frequently classified as *P. communis* but, like the apple, is self-incompatible and crosses readily with related species. There have been many recent contributions from other species (see Knight, 1963) and probably very many more unrecorded and it is therefore an oversimplification to consider cultivated pears to be merely clones of *P. communis* (see also the following section).

3 Early history (Fig. 72.1)

It is generally accepted that the primary centre of origin of both *Malus* and *Pyrus* is within the region which includes Asia Minor, the Caucasus, Soviet Central Asia, and Himalayan India and Pakistan (Wilcox, 1962); however, some authors (cited by Knight, 1963) claim a more easterly origin, even as far east as western China in the case of *Pyrus*. The secondary centres of origin of the individual species within the two genera are either within the primary centre of origin (e.g. *M. pumila*, *M. trilobata*) or to the east (e.g. *M. baccata*, *M. ioensis*, *P. ussuriensis*) or to the west (e.g. *M. florentina*, *P. longipes*, *P. nivalis*). No species in either genus has a natural distribution which includes areas both east and west of the primary centre. The development of species has been primarily related to physical isolation in a range of environments. Nearly always the isolation has been so recent that effective hybridization between species within both genera is still relatively easily achieved.

Evreinoff (see Knight, 1963) considers that *M. sylvestris*, and *M. pumila*, were the major ancestral species of modern apples but with *M. spectabilis*, *M. prattii*, *M. baccata*, *M. prunifolia* and *M. astracanica* as significant additional contributors. He considers *P. communis*, *P. nivalis* and *P. serotina* to be major contributors to the cultivated pear, with *P. syriaca*, *P. ussuriensis* and *P. longipes* significant minor donors. Other authors have implicated several additional wild forms in the origin of modern cultivars (Knight, 1963).

Unlike *Malus* (with section CHLOROMELES in North America), *Pyrus* did not spread to the New World until after its introduction by European settlers.

Apples and pears have existed in Europe from prehistoric times and have been cultivated in both Europe and western Asia since at least earliest historical times. Cato noted several varieties of apple and the Elder Pliny listed 22, while both Homer and Pliny recorded the names of many pears.

Early records indicate that Pearmain and Costard type apples were important in England by the thirteenth century. Pippin types probably became important in England later, following introductions from continental Europe in the sixteenth century.

4 Recent history

The Abbé d'Hardenpont and Van Mons bred pears on a large scale in Belgium during the seventeenth century and to a lesser, but nevertheless successful, extent they were also bred in France and England during the same period. Apple and pear cultivars and seedlings derived from them were taken to North America by the early settlers from Europe. At this time apples and pears were usually propagated in Europe by grafting. In North America, by contrast, initial distribution was mostly by seed. As a result,

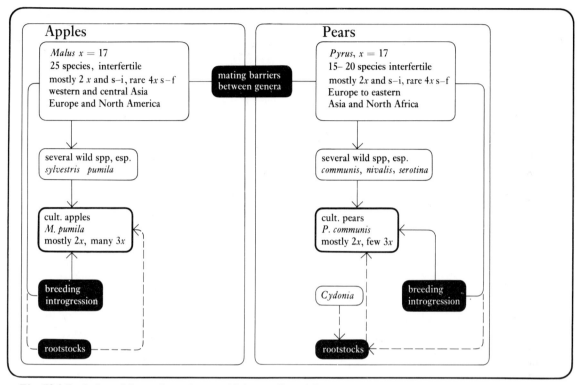

Fig. 72.1 Evolution of the apples and pears, *Malus pumila* and *Pyrus communis*.

the genetic diversity accumulated in North American orchards soon became considerably greater than in their European counterparts.

Subsequent selection in North America gave rise to a range of local varieties distinct in many respects from those grown in Europe. Even now, few European apple varieties are grown in North America, although several American varieties (Golden Delicious, Jonathan and Delicious) have become important in Europe in recent years. With pears the trend has been the reverse, with European pears widely grown in North America but with no American pears well established in Europe.

North American apple and pear breeders initiated the use of wild species in the latter half of the nineteenth century, mainly to provide better winter hardiness and also fireblight resistance in pear. The immediate effect of using wild species was to reduce fruit quality but this problem has largely been overcome by several generations of backcrossing.

Taking the world as a whole, varieties originating in North America or Europe are the important commercial types. Thus, for example the very promising new commercial apple and pear industry of Himachal Pradesh, in northern India, is based almost exclu-

sively on such varieties.

All important apple and pear varieties are grafted on special rootstocks since they cannot be readily vegetatively propagated on their own roots. East Malling Research Station has been a leader in the breeding of dwarfing precocious, high yielding, clonally propagated rootstocks for apple and pear. These rootstocks (especially the range of apple rootstocks) are widely used in modern orchards throughout the temperate fruit-growing areas of the world. Similar research programmes have been initiated in other countries. Despite the great success of clonal rootstocks, many growers, including a high proportion in North America, still depend on invigorating but very variable seed propagated rootstocks.

Many *Malus* species have been involved in breeding rootstocks. An example is *Malus robusta* 5 (*M. baccata* × *M. prunifolia*) which is widely used in the colder parts of Canada. For pears, wild species have also been used as rootstocks because of their winter hardiness and adaptation to a range of environments. Unlike those of apples, however, the pear species rootstocks are normally seed-propagated. Such rootstocks are widely used in North America, whereas clonal *Cydonia* selections have been the most widely used rootstocks for pear in Europe. Nowhere in the world have vegetatively propagated *Pyrus* rootstocks been successfully used on a larger scale, because they root only with difficulty (except during the juvenile period); however, some recent South African selections have been showing promise.

5 Prospects

The current trend is towards the use of a very few major varieties throughout the world, with an associated loss of the genetic diversity which was present 100 years ago when most important apple and pear growing countries grew hundreds or even thousands of local varieties, with only a sprinkling of foreign types. The loss of each local type reduces the scope for further evolution by breeding. Fruit collections such as the National Fruit Trials in Kent, England, will, if properly supported, play an increasingly important role in helping to retain for future use the remainder of such genetic diversity.

Many of the major varieties are closely related. Brown (1975) has drawn attention to some of the limitations inherent in attempting to breed from related apple cultivars. There is a prospect that the production of improved apple cultivars will first stagnate and then ultimately decline (with the advent of new hazards or new commercial requirements) if apple and pear breeding is restricted to closely related plants.

The evolution of improved cultivars will also be very seriously limited if the present trend towards the elimination of stands of wild species continues without steps being taken to preserve, for future utilization, adequate samples of representative types. Recent work at East Malling and at other research stations has shown the advantages to be gained by using wild *Malus* species in breeding programmes (Watkins, 1974*a*, *b*, 1975).

Future trends in breeding will increasingly emphasize fruit quality factors such as appearance, flavour, storage, shelf life and handling potential. Yield will also be of great importance in the immediate future although, in the long run, it will also need to be associated with suitability for mechanical harvesting. Environmental considerations will also increase the premium attached to breeding for levels of pest and disease resistance adequate to obviate chemical sprays. Evolution, through controlled breeding, to keep pace with these requirements will be dependent on the availability of a comprehensive gene pool (including both varieties and well chosen representatives of species) and on having a clear understanding of the genetic factors controlling economically important

characters (Alston, 1971–75; Alston and Watkins, 1975; Watkins, 1971). Possibilities also exist to develop varieties which are completely novel in other respects such as method of propagation and tree shape (Watkins and Alston, 1973).

In the future, techniques may be developed which would make it possible to produce triploid apples and pears from diploid × diploid crosses by producing unreduced gametes readily, rather than by chance and very infrequently, as at present. Given this potential, it would be possible to add a haploid set of chromosomes from a wild species (or an early generation hybrid) to a complete diploid set of chromosomes of a commercial variety, thus shortening breeding time. Such a technique would have the added advantages of making it possible to combine major genes and polygenes without the almost insuperable problem associated with the transfer of both types of genes in a lengthy backcrossing programme (Watkins, 1975).

Judging by the most advanced breeding programmes and the steady increase in the number of species being utilized, it seems likely that almost all of the *Malus* and *Pyrus* species will be involved in the evolution of future varieties and/or rootstocks.

6 References

Alston, F. H. (1971–75). Integration of major characters in breeding commercial apples. Early stages of pear breeding at East Malling. *Proc: Eucarpia Top Fruit Breed. Symp., Angers, 1970, Canterbury, 1973.*

Alston, F. H. and **Watkins, R.** (1975). Apple breeding at East Malling. *Proc. Eucarpia Top Fruit Breed. Symp., Canterbury, 1973.*

Brown, A. G. (1975). The effect of inbreeding on vigour and length of juvenile period in apples. *Proc. Eucarpia Top Fruit Breed. Symp., Canterbury, 1973.*

Knight, R. L. (1963). Abstract bibliography of fruit breeding and genetics to 1960: *Malus* and *Pyrus. Tech. Commun. Commonw. Bur. Hort. Plantn Crops*, **29**, pp. 535.

Rehder, A. (1954). *Manual of cultivated trees and shrubs.* New York, 2nd edn.

Watkins, R. (1971–75). Fruit breeding methodology: major gene and quantitative genetics. Rootstock breeding at East Malling. *Proc. Eucarpia Top Fruit Breed. Symp., Angers, 1970. Canterbury, 1973.*

Watkins, R. (1974*a*). Fruit breeding. *Rep. E. Malling Res. Stn, 1973*, 121–30.

Watkins, R. (1974*b*). Tree fruit breeding techniques at East Malling. *Proc. 2nd General Congr. Soc. Advancement Breed. Res. Asia Oceania, Delhi,*

Watkins, R. and **Alston, F. H.** (1973). Breeding apple cultivars for the future. In *Fruit, present and future*, **2**, 65–74, Royal Horticultural Society, London.

Wilcox, A. N. (1962). The apple: I Systematics. *Handbuch der Pflanzenzüchtung*, **6**, Berlin, 2nd ed., 637–45.

Raspberries and blackberries

Rubus (Rosaceae)

D. L. Jennings
Scottish Horticultural Research Institute
Dundee Scotland

1 Introduction

Raspberries and blackberries have been favourite garden fruits in Europe and North America for several centuries and they have now become important commercial crops, supplying some 100,000 tons of fruit annually for jam making, canning, freezing, yoghurt and flavourings. Commercial red raspberry production is particularly concentrated in eastern Scotland, eastern Europe and western North America, notably Oregon, Washington and British Columbia, while black raspberry production is limited to America. Commercial blackberry production has been neglected in Europe but forms a major part of the fruit-growing industry of Oregon, USA.

2 Cytotaxonomic background

Three sections of the genus *Rubus* contain cultivated fruits. They are: Idaeobatus (including raspberries) in which the ripe fruits separate from the receptacle; Eubatus (including blackberries) in which the drupelets of the ripe fruits adhere to the receptacle and both detach from the calyx; and Cylactis, including the herbaceous north-circumpolar or alpine species. A fourth group, Anoplobatus (including the flowering raspberries), has also been used in breeding. The basic chromosome number is 7.

Cultivated raspberries originate from three diploid species or subspecies: *Rubus idaeus* or *R. idaeus vulgatus*, the red raspberry of northern Europe and Asia; *R. strigosus* or *R. idaeus strigosus*, the red raspberry of North America, and *R. occidentalis* the black raspberry, also native to North America, but generally with a more southerly distribution than *R. strigosus*. Wild forms of both *R. idaeus* and *R. strigosus* are typical outbreeders. Fertilization is controlled by a multi-allelic oppositional type of gametophytic in-

compatibility system (Keep, 1968) and they are therefore highly heterozygous and show considerable inbreeding depression. Domestication involved growing in clonal monoculture and therefore demanded the selection of self-compatible forms, but the unilateral interspecific incompatibility shown in crosses with *R. occidentalis* suggests that only the pollen has changed while the style has retained its activity.

Three groups of blackberries have been domesticated (see Clausen *et al.*, 1945). First, the European blackberries extend over the whole of Europe and much of Asia. They include a very large number of polyploid forms, mostly tetraploid, which can conveniently be regarded as a single population characterized by morphological diversity and ecological differentiation. The only surviving primary diploids are *R. tomentosus* and *R. ulmifolius*. Second, there are the eastern American erect blackberries and trailing dewberries which include a large number of polyploid forms and also five primary diploids: *R. allegheniensis*, *R. argutus*, *R. setosus*, *R. cuneifolius* and *R. trivialis*. Third, the western American trailing blackberries are geographically separated by the prairies and rockies and have much higher chromosome numbers, 56 and 84 being the most common. Their origin is least understood, because there is no trace of lower ploidy levels, but the area shows a considerable diversity of some of the other *Rubus* sections and it seems likely that the early progenitors of this blackberry group combined genomes from several of these. They probably have the most complex origin of all the blackberries, since grade of polyploidy is a significant index of evolutionary activity.

Most of the diploid blackberries studied have proved self-incompatible; polyploidy seems always to have led to loss of this system, though development subsequent to polyploidy has differed in the several groups. Species of the European and eastern American groups used to be regarded as either facultative or obligate apomicts, but recent evidence casts doubts on claims that apomixis occurs through apospory or dispory. It turns out that European blackberries have a versatile reproductive system in which all the gametes are formed by meiosis; some egg cells may be fertilized and give sexual offspring but others are 'diploidized'. In *R. caesius* this occurs by fusion of two haploid nuclei produced by division of the haploid egg nucleus; there are several unusual features of this process, which begins only after a pollen tube has entered a synergid nucleus (Gerlach, 1965). Variations, with or without fertilization, give a wide range of chromosome types whose survival is largely determined by the pollen used, high-ploidy pollen favouring the development of non-sexual and high-ploidy genotypes. Allopolyploidy is the rule in this group, and, though the system described should lead to homozygosity of the chromosomes within each of the constituent complexes, differences persist between the complexes and variation is released slowly by occasional association between them. The breeding system clearly represents a successful restriction of full sexuality and is not the evolutionary cul-de-sac associated with apomixis. In contrast, the western American blackberries responded to polyploidy and the resultant loss of incompatibility by establishing dioecy and maintaining a predominantly sexual breeding system.

3 Early history

In raspberries, domestication involved the selection of wild forms with fewer, stronger canes, stronger fruiting branches and larger fruits. It preceded that of blackberries, particularly in Europe. An early reference to them in Turner's Herbal of 1548 notes that 'they growe in certayne gardines in Englande', but distinct cultivars were not recorded until late in the eighteenth century. About this time cultivars of European origin were first grown in America, where their superior fruits won them preference over the indigenous kinds, which did not appear in cultivation until later. Cultivated forms of black raspberry appeared about 1830 but were preceded by their natural

Fig. 73.1 Evolution of the cultivated forms of *Rubus*.

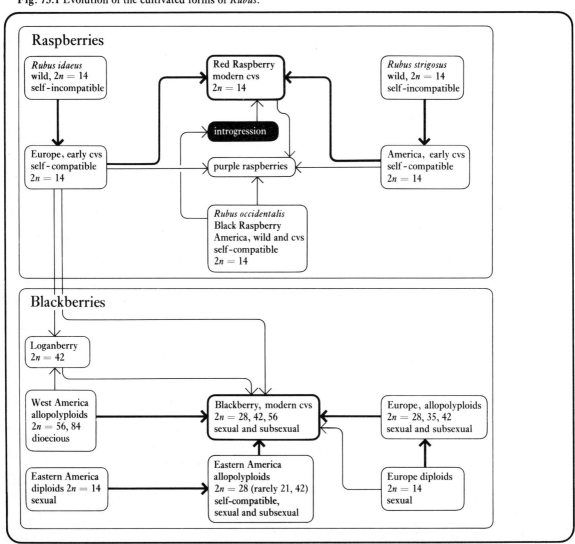

hybrids with the European red raspberry, known as purples or purple-cane hybrids. A major event in the evolution of the raspberry occurred with the advent of natural hybrids between the European and American red raspberries. As well as showing hybrid vigour, these combined the fruit qualities of European kinds with the hardiness and heat and drought tolerance of the American ones. Two famous ones were Cuthbert discovered in a New York garden about 1865, and Preussen discovered in a German garden about 1915. These later figured prominently in controlled breeding.

Selection for large-fruited variants also led to the selection of autotetraploid raspberries. They were all autumn-fruiting kinds and mostly originated in southern areas, particularly in France. Their selection clearly depended on the occurrence of high temperatures which maximize fruit-set in these sub-fertile genotypes, and their success was short-lived. However, autotetraploidy has played a role in the evolution of the Siberian raspberry *R. sachalinensis*, which is probably an autopolyploid derivative of *R. strigosus*.

It seems certain that, when man first cleared the forests, he both initiated an expansion of blackberry populations and also greatly accelerated evolutionary change. Clearance allowed previously isolated species to come together and gave opportunity for extreme hybridization; successful new combinations became fertile through polyploidy, were spread widely in the droppings of birds and were then maintained within narrow limits by their unusual reproductive system. The eastern American blackberry flora was probably mostly diploid until the time of settlement but a much longer time scale has operated in Europe. The culture of blackberries was late to start in Europe, possibly because of the abundance there of hedgerow fruit, but, began in the USA between 1840 and 1860, 18 cultivars being listed by 1867. Among these was the Evergreen or Cutleaf blackberry (*R. laciniatus*), which was introduced from central Europe about 1850; Himalaya Giant (*R. procerus*) was added from southern Europe about 1880. These two escaped from cultivation, rapidly became naturalized in western America and hybridized with the indigenous flora. They are both still highly successful commercial blackberries in Europe and America; indeed, a thornless mutant of *R. laciniatus* found wild about 1930 has now become America's highest yielding cultivar.

The western blackberries were popular with early settlers; the dioecious habit precluded domestication though their natural hybrids with other forms were successful. By far the most successful of these was the hexaploid Loganberry which Judge Logan discovered in his Californian garden about 1881; its origin has since been attributed to crossing between the octoploid Californian blackberry (*R. vitifolius*) and an unreduced gamete of the European raspberry.

4 Recent history (Darrow, 1966; Jennings *et al*, 1973)

The early and recent phases of evolution of raspberries and blackberries are not readily separated. Improvements discovered in nature and others produced by controlled breeding overlap in time and both are recent. Indeed, the Evergreen blackberry (*R. laciniatus*) and the Loganberry are examples of chance discoveries made nearly a century ago and not yet bettered by controlled breeding. In general, what began by accident was pursued and improved upon by intent. Further crosses between red and black raspberries were made and considerable crossing was done between the European and American forms of red raspberry. Later, rapid degeneration of successful cultivars through virus infection prompted breeding for resistance to the aphid vectors of these diseases and effective major-gene controlled resistances to both the main European and American aphids were found. Another advance was the development of raspberries adapted to the southern states of the USA by using Idaeobatus species like *R. parvifolius*, *R. biflorus* and *R. kuntzeanus* to contribute genes for resistance to leaf diseases and tolerance of heat and drought.

Breeding with the western American blackberries (Waldo, 1968) has attempted to combine their excellent flavour with a perfect-flowered and more productive growth habit. Both natural hybrids and hybrids obtained from controlled crosses involving the introduced forms of *R. laciniatus*, *R. procerus* and Loganberry have been used to achieve this. Spinelessness has been a major objective of all blackberry breeders, but nearly all of the many spine-free mutants found are chimeral and confined to the outer layer (I) of the meristem. These are inaccessible for breeding, but the diploid *R. rusticanus* var. *inermis* (a spine-free form of *R. ulmifolius*) has been a valuable source; when it was crossed in England with tetraploid *R. thyrsiger*, an unreduced gamete functioned and successful spine-free cultivars from this source have been developed in the USA. Another requirement, especially for northern Europe, is more rapid ripening to give earliness. The western American blackberries have this quality, but the only early European cultivar is Bedford Giant, which is a blackberry–raspberry hybrid and probably owes its earliness to raspberry germplasm.

5 Prospects

Natural evolution in the wild has shown the wide range of germplasm combinations that are possible in *Rubus* and breeders have found that most gene transfers can be achieved without specialized techniques. Current breeding therefore aims to bring together the very considerable range of available germplasm. In the short-term, the objectives are to breed raspberries and blackberries which meet the specialized requirements of processing and are capable of being machine-harvested. In raspberries this means combining the fruit size of *R. idaeus*, the firm fruit texture of *R. occidentalis*, the bright fruit appearance of *R. strigosus* and the disease and pest resistance of several species. In blackberries it means combining the erect growth habit of *R. argutus*, the flavour and early ripening of the western American blackberries and the spineless-ness of *R. rusticanus*. Other objectives include transferring genes for specific yield components (for example, floriferous laterals from *R. cockburnianus*) or resistance to specific pests and diseases (for example, resistance to beetle from *R. phoenicolasius* and *R. coreanus*). It may also be possible to reduce maintenance costs by restructuring the raspberry into a self-supporting plant which preferably bears all its crop on its first year's canes (Knight and Keep, 1966). This will involve combining raspberry dwarfing genes with genes for stout branched canes from *R. crataegifolius*; also assembling various sources of genes for autumn fruiting found both in European and American raspberries and in the Anoplobatus and Cylactis sections of the genus. Improvement of the Cyclactis group itself is also being attempted. Wild forms of *R. arcticus* and *R. stellatus* (the arctic raspberries) and *R. chamaemorus* (the Cloudberry) are popular in Scandinavia, especially for making liquors. All three are being domesticated and other improvements are being sought by crossing them with *R. idaeus* (Larsson, 1969).

6 References

Clausen, J., Keck, D. D. and Hiesey, W. M. (1945). Experimental studies on the nature of species. II. Plant evolution through amphiploidy and autoploidy with examples from the Madiinae. *Carnegie Inst. Wash. Pub.*, **564**.

Darrow, G. M. (1967). The cultivated raspberry and blackberry in North America – breeding and improvement. *Amer. hort. Mag.*, **46**, 203–18.

Gerlach, D. (1965). [Fertilisation and autogamy in *Rubus caesius*]. *Biologisches Zentralblatt*, **84**, 611–33.

Jennings, D. L., Gooding, H. J. and Anderson, M. M. (1973). Recent developments in soft fruit breeding. In *Fruit, present and future*, London (Royal Horticultural Society) **2**, 121–35.

Keep, Elizabeth (1968). Incompatibility in *Rubus* with special reference to *R. idaeus* L. *Can. J. Genet. Cytol.*, **10**, 253–62.

Keep, Elizabeth (1972). Variability in the wild raspberry. *New Phytol.*, **71**, 915–24.

Knight, R. L. and Keep, Elizabeth (1966). Breeding new soft fruits. In *Fruit, present and future*, London (Royal Horticultural Society), **2**, 98–111.

Larsson, E. G. K. (1969). Experimental taxonomy as a base for breeding northern *Rubi. Hereditas*, **63**, 283–351.

Waldo, G. F. (1968). Blackberry breeding involving native Pacific coast parentage. *Fruit Var. Hort. Dig.*, **22**, 3–7.

Quinine

Cinchona spp (Rubiaceae)

A. M. van Harten
Department of Plant Breeding
Agricultural University Wageningen
The Netherlands

1 Introduction

The bark of several species of the genus *Cinchona* contains quinine, an alkaloid noted for its anti-malarial action. The discovery of this characteristic of *Cinchona* plants, indigenous to the Andes, is hidden in the past. According to Cárdenas (1969), quinine, the common name of the plant and the product, was first introduced to Europe around 1640.

Initially, the whole world supply came from bark collected from wild trees in the Andes. Because of the apparent danger of extinction of quinine in its centre of origin, Blume proposed in 1829 to establish plantations elsewhere, which action in fact the Dutch and British undertook in their colonies in the Far East after 1850. Production in the then Dutch East Indies increased to such extent that, at the outbreak of the Second World War, more than 90 per cent of the world supply of quinine, equalling more than 10 million kg of bark, came from there.

Following the wartime discovery of synthetic anti-malarial drugs, the role of natural quinine diminished considerably, but there have been periods of revival, notably during later wars. At present there is still a rather important demand for cinchona bark.

It is noteworthy that some other alkaloids in the bark (such as quinidine) also have interesting pharmaceutical properties. The importance of these is increasing; for example, in treating infections by *Plasmodium vivax* (especially in the case of races resistant to synthetic drugs) and in treating certain heart diseases.

2 Cytotaxonomic background

The taxonomic situation within the genus *Cinchona* is extremely confusing. Most references are old (e.g. the extensive work of Weddell in 1842) and no recent revision of the genus, covering the whole area in which *Cinchona* occurs, is available. Some work, however, has been done on Peruvian (Hodge, 1944, 1950) and Bolivian (Stanley, 1931) species. Over the years, hundreds of specific names have been proposed and published, but many of them are undoubtedly synonyms; and some must refer to hybrids, which are numerous, as a result of easy natural hybridization. Differences in opinion between 'splitting' and 'lumping' taxonomists have only served to increase the confusion.

Cinchona is a relatively small genus in the large family Rubiaceae. The latter contains about 500 genera with 6,000 to 7,000 species, all tropical. Lanjouw *et al.* (1968) report that the genus *Cinchona* has 15 species, all South American, and that the economically most important species are *C. officinalis* and *C. pubescens* (syn. *C. succirubra*). Purseglove (1968) mentions *C. calisaya*, *C. ledgeriana* (probably a variety of *C. calisaya* or a hybrid), *C. officinalis* and *C. succirubra* (probably belonging to *C. pubescens*). Other authors consider *C. calisaya* as belonging to *C. officinalis*.

All the species so far investigated are diploids with $2n = 2x = 34$. They are small trees or shrubs indigenous to the rain forests on the eastern Andean slopes, on both sides of the equator, ranging in altitude from a few hundred metres above sea level to more than 3,000 m. The distance between the most northern and most southern habitats is more than 1,750 km. According to Purseglove (1968), *C. calisaya* and *C. ledgeriana* occur in southern Peru and Bolivia at 1,200–1,700 m; *C. officinalis* is found from Colombia to northern Peru at higher altitudes and *C. succirubra*, a very hardy species, within a wide range of altitudes from Costa Rica to Bolivia.

3 Early history

It is not certain whether the Indians of the Andes were acquainted with the use of quinine before the arrival of Europeans. Nor is it known exactly how the first introductions (*C. succirubra* and *C. officinalis*) to Europe around 1640 took place. The French botanist De la Condamine was, in 1738, the first European to see quinine plants alive in their natural stands in Ecuador. He also published the first scientific description (Cárdenas, 1969). In 1742 Linnaeus established the genus *Cinchona* and, in 1753, described *C. officinalis* in the *Species Plantarum*.

Originally, the selection of superior bark was done empirically on properties such as bark colour and texture. From 1820, chemical analyses were performed, which indicated that calisaya bark generally contained the highest content of quinine. Weddell was the first to obtain seeds (in Bolivia in 1840). His *calisaya* material was brought by way of Paris and London to Africa and the Malayan archipelago, but the introduction failed. In 1852, Hasskarl collected in Peru for the Dutch government. After many initial problems, the introduction to Java was successful but the bark quality proved to be only average and the content of alkaloids relatively low. In 1859 the British organized an expedition headed by Markham. Part of his material and material from later collectors, such as Pritchett and Cross, led to the establishment of plantations in India and Ceylon.

The most important collection however was made by Ledger and his assistant Miguel, who both lived in Puno, Peru, and collected (illegally) in Bolivia. After offering his material to the British, who were not interested, Ledger sold part of it to the Dutch in 1865. This material, later called *C. ledgeriana*, but very probably either a variety of *C. calisaya* or closely related to it, proved to contain an alkaloid content much superior to that of all other accessions known until then. In comparison with the previous 1–7 per cent of quinine, 11 years after the *ledgeriana* introduction in Java, 52 trees with 13–25 per cent of quinine were available. The trade in quinine quickly became more and more concentrated outside South America, notwithstanding laws forbidding exportation of plants or seeds from this region. Countries like Peru and Bolivia, where no plantations had been established, despite suggestions to do so (e.g. by Markham in 1863), were thus unable to benefit from their natural wealth. In fact, the Dutch obtained a position of virtual monopoly which lasted until the Second World War.

From an evolutionary point of view, nothing of significance happened during these early phases. Like rubber, the crop consisted of a diversity of wild plants brought directly into cultivation.

4 Recent history

Up to now, detailed information on *Cinchona* breeding has been available only from work in the previous Dutch East Indies (summarized by van Harten, 1969).

After some initial work with other species, attention was concentrated in the East Indies on: *C. ledgeriana*, because of its high quinine content; and *C. succirubra*, for its excellent characters as a rootstock.

Four to seven years are needed before a tree flowers and produces seeds. Heterostylism is common in *Cinchona*. Cross-pollination between the microstylic and macrostylic flowers occurs mainly via insects. Self-sterility and protandry are also common phenomena favouring cross-pollination. Large-scale artificial pollination is possible (Ebes, 1949). Thus the crop is substantially outbred.

Breeding efforts in the Dutch East Indies by government stations were concentrated for a long time on the selection of trees with high contents of quinine. This work was very successful but, on the other hand, the production of trees with improved bole shape, thicker bark and other useful agronomic characters was not undertaken in sufficient measure. Private planters however were working more on such aspects and gradually an agreement was reached. The breeding scheme adapted was analogous to that developed in rubber and based on the outbreeding habit and potential for vegetative propagation (on *succirubra* seedling stocks). It may be summarized thus: seedlings; clonal propagation of selected mother trees in seed orchards; improved seedlings; cycle repeated.

This system led to considerable improvement within a few decades. As a general criticism, it should be mentioned that insufficient attention was paid to testing phenotypic influences and to the mutual influence of rootstock and scion. Moreover, no sound scientific basis was established for the generally accepted assumption that a positive correlation should exist between the properties of parents and their seedling progeny. Too much attention was paid to the vegetative descendants of good mother-trees. Ebes (1949, 1951) was the first to apply the more modern views. It may be interesting to note that breeding work in the Dutch East Indies led, unintentionally, to the selection of types especially suitable for poor soils.

The exodus of the Dutch from the Far East more or less coincided with a general lack of interest in *Cinchona* as a result of the collapsed world market after the Second World War. The plantations were neglected and breeding research came practically to a standstill. When world demand went up again, Java (now Indonesia) was unable to regain its previous position. More modern plantations had been started during the war in other parts of the world (e.g. in Middle and South America) and quickly extended their activities. In addition to Indonesia, India and Sri Lanka (Ceylon), plantations are reported to exist in or introductions made to Guatemala, Costa Rica, Mexico, Jamaica, Réunion, Martinique, Puerto Rico, Brazil, Bolivia, Peru, Ivory Coast, Nigeria, Congo, Uganda, Tanzania, Kenya, Madagascar, Hawaii and Australia.

5 Prospects

The future of the crop is uncertain but it will surely never reach the same importance as in the past. However, there still seems to be a rather considerable market; some new applications of medical value have been found and more may be discovered in the future. From this point of view, it seems important to study further, and to conserve, the wild stands of *Cinchona* in South America and to continue breeding work on types with the desired medicinal and agronomic characters. Were serious breeding work to be undertaken again it is to be noted that the genetic base of the crop is very narrow and most of the wild species yet unexploited, indeed largely unknown as to breeding potential.

6 References

Cardenas, M. (1969). *Manual de plantas economicas de Bolivia.* Cochabamba, Bolivia.
Ebes, K. (1949). Artificial pollination of cinchona. *Report West Java Experiment Station* (unpubl.).
Ebes, K. (1951). Improvement of stemshape by breeding. *Ned. Boschb. Tijdschr.*, **22**, 1–7.
Harten, A. M. Van (1969). Cinchona (*Cinchona* spp.). In F. P. Ferwerda and F. Wit (eds.), *Outlines of perennial crop breeding in the tropics.* Misc. Papers, 4, Landbouwhogeschool Wageningen, 111–28.
Hodge, W. H. (1944). Sinopsis de la Cinchonas en el Peru. *Bol. Mus. Hist. Nat. 'Javier Prado', Peru.*
Hodge, W. H. (1948). Notes on Peruvian Cinchonas 1. *Bot. Mus. Lflt. Harvard Univ., Cambridge*, U.S.A.
Lanjouw, J., *et al.* (1968). *Compendium van de Pteridophyta en Spermatophyta.* Utrecht, The Netherlands.
Purseglove, J. W. (1968). *Tropical Crops 2, Dicotyledons.* London.
Stanley, C. (1931). *The Rubiaceae of Bolivia.* Chicago, USA.

Coffees
Coffea spp. (Rubiaceae)

F. P. Ferwerda
75 Bennekomse Weg Wageningen
The Netherlands; *formerly* Institute of Plant Breeding Wageningen The Netherlands

1 Introduction

Among the non-alcoholic stimulating beverages coffee ranks first. This widely appreciated mild stimulant is obtained from the roasted beans (seeds) of, mainly, two species of the genus *Coffea*: *Coffea arabica* and *C. canephora*. Two or three other less important species contribute only a small fraction of the total coffee crop. The economic species of coffee are shrubs or small trees up to 5–8 m high when unpruned. Coffee is mostly propagated from seed. Asexual propagation by means of grafting or by rooting softwood cuttings is also possible and practicable. Coffee species are tropical by origin and cultivations are all in warm countries wherever suitable conditions of climate and soil are found. *Arabica* prefers the relatively cool mountain slopes.

The major producers of *arabica* coffee are South and Central America, notably Brazil. The bulk of *canephora* (commonly called Robusta) coffee is produced in equatorial West and Central Africa. The world crop in 1971 amounted to approximately 5 Mt, of which, roughly, 80 per cent consisted of *arabica*.

2 Cytotaxonomic background

The genus *Coffea* is fairly large and comprises about 90 species of which only five or six are relevant to the evolution of the crop. For general systematics see Chevalier (1947). Africa is the home of most coffee species. Three of the four sections into which the genus is divided (Eucoffea, Argocoffea and Mascarocoffea) are native to Africa and Madagascar; the fourth, rather small, section (Paracoffea) is endemic in southeast Asia.

The basic chromosome number is $n = 11$. The most important species, *Coffea arabica*, is the only polylpoid yet known in the genus. It has $2n = 4x = 44$ and is self-fertile. The other members of the Eucoffea section, insofar as their chromosome numbers have been determined, are all diploids with $2n = 22$; those of which the breeding habits have been studied are pronouncedly self-incompatible. Compatibility relations are probably governed by an oppositional S-allele system.

Segregation of such genes as are known in *arabica* coffee is disomic. This fact, and the meiotic behaviour of some interspecific hybrids, strongly suggest alloploidy (Meyer, 1969; Kammacher and Capot, 1972). Its ancestry is still controversial; there are indications that *Coffea eugenioides* may be one of its ancestors (Carvalho *et al.*, 1969).

3 Early history (Fig. 75.1)

The evidence points to the essential early steps in *arabica* evolution having taken place in south western Ethiopia, the area that is still the primary centre of diversity of the crop. The oldest recorded use of coffee is that of chewing leaves or beans to dispel fatigue, hunger and pain. Human selection was presumably in favour of plants that had the strongest stimulating effect. In a later phase, chewing was replaced by the brewing of beverages. The habit of coffee drinking developed in Arabia in the fifteenth and sixteenth centuries and spread, in the course of the seventeenth century, via Turkey to western Europe and from there to the Western hemisphere.

In very early days, *arabica* seeds or plants must have been taken from Ethiopia to Yemen. The first recorded plantations of *arabica* coffee seem to have been established in Yemen in the fourteenth or fifteenth century. From there, small numbers of *arabica* plants were, in the seventeenth and eighteenth centuries, transported to the tropical countries which nowadays are known as the main areas of cultivation. The intricate migrations, as far as they can be retraced, have been summarized by Klinkowski (1947) and Ukers (1948), among others.

Arabica coffee first migrated eastward, from the Yemen to the Malabar Coast of India, thence to Ceylon and Indonesia, where it arrived in the last decade of the seventeenth century and gave rise to a prosperous cultivation. One *arabica* plant shipped to Amsterdam in 1706 was raised in the botanical garden of that city. Some of its seedling offspring migrated via the Jardin des Plantes in Paris to Martinique and thence to South America. Other *arabica* material, taken by the French from Yemen to Réunion (Bourbon) and from there to the Caribbean, ultimately also came to South America.

It is obvious that in all these introductions only small numbers of plants were involved. As a consequence, the *arabica* cultivations in southeast Asia and those in South and Central America rest on a very narrow genetical base. In recent years, attempts have

Fig. 75.1 Distribution of the cultivated coffees, *Coffea arabica* and *C. canephora*.

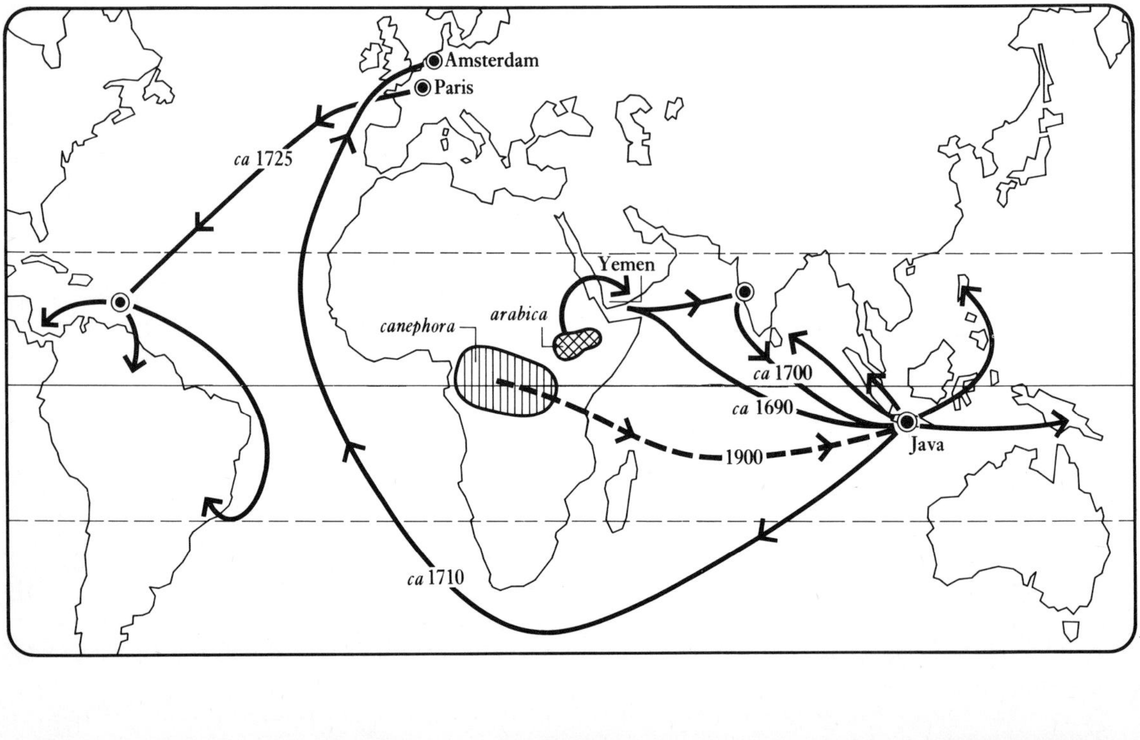

been made to remedy this situation by exploring the genetic variability still available in the primary centres of diversity and by preserving valuable types in gene banks.

The diploid coffees of which *Coffea canephora* is the principal economic representative, all have their primary centres of diversity in central and western equatorial Africa and in Madagascar.

Canephora coffee was found in a state of semi-cultivation when the first explorers reached central Africa and some crude selection may already have been made at that time. The species was introduced into southeast Asia, notably Indonesia, about the turn of the century. Its tolerance of *Hemileia vastatrix* made it a valuable substitute for *arabica* which had, in the last quarter of the nineteenth century, been virtually wiped out by coffee leaf rust in the lower altitudes of Java and Sumatra.

Initially, only small numbers of plants were introduced. The first *canephora* to reach Java came in 1900 from Zaïre (formerly the Belgian Congo); this consignment consisted of only 20 plants but it was subsequently supplemented by several consignments of seeds collected in various places.

On a smaller scale, seeds of economically less important species such as *Coffea liberica*, *C. dewevrei* var. *excelsa* and *C. congensis* were introduced into southeast Asia and into South America.

4 Recent history (Fig. 75.2)

Arabica breeding began in Brazil and India about 1925, in Tanzania in 1934 and in Kenya in 1944. The first stages of *canephora* breeding date back to 1908 in Java, 1934 in Zaïre and 1945 in the Ivory Coast. Breeding and selection in the autogamous tetraploid *arabica* and in the outbreeding diploid *canephora* have developed along different roads.

In *arabica*, the methods followed are those characteristic of autogamous crops. The main criteria of selection have been yielding capacity, size and shape of the bean and, in some instances, organoleptic properties. In Brazil, much attention has been given to genetical and cytological analysis, resulting in a wealth of fundamental data which will in the future prove increasingly useful.

Two devastating diseases have claimed a large share in the breeding effort of the past decades: *Hemileia vastatrix* the coffee leaf rust; and *Colletotrichum coffeanum*, the coffee berry disease that makes the berries mummify.

Hemileia, after its first appearance in Ceylon in 1869, has gradually invaded all the *arabica*-producing countries of the eastern hemisphere. Latin America remained free of the disease until, in 1970, *Hemileia* was found in southern Brazil and, soon afterwards, also in other parts of the vast Latin American *arabica* area. Coffee berry disease has become serious since

Fig. 75.2 Relationships of the cultivated coffees, *Coffea*.

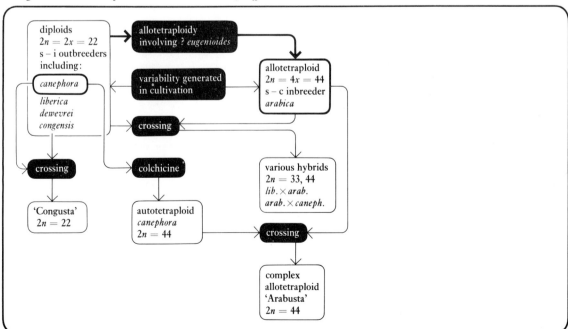

the 1950s, notably in the East African *arabica* areas.

In India and East Africa much effort has been spent on transferring *Hemileia* resistance from source material from Ethiopia and Timor into local strains. Types resistant to coffee berry disease have also been found and breeding programmes aiming at working up resistances into existing cultivars have been initiated, for example, in Kenya.

In the allogamous *canephora* coffee a different breeding plan has been developed. It consists of extensive crossing among large arrays of parents, followed by selection of outstanding F_1 families which can easily be recreated by means of bi-clonal or multi-clonal seed orchards. This method has worked well, resulting in yield increases up to 100 per cent in comparison with the initial material. Further progress is possible, provided a sufficiently broad genetic base can be maintained. Biometrical studies may show the way to some sophistication of mating patterns and selection procedures.

Clonal selection in *canephora* has received some attention in Indonesia and central Africa (Carvalho *et al.*, 1969; Dublin, 1967). Quicker results may be obtained thus than by breeding proper as any promising individual possessing resistance or excelling in organoleptic properties can be instantly fixed and propagated as a clone.

Interspecific hybridization received only limited attention before the late war and was then usually based on spontaneous hybrids. Only in the last 20–30 years has mutual crossability of various species been investigated systematically (Carvalho and Monaco, 1968) and hybridization projects on a cytogenetical basis have been carried out. A few of the old spontaneous hybrids (e.g. Congusta (*congensis* × *robusta*) and some *liberica–arabica* hybrids) have been introduced on a limited scale into commercial planting. Recently, amphidiploid Arabusta hybrids have been obtained by crossing *arabica* with colchicine-induced tetraploid *robusta* (= *canephora*) (Capot, 1972). They offer interesting prospects with regard to *Hemileia* (and possibly also *Colletotrichum*) resistance and improved organoleptic properties. As they do not breed true from seed, they must be propagated vegetatively.

5 Prospects

In future, the objectives of *arabica* breeding will be, to a large extent, dominated by disease resistances, notably to *Hemileia* and *Colletotrichum*. Considering the numerous physiological races in *Hemileia* (more than 25 are known) and numerous resistance genes (at least six) it seems an endless task to achieve vertical resistance. For field (horizontal) resistance, prospects seem better. This goal can probably be achieved in the next decade. The remarkable *Hemileia* resistance shown by some interspecific *arabica* hybrids deserves closer investigation; it might offer another approach to combating leaf rust.

In the *canephora* group, breeding will probably be continued along the lines followed thus far. Biometrical studies may possibly permit some sophistication of mating patterns and selection procedures.

As in other crop plants, coffee breeding is dependent on the conservation of genetic resources. A wealth of genetically highly diverse material is still available in the wild or under primitive agricultural conditions but it is already threatened by the cutting of indigenous forests and the expansion of pasture. There have been some recent collecting expeditions and there will no doubt be more.

International cooperation is indispensable. Collections have to be maintained as living museums, as prolonged seed storage is not feasible.

6 References

Capot, J. (1972). L'amélioration du café en Côte d'Ivoire. Les hybrides 'Arabusta'. *Café, Cacao, Thé*, 16, 3–18.

Carvalho, A. and Monaco, L. C. (1968). Relaciones genéticas de especies seleccionadas de *Coffea*. *Café (Lima)* 9, 1–19.

Carvalho, A. *et al* (1969). Coffee. In F. P. Ferwerda and F. Wit (eds). *Outlines of perennial crop breeding in the tropics. Misc. Paper* 4 Landbouwhogeschool, Wageningen, Netherlands, 189–215.

Chevalier, A. (1947). Les caféiers du globe III. *Encyclopédie Biologique*, (Paris), 27, pp. 356.

Dublin, P. (1967). L'Amélioration du caféier robusta en République Centrafricaine: dix années de sélection clonale. *Café, Cacao, Thé*, 11, 101–38.

Kammacher, P. and Capot, J. (1972). Sur les relations caryologiques entre *Coffea arabica* et *C. canephora*. *Café, Cacao, Thé*, 16, 289–94.

Klinkowsky, M. (1947). Die Wanderungswege des Kaffeebaumes. *Züchter*, 17/18, 247–55.

Meyer, F. G. (1969). The origin of arabica coffee. *Proc. XI intl. bot. Congr., Seattle, USA*, 146.

Monaco, L. C. and Carvalho, A. (1969). Coffee genetics and breeding in Brazil. *Span*, 12, 70–7.

Ukers, W. H. (1948). The romance of coffee. *Coffee and Tea Trade J.* (New York).

Citrus

Citrus (Rutaceae)

J. W. Cameron and R. K. Soost
University of California Riverside USA

1 Introduction

Citrus is grown throughout the world in tropical and subtropical climates where there are suitable soils and sufficient moisture to sustain the trees and not enough cold to kill them. The producing regions occupy a belt approximately 35° north and south of the equator. The main commercial areas are in the subtropical regions at latitudes more than 20° north or south of the equator. The total planted area of the 49 citrus producing countries is estimated at slightly over 1·6 Mha. The Mediterranean and North and Central America contain about 80 per cent of the commercial plantings and the remaining 20 per cent is distributed in the Far East (10%), South America (6%) and in other southern hemisphere countries (4%), including South Africa and Australia. Oranges constitute over 75 per cent of production, with lemons and grapefruit each accounting for approximately 10 per cent (Burke, 1967).

The genus *Citrus* has a great range of variability. Among the smallest fruits are the limes, which scarcely exceed 3 cm, while the pummelo may attain a diameter of 30 cm. Fruit rind colour ranges from the yellow-green of the limes to the red-orange of some mandarins, and shape varies from oblate to pyriform. At maturity, fruits of some cultivars are high in acid while others have almost none. All species of *Citrus* are evergreen, but the related genus *Poncirus* is deciduous. Altogether, there is much variability within the genus with which the breeder can work, and closely related genera provide an even wider range of characters.

2 Cytotaxonomic background

Citrus and its nearer relatives are represented by 28 genera in the tribe Citreae, of the orange subfamily Aurantioideae. Six of these genera, including *Citrus*, comprise the true citrus fruits, which are evergreen trees with highly specialized pulp vesicles in the fruit. The most closely related are *Citrus*, *Fortunella* (the kumquat) and *Poncirus*. *Poncirus*, a monotypic genus, has trifoliate, deciduous leaves and is extremely cold-resistant; it is an important rootstock. The remaining three genera are *Microcitrus*, *Eremocitrus* and *Clymenia*, the last being probably the most primitive, but imperfectly known.

The taxonomy of *Citrus* is confused. The system of Swingle (1967) established 16 species while, by contrast, Tanaka (1954) proposed 145 species and, later, 157. This lack of agreement reflects two familiar problems: what degree of difference justify species status and whether supposed hybrids among naturally occurring forms should be assigned species rank even though hybrids of known origin are often classed as cultigens. A compromise classification including 36 species has been proposed by Hodgson (1961). The *Citrus* forms of primary economic importance have many characters in common and are widely interfertile; their hybrids are also often fertile. Sterility and incompatibility occur but are not primarily determined by species. Asexual seed reproduction (nucellar embryony) is prominent but neither is it species-limited and its degree can easily be changed by hybridization. However, it has caused much confusion in understanding the nature of recurring genotypes after seed propagation.

The basic chromosome number is $x = 9$ in all *Citrus* species and related genera so far examined. These include at least 13 genera, and well over 70 forms within *Citrus* (Krug, 1943). The great majority are diploid; only a few polyploids have been identified in nature, including a tetraploid *Fortunella hindsii* and a tetraploid *Triphasia*, but many others have been produced experimentally.

3 Early history

The original distribution of the Aurantioideae was limited to the Old World. *Citrus* and its relatives arose in southeast Asia, the main centre apparently being eastern India. Relationships can be traced through the East Indies, Australia, central China, Japan and even Africa, to which four genera are endemic. Tanaka (1954) proposed a line of demarcation in southeast Asia which separated areas of probable development and spread of some *Citrus* species and relatives. This line runs south-eastwards from northern India, passing north of Burma and south of the island of Hainan. To the southwest is Tanaka's subgenus Archicitrus, within which the lemon, citron and orange probably developed. To the east and northeast is Metacitrus, which shows relationships to present-day mandarins, *Poncirus* and *Fortunella*.

No dates can be assigned to the domestication of

citrus in southeast Asia and no wild ancestors have been identified. The limited information available about the history of the principal horticultural forms is summarized in Table 76.1, together with data on reproductive behaviour and usefulness as rootstocks. It will be seen that one form (grapefruit) originated in the New World and that only two groups regularly reproduce sexually. The ease of hybridization among groups is indicated by the existence of many hybrids of horticultural importance.

Table 76.1 Principal horticultural groups of *Citrus*. (Data from: Webber, 1967; Swingle, 1967; Frost and Soost,1968).

Name*	Approximate dates of first written records		Type of embryo	Forms prominent as rootstocks
	Europe	New World		
1 Citron	300 B.C., Persia	—	Sexual	—
2 Lemon	12th–13th C.	1493, Haiti	Partly nucellar	Rough lemon (hybrid)
3 Lime	13th C.	—	Partly nucellar	Palestine sweet
4 Pummelo	12th C.	17th C., Barbados	Sexual	—
5 Sour orange	11th C.	—	Highly nucellar	Many
6 Sweet orange	15th C.	1493, Haiti	Highly nucellar	Many
7 Mandarin	1805	Mid 19th C., USA	Some cvs. nucellar, some sexual	Cleopatra
8 Grapefruit	—	18th C., West Indies	Highly nucellar	—

* Other group names, representing hybrids, include tangor = 6×7; tangelo = 7×8; orangelo = 6×8; citrange = 6×*Poncirus*.

4 Recent history

The first systematic citrus breeding was begun in 1893, by the US Department of Agriculture in Florida; a similar programme was begun at the University of California, Riverside, in 1914. Efforts by both groups have continued up to the present, and the USDA has extended its studies at other locations (Cameron and Frost, 1968). Other breeding studies have been carried out in Java, the Philippines, Italy, Japan and southern Russia, but most of them were seriously interrupted by the late war. In Japan, in particular, new research is presently underway (Nishiura, 1964). Citrus breeding requires long-term commitment and few institutions have given it sustained support. Breeding relationships are summarized in Fig. 76.1.

Citrus is capable of self-pollination since the male and female flower parts mature at the same time and the pollen is sticky rather than wind-blown. However, bees actively visit the flowers and, in mixed plantings, much cross pollination can take place; also self-incompatibility and inbreeding depression occur.

The interfertility of many forms, and even of *Citrus* with *Poncirus* and *Fortunella*, makes hybridization relatively easy unless other factors interfere (Frost and Soost, 1968). Genetic sterility is present in many oranges, such as Washington Navel, and in numerous cultivars of other species. In the lemon and lime there are high percentages of staminate flowers, which serve only as pollen sources. Intergeneric progenies vary greatly in vigour and fertility. Little is known about the behaviour of subsequent generations, although some backcross progenies have been obtained.

Self-incompatibility is present, determined by a series of oppositional alleles of poorly known distribution. Only in the pummelo are the alleles known to be widespread (Soost, 1964). For other than cultivated species, incompatibility has not been studied. Parthenocarpy occurs in some cultivars (e.g. Washington Navel orange and Satsuma mandarin) so that fruit set can sometimes occur without pollination.

Nucellar embryony has been a major obstacle to citrus breeding, although it is valuable for the reproduction of genetically uniform plants for rootstocks. Several common forms, including the orange, mandarin and grapefruit, include cultivars which bear embryos that are nearly all asexual, although related cultivars produce varying proportions of sexual ones. Two species, the pummelo and the citron, are apparently completely sexual. Inheritance of the character has proved relatively simple. Completely sexual parents have always produced offspring which are themselves completely sexual, while asexual × sexual crosses can produce both types. Progeny which are sexual have also been obtained when both parents were at least partially asexual. Many new sexual hybrids, useful for breeding, have recently been produced.

A long period of juvenility occurs in citrus seedlings, whether they arise sexually or asexually. Juvenility is evidenced by thorniness, upright growth and slowness to fruit; it is not easily modified by propagation on rootstocks.

Citrus is generally grown on rootstocks and breeding programmes for both rootstocks and scions have been carried on. For scion improvement, fruit size, flavour and season of ripening, as well as tree vigour, are critical. For fresh use, fruit shape and colour, number of seeds and storage life are also important. For juice concentrates, high solids and good colour and flavour are sought. Fruit size has often been a limiting character. Therefore the large-fruited pummelos have frequently been used as parents with, for example, mandarins and oranges. Hybrids of good size and flavour have been obtained

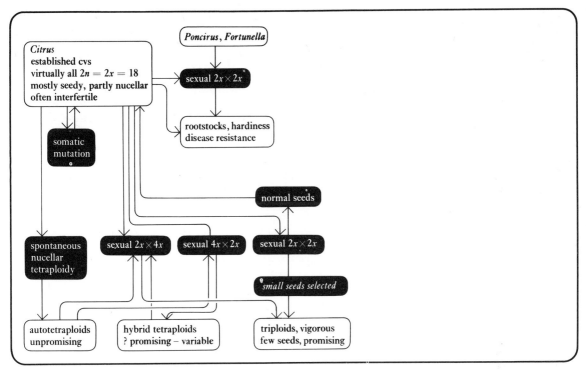

Fig. 76.1 Breeding behaviour of citrus fruits.

but high seed number is a persistent problem. Seediness is common in most diploid progeny and hybrids which are infertile often give low fruit set. Occasionally, parthenocarpy provides for good fruit set in a seedless selection.

Fruits with different (especially early or late) maturity are always of interest and a wide range of ripening periods can be obtained from parents ranging from early-ripening mandarins to late oranges such as Valencia. A low-acid pummelo has produced mild, early-ripening progeny, due to its effect on acidity. However, hybrids with a non-acid orange as one parent have not so far shown any such effect. High levels of soluble solids, with excellent flavours, have frequently been obtained among mandarin hybrids. Some *Citrus* selections tolerate higher maximum temperatures than others and certain hybrids sunburn or fail to set fruit in hot desert areas. Little is known about inheritance of resistance to pests. One study of California red scale showed that hybrids from susceptible parents such as pummelo and grapefruit were significantly more susceptible than hybrids from resistant mandarin parents. In some other studies, the sexual pummelo species has imparted marked vigour to its hybrids with other species.

Mutation is prominent in citrus, and can appear as limb sports, fruit sectors or whole plants from nucellar seeds. Several valuable cultivars, including the Washington Navel orange (parthenocarpic), the Marsh grapefruit and Shamouti orange, arose by mutation. Periclinal chimeras are rather common, one of the most interesting being the Thompson pink grapefruit.

During the last twenty years several new scion cultivars from controlled hybridization have been released in the United States. They are principally mandarins and mandarin hybrids. New cultivars of true oranges have been much more difficult to produce.

The first breeding work in Florida included crosses of *Citrus* with *Poncirus* to produce cold-hardy rootstocks. Several hardy hybrids were obtained. Additional crosses with *Poncirus* have since been made; in Japan the genus has been crossed with other cold-tolerant cultivars such as Yuzu and Natsudaidai. Crosses for cold-hardy scion types have been made, involving the Satsuma mandarin, Meyer lemon, *Fortunella* and, again, Natsudaidai. Evaluations are still underway.

Resistance to pests and diseases can depend on the scion, the rootstock or both. The most important diseases of citrus include *Phytophthora* root rot; citrus canker, the viruses of tristeza, psorosis and exocortis, and the mycoplasma causing stubborn disease. The citrus nematode and the burrowing nematode are also injurious in some areas. Since *Poncirus* is relatively resistant to *Phytophthora*, tristeza and the citrus nematode, it has often been used in crosses with *Citrus* for new rootstocks. Individual hybrids have shown high resistance to one or more of these maladies and often impart more growth vigour to their scions than does *Poncirus*. Tristeza is dependent on rootstock-scion interaction and the disease has at times destroyed large areas of citrus, principally trees on sour orange rootstock. Among hybrids which have provided tolerance is the widely used Troyer citrange (sweet orange × *Poncirus*) produced in 1909. In most breeding studies involving *Poncirus* as a male parent, use can be made of its trifoliate leaf as a marker, since this character is regularly dominant over the simple leaf of *Citrus*.

Hybrid rootstocks tolerant of the citrus nematode have been produced but tolerance is complicated by the occurrence of nematode biotypes. Breeding for tolerance to the burrowing nematode is underway. Stubborn disease is serious and wide-spread and injures many scion types regardless of rootstock; no breeding studies have been reported. A study on response to salinity among F_1s involving Rangpur lime and certain mandarins as parents was made by the USDA; a wide range of injury levels was observed.

Many spontaneous tetraploids in *Citrus* and *Poncirus* have occurred as nucellar seedlings, in frequencies as high as 6 per cent. Hybrid tetraploid seedlings seldom occur from crossing diploids. Tetraploids grow more slowly and are more compact and generally less fruitful than comparable diploids. Leaves are broader, thicker and darker. Fruits usually have thicker rinds, larger oil glands and less juice than diploids. Seed number depends on cultivars. Tetraploids have little commercial potential but they are valuable for the production of triploids.

Triploids frequently arise as sexual seedlings. They can be systematically produced by crossing tetraploids with diploids. However, production by crossing $4x$ (female) by $2x$ (male) is often not feasible for want of sexual $4x$ parents. The reciprocal cross ($2x \times 4x$) produces many tetraploid individuals (Esen and Soost, 1973). These occur because of the frequent production of diploid megagametophytes as well as the failure of most triploid embryos. The frequency of diploid megagametophytes is cultivar-dependent; of five cultivars tested, the highest percentage was 20. Failure of survival of triploid embryos is associated with failure of the endosperm and is related to the 3:4 embryo to endosperm ploidy ratio. The preferential survival of tetraploid seeds is associated with their 2:3 ploidy ratio. Their occurrence provides new hybrid tetraploids useful as seed parents.

The frequent occurrence of diploid megagametophytes also can be utilized for the production of triploids from $2x \times 2x$ crosses. Triploid seeds are much smaller than sister diploid seeds. Unlike triploids from $2x \times 4x$ crosses, their embryos and endosperms develop normally, except for size. By germinating the small seeds preferentially, recovery of triploids can be increased. Triploids are generally more vigorous than tetraploids. Fruitfulness has been highly variable: some triploids set well, while others are very unproductive. All have few or very few seeds. Triploids show promise for commercial use when selection for high yield is successful. Aneuploids have been weak and slow-growing.

5 Prospects

Advances in breeding and genetics by conventional methods will continue to be slow because of quantitative inheritance of most characters, long generation time and nucellar embryony. However, the range of strictly sexual parents has been increased and more can now be developed by breeding. Sexual tetraploids will greatly facilitate production of triploids. Several biochemical characters may have promise for separation of sexual and asexual seedlings (e.g. Pieringer et al., 1964).

Soil-borne pests and diseases, including diseases dependent on scion-rootstock interaction, will influence breeding efforts for rootstocks. Although partial resistance to several maladies exists, combining of resistance factors is needed. Continued screening for better sources of resistance is also essential. The genetic base of the major cultivars of important species is narrow and probably will remain so. Wide genetic diversity from other species and genera can only be introduced over a long period.

In vitro culture promises to contribute to genetic advance through embryo culture (Rangan et al., 1968), somatic hybridization, mutant cell selection, haploid cell culture and DNA incorporation. Sexual embryos that would not survive to maturity have been recovered by embryo culture. Somatic hybridization might eliminate the juvenile period. A shorter life cycle would permit a major increase in rate of genetic and evolutionary change. Diploidization of haploid cells would produce homozygous lines. Selection among cells in culture for specific mutants could also accelerate genetic advance.

6 References

Burke, J. H. (1967). The commercial citrus regions of the world. In Reuther, Webber, and Batchelor (eds.), *The citrus industry*, Berkeley, 1, 40–189 (revised).

Cameron, J. W. and Frost, H. B. (1968). Genetics, breeding, and nucellar embryony. In Reuther, Batchelor, and Webber (eds.), *The citrus industry*, Berkeley, 2, 325–71 (revised).

Esen, A. and Soost, R. K. (1973). Seed development in *Citrus* with special reference to $2x \times 4x$ crosses. *Amer. J. Bot.*, 60, 448–62.

Frost, H. B. and Soost, R. K. (1968). Seed reproduction: development of gametes and embryos. In Reuther, Batchelor, and Webber (eds), *The citrus industry*, Berkeley, 2, 290–324 (revised).

Hodgson, R. W. (1961). Taxonomy and nomenclature in citrus. In Price (ed.), *Intnl Org. Citrus Virologists, Proc. 2nd Conf.*, Gainesville, Florida, 1–7.

Krug, C. A. (1943). Chromosome numbers in the subfamily Aurantioideae with special reference to the genus *Citrus*. *Bot. Gaz.*, 48, 602–11.

Nishiura, M. (1964). Citrus breeding and bud selection in Japan. *Proc. Florida State hort. Soc.*, 77, 79–83.

Pieringer, A. P., Edwards, G. H. and Wolford, R. W. (1964). The identification of citrus species and varieties by instrumental analysis of citrus leaf oils. *Proc. Amer. Soc. hort. Sci.*, 84, 204–12.

Rangan, T. S., Murashige, T. and Bitters, W. P. (1968). *In vitro* initiation of nucellar embryos in monoembryonic *Citrus. Hort. Sci.*, 3, 226–7.

Soost, R. K. (1964). Self-incompatibility in *Citrus grandis. Proc. Amer. Soc. hort. Sci.*, 84, 137–40.

Swingle, W. T. (revised by Reece, P. C.) (1967). The botany of *Citrus* and its wild relatives. In Reuther, Webber, and Batchelor (eds.), *The citrus industry*, Berkeley, 1, 190–430 (rev. edn).

Tanaka, T. (1954). *Species problem in citrus*. Japanese Society for the Promotion of Science. Tokyo.

Webber, H. J. (revised by Reuther, W. and Lawton, H.) (1967). History and development of the citrus industry. In Reuther, Webber and Batchelor (eds), The *citrus industry*, Berkeley, 1, 1–39 (revised).

Peppers
Capsicum (Solanaceae)

C. B. Heiser, Jr.
Indiana University Bloomington
Indiana USA

1 Introduction

The chili or ají peppers, native to the New World tropics, are now widely cultivated for use as spices or vegetables in the temperate zones as well as the tropics. They were the Americas' most important contribution to the world's spices. Chili powder, red and cayenne peppers, Tabasco, paprika, sweet or bell peppers, and pimientos (not to be confused with *Pimenta dioica*) are all derived from the pod-like berry of various species of *Capsicum*. The plants are shrubby perennials, although usually grown as herbaceous annuals in the temperate zones. Propagation is by seeds. The pungency of peppers is due to capsaicin, the vanillyl amide of isodecylanic acid, contained in the placenta. A single dominant gene controls the presence of pungency. Since there are various degrees of pungency, apparently there are modifiers of the major gene; the environment has also been claimed to have an effect. Sweet, non-pungent, peppers, widely used in the immature or green stage as a vegetable for stuffing or for salads, are more appreciated in the temperate zones than in the tropics. Peppers are good sources of vitamins, particularly ascorbic acid in the raw sweet sorts, and vitamin A in the dried pungent kinds. In addition to their use as food or condiment, peppers still have some use in medicine and some are valued as ornamentals.

2 Cytotaxonomic background

Although earlier in this century only one or two species of cultivated peppers were recognized, there now seems to be some agreement that four or five species are involved (Heiser and Smith, 1953). The names used here follow Heiser and Pickersgill (1969); some European workers have advocated other names for two of the species (see Terpó, 1966). In addition to these domesticated species, there are some 20 wild species, mostly confined to South America. All of the

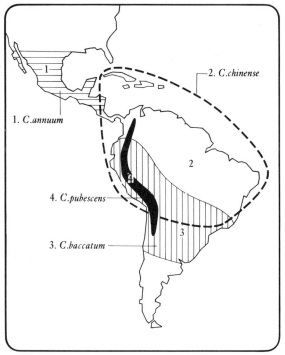

Fig. 77.1 Distribution of *Capsicum* peppers at time of European discovery.

following species show great diversity in the shape, size and colour of the fruits.

1 *Capsicum annuum* is by far the most widely cultivated and economically the most important species today. It includes the sweet peppers as well as most of those that are dried for hot peppers, chili powders and paprika. The domesticated forms are assigned to *C. annuum* var. *annuum* and the wild or weedy types to *C. annuum* var. *minimum*. The latter is distributed from the southern United States to northern South America. Several lines of evidence indicate that domestication first occurred in Middle America and karyotype analysis suggests that this was in Mexico (Pickersgill, 1971). This species is characterized by blue anthers, milky white corollas, inconspicuous calyx lobing and peduncles one at a node.

Attempts have been made, with varying degrees of success, at cultivar classification of *C. annuum*, most recently by Terpó (1966). These classifications have generally failed to include the large number of cultivars of tropical America.

2 *C. baccatum* is little cultivated outside parts of South America. The domesticated sorts are classified as *C. baccatum* var. *pendulum* and the wild types

as var. *baccatum* (Eshbaugh, 1970). The wild variety is largely confined to Bolivia and surrounding areas and it is probable that cultivation was initiated somewhere in this area. Although once confused with *C. annuum*, this species is readily distinguished from it by the yellow, tan or brown spots on the corolla, and prominent calyx teeth.

3 *C. frutescens* has a wide distribution as a wild, weedy or semi-domesticated plant in lowland tropical America and, secondarily, in southeastern Asia. The cultivar Tabasco is the only member of this species commonly in cultivation outside the tropics. The species is characterized by blue anthers, milky-, greenish- or yellowish-white corollas, and by usually having some nodes with two or more peduncles.

4 *C. chinense* is also widespread in tropical America and is the most commonly cultivated species in the Amazon region. Some varieties of this species grown in Africa are reportedly the most pungent of all the peppers. A constriction below the calyx is the only morphological character that separates *C. chinense* from *C. frutescens*. Clearly *C. chinense* and *C. frutescens* are very closely related and probably the two should be combined into one species, in which case the name *C. frutescens* takes preference. No wild type of *C. chinense* is known and it seems likely that the progenitor is the wild type of *C. frutescens*.

5 *C. pubescens*, the rocoto, is a highland species widely grown in the Andes. It is also in cultivation in a few places in highland Mexico and Central America but its entry there may well be post-Columbian (Heiser, 1969). No wild ancestral type is known but *C. pubescens* shows affinities with the wild South American species, *C. eximium*, *C. cardenasii* and *C. tovari*, one or other of which could conceivably be the progenitor. *C. pubescens* is morphologically the most distinct of the cultivated species, set apart from the others by a number of features, including dark, rugose seeds (the other species have straw-coloured, more or less smooth seeds). The fruit flesh of *C. pubescens* is considerably thicker than that of the other pungent peppers.

All the cultivated species, as well as the wild ones for which counts have been reported, are diploids with $2n = 2x = 24$. Karyotypes for several species are given by Ohta (1962). Hybrids have been obtained in most combinations for the first four species listed above and these show varying degrees of fertility. *Capsicum pubescens* appears to be well isolated genetically from the other cultivated species. A considerable number of hybrids have also been made between various wild species and the cultivated ones. All the cultivated species are self-compatible and selfing appears to be the general rule. Some wild species, however, have been reported as self-incompatible and others have long styles that probably promote out-crossing. Birds are probably the chief agents for the dispersal of the fruits of the wild species, which are sometimes called 'bird-peppers'.

3 Early history

Primitive man in America must have found wild peppers an interesting addition to his diet and they must have been eagerly sought. They are still collected in some places today and make their appearance in markets in parts of Latin America. Pepper seeds have been reported archaeologically from early levels (before 5000 B.C.) at Tehuacán, Mexico, and probably came from wild plants of *C. annuum*. Domesticated forms of this species are represented archaeologically in Mexico before the beginning of the Christian era. In archaeological deposits in coastal Peru, cultivated types of *C. baccatum* are found dated around 2000 B.C. and, in later levels on the coast, *C. frutescens* makes its appearance. Thus, pepper cultivation appears to be fairly ancient in the Americas (Pickersgill, 1969a, b). It seems most likely that pepper cultivation began independently in several areas employing different wild species. It is possible that, after an initial domestication of one species, stimulus-diffusion led to the attempt to cultivate other wild species in different areas. Domestication resulted in change, particularly in the fruits. The small, erect, deciduous red fruit of the wild type was replaced by larger fruits, often pendent, non-deciduous and having a variety of fruit colours in addition to red. There was, in fact, a remarkable parallel series of fruit types produced in the various cultivated species. Sweet types also early became known but these have assumed some importance only recently. Apparently, there was also a change with cultivation from outcrossing to inbreeding. Peppers became prized plants among the Indians, often held in second place only to the major staple, maize or manioc. Peppers also played an important role in religious ceremonies and legends among many Indian cultures.

4 Recent history

The first European record of the peppers is Peter Martyr's letter of 1493 in which he stated that Columbus had found peppers more pungent than those (i.e. *Piper nigrum*) from the Caucasus. Following Columbus's voyages, many peppers were introduced to Europe, and these, unlike their relatives, the Irish potato and tomato, found almost immediate acceptance. A great many peppers were described by the herbalists in the sixteenth and seventeenth centuries. The peppers also soon made their way to other parts of the world and were eagerly adopted, particularly in India, which today is one of the leading countries in the export of chilies. Peppers were apparently not among the early garden crops in the United States but today the crop is extremely important there, being valued in excess of $44 million in 1970. Peppers still retain their importance in their homeland where most of the peppers produced are consumed locally.

5 Prospects

Breeding of peppers has been rather similar to that of other solanaceous fruit crops. They are tolerant of inbreeding and conventionally bred as pure lines. Although peppers are perhaps less subject to fungus and insect attacks than many plants, it has still been necessary to devote considerable effort to the breeding of resistant varieties, particularly against virus diseases. More of the breeding work has concentrated on sweet types than on pungent ones. Polyploids have been produced in *C. annuum* but at present are of no commercial importance. Haploids are known and have some potential in plant breeding. Cytoplasmic male sterility was discovered by P. A. Peterson in 1958. Today, F_1 hybrids of a number of sweet types are available to the grower. The genetic base of the peppers is not as narrow as that for many crops, and a large number of yet unexploited varieties still exist in the tropics. Seed banks have now been established in some countries but much more work of this nature needs to be undertaken.

6 References

Eshbaugh, W. H. (1970). A biosystematic and evolutionary study of *Capsicum baccatum* (Solanaceae). *Brittonia*, **22**, 31–43.

Heiser, C. B. (1969). Systematics and the origin of cultivated plants. *Taxon*, **18**, 36–45.

Heiser, C. B. and **Pickersgill, B.** (1969). Names for the cultivated *Capsicum* species (Solanaceae). *Taxon*, **18**, 277–83.

Heiser, C. B. and **Smith, P. G.** (1953). The cultivated *Capsicum* peppers. *Econ. Bot.*, **7**, 214–27.

Lippert, L. F., **Smith, P. G.** and **Bergh, B. O.** (1966). Cytogenetics of the vegetable crops. Garden pepper, *Capsicum. Bot. Rev.*, **32**, 24–55.

Ohta, Y. (1962). [Karyotype analysis of *Capsicum species*]. (Japanese with English summary.) *Seiken Ziho*, **13**, 93–9.

Pickersgill, B. (1969a). The archaeological record of chili peppers (*Capsicum* spp.) and the sequence of plant domestication in Peru. *Amer. Antiq.*, **34**, 54–61.

Pickersgill, B. (1969*b*). The domestication of chili peppers. In P. J. Ucko and G. W. Dimbleby (eds.), *The domestication and exploitation of plants and animals*, pp. 443–50. London.

Pickersgill, B. (1971). Relationships between weedy and cultivated forms in some species of chili peppers (genus *Capsicum*). *Evolution*, **25**, 683–91.

Terpó, A. (1966). Kritische Revision der wildwachsenden Arten und der kultivierten Sorten der Gattung *Capsicum*. *Fedde Repert.*, **72**, 155–91.

Tomato

Lycopersicon esculentum (Solanaceae)

Charles M. Rick
University of California Davis USA

1 Introduction

The tomato is an annual vegetable crop of widespread culture and popularity. In the USA the crop has recently had an annual value of about $600 million, ranking second only to potatoes in importance amongst vegetables. Elsewhere, its popularity varies but the areas are few in which it is not used in one form or another. By virtue of its many attributes, the tomato is also a favourite subject for research in physiology and cytogenetics.

The reasons for the tomato's popularity are various. It supplies vitamins (particularly C) and a variety of colour and flavour to the diet. Its many forms are adapted to a wide range of soils and climates, although always demanding a warm season and well-drained soil. Its culture extends from the tropics to within a few degrees of the Arctic Circle. Wherever length of season permits, the tomato is grown in the open; elsewhere, and in the off-season, it is often cultured in glasshouses or other protective structures. A favourite of home gardeners, the crop is also grown commercially, even on a large scale, single plantings of 200 ha being not unusual. Sometimes, the tomato is managed in the same fashion as a row crop and harvested mechanically in a single operation; at the other extreme, it may be trained on trellises and a single planting grown throughout most of the year and harvested continuously by hand. Under the latter régime, astounding yields may be realized although, for general field culture, harvests of 50 t/ha are not unusual.

2 Cytotaxonomic background

The tomato is one of the 8–10 closely interrelated species of the genus *Lycopersicon*, which is distinguished from the most closely related (and probably ancestral) genus *Solanum* by the pollen-shedding characteristics of its anthers. The latest complete systematic treatment is that of Muller (1940), a monograph that is useful but outdated by recent biosystematic research. Most of the species are short-lived perennial herbs but can be grown to flowering and fruiting in five months or less. All species are native to western South America; the wild form of *L. esculentum*, var. *cerasiforme*, is also found in Mexico, Central America and other parts of South America, as well as the subtropics of the Old World, but it is weedy and its true native range is obscured by its aggressive colonizing tendencies. The chromosomes of all species are alike in number ($2n = 2x = 24$) and morphology, and even the cytology of inter-specific F_1 hybrids yields little or no evidence of structural differences. Rare instances of spontaneous autopolyploidy have been recorded.

L. esculentum and its near relatives are self-fertile. The former is outcrossed to a considerable extent in its native region and certain other subtropical areas but, elsewhere, is almost completely self-pollinating (Rick, 1950). In contrast, the other species are self-incompatible, showy-flowered and consequently entirely outcrossed (by various species of bees).

L. esculentum can be hybridized with, and thereby receive genes from, all other species, with varying degrees of difficulty. Numerous desired characters have thereby been transmitted to, and incorporated in, acceptable cultivars.

3 Early history

The tomato was already a well-developed cultigen in the New World at the time of the conquests. It was taken to Europe in the sixteenth century and later disseminated thence to many areas of the world. This pattern of migration is matched by those of many other cultivars of American origin. These facts are established beyond dispute; however, we stand on less firm ground regarding other aspects of the origin and early history of the cultivated tomato. Our under-standing of these matters has been modified slightly, if at all, since the publication of Jenkins's (1948) critique. Thus, despite earlier wide acceptance of Peru as the centre of domestication, the bulk of the historical, linguistic, archaeological and ethnobotani-cal evidence favours Mexico, particularly the Vera Cruz–Puebla area, as the source of the cultivated tomatoes that were first transported to the Old World. Mexico was also considered the most likely centre of domestication in spite of the undisputed distribution of the genus in South America. To a large extent, these conclusions are based on fragmentary or negative evidence. The archaeological data, for example, does not positively favour any particular region but the dearth of representations of the tomato in Peruvian

artefacts is considered significant in view of the pen-chant there, of certain pre-Columbian cultures, for depicting their cultigens on pottery. The poor preservation qualities of tomato plant parts undoubt-edly obstructs progress in detecting remains of either fossil or archaeological nature.

The genetic evidence is scarcely more illuminating. Jenkins (1948) stresses the great morphological variabi-lity in the cultivated tomatoes of Mexico; yet this level of variation can be matched by that of Central America (Lehmann and Schwanitz, 1965) and coastal Peru (Rick, 1958, and unpublished). But, as many re-searches reveal, genetic variability *per se* is not very diagnostic in determining centres of origin. Far more critical would be genetic distance as determined by the degree of genetic differentiation between modern cultivated tomatoes and their counterparts in putative areas of origin. Unfortunately, again, the data are not critical, although they do reveal that the tomatoes introduced by the Spanish explorers (which furnished the entire base for modern cultivars) represent only a minute portion of the germ plasm available in the same and closely related species. In respect of the distribution of alleles at the *Ge* (gamete eliminator; Rick, 1971) and *Prx* (peroxidase allozymes; Rick *et al.*, 1974) loci, the currently available European and US cultivars (derived almost entirely from the first New World introductions) are exceedingly homogeneous. The few deviations from the 'normal' genotype appear in tiny enclaves, which might represent either *de novo* appearance by mutation or (probably less likely) ancestry in western South America. In these assays, the greatest overall similarity in allele frequency is with cultivars and var. *cerasiforme* from Mexico and Central America. The differing alleles found in the Ecuadorean and Peruvian cultivars are also common to var. *cerasiforme* and *L. pimpinellifolium* from their respective regions. The latter information does not discriminate between hypotheses of filial or intro-gressive relationships.

Whatever the geography of domestication of the tomato, its immediate ancestor was probably var. *cerasiforme*, the common, weedy, wild counterpart of the same species, as concluded by Jenkins (1948) and still widely accepted today. The latter bears greater genetic resemblance to the cultivated tomato than the only other likely candidate, *L. pimpinellifolium*, which is probably a by-product rather than a member of the stem-line of the crop. Although these taxa differ considerably from each other in size characteristics, the magnitude of genetic difference does not appear to be great. Thus, many breeding experiments have succeeded in restoring the large fruit size of *esculentum*

269

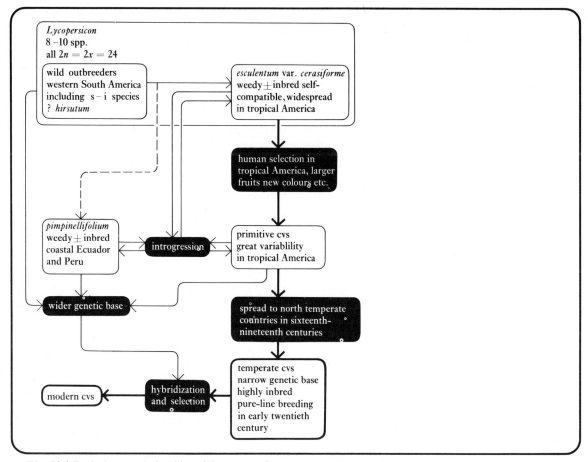

Fig. 78.1 Evolutionary relationships of the tomato, *Lycopersicon esculentum*.

cultivars after a cross with *L. pimpinellifolium* followed by only three generations of backcrossing and selection. Further, Stubbe (1971) has demonstrated that it is possible, by successive steps of induced mutation and selection, to increase fruit size in *pimpinellifolium* to a level comparable with that of smaller fruited cultivars. Similar progress was reported in reducing fruit size of the *esculentum* cultivars toward that of *L. pimpinellifolium*. If such progress can be achieved experimentally within about two decades, it is certainly conceivable that primitive man could have effected the smaller increase in fruit size between var. *cerasiforme* and the cultivars in the course of millenia.

4 Recent history

In the evolution of cultigens, improvement and utilization are undeniably interdependent. In the case of the tomato, fears of toxicity long restricted consumption. Such notions, based upon the presence of poisonous glycoalkaloids in the foliage and fruits of other members of the nightshade family, were dispelled first in the Mediterranean region and later in northern Europe and North America but persisted well into the twentieth century. The history of utilization, particularly as affected by such superstitions, is carefully documented in McCue's (1952) annotated bibliography.

Prior to the 1920s, tomato improvement depended largely upon selection of chance variants that originated as a result of mutation, spontaneous outcrossing or recombination of pre-existing genetic variation. Such selection certainly succeeded in developing cultivars adapted to such widely differing environments as the long warm season in Italy, the short season in outdoor culture in northern Europe, greenhouse production in Britain, the mesophytic conditions of the eastern and midwestern USA and the warm, arid districts of the western USA.

Much more rapid progress in tomato breeding and utilization has come about during the past 50 years.

Application of modern methods resulted in the breeding and introduction of hundreds of cultivars adapted to widely differing conditions and intended for diverse purposes as fresh vegetables and for various forms of preservation. The greatly accelerated rate of varietal introduction has led to rapid replacement of cultivars; recently, few cultivars have held supremacy for as long as 10 years.

The normal mode of reproduction and the ease of manipulation of the flowers for controlled matings make the tomato ideal for application of the traditional 'Minnesota' method: single hybridization followed by pedigree selection is highly efficient for achieving desired combinations of parental characteristics. Certain dominant characters can also be combined and heterosis exploited in F_1 hybrid cultivars. A complete list of improvements attained during this period would be too long for presentation here; instead, the following list of principal categories of achievement is submitted.

1. Increased yields by way of larger fruit size and increased fruit number. The ability of flowers to set fruit has been improved and, in certain types, a more concentrated fruit set achieved.

2. Improved fruit quality: shape, texture, colour, flavour; uniformity in all characteristics.

3. Plant habit modified to facilitate cultural and harvest operations. The most significant change has been the discovery and exploitation of determinate (restricted) growth conditioned by the *sp* gene.

4. Improved handling and storage durability (but sometimes to the detriment of culinary quality).

5. Pest resistance. The list of accomplishments includes resistance to various species of insects, viral and fungal parasites, and nematodes.

During the last 50 years of intensive plant breeding, floral structure has evolved so as to improve fruit setting ability. Intentionally or otherwise, the position of the stigma has been markedly changed (Fig. 78.2). Like their closest wild relatives, the cultivars from Latin America tend to have well exserted stigmas. If such types are cultivated in foreign areas lacking the appropriate pollinating insects, fruit setting is greatly diminished. Thus, when the tomato was first introduced to Europe, an intense selection would have been expected for less exserted stigmas. In glasshouse culture, the selection would have been even stronger because insects are excluded as well as wind-induced vibrations, which can aid in pollination. As a result, the stigma of most cultivars became fixed at the mouth of the anther tube. During the past decade stigma position was shifted again, this time to an even lower level, well within the anther tube (Rick and Dempsey, 1969). This change, which further improved self-pollination and consequent fruit set and practically eliminated outcrossing, has been incorporated in the leading California cultivars.

Perfection of completely mechanized harvest has had a strong impact on the recent evolution of the crop. Although applied primarily to the crop grown for processing, the method is also being adapted for harvest for the fresh fruit market. The impact of this development is manifest in the records of the 1973 crop year: approximately 80 per cent of the US pack and a substantial part of world production was thus produced. Mechanized harvest requires careful integration of plant breeding, design of the harvester, crop management and cannery operations. Success in developing this procedure therefore hinged on co-operation of specialists in all of these fields, on the

Fig. 78.2 Change in style-length in the evolution of the tomato, *Lycopersicon esculentum*.

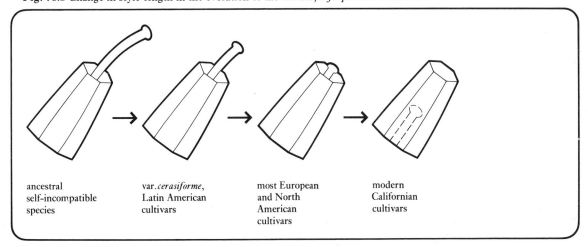

ancestral
self-incompatible
species

var. *cerasiforme*,
Latin American
cultivars

most European
and North
American
cultivars

modern
Californian
cultivars

cultivation of large acreages of relatively flat terrain and on California's desert summer climate that permits production without staking, accumulation of maturing fruits on the plant until an economic proportion are ripe and a long growing season allowing direct seeding and successive plantings.

Tomato F_1 hybrids have been investigated for at least 50 years, and numerous F_1 cultivars have been introduced to the trade. Advantages of such hybrids are increased earliness, improved quality and yield, as well as combinations of desirable dominant parental alleles, particularly monogenic disease resistances. Capable of setting fruit under adverse conditions, F_1 cultivars tend to be popular with home gardeners. Their use in commercial plantings depends on economic considerations such as benefits received as against cost and other problems of seed production. The extent of F_1 plantings therefore varies according to country and district; they constitute at least 95 per cent of commercial plantings in Japan in contrast to less than 5 per cent in the USA.

Progress in tomato breeding depends to some extent upon an understanding of the inheritance of desired characters, linkage relationships and cytogenetics of interspecific hybrids. It is therefore noteworthy that the tomato ranks among the best known crop plants in terms of its genetics and cytogenetics, a situation that owes partly to its natural advantages for such research: (a) short life cycle, (b) ecological versatility, (c) high seed yields, (d) self-pollinating mechanism, yet amenability to controlled hybridization, and (e) diploidy. Some 250 genes have been allocated to their respective chromosomes (mostly located with reasonable accuracy) and chromosome arm contents have been well established by cytogenetic identification of centromere positions (Rick, 1974). This research has also been expedited by the coordination of geneticists' efforts and the liaison between geneticists and breeders promoted by the Tomato Genetics Cooperative. As examples of the reciprocal benefits, much useful germ plasm, including numerous male-sterile mutants and derivatives of species hybrids, has been contributed to breeding programmes, whilst the latter have provided many mutants and other materials of great interest to tomato geneticists.

5 Prospects

In the immediate future we shall undoubtedly witness a continuation of present trends and the application of traditional methods, notably of intraspecific hybridization and selection as a standard breeding method. The use of F_1 hybrid cultivars is likely to increase as a result of appreciation of their attributes and efforts

to economise in seed production and seeding of the crop. Desired characters, not otherwise attainable, are sought in induced mutants and other species. Large resources of germ plasm are available and yet unexploited opportunities (for example, the high water economy of *Solanum pennellii*) await investigation (Rick, 1973).

Innovative methods of great promise are attracting the attention of tomato specialists. The dream of breeding this species by way of autodiploids seems closer to realization, now that methods of microspore culture are being perfected. So far, the method has met with rather limited success in *L. esculentum* but it is reasonable to expect that formulae for the consistent culturing of haploid embryos from microspores will soon be found. By permitting homozygotization in a single generation, this method presents enormous advantages, not only in time, but also in segregation that is vastly simpler at the *n* rather than at the *2n* level, in utilizing induced mutation, in cell hybridization and in other respects.

Another goal that seems within reach is single-cell culture and cell hybridization. Methods that have succeeded in tobacco species should be applicable to tomatoes to permit wider crosses than are now attainable by normal hybridization procedures.

Many objectives are being sought currently in tomato breeding programmes. Efforts are being concentrated on the analysis of flavour constituents, on their genetic control and on the utilization of such knowledge to improve quality. Total yield, although a *sine qua non* for plant breeding in general, may receive less attention, in as much as the yield potential of existing cultivars is already very high. The current emphasis on mechanized harvest will undoubtedly have a strong impact on the direction of breeding efforts. The objectives being sought are compact plant habit, concentrated fruit set, durability of fruits and retention of fruits on trusses to reduce losses from shattering. The appearance of new pests and new strains of existing ones will require constant surveillance and breeding resistance thereto. Resistance is badly needed to certain diseases, notably bacterial wilt (*Pseudomonas solanacearum*), that currently impede tomato production in the wet tropics.

6 References

Jenkins, J. A. (1948). The origin of the cultivated tomato. *Econ. Bot.*, **2**, 379–92.

Lehmann, C. O. and **Schwanitz, F.** (1965). Ein Beitrag zur Kenntnis der Formenmannigfaltigkeit der Kulturtomaten (*L. esculentum*). *Kulturpfl.*, **8**, 545–85.

McCue, G. A. (1952). The history of the use of the tomato:

an annotated bibliography. *Ann. Missouri bot. Gdn.*, **39**, 289–348.

Muller, C. H. (1940). A revision of the genus *Lycopersicon*. *US Dep. Agric. misc. Publ.*, **382**, pp.

Rick, C. M. (1950). Pollination relations of *Lycopersicon esculentum* in native and foreign regions. *Evolution*, **4**, 110–22.

Rick, C. M. (1958). The role of natural hybridization in the derivation of cultivated tomatoes of western South America. *Econ. Bot.*, **12**, 346–57.

Rick, C. M. (1971). The tomato *Ge* locus: linkage relations and geographic distribution of alleles. *Genetics*, **67**, 75–85.

Rick, C. M. (1973). Potential genetic resources in tomato species: clues from observations in native habitats. In A. M. Srb (ed.), *Genes, enzymes, and populations*. New York, 255–69.

Rick, C. M. (1974). *The tomato*. In King, R. C. (ed.), *Handbook of genetics*, New York, 247–80.

Rick, C. M. and Dempsey, W. H. (1969). Position of the stigma in relation to fruit setting of the tomato. *Bot. Gaz.*, **130**, 180–6.

Rick, C. M., Zobel, R. W. and Fobes, J. F. (1974). Four peroxidase loci in red-fruited tomato species: genetics and geographic distribution. *Proc. Natnl. Acad. Sci. USA*, **71**, 835–9.

Stubbe, H. (1971). Weitere evolutionsgenetische Unterschungen in der Gattung *Lycopersicon*. *Biol. Zbl.*, **90**, 545–59.

Acknowledgment. The author's share of the research reported here was supported in part by grants from the National Science Foundation and National Institutes of Health, USPHS.

Tobacco

Nicotiana tabacum (Solanaceae)

D. U. Gerstel
North Carolina State University
Raleigh NC USA

1 Introduction

Sixty-four recognized species of the genus *Nicotiana* are found in the Americas, Australia and a few islands of the South Pacific. Because of the intoxicating effect of the alkaloid nicotine they contain, about ten species were employed by aborigines for ritualistic, medicinal and perhaps hedonistic purposes; for example: *N. tabacum* in South and Central America; *N. bigelovii*, *N. attenuata* and *N. trigonophylla* in western North America; *N. rustica* east of the Mississippi, in northern Mexico and the West Indies; and *N. benthamiana*, among others, in Australia. The natural ranges of most or all of the species mentioned were extended by cultivation before the arrival of white men. They were used as chew or snuff or were smoked in pipes, cigars and reed-cigarettes. In modern times, *N. rustica* has been utilized also as a source of nicotine for insecticides and of citric acid. A few species serve as ornamentals (*N. alata*, *N. sanderae*, *N. glauca*). *Nicotiana tabacum* is by far the most important species in modern agriculture and international trade. Estimated world production of this species for 1973 was 4,750 million kg, grown throughout the world, except for the arctic and near-arctic zones.

It is characteristic of cultivated species in general to be polymorphic and this is true also for cultivated species of *Nicotiana*, e.g. *N. bigelovii*, *N. rustica* and *N. tabacum*. *Nicotiana tabacum*, in particular, is very variable and numerous commercial types have been distinguished. These differ widely in morphological, physical and chemical characteristics. The most important are: flue-cured tobacco (named after the metal flues which distribute heat in the curing barns); burley, a thin-leafed light-coloured tobacco; aromatic or oriental; and air-cured tobaccos. Tobacco types are not used interchangeably to any extent and specific types have specific uses. Thus, flue-cured and aromatic mainly go into blends for cigarettes; and cigar-wrapper, an air-cured type, is limited to the use indicated by its name.

2 Cytotaxonomic background

Species of *Nicotiana* played an important role in early botanical studies of the contribution made to the off-spring by male and female parents. In the middle of the eighteenth century Kölreuter investigated hybrids between *N. rustica* and *N. paniculata* and their progenies. *Nicotiana* has remained a favoured subject of genetical work because of the ease of manipulation of the flowers, the large number of seeds produced and a crossing range extending from high fertility to absolute genetic, chromosomal or cytoplasmic barriers to interspecific hybridization.

An early analysis of the heredity of oppositional self-incompatibility alleles was performed by East and Mangelsdorf (1925) with *N. sanderae*, a horticultural form derived from hybrids between *N. alata* and *N. forgetiana*. Since then, a few additional diploid self-incompatible species have been found. However, none of the extensively cultivated species are self-incompatible and *N. tabacum* and *N. rustica* are mainly self-fertilizing.

A very large number of intra- and intersectional hybrids has been produced and meiosis has been studied in most of them in order to establish phylogenetic relationships (Kostoff, 1943; Goodspeed, 1954) and to provide a background for interspecific gene transfers in breeding work. A cytotaxonomic treatment of the genus was given by Goodspeed (1954) who divided it into three subgenera and 14 sections. The basic chromosome number is $x = 12$ and most of the present-day species have $2n = 2x = 24$. Allopolyploidy has played an important role in the evolution of the genus and nine American species have $2n = 4x = 48$. Deviating numbers exist in South America ($x = 9, 10$) as well as in Australia ($x = 16–23$).

Nicotiana tabacum ($4x = 48$) is cytogenetically the most extensively studied species. An ancestor of present-day *N. sylvestris* (section Alatae) and a member of the Tomentosae section (either *N. tomentosiformis* or *N. otophora* or a form antecedent to these) are thought to be the diploid parents of *N. tabacum*. The letters S and T are used to designate the two genomes of the amphidiploid, so *N. tabacum* is SSTT. Goodspeed (1954) thought that *N. otophora* was closer to *N. tabacum* than was *N. tomentosiformis* because the ranges of *N. otophora* and *N. sylvestris* overlap in the Salta region on the eastern slope of the Andes mountains. Furthermore, synthetic *N. sylvestris* × *N. otophora* amphidiploids are male and female fertile and resemble

N. tabacum. The writer believes that *N. tomentosiformis* may be closer to *N. tabacum*, even though the present distributions of *N. sylvestris* and *N. tomentosiformis* are disjunct (being separated by about 3° latitude) and the amphidiploid is highly female sterile. The flowers of amphidiploid *N. sylvestris* × *N. tomentosiformis* resemble those of *N. tabacum* more closely than those of 4x (*N. sylvestris* × *N. otophora*). The corolla lobes of the latter fold back at maturity and have a pronounced bilateral symmetry, while the flowers of 4x (*N. sylvestris* × *N. tomentosiformis*) are like those of *N. tabacum* in having flat lobes and in being nearly actinomorphic. Other evidence stems from a comparison of gene segregations of the synthetic allopolyploids 6x (*N. tabacum* × *N. tomentosiformis*) and 6x (*N. tabacum* × *N. otophora*) which possess the genome formula SSTTTT. In the former, backcross segregation ratios for genes located in the T genomes approximate to those of an autotetraploid, indicating close similarity of the T genomes of *N. tabacum* and *N. tomentosiformis*. In contrast, the wider ratios from 6x (*N. tabacum* × *N. otophora*) demonstrate a reduced homology between the T genomes of *N. tabacum* and *N. otophora* (Gerstel, 1963). Finally, isozyme studies by Sheen (1972) support the view that *N. tomentosiformis* is more closely related to *N. tabacum* than is *N. otophora*.

Chromosome pairing in undoubled F_1 hybrids between *N. sylvestris* and Tomentosae species is very low and multivalent formation in their amphidiploids is near zero. Therefore, a mechanism suppressing homeologous associations like the one found in chromosome 5B of the polyploid wheats is not required in *N. tabacum*; differential affinity assures regularity of meiosis here (Gerstel, 1963). The alternative view, that the S and T genomes of the ancestral species were once more similar than they are now and that a mechanism to prevent homeologous associations was necessary in the early amphidiploid, appears less probable. Haploid *N. tabacum* exhibits even less chromosome pairing than is found in the F_1 hybrids of the putative parents; the most likely explanation is that further divergence ('diploidization') of the S and T genomes in the amphidiploid has taken place after *N. tabacum* originated.

The situation in *N. rustica* ($2n = 4x = 48$, from *N. paniculata* × *N. undulata*) appears to be similar, but cytological evidence is scanty.

A recent survey of the genus for locations of heterochromatin (Merritt, 1974) has brought to light another interesting problem concerning *N. tabacum* and *N. rustica*. All the *Nicotiana* species studied have small heterochromatic knobs but only a few species, scattered through the three subgenera, possess in addition large blocks of heterochromatin. One of the ancestral species of each of *N. tabacum* and *N. rustica* have such blocks. This is true for *N. otophora* and *N. tomentosiformis*, both of which possess several large terminal or subterminal heterochromatic blocks. *Nicotiana paniculata*, ancestral to *N. rustica*, also possesses a number of blocks of heterochromatin. The other ancestral species, *N. sylvestris* and *N. undulata*, respectively, have only very small knobs, as is the case with the two amphidiploids, *N. tabacum* and *N. rustica*. The puzzling question of what caused the disappearance of the large blocks during the evolution of the amphidiploids will be solved only after the function or functions of heterochromatin are better known than they are today. The alternative, that the heterochromatic blocks appeared in the diploids only after the origin of the amphidiploids, appears less probable.

Differentiation of the cytoplasm has played a role in the evolution of the genus, as is shown by the frequent occurrence of male sterility when the cytoplasm of one species is combined with the chromosomes of another. (This has also been observed in other genera, e.g. in the Triticinae and in *Gossypium*.) Thus, Chaplin (1964) introduced the chromosomes of *N. tabacum* into the cytoplasm of six other *Nicotiana* species by means of backcrosses to *N. tabacum*. The chromosomes of the non-recurrent parent were eliminated by this procedure. The effects of the cytoplasm are specific; the various cytoplasms produce very distinctive morphological alterations of the male organs and sometimes of the corollas also. Sometimes, male fertility can be restored by reintroducing specific chromosomes from the non-recurrent parent (e.g. Burk and Mann, 1970).

3 Early history

The earliest documented evidence of the use of tobacco is a bas-relief from a temple at Palenque in the State of Chiapas, Mexico, dated A.D. 432, showing a Mayan priest blowing smoke through a tubular pipe during a ceremony. The tobacco used was probably *N. tabacum*. Another find consisted of loose tobacco and pipe dottle left by the cave-dwelling Pueblo Indians of northern Arizona. These remains, dating from approximately A.D. 650, have been shown by chromatographic and spectrophotometric analyses to contain nicotine; presumably their source was *N. attenuata*. Little other archaeological evidence of the origin and early uses of tobacco has been found.

Nicotiana tabacum and *N. rustica* are the only Nicotianae remaining in cultivation today. Probably, neither

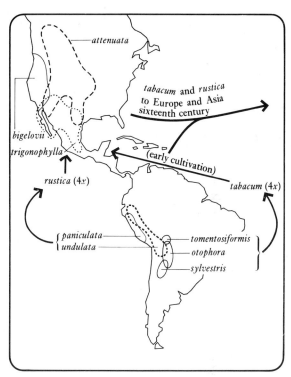

Fig. 79.1 Evolutionary geography of the cultivated tobaccos. Based on data of Goodspeed 1954.

of these two amphidiploids exists in the wild state, even though both have become naturalized in various places. *Nicotiana rustica* persists in the Bolivian, Peruvian and Ecuadorian highlands in ruderal locations coinciding, in part, with the areas of distribution of the progenitor species. Wild *N. tabacum* is also unknown but this species has been collected at locations where, in Goodspeed's (1954) opinion, it was an escape from cultivation. We may assume that these species originated in the areas of distribution of their diploid ancestors.

Since *N. rustica* and *N. tabacum* are not known as wild plants, we are faced with the problem as to which came first: human utilization of an ancestral species (as in the case of hexaploid wheat) or the hybridization which gave rise to the amphidiploids. In the case of *N. tabacum*, some details are known about the alkaloids which show that an additional problem exists.

The valued alkaloid of tobacco is nicotine, produced in the roots and translocated to the leaves (Dawson, 1942). But the wild parents of *N. tabacum* (i.e. *N. sylvestris* and the Tomentosae species) both possess a dominant 'converter' gene which, by enzymatic action, demethylates nicotine into undesirable nornicotine in the leaves. For this reason, these species may have been useless, as must have been, *a fortiori*, the original amphidiploid, for it combined the 'converter' genes of both its ancestors. The Indians hardly could have circumvented this problem by using green leaves (either by making extracts or by chewing) because the converter gene from *N. tomentosiformis* acts even in the green leaves (Wernsman and Matzinger, 1968).

Since the converter gene of *N. sylvestris* is weaker and acts later, it is possible that this species was the first to be used. In fact, Spegazzini (cited by Kostoff, 1943) claims that the Indians used the leaves of this diploid, though Goodspeed (1954) does not seem to think so. If they did, they may have switched later to the amphidiploid, provided it came from hybridization with a form of a Tomentosae species which did not have a strong converter allele. Further evidence is needed to test this supposition.

Wild species possess means of disseminating their seeds or other propagules. Seed-propagated cultivated species generally do not scatter their seeds. This is also true of the Nicotianae; the wild species have dehiscent capsules while many races of *N. tabacum*, *N. rustica* and *N. bigelovii* have capsules which remain closed at maturity (Goodspeed, 1954). Undoubtedly this is a result of human selection which eased the process of seed collection.

The early history of tobacco culture by European settlers was described by Garner *et al.* (1936). When Spanish explorers landed early in the sixteenth century in Yucatan they found *N. tabacum* under cultivation by the Indians. Soon the conquerors themselves founded plantations of this species throughout the Caribbean and as far as Venezuela and Brazil, establishing a number of morphologically distinct types in the process. The product found a rapidly expanding market throughout Europe and in many parts of Asia. Cultivation of the crop in these parts followed not far behind.

4 Recent history

Tobacco culture was begun in Virginia in 1612 by John Rolfe. The plant grown by the local Indians was *N. rustica* but this species, with its harsh and irritating smoke, was replaced early by *N. tabacum* because of European demand for 'Spanish' leaf. Since *N. tabacum* is self-pollinated, Rolfe's planting of 'Orinoco' was probably a heterogeneous mixture of inbred lines. The practice of the early growers to produce their own seed soon resulted in a great variety of types. Mechanical mixture of seed, mutation, natural crossing and a system of mass selection may all have served to increase diversity. Little is known about the times and

places of origin of the diverse types. For instance, among oriental tobaccos grown around the eastern Mediterranean and the Black Sea, there exists a prodigious variety of forms which, however, share certain features such as the low stature, small leaves and high aroma that distinguish oriental tobaccos from all other types. It is not known from where in the Americas these very distinct and yet diverse forms came, whether from one or many locations. The story of the many distinct tobaccos of India may not be very different. Many authors have commented on the genetical plasticity of the species; no doubt soil type, climate and cultural and curing practices also contributed greatly to the diversity.

An account of the history of tobacco breeding in the United States has been given by Clayton (1958). At the beginning of this century, attempts were made to increase yield by hybridization of cultivars followed by limited selection. These efforts ended in failure, largely because the manufacturers perceived a change in taste and raised long-persistent objections to hybridization. The next phase, starting around 1920, saw breeding for disease resistance (particularly for black root rot) by intervarietal crossing, which was successful. The use of interspecific hybrids started with the transfer of resistance to tobacco mosaic virus by means of a substitution of an intact chromosome from *N. glutinosa* for a *N. tabacum* chromosome by Holmes (1938). By repeated backcrossing, the alien chromosome was broken up; later, mosaic resistance was transferred to other varieties. Efforts to obtain resistance to other diseases continued and were successful in several instances, using both intra- and interspecific sources of resistance. An outstanding example of the latter is resistance to blue mould. This disease reached epidemic proportions in Europe around 1960. Control was achieved by transferring resistance from the Australian species, *N. debneyi*. Intensive search of world collections of *Nicotiana* for simply-inherited types of resistance to pathogens (surveyed by Burk and Heggestad, 1966) and to insects is still in progress. It should be recognized, however, that monogenically controlled resistance is more likely to break down under an attack by new pest biotypes than polygenic resistance; but the latter is harder to identify and to transfer.

Extensive biometrical studies have been executed in *Nicotiana* (partial review by Smith, 1974). There is wide agreement that there is a predominance of additive genetic variance and little dominance variance in *N. tabacum*. Little heterosis or inbreeding depression for several characters has been found. Consequently, accumulation of favourable homozygous genes would be more effective than use of first generation hybrids. High degrees of hybrid vigour were, however, obtained by workers in the USSR, Poland, Israel and other countries, especially when crosses were made between varieties of diverse origins. Matzinger and Wernsman (1967), who listed some of these studies, observed heterosis in crosses of *N. tabacum* with the progenitor species, particularly with the Tomentosae. These authors are attempting to broaden the genetic base of flue-cured tobacco by introgression from these species.

5 Prospects

Two developments which may lead to a modification of breeding plans and objectives are of great concern to the breeder; these are agricultural mechanization and the emphasis on health problems associated with smoking. Methods of machine harvest vary widely between types. In burley, for example, entire plants are cut at the base while, in flue-cured tobacco, leaves are harvested individually. In either case, uniformity of maturity is desirable and both chemical treatments and more uniformly ripening genotypes are under investigation. Several morphological aspects and physical properties may also require modification. Uniformity of seed germination and seedling growth need improvement to aid mechanized transplanting from seedling beds to the field.

The health problem has produced a call for low nicotine content – which is easily achieved by selection. As a side effect, yield may be increased because of a negative correlation between nicotine content and yield. Investigation of other chemical components may lead to the identification and ultimate reduction of constituents considered to be a health hazard.

Recently, haploid plants have been obtained by culturing pollen grains, a technique which at this time is further advanced in *Nicotiana* than in any other genus (Smith, 1974). The use of haploids can avoid the complications resulting from dominance in genetical studies and derivatives with doubled chromosome numbers are completely homozygous, at least initially. The value of the technique for tobacco improvement remains to be assessed, however.

Another recent achievement consists of the isolation and fusion *in vitro* of protoplasts, by which means Carlson *et al.* (1972) have produced an amphidiploid hybrid, *N. glauca* × *N. langsdorffii*. This method may promise the production of hybrids which cannot be obtained by sexual means.

Finally, recent studies of photorespiration merit attention. They indicate that low rates of photorespiration can bring about large increases in net photosynthesis and production of dry matter. Such low rates of photorespiration characterize certain lines of *N. tabacum* and are heritable (Zelitch, 1971); they may therefore be made the subject of selection.

6 References

Burk, L. G. and **Heggestad, H. E.** (1966). The genus *Nicotiana*: a source of resistance to diseases of cultivated tobacco. *Econ. Bot.*, **20**, 76–88.

Burk, L. G. and **Mann, T. J.** (1970). Onset, prevention and restoration of male-sterile flower anomalies in tobacco. *J. Hered.*, **61**, 142–6.

Carlson, P. S., Smith, H. H. and **Dearing, R. D.** (1972). Parasexual interspecific plant hybridization. *Proc. natn. Acad. Sci. USA*, **69**, 2292–4.

Chaplin, J. F. (1964). Use of male-sterile tobaccos in the production of hybrid seed. *Tob. Sci.*, **8**, 105–9.

Clayton, E. E. (1958). The genetics and breeding progress in tobacco during the last 50 years. *Agron. J.*, **50**, 352–6.

Dawson, R. F. (1942). Accumulation of nicotine in reciprocal grafts of tomato and tobacco. *Am. J. Bot.*, **29**, 66–71.

East, E. M. and **Mangelsdorf, A. J.** (1925). A new interpretation of the hereditary behavior of self-sterile plants. *Proc. natn. Acad. Sci. USA*, **11**, 166–71.

Garner, W. W., Allard, H. A. and **Clayton, E. E.** (1936). Superior germ plasm in tobacco. *Yearbook of Agriculture*, US Dept. of Agriculture, Washington, USA., 785–830.

Gerstel, D. U. (1963). Evolutionary problems in some polyploid crop plants. *Hereditas*, suppl. **2**, 481–504.

Goodspeed, T. H. (1954). *The genus Nicotiana*. Waltham, Mass., USA.

Holmes, F. O. (1938). Inheritance of resistance to tobacco mosaic in tobacco. *Phytopathology*, **28**, 553–6.

Kostoff, D. (1943). *Cytogenetics of the genus Nicotiana*. Sofia, Bulgaria.

Matzinger, D. F. and **Wernsman, E. A.** (1967). Genetic diversity and heterosis in *Nicotiana*. *Züchter*, **37**, 188–91.

Merritt, J. F. (1974). The distribution of heterochromatin in the genus *Nicotiana* (Solanaceae). *Am. J. Bot.*, **61**, 982–94.

Sheen, S. J. (1972). Isozymic evidence bearing on the origin of *Nicotiana tabacum* L. *Evolution*, **26**, 143–54.

Smith, H. H. (1974). Nicotiana. In R. C. King (ed.), *Handbook of genetics*, **2**, New York and London.

Wernsman, E. A. and **Matzinger, D. F.** (1968). Time and site of nicotine conversion in tobacco. *Tob. Sci.* **12**, 226–8.

Zelitch, I. (1971). *Photosynthesis, photorespiration and plant productivity*. New York and London.

Eggplant

Solanum melongena (Solanaceae)

B. Choudhury
Indian Agricultural Research Institute
New Delhi India

1 Introduction

The eggplant, brinjal or aubergine is cultivated as a vegetable throughout the tropics and, as a summer annual, in the warm subtropics. It is highly productive and usually finds its place as the 'poor man's crop'. In India, it is an important article of diet consumed in a great variety of ways and it also has considerable medical uses (Kirtikar and Basu, 1933).

2 Cytotaxonomic background

Solanum is a very large genus that is mostly American in distribution. Among the 22 Indian species there is a group of five related ones, all prickly and all diploids with $2n = 2x = 24$, namely: *melongena*, *coagulans*, *xanthocarpum*, *indicum* and *maccanii*. Occurrence of polyploidy within *S. melongena* (with small and poor fruit formation) has also been reported. *Solanum melongena* produces fertile hybrids in crosses with *S. incanum* (often considered synonymous with *S. coagulans*). The cross *S. xanthocarpum* × *S. melongena* is successful when the former is used as the pistillate parent but the resulting seeds are inviable. There are variable reports regarding the compatibility of *S. indicum* to *S. melongena* and *S. xanthocarpum*. *Solanum indicum* is a highly variable species and further studies regarding the breeding behaviour of different forms of *S. indicum* with other species are necessary to determine affinities between the species. Taking compatibility as a criterion in establishing affinities, it appears that *S. melongena* is more closely related to *S. incanum* than to any other species. Two other species, *S. aviculare* and *S. khasianum*, having the same chromosome number, have gained importance as sources of solasidine used for synthesis of steroid hormones. *Solanum khasianum* crosses with *S. melongena* only when the latter is used as the female parent. The fruits formed by the hybrid have shrunken seeds.

Wild *S. melongena* is an armed perennial shrub/herb with bitter fruits and it occurs in India (Bhaduri, 1951). There have been suggestions of an African origin of the crop (Sampson, 1936) but these arose from confusion as to the taxonomy and distribution of wild relatives. Certainly, there are spiny African Solanums but there now appears to be no evidence that *S. melongena* is native there.

3 Early history

That the crop is old in India is clear from an extensive list of old names but no date for domestication can be suggested. Since wild forms are spiny and bitter it seems clear that early selection must have been directed against these characters and in favour of large fruit size.

On the basis of a study of the distribution of variability, Vavilov (1928) regarded the crop as being of Indian origin and added that some secondary variability had also developed in China. It has been known in China since the fifth century B.C.

The crop is certainly extremely variable in India but the several formal botanical classifications of the species which have been proposed do not seem very helpful, as they are sometimes based in part on freak forms. The species has perfect flowers and is self-compatible and an inbreeder. The amount of natural cross pollination varies with the variety and environment, the average being about 6–7 per cent.

The eggplant was taken to Africa by Arab and Persian travellers before the middle ages in Europe. First records in Europe were in the fifteenth century (Hedrick, 1919, citing Sturtevant) but the crop did not become generally known there until the seventeenth century. The early European name, eggplant, suggests that first introductions were small fruited; an early Italian name, *melazana*, meaning mad apple, later became *melongène* and Linnaeus's Latin, *melongena* (Heiser, 1969). The French name, *aubergine*, comes from the Arabic by way of the Spanish *berenjena*.

4 Recent history

Eggplant breeding has mainly been carried on in India and Japan and, to some extent, also in the USA and Europe. The principal methods used for improvement of the crop are selection from inbred lines and intervarietal crosses. The occurrence of considerable hybrid vigour in eggplants was recorded as early as 1892 by Munson in the USA and in 1926 in Japan by Nagai and Kida. Hybrid eggplants are now commonly used in many countries, especially in Japan. Though male-sterile lines have been detected, F_1 hybrid seed is still produced by emasculation and hand pollination.

5 Prospects

Breeding for disease resistance will no doubt continue to be important, especially in connection with *Phomopsis*, *Verticillium* and bacterial wilts and little leaf disease. Breeding for resistance to insects like shoot and fruit borer (*Leucinodes orbonalis*), jassids and aphids and root knot nematodes should also receive proper attention.

Several related species might offer opportunities for interspecific transfer of disease, insect and nematode resistance, but the crosses have so far proved difficult or impossible. The *Solanum* spp., *nigrum*, *khasianum*, *sisymbrifolium*, *integrifolium* and *gilo*, are possibilities. There are also prospects of improving protein and vitamin content.

6 References

Bhaduri, P. N. (1951). Inter-relationship of non-tuberiferous species of *Solanum* with some consideration on the origin of brinjal. *Indian J. Genet. Pl. Br.*, 2, 75–86.

Hedrick, U. P. (1919). Sturtevant's notes on edible plants. *N.Y. Dep. Agric. ann. Rep.*, 27, 212, pp. 685.

Heiser, C. B. (1969). *Nightshades, the paradoxical plants.* San Francisco, USA.

Kirtikar, K. R., and Basu, B. D. (1933). *Indian medicinal plants*, 3, 1757–9.

Munson, W. M. (1892). Notes on eggplants. *Maine agric. Expt. Sta. ann. Rep.*, 1892, 76–89.

Nagai, K. and Kida, S. (1926). Experiments on hybridization of various strains of *Solanum melongena*. *Jap. J. Genet.*, 4, 10–30.

Sampson, H. C. (1936). Cultivated crop plants of the British Empire and the Anglo-Egyptian Sudan. *Bull. misc. Inf. Roy. bot Gdn Kew add. Ser.*, 12, 159.

Vavilov, N. I. (1928). Geographical centres of our cultivated plants. *Proc. V Intnl Congr. Genet.*, New York, 342–69. 342–69.

Potatoes

Solanum tuberosum (Solanaceae)

N. W. Simmonds
Scottish Plant Breeding Station
Edinburgh Scotland

1 Introduction

'Of the eight food crops which may be regarded as staple in the broadest sent of the word, seven are grain . . . The other, produced in greater quantity, though on a smaller area than any one of the grain crops is a tuber crop.' (Burton, 1966). The potato is thus a major foodstuff even though it is not, by reason of relatively low dry matter content, as important for human nutrition as the crude comparison with cereals would suggest. The world crop in 1959 was estimated as 279 Mt. Potatoes are cultivated in all cool countries, with the USSR and eastern Europe as leading producers.

International trade (except in seed tubers) is negligible and the crop is always consumed more or less locally. Dietetically, the potato is remarkably good and the protein is of excellent quality, a fact of considerable historical significance.

2 Cytotaxonomic background

The tuber-bearing Solanums are one relatively small group of a very large genus. (For general systematics, see Correll, 1962.) They are classified in Subgenus Pachystemonum, section Tuberarium, subsection Hyperbasarthrum. About 170 species are included, of which fewer than ten are relevant to the evolution of the crop. The basic chromosome number is $x = 12$. Among the wild forms rather less than half occur in north and central America (to about 40°N); they include diploids, allotetraploids and allohexaploids ($2n = 24, 48, 72$) and have, at most, marginal bearing on cultivar history, a few having been used in potato breeding. The majority of species are South American (to about 45°S) and most of these are diploids. It is from a few of these diploids ($2n = 24$) that the cultivars derive, though one of the rare tetraploids (*Solanum acaule*) has also contributed (Figs. 81.1 and 81.2).

A conspicuous biological feature of the wild

potatoes is the contrast between very variable, outbred diploids and the less variable inbred polyploids; the former have a gametophytic (oppositional) *S*-allele incompatibility system (like *Nicotiana*) (with, possibly, a more complex two-locus system in some central American species) and are highly intolerant of inbreeding; the polyploids are self-compatible and often (usually?) self-pollinated. It may be that polyploids have been selected at the limits of geographical and altitudinal range by reason of their potential for inbreeding and the point deserves further study.

3 Early history (Fig. 81.2)

Wild potato tubers are all bitter to the taste and contain potentially toxic amounts of various steroidal alkaloids. The first step in the evolution of the crop must have been the recognition and selection at the gathering stage of clones that were less bitter (and therefore potentially less toxic) than usual. Wild tubers are still gathered and eaten in various places in South and Central America. The first step was therefore the emergence of alkaloid-free diploids which could safely be eaten in quantity. When this happened is not known but the general historical–archaeological context suggests the period 2000–5000 B.C., con-

Fig. 81.1 Distribution of the tuber-bearing species of *Solanum*.

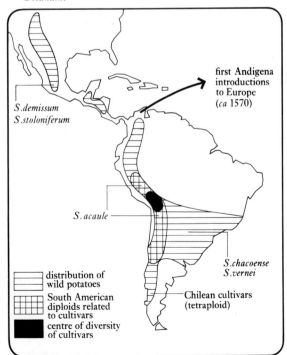

distribution of wild potatoes

South American diploids related to cultivars

centre of diversity of cultivars

S.demissum
S.stoloniferum

first Andigena introductions to Europe (*ca* 1570)

S. acaule

S.chacoense
S.vernei

Chilean cultivars (tetraploid)

currently with the domestication of the llama (Jensen and Kautz, 1974). The wild species concerned is/are not known with certainty but, on systematic grounds, several candidates in the series Tuberosa have been proposed; of these *S. brevicaule*, *S. leptophyes*, *S. canasense*, *S. soukupii* and *S. sparsipilum* are perhaps most frequently mentioned but others (e.g. *S. vernei*) have their proponents. Since related wild species hybridize readily enough with each other and with cultivated diploids, several wild forms could have been implicated, either before or after the primary domestication (see Ugent, 1970).

Whatever wild species was/were concerned, the cultivated diploids developed a great deal of new variability in foliage characters and in tuber shapes and colours. They have been classified into a number of species (with *S. stenotomum* as the most important) but they are all readily intercrossed, there are no conspicuous discontinuities and they can be treated for the present purpose simply as 'cultivated diploids'. They retain the self-incompatibility and outbreeding habits of their putative ancestors (Dodds, 1965).

The area of domestication may be assumed to be where the wild and cultivated diploids are still present and most variable, namely in the high plateau of Bolivia–Peru, in the general region of Lake Titicaca (Fig. 81.1). The diploids are still there but have been largely superseded by cultivated tetraploids and a few triploids. The tetraploids are variously referred to as *Solanum andigena*, *S. tuberosum* subsp. *andigena* or Group Andigena. They are autotetraploids, though segmental allopolyploidy is favoured by some workers (review in Howard, 1970) and they display the same great range of variability of habits, leaf characters, and tuber shapes, colours and sizes as their diploid ancestors (with which they share common *S*-alleles – Dodds, 1965). Andigena potatoes are self-compatible (cf. Lewis's rule) but intolerant of inbreeding and often rather infertile.

The general outcome of the first phases of potato evolution was the diffusion through highland South America of a great complex of diploid, autotriploid and autotetraploid potatoes with a centre of variability in Peru–Bolivia. This much was first revealed by the pioneering Russian expeditions in the 1920s. Secondary outcomes were the establishment in Chile, from migrant Andigenas, of a group of tetraploids adapted to high latitudes; and the occurrence in the central Andes of a few allotriploids (*S. juzepczukii*, $2n = 3x = 36$) and allopentaploids (*S. curtilobum*, $2n = 5x = 60$) derived from hybridization with the wild Andean tetraploid, *Solanum acaule* (Fig. 81.2).

4 Recent history

General accounts of the history of potatoes in Europe will be found in Salaman (1949), Hawkes (1967) and Simmonds (1969). The first recorded European contact with the potato was in 1537 in the Magdalena valley. The Spanish invaders became familiar with the crop and it was probably about 1570 that a Spanish ship first introduced potatoes to Europe. Legends notwithstanding, Raleigh and Drake had no hand in the introduction. From Spain, potatoes were widely spread round Europe by the end of the century and were repeatedly the object of writings and drawings by the herbalists. All the evidence goes to show that there were few initial introductions and that they were all Andigena types. Probably the first introduction to Britain (about 1590) was independent of the earlier Spanish introduction. The first potatoes to reach North America came from Europe about 1621. A source in the northern Andes for the first introduction to Europe seems very likely.

The potatoes of the central Andes were adapted to the prevailing short days of those latitudes; they tuber very late or not at all in the long days of a north temperate summer. Andean potatoes are therefore ill-adapted to Europe and, indeed, it was nearly 200 years before the crop began to have any significant agricultural impact in its new home. By the late eighteenth century, clones adapted to long days had emerged. This was no doubt accomplished simply by selecting for earliness and size of crop among seedlings raised from open-pollinated berries but there must have been a considerable element of natural selection. That Andigena has the genetic capacity to respond to such selection has been demonstrated experimentally (see below). Thus the potato is one of a considerable list of crops in which a day-length bottleneck had to be passed before adaptation to a new environment was possible.

In South America the Chilean tetraploids (a local offshoot of the Andigena Group) are also adapted to long days (lat. 45°S) and the early Russian workers argued therefore that Chile was probably the source of the European populations. History is against this view and Andigena, anyway, is capable of adaptation. A

Fig. 81.2 Evolution of the cultivated potatoes, *Solanum tuberosum*.

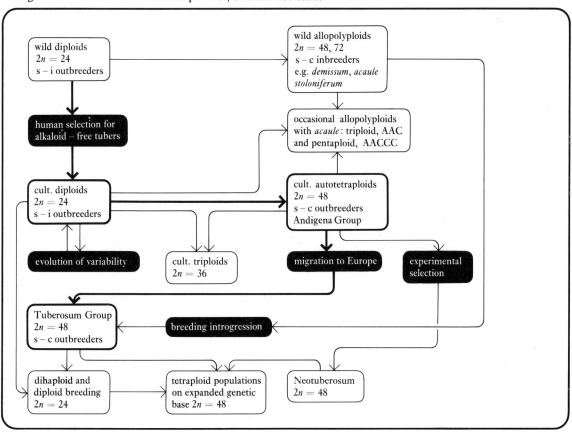

Chilean variety, however, was probably important in a later phase of potato breeding (Rough Purple Chile introduced to the USA *ca*. 1850).

The nineteenth century in Europe and North America saw rapid and sustained progress in potato breeding and modern varieties emerged by the end of the century. The products (*S. tuberosum* subsp. *tuberosum*, informally Group Tuberosum) differed from their Andigena ancestors not only in day-length response but also in having smaller haulms and fewer flowers (largely a correlated response), larger leaves, shorter stolons, fewer, larger and smoother tubers, less pigmentation and enhanced disease resistance. The changes were indeed dramatic. Opinions differ as to how long the transformation took; some authors would put the start about 1700, others nearer 1800. Certainly, the major change was accomplished in between 100 and 200 years.

Potato blight (due to *Phytophthora infestans*) was a significant episode from the 1840s. Besides the tremendous social and historical impact of the disease (Salaman, 1949) it imposed a major new selective factor on the potatoes of the time. Some resistance became available in the 1870s but the episode must be presumed still further to have narrowed the genetic base.

The modern potatoes which emerged around the end of the nineteenth century were all founded genetically upon the original Andigena introductions supplemented by (probably few) later accessions (such as Rough Purple Chile, mentioned above); wild species had no part in the story. The first interspecific crosses in potatoes were tried about 1850 but it was not until this century that wild species were extensively used in potato breeding. Essentially, the procedure has been to cross cultivars by species having desirable characters (virtually always disease resistance) and backcross to cultivars. By this means breeding stocks have become somewhat introgressed by: *S. demissum* (race-specific hypersensitivity to blight); *S. stoloniferum*, *S. chacoense*, *S. acaule* (resistances to viruses); *S. multidissectum*, *S. spegazzinii*, *S. kurtzianum*, *S. oplocense* and *S. vernei* (resistances to eelworms). Some of these programmes may ultimately be successful but none has yet made a significant contribution to potato improvement. In the late 1960s, only four varieties out of 30 (the top ten each from Holland, USA and UK) bore any genetic contribution from wild species (all from *S. demissum*) and the leading varieties were all pure Tuberosum. The average age of the leading varieties (more than 10% of acreage in each country) was over 50 years (Howard, 1970). The broad picture that emerges therefore is of slow progress on a narrow genetic (Tuberosum) base supplemented by some (yet ineffectual) introgression by several wild species.

Contemporary breeding methods are dominated by the mating system of the crop. As a (more or less concealed) outbreeder, tetraploid potatoes show inbreeding depression and a good deal of erratic sterility; they are highly heterozygous but promising clones are instantly 'fixed' by vegetative reproduction. There is yet little biometrical–genetic information about the control of economic characters; what there is, and general experience, would suggest a fairly large measure of specific combining ability. Extensive crossing among large arrays of parents is indicated, followed by selection of outstanding (presumably heterotic clones). The general pattern is of generationwise assortative mating.

The objectives of potato breeding have been, to a perhaps unusual extent, dominated by disease resistances (as the list of wild species above indicates). Of the diseases, blight has received by far the largest share of the effort; by the use of R (hypersensitivity) genes from *S. demissum*, success seemed near in the early 1930s but general failure ensued because the fungus rapidly evolved new specificities. Even multiple R-genes proved to be ineffective. By the late 1950s the Rockefeller workers in Mexico had shown that the use of polygenic 'field-resistance' ('horizontal resistance') was feasible and most potato breeding programmes now ignore R-genes. No excellent varieties with high field resistance have yet been bred but this goal will probably be achieved in the next few years. The tetraploid cultivars seem to have the genetic capacity needed and wild species are probably not necessary.

5 Prospects

The next decade will probably see considerable changes in potato varieties because: first, there is an increasing demand for clones specialized for processing uses throughout western Europe and North America; and, second, several disease resistances (notably to blight, viruses and eelworms) now being worked up in breeding stocks should emerge into practical use. None of this, however, implies any significant evolutionary change because the base population is still the Tuberosum Group supplemented by a few backcrossed resistances from wild species or primitive cultivars. Nor are significant changes in breeding methods likely though, no doubt, biometrical studies will permit some sophistication of mating patterns and selection procedures.

Of much greater evolutionary potential is the attempt now being made in Britain and the USA to broaden the genetic base of potato breeding by means of the Andigena Group (review in Simmonds, 1969). This approach was prompted by the observation (see Section 4, above) that north temperate potatoes were founded on a narrow genetic base, that they had made very rapid progress under selection in the nineteenth century and that breeding in this century seemed to be rather unsuccessful. Accordingly, a mass selection experiment using South American, unimproved Andigena stocks was started in 1959 with the object of re-creating the Tuberosum Group. Methods used were similar to those practised by nineteenth century breeders. Excellent progress was apparent in five years and in ten years (five generations) the selected population approximated to Tuberosum in performance and showed heterosis in inter-group crosses. The experiment both supports the conventional view as to the mode of evolution of the Tuberosum Group and provides wholly new potato breeding stocks on a fairly massive scale. The US experience has been similar to the British. Neo-tuberosum potatoes (as such adapted Andigenas may conveniently be called) are now entering several breeding programmes and the first products may emerge in a decade or so. Neo-tuberosum clones are expected to contribute yield heterosis, processing qualities and a useful array of disease resistances.

Potato breeding so far has almost entirely been directed towards the needs of temperate (mostly north temperate) countries. The last 20 years has seen a trend towards widening still further the range of adaption of the crop by extending it, first, to low latitudes at middle elevations (e.g. Kenya, India, Central and South America) and second, perhaps, even to the lowland tropics (Simmonds, 1971). Basic materials have so far mostly been Tuberosum stocks of north temperate origin but the breeding potential of Andigena and Neo-tuberosum in these circumstances must be very great.

Finally, we must consider the potential of diploids and dihaploids ($2n = 24$). The latter have been produced in quantity from tetraploids pollinated by diploids (Peloquin et al., 1965; Howard, 1970). They represent, in effect, a sample of female gametes. Most, as expected of an autotetraploid outbreeder, are weak and infertile but crosses with cultivated diploids can yield remarkably vigorous and productive seedlings. Dihaploids, however, will very probably have their uses in synthesizing special parents (e.g. tetraploids multiplex for disease resistance genes). As a complement to the dihaploids, at least one cultivated diploid population adapted to north temperate conditions is being developed. The outcome should be a diploid-

dihaploid gene pool utilized for genetic studies and the synthesis of special tetraploid parents.

6 References

Burton, W. G. (1966). *The potato*. Wageningen, The Netherlands.

Correll, D. S. (1962). *The potato and its wild relatives*. Renner, Texas.

Dodds, K. S. (1965). The history and relationships of cultivated potatoes. In J. B. Hutchinson (ed.), *Essays in crop plant evolution*, London, pp. 123–41.

Hawkes, J. G. (1967). The history of the potato. *J. Roy. hort. Soc.*, 92, 207–24, 249–62, 288–302, 364–65.

Howard, H. W. (1961). Potato cytology and genetics. *Bibliogr. Genet.*, 19, 87–216.

Howard, H. W. (1970). *Genetics of the potato*. London.

Jensen, P. M. and Kautz, R. R. (1974). Preceramic transhumance and Andean food production. *Econ. Bot.*, 28, 43–55.

Peloquin, S. J., Hougas, R. W. and Gabert, A. C. (1965). Haploidy as a new approach to the cytogenetics and breeding of *Solanum tuberosum*. In Riley and Lewis (eds.), *Chromosome manipulations and plant genetics*. Edinburgh and London.

Salaman, R. N. (1949). *The history and social influence of the potato*. Cambridge.

Simmonds, N. W. (1969). Prospects of potato improvement. *Ann Rep. Scott. Pl. Br. Sta.*, 48, 18–38.

Simmonds, N. W. (1971). The potential of potatoes in the tropics. *Trop. Agriculture Trin.*, 48, 291–9.

Swaminathan, M. S. and Howard, H. W. (1953). The cytology and genetics of the potato (*Solanum tuberosum*) and related species. *Bibliogr. Genet.*, 16, 1–192.

Ugent, D. (1970). The potato, *Science*, N.Y., 170, 1161–6.

82

Kola

Cola spp. (Sterculiaceae)

C. L. M. van Eijnatten
Department of Crop Science Faculty of
Agriculture University of Nairobi Kenya

1 Introduction

Of the approximately 50 *Cola* species, one, *Cola nitida*, is an important economic crop in the rain forest area of West Africa; another, *Cola acuminata*, is a minor crop in the rain forest and derived savannas of the eastern part of West Africa. The remaining edible *Cola* species are rarely cultivated, but are retained when vegetation is cleared for farming purposes.

The kola tree is cultivated for its seeds. After the removal of the fleshy seed coat from the seed, the embryos are left, commonly called kola 'nuts'. The nut is consumed for its stimulant effect, as it dispels sleep, thirst and hunger. The principal stimulating alkaloid component is caffeine. Its use is closely interwoven with local customs.

The kola tree may grow up to a height of 20 to 25 m but is more usually from 8 to 12 m high. It develops a characteristic dome-shaped canopy which reaches almost to ground level, unless planted close to other trees. The tree is evergreen with leathery leaves, many of which may absciss under exceptionally dry conditions. When all leaves are lost the tree usually dies back. This behaviour restricts the kola to the rain forest areas or areas where water is otherwise continuously available.

Cola nitida, the common kola nut, is spread over the whole West African rain forest area from Sierra Leone to Cameroon, whilst *C. acuminata* is more common in the area from Dahomey to Angola. Both species were introduced into South America during the period of the Slave Trade. *Cola nitida* is widely spread in the Caribbean Islands, whilst *C. acuminata* occurs on the Brazilian coast.

The world production of kola nuts was estimated at 175 kt in 1966, of which 3 kt came from South America and 172 kt from West Africa, with Nigeria in the forefront with 120 kt.

2 Cytotaxonomic background

The genus contains some 50 species of which the taxonomy has been traced by Bodard (1962), Chevalier and Perrot (1911) and Schumann (1900). The subgenus *Cola* (Bodard) or *Eucola* (Chevalier) includes the edible kolas. The basic chromosome number is $x = 10$. *Cola nitida* and *C. acuminata* are tetraploids with $2n = 4x = 40$ (Bodard 1962).

Pollination of *C. nitida* by *C. acuminata* or vice versa leads to seed set, but the resulting plants are invariably sterile. Abnormal meiotic pairing of chromosomes has been observed and this is a possible explanation of the high frequency of infertile kola trees in southern Nigeria, where the two species meet.

3 History

The two main species evolved in separate regions within West Africa. *Cola nitida* was restricted to the rain forest areas west of the dry coastal savanna region of southeastern Ghana, Togo and Dahomey. *Cola acuminata*, on the other hand, occurred to the east of this ecologically unsuitable area and was distributed mainly in Nigeria, Cameroon and Angola.

Both species were spared by the early inhabitants of these areas rather than taken into cultivation. The domestication of *C. nitida* took place during the last 100 years, particularly in Nigeria, outside its original area of distribution, and, to a far lesser extent, also for *C. acuminata*.

Within *C. nitida* nut colour varies from a creamy white, through pink to very dark purple. There is strong social preference for white nuts. This is the explanation of the fact that new kola farms established in Nigeria in the last 40 to 50 years have a preponderance of white nuts, introduced, according to the village elders, from Ghana. Earlier plantations must have been sown with the more common purple or red types. The selection of white nuts has been associated with selection for smaller trees, not exceeding heights of 6 to 8 m at the age of 30 to 40 years.

Kola is insect-pollinated, and almost 90 per cent of the trees are self-incompatible; cross-incompatibility is also common and the genetical situation, which has not been analysed, seems to be reminiscent of that in the related crop, *Theobroma*. These features are leading to the development of a secondary centre of diversity in southwestern Nigeria, where extreme variations occur in yielding ability of individual trees. Commonly, those trees which regularly yield well (several thousand nuts) are chosen for the establishment of new kola farms. The tremendous variation in yielding ability of individual trees becomes apparent, when one notes that the average yield per tree lies around 231 nuts (or 5 to 6 fruits) whilst individuals may produce up to 10,000 nuts (or 250 fruits).

4 Prospects

First steps in kola breeding were taken by Russell (1955), Dublin (1965) and Van Eijnatten (1969*a*, *b*). The approach was by vegetative propagation of proven high yielding trees in order to exploit existing potential quickly (Van Eijnatten, 1973). Much of the high-yielding material available in Nigeria in the 1960s was interpollinated and progeny planted out to obtain information on the combining ability of the parent trees and to identify exceptionally favourable combinations. This work will, in time, provide new materials for vegetative propagation and also open the way towards the use of second-cycle seedlings among the progeny of favourable combinations.

When the use of clones shall have become widespread, the variability available in cultivated material will slowly diminish, although the likelihood that seedling trees will rapidly be removed is not great. Nevertheless, it is of importance to collect material of interest from the point of view of disease resistance, insect resistance (stem borer, kola weevil), low stature and other desirable features. Such materials can easily be clonally maintained.

The cultivation and hence the genetic improvement of kola has been receiving attention only for the last 20 years. This is a short period for a tree crop and changes in objectives and methods of breeding are likely only to develop over a longer period of time.

5 References

Bodard, M. (1962). Contribution à l'étude systematique du genre *Cola* en Afrique occidentale. *Annales Fac. Sci. Univ.* Dakar.

Chevalier, A. and **Perrot, E.** (1911). *Les végétaux utiles de L'Afrique tropicale francaise*, fasc. VI. *Les kolatiers et les noix de kola.* Paris.

Dublin, P. (1965). Le colatier (*C. nitida*) en République Centrafricaine: culture et amélioration. *Café, Cacao Thé*, **9**, 97–115, 175–91, 294–306.

Eijnatten, C. L. M. van (1969*a*). Kola, its botany and cultivation. *Dept. Agric. Research, Royal Tropical Institute, Amsterdam, Comm.* **59**, pp. 100.

Eijnatten, C. L. M. van (1969*b*). Kolanut (*Cola nitida* and *C. acuminata*), In Ferwerda F. P. and Wit F. (eds), *Outlines of perennial crop breeding in the tropics.* Misc. Paper, **4**, Agric. Univ., Wageningen, 289–307.

Eijnatten, C. L. M. van (1973). Kola, a review of the literature. *Trop. Abstr.*, **28**(8), 1–10.

Russell, T. A. (1955). The kola of Nigeria and the Cameroons. *Trop. Agric. Trinidad*, **32**, 210–40.

Schumann, K. (1900). Uber die Stammpflanzen der Kolanuss. *Tropenpfl.*, 219–23.

Cacao

Theobroma cacao (Sterculiaceae)

F. W. Cope
University of the West Indies Trinidad

1 Introduction

Cacao is a crop of strictly tropical cultivation, the current annual production of which is around $1 \cdot 5$ Mt of dried, fermented beans, valued in round figures at £1,000 million. Consumption of cocoa beans, for the manufacture of drinking cocoa and chocolate, is today exceeding the supply, and seems likely to increase. World production in 1900 was approximately 100 kt, reaching the million-ton level for the first time in 1960.

Africa produces about two-thirds of the world crop, the rest being derived from South and Central America and the West Indies, with minor productions in Asia and the Pacific region. European countries consume more than half of the world's production of cacao, with American countries accounting for another 40 per cent. Consumption in the producing countries is quite small, the product of cultivation being mostly processed and consumed in temperate countries.

The fat content of the bean is high. Its content in drinking cocoa is reduced by about a half (to around 25%), the fat that is extracted being added to other beans to make chocolate, or sold as cocoa butter. Cocoa and chocolate products are almost always derived by blending beans from different sources. The beans, following extraction from the fruit, need to undergo a fermentation process before the chocolate aroma develops when the beans are later roasted. Fermentation temperatures kill the bean and death is followed by complex chemical reactions in the cotyledons which lead to the formation of chocolate flavour precursors.

General accounts of the crop have been presented by Hardy (1960) and Urquhart (1961).

2 Cytotaxonomic background

Theobroma cacao is the only economically important species of the genus which, according to Cuatrecasas (1964), comprises 22 species, all confined to tropical America. He recognizes six sections, with *T. cacao* alone representing one section. The *Theobroma* species all appear to be normal diploids $2n = 2x = 20$; the chromosomes are small, the maximum length of mitotic chromosomes in *T. cacao* being 2μ. Species crosses within sections often give fertile hybrids; but crosses between *T. cacao* and the other species do not give viable progeny, despite the fact that fruit enlargement is often seen when *T. cacao* is used as female. Little of breeding value, however, appears to be offered by the other *Theobroma* species.

Cacao is extremely variable. Cuatrecasas recognized two subspecies, namely, *T. cacao* subsp. *cacao* and *T. cacao* subsp. *sphaerocarpum*, the former being subdivided into four forms. All subspecies and forms interbreed to give fully fertile F_1 hybrids.

The cocoa trade recognizes three main types of product, as follows:

1   *Criollo cocoa:* the beans are large, plump and round in cross-section, with the cotyledons white or very pale violet in the fresh bean (recessive character), giving a chocolate product on roasting which is notably lacking in astringency. Fermentation times after harvest are short (sometimes no more than a day). The soft fruit wall is usually markedly 5-angled, with a warty surface. Criollo cocoa is rated the highest in quality but yields are low and the tree lacks hardiness; today only a little is available.

2   *Amazonian forastero:* the beans are flattened, with the cotyledons showing violet pigmentation in the fresh bean, giving a roasted product that may be astringent. Fermentation times range between a few days to a week or more. The hard pericarp is not usually conspicuously angled or warty. Forastero cocoas are hardy and often high-yielding and today supply the bulk of the world's cocoa.

3   *Trinitario cocoa:* may be regarded as a variable hybrid population of criollo and forastero ancestry. Cotyledon colour, bean size and shape, pod-wall characters and astringency are highly variable. The trade refers to it as a 'fine cocoa'.

Cheesman (1944) placed the centre of origin of cultivated *T. cacao* on the eastern equatorial slopes of the Andes. This area of South America appears to embrace the widest range of variability as recorded by Pound (1938) on his collecting expeditions in the Amazon basin. The centre of cultivation, however, is undoubtedly Central America, for the crop has been under cultivation there for more than 2,000 years. Purseglove (1968) believed that cacao was probably not truly wild in Central America. Cuatrecasas (1964) thought, however, that *T. cacao* in early times became spread throughout the central part of Amazonia and

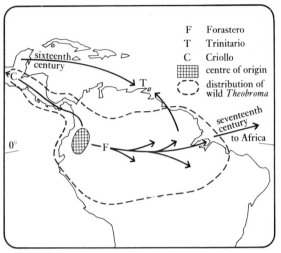

Fig. 83.1 Evolution of cacao, *Theobroma cacao*.

northwards into southern Mexico, with the two areas developing cacao populations geographically separated by the Panama isthmus.

Outbreeding in cacao, a very prolific producer of small flowers adapted to pollination by small insects, is promoted by a pollen incompatibility mechanism that is unique in the angiosperms. Materials collected in or near the suggested centre of origin seem to be uniformly self-incompatible (SI) but, away from this centre, we may find self-compatible (SC) trees. The *S*-locus shows multiple allelomorphism, the *S*-alleles bearing dominance and independence relationships towards one another (Knight & Rogers, 1955). Two SI trees will also be cross-incompatible (CI) if: (*a*) they both carry the same dominant *S*-allele; or (*b*) one has a dominant *S*-allele which in the other is accompanied by an allele independent of it; or (*c*) both carry independent *S*-alleles, one of which is common. Pollen from the SC tree appears able to set fruit on any other cacao tree; and any pollen will bring about fertilization and fruit set on the SC tree.

SI trees homozygous for an S-allele are known and progeny derived from a compatible cross between SI parents show varying degrees of cross-incompatibility with one or both parents. Control of the incompatibility reaction in *T. cacao* appears, then, to be of the sporophytic type, as postulated by Knight & Rogers (1955). That this mechanism is not a standard one however is evident from the fact that pollen germination and pollen-tube growth in SI and CI matings are uninhibited, so that pollen-tubes enter the ovules and liberate their male gametes into the embryo-sacs. Then, according to the *S*-genotype(s) of the parent(s), male and female nuclei do, or do not, fuse according to their *S*-allele contents. Counts of ovules in incompatibly-pollinated ovaries showing: (*a*) non-fusion of the first male gamete with the egg nucleus and non-fusion of the second male gamete with the polar nuclei ('non-fusion ovules'); and (*b*) normal syngamy ('normal ovules') yield the predictable ratios 1:3, 1:1 or 1:0 (Cope, 1962*a*). Abscission of the flower takes place at a well-defined abscission layer about 72 hours after pollination whenever the ovary contains a number of ovules in which syngamy has not taken place. Crosses and selfs involving SC trees and compatible crosses between SI trees give rise only to ovules showing normal gamete fusion.

Cacao (and possibly other *Theobroma* species) appears, then, to combine features of sporophytic control with those of gametophytic control of incompatibility (which suggests that the *S*-locus acts both before and after meiosis) and this peculiar combination could be related to the reproductive biology of the tree. The commonest *S*-genotype in cacao is heterozygous for alleles of different dominance status; self-pollinations on such genotypes give an incidence of about 25 per cent non-fusion and 75 per cent normal ovules in the ovary. The ratio of non-fusion to normal ovules is, of course, subject to sampling effects and it may therefore sometimes happen that a selfed ovary will contain so few non-fusion ovules that it develops into a fruit by avoidance of the abscission process. Fruit-setting from self-pollination on SI cacao trees is known to occur with a very low frequency; and selfed progeny can be obtained from an SI tree by using a mixture of its own pollen with pollen from a cross-compatible source (homozygous for a dominant marker). The SI tree which grew isolated in the Amazon forest would not then be completely non-reproductive, for the occasional success of self-pollination could result in the setting up of a small breeding community in which certain cross-compatibilities between parents and offspring and between sibs would exist. Incompatibility in *T. cacao* is totally unrelated to the conservation of female gametes, a feature in which it again contrasts with other angiosperms which have incompatibility mechanisms.

Studies made by Cope (1962*a*) on the inheritance of the SI and SC conditions in materials of Trinitario and Amazonian forastero origins suggest that the *S*-locus acts in complementary fashion with two other, independent loci, designated A,a and B,b. These two other loci are thought to produce, again by complementary action, an incompatibility precursor substance which is given highly specific properties by the *S*-alleles present. In SI trees heterozygous for *S*-alleles of different dominance status, the gametes receiving the dominant allele are fully activated against fusion with other gametes carrying the same alleles as a dominant; in trees heterozygous for independent *S*-alleles, each gamete type becomes activated against fusion with other gametes of the same genotype; and, where a tree is homozygous for an *S*-allele, all gametes are activated against fusion. Gametes carrying recessive alleles are not activated. Very likely, the fusion or non-fusion of gametes involves the cytoplasm associated with the gamete nuclei. The SC condition (a true-breeding recessive character) can arise by (*a*) homozygosity at the *S*-locus for an amorph of the *S*-series; or by (*b*) homozygosity for the recessive a and b alleles; or (*c*) both recessive conditions combined. Gametes produced by the SC tree all lack non-fusion properties, so that pollinations involving SC trees do not give rise to non-fusion ovules.

Other studies (Cope, 1962*b*), of a theoretical nature, suggest that a large, randomly mating population of SI types can attain equilibrium in respect of *S*-gene and *S*-genotype frequencies. If the *S*-alleles present in the population have simple dominance and recessive relationships, the equilibrium state would have the *S*-phenotypes represented in equal proportions; further, the *S*-gene frequencies would be calculable, with the top dominant having the lowest frequency and the bottom recessive the highest. But, once the population came to contain individuals which, as a result of mutation, were SC, then the selective advantage of recessivity becomes so pronounced that in the absence of a strong outbreeding advantage, SC would tend to displace SI genotypes in the population.

3 Early history

Man has made use of cacao in the New World for possibly 2,000 years. Historical records show that the criollo type was well known to the lowland Amerindian inhabitants of Central America, principally the Maya group. According to Hardy (1960), our words 'cacao' and 'chocolate' derive from pairs of Mayan words.

The inhabitants of Central America believed that the cacao tree was of divine origin, the seeds having come from Paradise; Linnaeus honoured this belief by naming the genus *Theobroma*, although the first reference to cacao in botanical literature appears to have been that of Charles de l'Écluse (Clusius) in 1605, who called it *Cacao fructus*.

Besides using cacao beans for the preparation of a beverage (made by boiling beans and ground maize together, with *Capsicum* pepper added), the Amerindians used them for currency, so that cacao was a valuable commodity. The Spaniards did not relish the beverage when they became acquainted with it and so did not place much value upon the tree; but later they learnt to make a palatable drink by adding sugar and vanilla to roasted and ground cocoa beans. Export of cocoa from Central America to Spain soon followed.

With cacao becoming an economically important commodity, the cultivation of the crop spread rapidly in the New World. Criollo material from Central America was planted in Venezuela and Trinidad in 1525. Then Jamaica, Haiti and the Windward Islands became important producers. In 1727 Trinidad suffered from the effects of the 'blast' (a disaster of unknown cause) which destroyed much of the criollo planting; some years later, forastero types were introduced as a replacement for the partially destroyed criollo, and hybridization between the two types generated the hybrid population which is now called 'trinitario'. The cultivation of cacao spread to Asia, and to Fernando Po and Principe in the Gulf of Guinea in the seventeenth century.

Nothing is known of the uses to which the inhabitants of the Amazon basin might have put the forastero types. Perhaps then, as now, their interest lay only in the acid-sweet pulp which surrounds the seeds.

4 Recent history

The history of cacao in the last hundred years has been one of increasing importance of Africa as a producer so that, today, the areas of early cultivation in the New World produce no more than some 30 per cent of the world's crop. Cacao was introduced into West Africa from Fernando Po by a farmer taking home a few pods, and from this very limited introduction of an Amazon forastero type (called West African amelonado) virtually all of the large West African cocoa industry grew up. The population is remarkably uniform and is SC.

Cocoa has doubtless been subject to mass selection ever since man became interested in it as a crop plant; the impact of such selection is impossible to assess since information on truly wild and undisturbed natural populations of *T. cacao* is not available. Departments of Agriculture in tropical countries, since the beginning of the twentieth century, often showed considerable interest in selecting mother-trees of good performance as sources of seed for extension of the crop and for replacement of poor bearers and missing trees in established cultivations; some Departments (e.g. that of Trinidad and Tobago) also investigated vegetative propagation methods for cacao and were able to propagate good mother-trees as buddings and grafts.

Since 1930, organized research on cacao in its many aspects has been conducted at several centres in the tropics. Improvement programmes have usually had increased productivity as their prime objective but resistance to or tolerance of pests and diseases have also been important. The crop is subject to certain viruses, particularly in West Africa ('swollen shoot disease'); fruits and vegetative parts can suffer heavy attack by *Phytophthora palmivora* throughout the tropics; *Marasmius perniciosus* is the fungus causing witches' broom disease in South America, often reducing crop very seriously by fruit infection; *Monilia roreri* produces pod rot in South America; *Ceratocystis fimbriata*, associated with *Xyleborus* borers, brings about rapid death of the tree in the New World. Capsid insects can cause serious loss of trees and crop in West Africa; the cocoa beetle (*Steirastoma* spp.) and thrips (*Selenothrips rubrocinctus*) are important pests in parts of America.

The early approach to yield improvement was to select phenotypes of superior yield for subsequent conversion to clones; successful methods of rooting semi-hardwood cuttings were evolved by Pyke in Trinidad and subsequently modified and improved in other territories. Clonal material coming from these programmes was often disappointing because: (*a*) large clone/environment interactions showed themselves; (*b*) most clones were no better than unselected material in disease and pest resistance; (*c*) the flavour derived from clonal beans, even in mixtures, was rarely acceptable to manufacturers; and (*d*) the rooted cuttings were expensive to produce and distribute.

When, later, deliberate hybridization was undertaken, the objective was often the identification of good recombinants which were then converted to clones. But, frequently, the hybrid progenies showed remarkable heterosis, so that the possibility of growing hybrid or synthetic varieties by seed became apparent. The last decade has therefore seen efforts directed towards the identification of parents of good combining ability, preferably also with resistance to various diseases.

Seed of hybrid varieties in cacao can be made cheaply where one or both parents is SI (review in Bartley and Cope, 1973). A crossing-block can be set up, in isolation from other cacao, with cuttings of the two parents, one or both of which could be SI. Where one parent is SI, fruits are taken only from this parent; but, if both parents are SI, fruits may be taken from all cuttings in the crossing-block. Cross-compatibilities permitting, more than two parents might be included in one crossing-block. Hybrid varieties could also be derived from parents of various degrees of inbreeding where it was found desirable to reduce the variability usually encountered in the progeny of non-inbred parents (Bartley, 1957). Natural cross-pollination within a crossing-block could also be employed to produce polycross seed and synthetics. Inbreeding programmes often form part of the breeding activities of research organizations, not only to breed parents with some degree of homozygosity for the production of hybrids but also to breed materials homozygous for such desirable traits as disease resistance. Only in rare circumstances would the incidence of self-incompatibility render inbreeding difficult or impossible.

Since cacao is a flavour crop, with utilization normally based on the blending of beans from different sources, the cacao breeder usually has to maintain traditional flavour in his new varieties. Where such a traditional flavour arises from heterogeneity (as in Trinitario cacaos), the breeder may need to produce a wide range of improved genotypes for planting in mixture. Bean size is another character that the breeder needs to pay attention to, since manufacturers do not like small beans.

Several cacao-collecting expeditions have been mounted in various parts of the American continent in the last 30–40 years. Pound's travels in South America in 1937 and 1943 have provided breeding centres throughout the tropics with valuable breeding materials. Collections made on later expeditions are still largely under assessment. Most centres show a commendable willingness to exchange materials.

5 Prospects

The demand for cocoa is likely to continue to rise and the prospects for efficient production seem good. The bulk of the world's cocoa is still produced from old and ageing plantings of unimproved seedlings and the trend will surely be towards the replacement of these by high-yielding hybrid and synthetic varieties, locally well adapted, responsive to intensive cultivation and yielding a product acceptable to the manufacturer. The present-day emphasis on meeting manufacturers' flavour requirements may diminish as knowledge of the components of chocolate flavour increases. Clonal propagation will, of course, remain useful – indeed essential – to the breeder but is certain to decline as a means of propagation in agricultural practice. Cacao is thus one of the very few crops in which a trend from seedlings to clones has been reversed. Clearly, the development, maintenance and dissemination of collections will be as important in the future as it has been in the past. Thus the recent trend towards widening the genetic base of the crop will be continued.

6 References

Bartley, B. G. (1957). Methods of breeding and seed production in cacao. *6th Mtg Interamerican Cacao Tech. Committee, Bahia, 1966.*

Bartley, B. G. and **Cope, F. W.** (1973). Practical aspects of self-incompatibility in *Theobroma cacao.* In R. Moav (ed.), *Agricultural genetics,* New York and Toronto, 109–34.

Cheesman, E. E. (1944). Notes on the nomenclature, classification and possible relationships of cacao populations. *Trop. Agriculture, Trin.,* **21,** 144–59.

Cope, F. W. (1962*a*). The mechanism of pollen incompatibility in *Theobroma cacao* L. *Heredity,* **17,** 157–82.

Cope, F. W. (1962*b*). The effects of incompatibility and compatibility on genotype proportions in populations of *Theobroma cacao* L. *Heredity,* **17,** 183–95.

Cuatrecasas, J. (1964). Cacao and its allies: a taxonomic revision of the genus *Theobroma. Contrib. US Nat. Herb.,* **35,** 379–614.

Hardy, F. (1960). *Cacao Manual.* I.A.I.A.S., Turrialba, Costa Rica.

Knight, R. and **Rogers, H. H.** (1955). Incompatibility in *Theobroma cacao. Heredity,* **9,** 69–77.

Pound, F. J. (1938). *Cacao and witchbroom disease in South America.* Trinidad.

Purseglove, J. W. (1968). *Tropical crops: dicotyledons* **2.** London.

Urquhart, D. H. (1961). *Cacao.* London, 2nd edn.

Jute

Corchorus spp. (Tiliaceae)

D. P. Singh
Jute Agricultural Research Institute
Barrackpore West Bengal India

1 Introduction

The two species of *Corchorus*, *C. capsularis* and *C. olitorius*, yield a bast fibre second in importance among the vegetable fibres only to cotton. By far the largest production is in the Ganges–Brahmaputra delta where *capsularis* exceeds *olitorius* in a ratio of about 3:1. The crop is produced by small farmers and needs a plentiful supply of water for retting. For a survey of economic botany see Purseglove (1968).

2 Cytotaxonomic background

Corchorus contains about 40 species and is pantropical in distribution. Most of the few species that have been counted, including the cultivated ones, have $2n = 2x = 14$. A few tetraploids ($2n = 28$) are also known. There is no information about cytotaxonomic relationships of the wild species. The two cultivated ones can be crossed, but only with great difficulty (Bhaduri and Bairagi, 1968). Swaminathan *et al.* (1961) reported translocations in the hybrid and Datta (1958) detected considerable differences in chromosome morphology between the two species.

There is much disagreement in the literature as to the 'natural' distribution of the two species. *Corchorus olitorius* is pan-tropical but may often be escaped rather than wild. Kundu (1951) regarded it as primarily African; *capsularis* has often been regarded as Chinese but Kundu thought the evidence for its being truly native in Indo-Burma was strong.

3 Early history

Corchorus olitorius has been used (hardly cultivated) as a minor vegetable in Africa and the Middle East for a very long time; hence, no doubt, some of the uncertainty as to where it is truly native. As fibres, both species were certainly domesticated in India in quite recent times. There is no Sanskrit word for jute and the name (a local Indian one) was first given general currency by the botanist Roxburgh, in Calcutta in the late eighteenth century (Kūndu, 1951). The domestication of both species as fibre crops was the outcome of a deliberate search for new fibres to replace hemp for cordage and sacking. Success came in the middle of the nineteenth century, after the Dundee spinners had learned (in 1838) how to spin the new material. It is, perhaps, surprising that the early history of such a recent crop is so poorly documented.

From the fact that wild *Corchorus* species are generally low, and branched, it may be inferred that early selection was directed towards the development of a tall, unbranched habit and, perhaps, a shorter, strictly annual, lifecycle. Cultivated plants may be up to 5 m tall: *capsularis* is regarded as ecologically the more adaptable (it tolerates deep flooding) but *olitorius* has the finer fibre.

4 Recent history

Both species are naturally more or less inbred and are tolerant of selfing. Crossing rates have been estimated in the range $0 \cdot 5 - 10$ per cent. Breeding began in Bengal in 1904 and successful pure lines of both species emerged in 1920. Later, mass-selection was successfully practised.

Jute plants fruit after fibre would normally have been harvested. Fibre yield can, however, be fairly well predicted from plant height and basal stem diameter by means of a selection index (Shukla and Singh, 1967).

The flowering process is day-length sensitive. In Bengal, early-sown *olitorius* flowers prematurely but a Sudanese strain (from a lower latitude) did not bolt under similar conditions and proved useful in breeding.

Fibre quality is of great importance in any breeding programme. Long, fine, fibres, smooth and not 'looped' are desired; fortunately, these characters all give evidence of useful heritability (Singh, 1971; Maiti, 1974). Some work on induced mutations has been reported (Singh *et al.*, 1973).

5 Prospects

No doubt breeding will continue upon established lines and there must yet be plenty of room for varietal improvement. Perhaps the most valuable advance would be a means of effecting recombination between *olitorius* and *capsularis*. As noted above, the cross is very difficult (Bhaduri and Bairagi, 1968) and extensive recombination has not yet been achieved. The allotetraploid, OOCC, might provide a useful route and might even be valuable in its own right. It has not yet been studied. The potential of wild *Corchorus* species for breeding is unknown; certainly there is a great deal of variability and the genus – like many genera of Tiliaceae – is highly fibrous.

6 References

Bhaduri, P. N. and **Bairagi, P.** (1968). Interspecific hybridization in jute (*Corchorus capsularis* × *C. olitorius*). *Sci. Cult.*, **34**, 355–7.

Datta, R. M. (1958). Chromosome studies in the genus *Corchorus. Ind. Agriculturist*, **2**, 120–2.

Kundu, B. C. (1951). Origin of jute, *Ind. J. Genet. Pl. Br.*, **11**, 95–9.

Purseglove, J. W. (1968). *Tropical crops. Dicotyledons*, **2**, London, 613–9.

Maiti, R. K. (1974). *Histomorphological studies of some long fibre crops in relation to yield and quality*. D.Sc. Thesis, Calcutta University.

Shukla, G. K. and **Singh, D. P.** (1967). Studies on heritability, correlation and discriminant function selection in jute. *Ind. J. Genet.*, **27**, 220–5.

Singh, D. P. and **Joseph, J.** (1970). Origin and evolution of cultivated jute (*C. capsularis* and *C. olitorius*). *Symp. Ind. Soc. Genet. on Crop Plant Evolution.*

Singh, D. P. (1971). Estimates of correlation, heritability and discriminant function in jute. (*C. olitorius*). *Ind. J. Hered.*, **2**, 65–8.

Singh, D. P., Sharma, B. K. and **Banerjee, S. C.** (1973). X-ray induced mutations in jute (*C. capsularis* and *C. olitorius*). *Genet. agr.*, **27**, 115–47.

Swaminathan, M. S., Iyer, R. D. and **Sulbha, K.** (1961). Morphology, cytology and breeding behaviour of hybrids between *Corchorus olitorius* and *C. capsularis. Curr. Sci.*, **30**, 67–8.

Carrot
Daucus carota (Umbelliferae)

O. Banga
Diedenweg 6 Wageningen
The Netherlands; *formerly* Institute of Horticultural Plant Breeding Wageningen

1 Introduction

Carrot is a cool weather plant. It is grown round the world in temperate climates in spring, summer and autumn and in sub-tropical climates during the winter. The carrot is not one of the most important vegetables, but still has a significant place. It serves as an ingredient in soups and sauces and in dietary compositions but also as a vegetable in itself or in combination with peas. It used to be grown for the production of carotene (from which the mucous membrane of the intestines can make vitamin A) but, following industrial synthesis of carotene (1947), the culture of carrots for this purpose declined.

2 Cytotaxonomic background

All cultivated carrots, together with the wild ones which are known in Europe, belong to the species *Daucus carota*, with $2n = 2x = 18$. Most wild *Daucus* forms are found in southwestern Asia and the Mediterranean, a few in Africa, Australia and America. All wild forms investigated in Asia, Asia Minor, Japan and the USA have the same chromosome number as their European relatives. Neither polyploidy nor structural changes in the chromosomes seem to have played a role in the differentiation of the species (Whitaker, 1949).

Daucus carota ssp. *carota* is the commonest wild form in Europe and southwestern Asia. The ssp. *maximus* is found in the Mediterranean and eastwards to Iran; *maritimus, commutatis, hispanicus* and *gummifer* grow in the western part of the Mediterranean, *fontanesii* and *bocconei* in the central part, and *major* in the Balkans.

The ssp. *carota*, as found in Afghanistan, is described by Mackevic (1929) as rather variable in many morphological characters. The roots vary in degree of ramification, fleshiness and colour. Some are white and some coloured in varying degrees by anthocyanin.

The cultivated *D. carota* intercrosses freely with ssp. *carota* and probably with all or most of the other wild forms. It is an outbreeder. No incompatibility has been observed, but selfing of individual flowers and also of individual umbels is practically prevented by protandry.

3 Early history

The western carotene carrot has been derived from the eastern anthocyanin carrot, and the latter has probably been developed from anthocyanin-containing forms of *D. carota* ssp. *carota* such as those found in Afghanistan. Forms with fleshier, smoother, less forked and better coloured roots must have been selected to provide the earliest cultivars. Their roots may have been shades of purple or violet, from light to almost black, and their shapes generally long-conic. According to Mackevic (1929), Afghanistan is the centre of diversity of the anthocyanin carrot so that this country is suggested as the primary centre of origin.

From written sources we know that the cultivated anthocyanin carrot, in most cases accompanied by a yellow mutant devoid of anthocyanin, spread westwards and eastwards. They can be traced in Asia Minor in the tenth and eleventh centuries, in Arab-occupied Spain in the twelfth century, in continental northwestern Europe in the fourteenth century and in England in the fifteenth century. Their introduction to China was in the fourteenth century and to Japan early in the seventeenth century.

The carrots that were grown in northwestern Europe before and during the sixteenth century were all purple or yellow and of long-conic shape. The purple type was at first the more popular, probably because its taste was somewhat milder. However, when cooked, the root exuded anthocyanin which gave soups and sauces a brownish purple colour. The yellow type therefore became preferred. In The Netherlands about 1600, selection was initiated to derive a more orange-coloured type from the yellow. The orange colour was gradually intensified and, early in the seventeenth century, the Long Orange had been established. From this variety, the finer so-called Horn carrots were selected which were, by 1763, differentiated into three orange varieties differing in earliness and size: the Late Half Long (the biggest), the Early Half Long and Early Scarlet Horn (the smallest). All present varieties of the western carotene carrot have been developed from these four closely related varieties, either by simple selection or by intercrossing different types (Banga, 1963).

Anthocyanin carrots are still grown in eastern countries, together with most of the western orange varieties.

Lycopene (the colouring substance of tomatoes) may also occur in carrots. Katsumata *et al.* (1966) found a relatively high content in the Japanese variety Kintoki. There are also white carrots which probably originated as mutants of the yellow. They had a limited popularity but are of very little importance now.

A notable point in the evolution of carrots in northern latitudes is that there has been an important change in photoperiodic reaction. When plants adapted to sub-tropical latitudes are grown in the northern summer they respond to long days by bolting before the root has thickened properly. Long-day tolerance must have been a consequence of prolonged rogueing of bolters. Simultaneously, no doubt, selection must have favoured types with increased sensitivity to, indeed the need for, vernalizing temperatures, because, after storage in a cool place during winter, such types will flower and seed more readily.

In the seventeenth and eighteenth centuries Holland played a leading role in carrot breeding, succeeded by France in the nineteenth century. Later, several other western countries and Japan made important contributions.

4 Recent history

Carrot breeding on a scientific basis started very recently. The methods previously employed were mass-selection or combined mass-pedigree selection. The outbreeding character of the crop and consequent inbreeding depression forbade the use of inbred lines and the development of a high degree of uniformity. Uniformity, however, is wanted for mechanical harvest and handling. This has encouraged the adoption of the F_1 hybrid approach, based upon male-sterile lines. By the use of a suitable cytoplasm and one or more genes, it has proved possible to facilitate production and maintenance of hybrids. But there are difficulties. First, part of the lack of uniformity of ordinary varieties is due to lack of uniformity of seed behaviour, which is associated with the fact that seeds of a higher order umbel start growth and ripen later than those of a lower order umbel; and, second, the large scale maintenance of hybrids has, so far, not been very successful because of loss of vigour of inbred components or lack of stability of sterile or maintainer lines. Nevertheless, the use of hybrids has started and is extending. In the USA, they are not yet dominant but their share is slowly increasing. In Japan, some hybrids are in commercial production. In Europe they are in an advanced experimental stage.

Regardless of breeding plans adopted, the objectives of carrot breeding are largely concerned with yield, shape, colour, earliness, top-root ratio, resistance to bolting and quality characteristics.

Yielding capacity depends primarily on four factors: root length, root shape, dry matter content and resistance to unfavourable conditions and pathogens. Yielding capacity increases with root length but root length is limited to about 25–30 cm by the requirement that harvesting and handling of the roots must be practicable without too much risk of damage. A cylindrical root weighs more than a conical one of the same length but the latter is less readily affected by unfavourable growth conditions. So the most popular varieties are slightly to rather conical. Generally, a fairly high dry matter content is preferred because this favours good storage and transport characters. In future, breeding for resistance to pathogens may assume greater importance.

Earliness can be achieved either by smaller size or higher growth rate or both. In the series: Amsterdam Forcing–Nantes–Berlikum, the first variety is the smallest and the earliest. On the other hand, the orange (but annual) North African variety, Muscade, is larger but nevertheless earlier than Amsterdam Forcing because of its higher growth rate.

The top-root ratio varies considerably between varieties. The carrot has a juvenile phase during which ripening is not possible. The length of this phase is longer in types with large tops and strong taproots, shorter in those with small tops and weak taproots. This situation has led to the evolution of large-top varieties in continental climates and of small-top varieties in more maritime environments where at least the first part of the growing season is cool. The reason is that the large-top types will, when grown under warm conditions from the start, make a sufficiently large vegetative frame before high temperatures stop growth and turn the balance toward ripening. Small-top varieties, on the other hand, will, under warm conditions from the start, stop growth and turn to ripening when they are so small that no yield of any consequence is possible (Banga and de Bruijn, 1968).

Between 1940 and 1945, high-carotene types with two or three times the ordinary carotene content were bred in the UK to minimize night blindness among aviators. As carotene is now synthetically manufactured and very high carotene roots tend to be rather dry, breeding programmes are now aimed, not at the carotene as such, but rather at an even colouring throughout the whole root in both core and flesh, in balance with the maintenance of some degree of tenderness.

5 Prospects

As has been shown above, the genetic base of modern carrots is narrow, being founded upon a few eighteenth century Dutch varieties. Eastern genetic material could be used to widen the base, as Japanese breeders have long used material from both sources. Much effort is still going into the development of F_1 hybrid varieties but the problems, both genetic (of stability of components) and physiological (of seed uniformity), are still formidable. There may be opportunities to use annual material to improve earliness under short day-lengths. Resistance breeding, already begun in some programmes, will surely be extended.

6 References

Banga, O. (1963). *Main types of western carotene carrot.* Zwolle.

Banga, O. and **De Bruijn, J. W.** (1968). Effect of temperature on the balance between protein synthesis and carotenogenesis in the roots of carrot. *Euphytica,* 17, 168–72.

Katsumata, H., Yasui, H., Matsuo, Y. and **Yamazaki, K.** (1966). Studies on premature bolting and carotene, lycopene content in carrot. *Bull. hort. Res. Sta. Japan, Ser. D,* 4, 107–29.

Mackevic, V. I. (1929). The carrot of Afghanistan. *Bull. appl. Bot., Genet. Pl. Br.,* 20, 517–57.

Whitaker, T. W. (1949). A note on the cytology and systematic relationship of the carrot. *Proc. Amer. Soc. hort. Sci.,* 53, 305–8.

Grapes

Vitis, Muscadinia (Vitaceae)

H. P. Olmo
University of California Davis USA

1 Introduction

The grapevine is a perennial, woody vine climbing by coiled tendrils. As a cultivated plant it needs support and must be pruned to confine it to a manageable form and to regulate fruitfulness. The fruit (a berry) is juicy and rich in sugar (15–25%), in roughly equal proportions of dextrose and levulose. It is the commercial source of tartaric acid and is rich in malic acid. Cultivation of the crop is largely concentrated in regions with a Mediterranean-type climate, with hot dry summers and a cool rainy winter period.

The world's vineyards occupy about 11 Mha. The principal product is wine, the mean annual output of which, 1964–68, was 280·9 Mhl. About 10 per cent of this production enters international trade. The biggest producers circle the Mediterranean, with Italy, France and Spain in the lead. The production of table grapes consumed as fresh fruit is about 6 Mt and the leading growers are Italy, Turkey, Bulgaria, the USA, Greece and Portugal. Production of raisins, largely sun-dried fruit of seedless cultivars, reached 0·9 Mt in 1970, the leading producers being the USA, Turkey, Greece and Australia.

2 Cytotaxonomic background

Lavie (1970) has summarized the cytotaxonomy of the family. *Vitis* contains about 60 species, but botanical knowledge is incomplete. This genus is unique amongst the 12 recognized in the family Vitaceae in having 38 very small somatic chromosomes that regularly form 19 bivalents at meiosis. Wild grapes can be divided into three geographical groups: American, Middle Asian and Oriental. North America, especially the southeastern and Gulf region of the USA, is particularly rich in *Vitis*. Bailey (1934) lists 28 species, but does not include Mexico.

Most other related genera, including *Muscadinia*, have $2n = 2x = 40$. Formerly classified as a section of *Vitis*, *Muscadinia* has only three known species, restricted to the southeastern USA and northeastern Mexico. The colonists of the Carolinas cultivated *M. rotundifolia* directly and its domestication dates from the latter part of the seventeenth century. Since this species does not hybridize naturally with sympatric species of *Vitis*, it represents an example of the domestication of a single species *in situ*. Though isolated in nature from *Vitis*, *M. rotundifolia* can be hybridized experimentally with *V. vinifera* and the cross has been explored as a means of improving the disease resistance of *vinifera* and the fruit quality of *rotundifolia*.

The cytogenetics of the F_1 (*vinifera* × *rotundifolia*) with 39 somatic chromosomes and its backcross derivatives have been studied in some detail (Patel and Olmo, 1955). Unlike hybrids between *Vitis* species (which are fertile), the intergeneric hybrids are highly or completely sterile, though occasional viable seeds are obtained in some combinations. At meiosis about 13 bivalents are formed, with univalents. The genomic formula of the F_1 hybrid is thus $13 R^r R^v + 7A + 6B$, in which 13 chromosomes of *vinifera* and *rotundifolia* are homologous enough to pair. The ancient basic chromosome numbers in the family are probably 5, 6 and 7. *Vitis* species are thus ancient secondary polyploids involving three basic sets in the combination $(6+7)+6 = 19$. *Muscadinia* species, on the other hand, are $(6+7) +7 = 20$. Both have undergone diploidization to give regular bivalent pairing.

The delimitation of species in *Vitis* has been extremely difficult if not altogether artificial. Taxonomists have been reduced to using such ephemeral characters as degree of hairiness of the young shoots or leaves. Many species are sympatric with one or more others and are extremely variable. This variability reaches its greatest expression in passing from uniform tropical to subtropical environments where great differences in rainfall exist within short distances and isolated communities of vines become differentiated. The populations represent ecospecies rather than species. All known *Vitis* species can be easily crossed experimentally and the F_1 hybrids are vigorous and fertile. Studies of natural populations indicate that hybridization has occurred and continues. Even though species cannot be delimited on degree of genetic isolation, the species classification remains useful because of its practical value in separating norms of variation that are important in breeding. For example, *riparia* is highly resistant to phylloxera but *vinifera* quickly succumbs; and *riparia* roots readily from dormant cuttings but *aestivalis* does not.

3 Early history

The estimated 10,000 cultivars of the Old World are thought to derive from the single wild species, *V. vinifera*, of Middle Asia, still found from northeastern Afghanistan to the southern borders of the Black and Caspian Seas. Legend and tradition favour ancient Armenia as the home of the first grape and wine culture. Small refuge areas isolated by the glacial epochs are found scattered in southern Europe, but their role in domestication is questionable. Negrul (1938) has proposed three principal groups of cultivars: *occidentalis*, the small-berried wine grapes of western Europe; *orientalis*, the large oval-berried table grapes; and *pontica*, intermediate types of Asia Minor and eastern Europe.

The fruit of wild *vinifera* (*sylvestris*) is palatable and the wine is of a quality comparable to that made from present cultivars. It was used *in situ* long before any settlement occurred. Domestication started when migratory nomads marked forest trees (usually poplar, pear, willow, plum or fig) that supported particularly fruitful vines. This was most often near watering holes serving their herds. Sparing these vines was associated with a spiritual taboo respected by other tribes as well. As sedentary agriculture developed and the mixed deciduous forest was cleared, fruit trees and vines were spared along boundary lines where irrigation ditches were developed and the vines were out of reach of grazing animals. Vineyards as such developed later when they could be protected by high mud walls from the ever-present sheep and goats, but this came as part of village settlement.

Cultivation of the wine grape was under way in the Near East as early as the fourth millennium B.C. There is no evidence of any cultivation west of Greece until the first millennium B.C. The products of the vine were exported westwards from very early times, to be followed later by the practices peculiar to viticulture and by domesticated varieties (Helbaek, 1959). The westward movement fanned out from Asia Minor and Greece, following the Phoenician sea routes. During the Roman period, the spread of the vine was associated with that of the Christian faith; wine is a necessary ingredient in the consecration of the Mass. In the Middle Ages, the Catholic monasteries throughout Europe were the guardians of select vineyards. A vineyard covered by the eruption of Vesuvius in A.D. 79 illustrates the practices recommended in the early Roman agricultural manuals of that era (Jashemski, 1973). The vine followed the main river valleys, the Danube, Rhône, Rhine, Tiber and Douro and, by A.D. 55, the northernmost vineyards were being established along the Moselle Valley in Germany. The *vinifera* grape was introduced to the New World at the time of discovery and later accompanied practically all the Spanish and Portuguese voyages of discovery and conquest (Fig. 86.1). The first recorded introduction to the east coast of the United States was in 1621 by the London Company but this was probably preceded by the Spanish landings in Florida. The most recent incursions of *vinifera* are in tropical countries; thus, in 1958, *vinifera* was introduced into

Fig. 86.1 Evolutionary geography of the grape, *Vitis vinifera*.

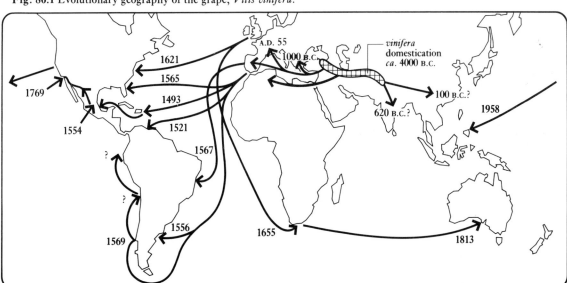

the Phillipines from California.

As the *vinifera* grape was introduced into zones beyond its natural range it often hybridized with native *Vitis* to produce new races better adapted to local environments. Thus, new American hybrids arose along the Atlantic seaboard as spontaneous seedlings which were prized as new varieties. The Alexander, Concord and Delaware are examples. In the Caribbean islands and Venezuela, introduced *vinifera* has introgressed with the tropical *caribea* to produce new vigorous races of Criollas that are disease resistant and tolerant of the climate, giving hopes of a grape culture where none could survive before. Some Japanese and Chinese cultivars are oriental species introgressed with *vinifera*.

Wine grapes are seedy, acid and juicy. Among the dessert grapes, seedlessness has evolved in a wide range of expression (Stout, 1936). On the one hand are cultivars that have very small, parthenocarpic berries such as the ancient Greek variety Black Corinth, dried to produce currants. The Kishmish or Sultanina, the most important raisin variety, is said to be stenospermocarpic, abortion of the seeds occurring soon after fertilization. Seedless varieties arise by somatic

mutation but are of different types and involve different genetic backgrounds (Olmo and Baris, 1973).

4 Recent history

Vitis species are dioecious or, occasionally, subdioecious when some male flowers transform to hermaphrodites. The sexual type is determined by three alleles. This is a primitive type of sex determination in which gross differentiation of sex chromosomes has not occurred (Negi and Olmo, 1971). The primitive hermaphrodite is Su^+Su^+. A dominant mutation, Su^F, suppresses ovary development to produce maleness. A recessive allele, Su^m, results in reflexing of the filament and in sterile pollen, to produce functional femaleness. In natural populations, males (Su^FSu^m) and females (Su^mSu^m) occur in equal numbers and cross-pollination by wind and bees occurs. The dominance relationship of the three alleles is $Su^F > Su^+ > Su^m$.

Practically all cultivars of Europe and the New World are hermaphrodites (Su^+Su^+ or Su^+Su^m) and self-pollinating. In Middle Asia, however, many are female. In a warm, dry climate and with close planting of mixed cultivars, cross-pollination is effective. As

Fig. 86.2 Evolution of the grapes, *Vitis* and *Muscadinia*.

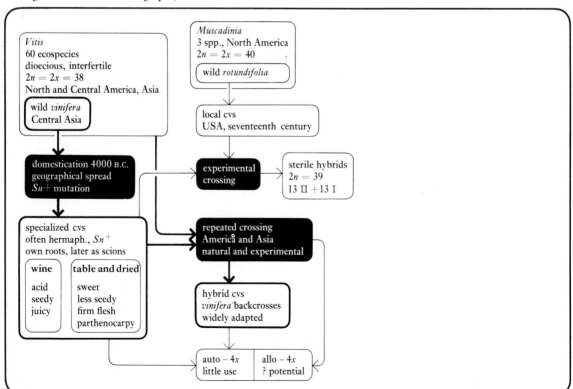

these female cultivars are moved to new areas and isolated, they are rapidly selected against because of poor fruitfulness. Female vines are useful in breeding, eliminating the need for emasculation (Levadoux, 1946).

No haploids ($n = 19$) have been reported in *Vitis* and the only aneuploids ($2n+2$) appear as rare aberrants. Autotetraploids were first described in 1929 and have been found to arise spontaneously in most cultivars. They are often periclinal chimeras; only two layers are involved in meristem differentiation, so that three types of tetraploid chimera have been found: 2–4, 4–4 and 4–2. The larger berry size attracted attention to the possibility of producing improved table grapes but, in general, the auto-tetraploids have poor cultural characteristics, being less fruitful, irregular in berry size and more fragile; also, the root system is weaker and tetraploids are better grafted on diploid rootstocks. However, a few tetraploid varieties have long been grown commercially in greenhouses where special attention to pollination and cultural factors is possible. Allo-tetraploids (Jelenković and Olmo, 1969) are more promising, since the undesirable features of the auto-tetraploids are not so evident and selection can proceed in a wider genetic base. Triploids are highly sterile but may be useful for vigorous rootstocks, especially if a wide range of resistance to soil pests is desired. Pentaploids are weak and useless.

Vinifera grapes were propagated from earliest times by cuttings or layering and remained relatively free of pests and diseases. However, vines began to die in French vineyards in 1860 and, in 1868, a root aphid, *Phylloxera*, was identified as the cause. This insect had been introduced from the USA where it lived as a natural symbiont on tolerant native vines. Within a few years, thousands of acres were ruined and, eventually, nearly all the vineyards of Europe were in trouble. It was noticed that American hybrids such as Isabella, Herbemont, Concord and others which had been introduced as exotics years before, showed some tolerance. The French government then sent specialists to the United States to discover the best sources of resistance. From Missouri and other areas of the midwest, thousands of cuttings and seeds were sent back to France for local selection. A few proved tolerant, were widely propagated and utilized as resistant rootstocks. Selections of *riparia*, *rupestris* and *berlandieri* proved most useful. From about 1880 onwards, interspecific hybrids were deliberately bred as rootstocks. Thus *berlandieri* was the best adapted to calcareous soils but rooted with difficulty, so it was crossed with *vinifera* to improve propagation. The

breeding of rootstocks was the first massive improvement programme, but the germ plasm of the *vinifera* scions remained intact.

Grafting, however, is expensive. Many breeders therefore set out to breed new vines that would combine resistance to phylloxera with fruit of good wine quality, the direct producer. After almost a century, this ideal still remains a dream and we speak now of 'French hybrids'. However, some of the hybrids proved valuable in other ways; for example, in having better resistance to fungus diseases and greater hardiness. In some areas they were better adapted than ordinary *vinifera* and produced wine of passable quality. They form the base of new wine industries in many parts of the world where *vinifera* is not well adapted. The starting point was a female vine, selection 70 (*rupestris* × *lincecumii*), sent by Jaeger from Missouri to Contassot in France in 1882. Contassot distributed open-pollinated seed to Couderc and Seibel who produced the first series of hybrids. The work continues and some of the more recent hybrids have germ plasm from as many as six American species, but backcrossing to *vinifera* is still practised to improve quality. For a summary of breeding programmes and accomplishments, particularly in Europe, refer to Neagu (1968).

Grapes are outbreeders. Cultivars are highly heterozygous and carry a heavy load of deleterious recessives. Inbreeding depression is severe so that, by the second or third generation, sterility usually ensues. The most successful breeding method is to maintain heterozygosity by crossing the best representatives of unrelated lines, resorting occasionally to closer mating to concentrate desirable combinations of characters.

5 Prospects

Cultivar improvement is increasingly directed toward disease and insect resistance. A high priority is given to virus resistance, since many of the world's oldest and most renowned vineyards are seriously menaced by soil-borne infections. Native species must be more thoroughly studied, screened and compared as sources of resistance. Allopolyploidy as a tool to produce larger berry size and better cultural features has been neglected in the quest for improved table varieties. We should see the use of native tropical species as a base in greatly extending the zone of commercial grape culture. As before, *vinifera* must be used to introduce high quality. Hardy clones having short growth cycles should further extend the range. A beginning has been made with oriental *amurensis* but American species can also be used. Selected female cultivars to obtain mass hybridization can be useful. Increasing need for mechanization of harvesting, pruning and other cultural methods will place new demands on the breeder. The long generation time from seed to fruit (three to five years) could, perhaps, be shortened by the use of biochemical methods of selection at the earliest possible stage of seedling development.

6 References

Bailey, L. H. (1934). Vites peculiares ad Americam borealem. *Gent. Herb.*, 3, 149–244.

Helbaek, H. (1959). Domestication of food plants in the old world. *Science, N.Y.*, 130, 365.

Jashemski, W. F. (1973). Large vineyard discovered in ancient Pompeii. *Science, N.Y.*, 180, 821–30.

Jelenkovic, G. and Olmo, H. P. (1969). Cytogenetics of *Vitis*. V. Allotetraploids of *V. vinifera × V. rotundifolia Vitis*, 8, 265–79.

Lavie, P. (1970). *Contribution a l'étude caryosystématique des Vitacées*. Thèse, Faculté des Sciences de Montpellier, 1, pp. 213.

Levadoux, L. (1946). Étude de la fleur et de la sexualité chez la vigne. *Ann. École Nat. Agr. Montpellier*, 27, pp. 89.

Neagu, M. M. (1968). Génétique et amélioration de la vigne. Rapport général. *Off. Int. Vigne et Vin. Bull.*, 41, 1301–37.

Negi, S. S. and Olmo, H. P. (1971). Conversion and determination of sex in *Vitis vinifera (sylvestris)*. *Vitis*, 9, 265–79.

Negrul, A. M. (1938). Evolution of cultivated forms of grapes. *C. R. Acad. Sci., U.S.S.R.*, 18, 585–8.

Olmo, H. P. and Baris, C. (1973). Obtention de raisins de table apyrènes. *O.I.V. intnl Symp., Cyprus*, 32–11.

Patel, G. I. and Olmo, H. P. (1955). Cytogenetics of *Vitis*. I. The hybrid *V. vinifera × V. rotundifolia. Amer. J. Bot.*, 42, 141–59.

Stout, A. B. (1936). Seedlessness in grapes. *N.Y. Agric. Exp. Sta. tech. Bull.*, 238, pp. 68.

Timber trees

R. Faulkner
Forestry Commission Northern Research Station Edinburgh Scotland

1 Introduction

Most coniferous forest tree species of economic importance are marketed as 'softwoods' and broad leaved species as 'hardwoods'. Softwoods are preferred by the major wood-using industries because they are light, easy to handle and work; also, many species have similar wood properties and are thus interchangeable. Exploitation is often simple because most species occur over large areas either in pure stands or in mixture with a limited number of other species. Most conifers are easy to cultivate as pure crops. Between hardwood species there is a greater diversity in wood qualities and specialized markets and end-products often have to be developed for each species. Many of the commercially valuable species are heavy and difficult to handle and transport and, for this reason, exploitation is often difficult. This is particularly true of tropical rain-forests in which only a few trees of commercial importance are found per hectare. Relatively few commercially important broad-leaved species grow well in monoculture.

Softwood timber is chiefly marketed as: sawnwood, for building and packaging purposes; roundwood, for mining timber and fencing; pulpwood, for the paper and board industries; and veneerwood, for plywood and decorative facings. Hardwood timbers are marketed for similar purposes, with most of the high quality grades going for high-class joinery and decorative veneer work.

International trade is considerable and world-wide 1971 log exports alone totalled 61·5 million m^3. The 1971 national statistics on wood production, exports, imports and *per capita* consumption have been summarized (Anon, 1974). Many highly industrialized countries now have enormous timber deficits, amounting, in the EEC countries alone, to an estimated 114 million m^3. *Per capita* consumption is rising rapidly and, in the USA, for example, it was expected in 1962 that pulpwood needs would rise threefold and combined timber products by 80 per cent between then and the year 2000. By the year 2015, the annual increase in world demand for industrial wood is expected to be 60 million m³ (FAO, 1967). It has been estimated that, at present, about 50 per cent of the world consumption of wood is still used as a fuel.

2 Cytotaxonomic background

Essential features are summarized in Table 87.1. All 19 of the most widely cultivated conifers are members of the Pinaceae. The cytotaxonomy of the conifers has not been intensively investigated; most are believed to be stable diploids. Polyploidy is rare, although some aneuploids, mixoploids and triploids have been reported. By contrast, polyploidy is common among the broadleaved species (Gustafson and Mergen, 1963). Most of the poplar 'cultivars' are vegetatively propagated clones derived from individual selections in natural forests or from artificial hybridizations, particularly within the Black and Balsam sections of the genus. Several commercially valuable triploid and tetraploid ($2n = 57$, 76) aspens and poplars are known.

All conifers and the broadleaved species listed in Table 87.1 are outbreeders and, with the exception of the *Eucalyptus* species and teak (which are insect-pollinated), all are naturally wind-pollinated. The majority are highly self-incompatible but, when self-pollination does occur, the seed normally contains a high proportion of abnormal embryos and the plants show inbreeding depression and chlorophyll deficiencies.

3 Early history

Many of the world's forest ecosystems have been influenced by man throughout the ages and fire has played a major role. Other important influences on forests and forestry have been grazing by domestic animals, agriculture and hunting (see Stern and Roche, 1974, for a historical review). Uncontrolled burning of forests to provide land for grazing animals was common from the Bronze Age through to the early nineteenth century in northern Europe. Other forests were cleared to provide charcoal for smelting, particularly from mediaeval times onwards to the latter part of the seventeenth century. Many remaining forests were over-exploited to provide timber for ship-building and general building work; by the fifteenth century Britain, for example, was already importing timber from the Baltic area and from eastern America by the late eighteenth century. Selection was no doubt often dysgenic, the best trees being used, the worst left to bear seed.

The cultivation of trees for wood production has been practised on a limited scale since ancient times.

Table 87.1 Principal cultivated timber tree species

Genus	Species	Vernacular	$2n$	Distribution
Softwoods				
Pinus (94 spp.)	*caribaea*	Caribbean pine	24	Central America, Cuba, Bahamas
($x = 12$)	*densiflora*	Japanese red pine	24	Eastern Asia
	elliottii	Slash pine	24	South/eastern USA
	patula	Mexican pine	24	Mexico
	pinaster	Maritime pine	24	South/west Europe, Italy
	radiata	Monterey pine	24	California
	resinosa	Red pine	24	North/eastern USA, Canada
	sylvestris	Scots pine	24	Eurasia
	taeda	Loblolly pine	24	East and south/eastern USA
Picea (31 spp.)	*abies*	Norway spruce	24	North and central Europe
($x = 12$)	*glauca*	White spruce	24	Northern North America
	mariana	Black spruce	24	Northern North America
	obovata	Siberian spruce	24	Northern Asia
	sitchensis	Sitka spruce	24	Coastal north/west America
Larix (10 spp.)	*decidua*	European larch	24	Europe
($x = 12$)	*kaempferi*	Japanese larch	24	Japan
Pseudotsuga (6 spp.) ($x = 13$)	*menziesii*	Douglas fir	26	Northwest America
Cryptomeria (1 sp.) ($x = 11$)	*japonica*	Japanese cedar	22	Japan
Araucaria (9 spp.)	*cunninghamia*	Hoop pine	?	Australia
Hardwoods				
Betula (28 spp.)	*pubescens*	Silver birch	28	Europe, northern Asia, North America
($x = 14$)	*verrucosa*	Silver birch	56	Europe, northern Asia, North America
Eucalyptus (2,800 spp.)	*camaldulensis*	Murray Red gum	22	Australia
($x = 11$)	*globulus*	Tasmanian Blue gum	?	Australia
	grandis	Flooded gum	?	Australia
Fagus (8 spp.) ($x = 12$)	*sylvatica*	European beech	24	Europe
Populus (31 spp.)	*deltoides*	Cottonwood	38	Eastern North America
($x = 19$)	*nigra*	Black poplar	38	Europe, western Asia
	tacamahaca	Balsam poplar	38	North America
	tremuloides	American aspen	38	North America
	trichocarpa	Californian poplar	38	Western North America
Tectona ($x = 12$)	*grandis*	Teak	24	Southeast Asia

Serious afforestation programmes probably started in Japan some 400 years ago and in Europe a century later. Large schemes were started in Australasia and North America after about 1870 and in South America and tropical Africa since the 1920s (Streets, 1962; Toda, 1974). Forest management systems which rely on natural regeneration have long been, and still are, practised in many countries, but on a diminishing scale.

Plant explorers of the nineteenth and early twentieth centuries introduced many tree species from northwestern America and eastern Asia into Europe, many of which have proved to be well suited to cultivation. Douglas fir (*Pseudotsuga menziesii*) and Sitka spruce (*Picea sitchensis*) are particularly noteworthy. *Pinus radiata*, which occurs naturally in a very limited area in California, was introduced into New Zealand in the 1850s and has since been extensively planted in Chile, Australia and Spain on a very wide range of sites, many having climates which differ markedly from that of the source.

4 Recent history

In the last 20 years extensive plantations of several species of *Eucalyptus* have been successfully established in the Mediterranean region, East and Central Africa and South America. Teak, which is indigenous to India, Thailand, Burma and Indo-China, has been extensively planted both within and outside its natural range in these countries and also in Sri Lanka, Malaysia, Sumatra, Borneo and Papua New Guinea. It is also successfully grown on a limited scale in Jamaica, Trinidad, Nigeria, Ghana and Tanzania.

The early success of useful exotic species quickly led, in many countries, to the establishment of comparative species trials. Since the 1920s these have usually been followed by comparative tests of trees derived from seed from various naturally occurring ecotypes. Many such 'provenance' tests have been supported on an international scale by the Food and Agriculture Organization in conjunction with the International Union of Forest Research Organizations (IUFRO). The Commonwealth Forestry Institute in Oxford has also played a major role in the organization and collection of seed and the design and establishment of experimental tests of many tropical and subtropical pine species (Burley and Nikles, 1972, 1973).

The current general position is that most countries have established large areas with trees raised from unselected seed sources. More recently, and on a smaller scale, plantations have been based on general 'provenance' collections within imprecisely defined seed collection regions; exceptionally, seed from specific ecotypes has been used. For some species, certain provenances have been outstandingly successful and many large-scale forest programmes are now based on very restricted parent sources. Apart from the southeastern part of the USA, northern Europe, New Zealand and Australia (where the Southern pines, Scots pine and Monterey pine respectively have been bred on

a considerable scale), tree breeding has yet to make a substantial impact on forestry practice.

Commercially orientated tree breeding programmes first started in America in 1924, with the object of producing hybrid poplars for pulpwood. Other programmes were started in Scandinavia during the 1930s, with emphasis on poplars, aspens, birches and indigenous conifers. Since 1950, many other countries have started breeding programmes, most of which are based on mass-selection and depend upon the outbreeding nature of the species concerned. Superior phenotypes are vegetatively propagated and subsequently established in clonal seed orchards from which genetically superior seed is expected. Concurrently, progeny tests, usually based on open-pollinated seed from the parent trees, provide a basis for selectively roguing the orchards for further improvement. Libby (1973) has discussed domestication strategies for forest trees and Shelbourne (1969) has summarized the more important tree breeding methods. Inter- and intraspecific hybridization is being investigated and some hybrids, such as the *Larix kaempferi* × *L. decidua* hybrids in Europe and the *Pinus rigida* × *P. taeda* hybrids in Korea are already in production.

Breeding goals vary but most programmes aim to improve growth-rate, thus leading to shorter rotations and shorter periods of investment; there is no doubt that enhanced growth rates can, in fact, be achieved. Others aim to improve stem-form, wood qualities for specific end-uses, disease and insect-pest resistance, cold- and drought-resistance. FAO (1963, 1970) has published useful general summaries of most aspects of tree·breeding.

5 Prospects

Strong action will undoubtedly be taken to conserve the germplasm of forest tree species in general and of particularly precious but endangered populations in particular. This is likely to be coupled with the establishment of more international seed collection centres and of pollen banks for longer-term germ plasm storage. There is already a growing trend for seed users to demand authoritative certificates of origin for both imported and home-collected seed and to demand that seed be collected from nationally registered phenotypically superior sources. Both national and international seed and plant certification schemes are in being for the registration and marketing of tested seed materials. Ultimately, genetically superior and tested cultivars will be artificially created for specific ecological conditions. The leading timber species are thus in transition to becoming truly cultivated plants, evolving in response to human selection.

There is a growing interest in the development of cheap ways of mass producing genetically superior trees by rooting cuttings or by cell-culture. If successful, extensive use will be made of superior clones, particularly for those species grown on very short (10- to 15-year) rotations, where financial losses from unpredictable catastrophes would be least. Such clonal forests will no doubt stimulate interest in the development of forests based on mixed rather than on single clones. Clonal forests are particularly attractive from the point of view of mechanization potential. The idea, incidentally, is not a new one: clonal poplars in Europe and clonal *Cryptomeria* in Japan have been grown for many rotations.

Improvements of other characters, such as resistance to atmospheric pollution, low demands for soil nutrients, tolerance of drought (or, alternatively, tolerance of soil wetness) will be sought, since there is an ever-growing tendency to locate forests on marginal agricultural land and towards extreme sites. For the mass-production of F_1 hybrid seed, suitable dioecious clones or the effective use of gametocides would be essential and will continue to be sought.

The high rates of growth obtainable in some tropical areas (which may exceed growth-rates under temperate conditions by as much as 200 per cent) will lead to greater emphasis being placed on speeding the breeding work in tropical countries.

6 References

Anon. (1974). 1974 World wood review. *World Wood*, 15, 19–99.

Burley, J. and Nikles, D. G. (eds.) (1972, 1973). *Selection and breeding to improve tropical conifers*. Oxford, 2 vols.

FAO (1963). Report on first world consultation on forest tree breeding. *Unasylva*, 18, (2–3), 1–141.

FAO (1967). Actual and potential role of man-made forests in the changing pattern of wood consumption. In *FAO world symposium on man-made forests and their industrial importance*. Rome, 1–50.

FAO (1970). Report on second world consultation on forest tree breeding. *Unasylva*, 24 (2–3), 1–132.

Gustafson, A. and Mergen, F. (1963). Some principles of tree cytology and genetics. *Unasylva*, 18 (2–3), 7–20.

Libby, W. J. (1973). Domestication strategies for forest trees. *Can. J. Forest Res.*, 3, 265–76.

Shelbourne, C. J. A. (1969). Tree breeding methods. *Tech. Pap. N.Z. Forest Serv.*, 55, pp. 43.

Stern, K. and Roche, L. (1974). *Genetics of forest ecosystems*. Berlin and New York.

Streets, R. J. (1962). *Exotic forest trees in the British Commonwealth*. Oxford.

Toda, R. (ed.) (1974). *Forest tree breeding in the world*. Tokyo.

Minor crops

P. M. Smith
Botany Department
University of Edinburgh Scotland

Introduction

In this appendix an attempt is made to summarize very briefly the available information on the evolution of some crops (or near-crops) which are either minor or so poorly documented that little can be said about them. Some, indeed, are far from being of minor importance to local economies and others clearly may turn out to have long and interesting histories when they are adequately studied.

Inevitably, the material of this appendix has had to be drawn largely from secondary and other fragmentary sources. To save space, several major works of general relevance which have been repeatedly consulted are listed here and are not cited in the text. They are:

Burkill, I. H. (1935). *Dictionary of the economic products of the Malay Peninsula*. London.

Chandler, W. H. (1950). *Evergreen Orchards*. Philadelphia.

Corner, E. J. H. (1940). *Wayside trees of Malaya*. Singapore.

Corner, E. J. H. (1968). *The natural history of palms*. London.

Darlington, C. D. and Wylie, A. P. (1955). *Chromosome atlas of flowering plants*. London, 2nd edn.

Hill, A. F. (1952). *Economic botany*. New York.

Kay, D. E. (1973). *Root crops*. Trop. Prod. Inst. London. (Crop and Prod. Digest, 2), pp. 280.

Kirby, R. H. (1963). *Vegetable fibres*. London.

Purseglove, J. W. (1968, 1972). *Tropical crops. Dicotyledons*, 1, 2; *Monocotyledons*, 1, 2. London.

Schery, R. W. (1954). *Plants for man*. London.

Smartt, J. (1975). *Tropical pulses*. London.

The arrangement of the text follows that of the book as a whole: entries are set out alphabetically by families and then by genera within families.

Thanks are due to several correspondents for kindly providing materials upon which entries were based, namely: E. Keep (gooseberry), J. W. Purseglove (millets), J. Smartt (legumes), W. B. Storey (*Macadamia*), T. W. Whitaker (cucurbits), and also to N. W. Simmonds for help and encouragement.

Mauritius hemp

Furcraea gigantea var. *willemettiana* (Agavaceae)

Several species of *Furcraea*, *Agave*-like plants with very large leaves, are grown in South America as local sources of fibre for sacking and small cordage. All the species are native to tropical America but the best known fibre plant in the genus is *F. gigantea* var. *willemettiana* ($2n = 2x = 60$) which is grown commercially in Mauritius and Madagascar. *Furcraea gigantea* is native to Mexico.

It is not known how var. *willemettiana* arose but it is likely to have been introduced to Mauritius, as was *F. gigantea* itself. Seed is rarely produced and commercial propagation is by bulbils. After being first grown as a hedge plant for delimiting properties, *F. gigantea* escaped and the resultant naturalized populations are also exploited for fibre.

Bow-string hemp

Sansevieria spp. (Agavaceae)

Several species of *Sansevieria* are grown for their strong, elastic leaf fibres, wild plants also being a source of supply. They are fleshy, xerophytic perennials with rosettes of sword-like leaves. The fibres are also used for making matting, hammocks and cords. *Sansevieria zeylanica* ($2n = 2x = 40, 42$) is native to Ceylon and is also cultivated there. *Sansevieria roxburghiana* ($2n = 2x = 40$) is used in India and has long been grown as a source of bowstrings. *Sansevieria guineensis* is a native of tropical Africa, now grown in Jamaica and central America. *Sansevieria zeylanica* has been widely introduced in the tropics and often becomes naturalized. Some species produce seed only rarely. Propagation by rhizomes or leaf cuttings is usual.

Cashew

Anacardium occidentale (Anacardiaceae)

Cashew ($2n = 42, x = ?$) is a native of tropical America, from Mexico to Peru and Brazil, also of the West Indies. Following introduction to the Old World tropics by early Spanish and Portuguese adventurers (fifteenth to sixteenth century), it has become widely naturalized in coastal areas. The fleshy pedicel and receptacle (cashew apple) and the seeds (cashew nuts) are edible, the latter after the achenes have been roasted. Flowers (male and hermaphrodite) are mainly cross-pollinated by a diversity of insects but many of the fruits fall before maturity. Floral biology is still poorly understood; there are stamens of three different lengths. Propagation methods include seed, grafting and air-layering. There is a great deal of variability in the crop but little breeding has been done.

Northwood, P. J. (1966). Some observations on flowering and fruit setting in cashew, *Anacardium occidentale*. *Trop. Agric., Trin.*, **43**, 35–42.

Pistachio

Pistacia vera (Anacardiaceae)

Pistacia vera ($2n = 2x = 30$) is a tree native in the Near East and western Asia. It has been cultivated in the Mediterranean area and western Asia for 3,000–4,000 years and has been successfully introduced as a cultigen to the southern United States. It is a dioecious evergreen with small apetalous flowers. The fruit is a drupe and the seed within is the pistachio nut.

Two centres of diversity have been indicated, in Turkey and Kirgizstan. Natural populations with large nuts occur in Turkmenia. Other tree crops show diversity in the same area. Few cultivars of pistachio exist, partly due to the breeding system and long life of the trees, the millennia of propagation by grafting and because the natural diversity has not been fully exploited. Many wild populations have been destroyed by forest clearance or grazing. In the Tajik SSR male pistachio trees differ widely in fertility. Male trees often shed pollen too soon and a poor crop results. Other *Pistacia* species (e.g. *P. terebinthus*) are sufficiently closely related to *P. vera* to act as alternative pollinators and as stocks on which young trees, male or female, may be grafted.

Soursop

Annona spp. (Annonaceae)

Several species of *Annona*, a genus of small neotropical trees, yield useful fruits. Fruits are syncarpous, with the carpels fused to the receptacle. Distorted fruits, characteristic of *Annona* spp., arise from failure of fertilization in one or several of the numerous carpels. Hand pollination improves fruit set; beetle pollination is the natural system. Cultivation is usually in gardens. Some species are used for medicinal or stimulating infusions (e.g. *A. glabra* ($2n = 4x = 28$) in Curacao, an advanced species on karyotype symmetry classification).

Soursop (*A. muricata*, $2n = 2x = 14$), is widely cultivated throughout the lowland New World tropics. It is very sensitive to cold and succeeds in Florida only in the warmest parts, which seem to be its northern limit. Custard Apple (*A. cherimolia*, $2n = 2x = 14$), is native to the highlands of Peru and Ecuador, and has long been established in Mexico, probably by introduction. Varieties differing in fruiting time have been established from seedlings. Said to be true breeding, they enable fruit to be produced for 6–7 months of the year in Mexico and parts of South America. Clonally propagated varieties are used in California. Custard Apple does well only at higher altitudes in the tropics. Hybrids of Custard Apple with *A. squamosa* (*q.v.*) do well in Florida, unlike *A. cherimolia* itself. Sugar Apple or sweetsop (*A. squamosa*, $2n = 2x = 14$) is known in parts of Asia as Custard Apple. An Indian origin for this crop can be discounted: it was certainly introduced to the East from the West Indies or South America, where it is native. It is widely grown at low altitudes throughout the tropics and sets fruit without hand pollination better than does *A. cherimolia*, with which it forms a hybrid, the Atemoya.

Maté

Ilex paraguariensis (Aquifoliaceae)

Yerba de Maté or Paraguay tea is a small, evergreen tree, native to the mountains of southern Brazil, Paraguay and northern Argentina; it is cultivated extensively throughout its native range. Trees of *I. paraguariensis* ($2n = ?4x = 40$), and possibly other species (e.g. *I. pseudobuxus*) are grown from seed for their leaves, which contain caffeine. Infusions of the prepared leaves are greenish and aromatic. It is a popular beverage in South America and there is a small export trade to Europe and North America. Wild trees are still used as a source of maté to supplement the plantation crop.

Ullucu

Ullucus tuberosus (Basellaceae)

Ullucu (also called melloco, timbos, etc.) is a minor tuber crop of the high Andes. The tubers demand very careful preparation to eliminate an unknown toxic component. It ranges from Colombia to northern Argentina and is quite variable. Wild forms of ullucu (or perhaps escapes) are known from 'cloudy, damp' parts of Bolivia; they have long rambling stems and small tubers and may relate to a new species described by Brücher (1967) as *Ullucus aborigineus* from cloud forest in Argentina. Cultivated clones are generally diploid ($2n = 2x = 24$) but triploids ($2n = 3x = 36$) have been reported. It is clonally propagated and seeds very rarely, perhaps never. An Andean origin (probably in the altiplano region) is certain but the history of the crop is unknown.

Brücher, H. (1967). *Ullucus aborigineus* sp. nov., die Wildform einer Andinen Kulturpflanze. *Ber. dtsch. bot. Ges.*, **80**, 376–81.

Leon, J. *et al.* (1958). Estudios sobre tubérculos alimenticios de los Andes. *Commun. Turrialba*, **63**, pp. 49.

Leon, J. (1967). Andean tuber and root crops; origin and variability. *Proc. intnl Symp. trop. Root Crops, Trinidad*, 1, 118–23.

Calabash

Crescentia cujete (Bignoniaceae)

Calabash fruits provide durable epicarps which may be polished or carved and used as containers (e.g. for calabash curare) or for musical instruments (maracas). The flowers are probably bat pollinated and are succeeded by the hard-shelled, gourd-like fruits. The trees are easily grown from seeds or cuttings. *Crescentia cujete* ($2n = 2x = 40$) is native to the West Indies and tropical America and has been widely introduced. Calabashes have been found in archaeological sites in Peru.

Morton, J. F. (1968). The calabash (*Crescentia cujete*) in folk medicine. *Econ. Bot.*, **22**, 273–80.

Durian

Durio zibethinus (Bombacaceae)

The Durian is a cauliflorous fruit tree of Malaysia and neighbouring areas. A liking for the powerful smell and taste of the ripe fruit is acquired by some people but elephants, tigers and monkeys seem especially attracted to it. The precise place of origin of durian ($2n = 2x = 14$) is uncertain but is likely to be in

303

Western Malaysia since the crop is well adapted and has relatives there and attempts to introduce it elsewhere have usually failed. Self-pollination is possibly the normal means of pollination since the flowers open for only a short time but bees and bats have been implicated as pollen vectors, presumably implying some crossing. Propagation is possible from seeds which are, however, of very short viability. Dispute surrounds the status of the 'wild' durian. Soegeng-Reksodihardjo (1962) asserts that it is truly wild, but Corner disagrees. Cultivated durian may have originated from *D. malaccensis* or from *D. wyattsmithii*, both of which are genuinely wild trees in Malaysia, but it is at least equally likely that the progenitor was wild *D. zibethinus*.

Soegeng-Reksodihardjo, W. (1962). The species of *Durio* with edible fruits. *Econ. Bot.*, **16**, 270–82.

Queensland arrowroot
Canna edulis (Cannaceae)

Canna edulis ($2n = 2x$, $3x = 18$, 27) is grown for its starchy rhizomes in parts of tropical America and the West Indies, Africa, Australia and Asia. In its native South America it has declined considerably from its former importance. Propagation is vegetative and only one variety seems to be known.

The crop can be associated with primitive agriculture only in tropical South America. It occurs as a wild plant in the Andes below about 2,800 m, from Venezuela to northern Chile, in the basins of the Amazon and Paraguay, and in the West Indies. In central America it is a weed. South America is the centre of diversity of the genus. Wild *C. edulis* appears to be the only ancestor of the cultivated canna, producing rather smaller rhizomes than the cultivated form.

The earliest known site of domestication (Huaca Prieta) is dated at 2500 B.C. The canna remains are very unlikely to be of wild origin, since the arid coastal strip of Peru would not have supported the plant. Gade (1966) believes that Colombia, which had an early root crop tradition, may have been the original place of domestication, whence the crop was disseminated to the western coast of South America and then to the West Indies with the Caribs. The Incas grew it on the banks of their irrigation ditches. Later, it was in use as a root starch by the Tainos of the Greater Antilles. Tous-les-mois is a recent West Indian name for the plant. The Arawaks, who probably took the canna into eastern South America, called it imocona. It was known in Europe in 1570, though as a variant of *C. indica*.

Cultivation in South America is now very restricted. In Peru, the crop is grown extensively in the upper parts of the valley of the Apúrimac River. Here cultivation seems to have survived the advent of better root crops because of the traditional use of baked *achira* (the Quechua name) in the Corpus Christi festival of nearby Cuzco (southern Peru).

Gade, D. (1966). Achira, the edible canna, its cultivation and use in the Peruvian Andes. *Econ. Bot.*, **20**, 407–15.

Spinach
Spinacia oleracea (Chenopodiaceae)

Spinach is an annual green vegetable grown in cool, moist climates. It is native to southwestern Asia but is now widely introduced in Europe, the Americas and the Far East. Spinach was not known as a potherb to the Greeks or Romans and it seems to have spread beyond its native area since their times.

S. oleracea ($2n = 2x = 12$) occurs in round-fruited and prickly-fruited varieties. Prickly fruits are probably a primitive character. Girenko (1968) emphasizes the distinctness of *S. oleracea* from its two diploid relatives (*S. tetrandra* and *S. turkestanica*) and describes primitive forms of *S. oleracea* from northwestern India and Nepal. The characters of round fruit, entire leaves and low pigmentation are features associated with cultivation and seem to have arisen independently in a number of different spinach-growing areas.

Cultivated spinach is normally dioecious with equal proportions of male and female plants. Many gradations of monoecism and hermaphroditism are known, enabling the selection of 'highly male' or 'highly female' lines. These lines can subsequently be used to produce 'hybrid' varieties.

Girenko, M. M. (1968). [Classification of spinach]. *Trud. priklas. Bot. Genet. Seleckc.*, **40**, 28–36. (In Russian.)
Janick, J. and Stevenson, E. C. (1955). Genetics of the monoecious character in spinach. *Genetics*, **40**, 429–37.

Chicory and Endive
Cichorium spp. (Compositae)

Chicory (*C. intybus*, $2n = 2x = 18$) is a perennial salad

plant, native to Europe, which has been cultivated at least since Greek and Roman times. It has become naturalized in parts of North America. Endive (*C. endivia*, $2n = 2x$, $3x = 18, 36$) is an annual or biennial salad plant, indigenous to the eastern Mediterranean area, although an Indian origin has been suggested. The leaves are blanched before eating. Endive was cultivated by the ancient Egyptians. It is thought to have originated from the cross, *C. intybus* × *C. pumilum*. Considerable development has gone into the chicory crop, particularly in Europe. Self-sterile lines are sought for bulk crossing programmes. Autotetraploid chicories have been produced and show wider variation than the diploid. They are considered to have breeding potential. Marketing chicory on a large scale has encouraged a search for uniformity of size, shape and time of maturation. In France, cross pollination of highly uniform Flambor strains is encouraged by placing beehives in the fields.

Artichoke
Cynara scolymus (Compositae)

The globe artichoke ($2n = 2x = 34$) is a native of the Mediterranean area and the Canary Isles. It is a tall, perennial thistle with large fleshy, edible capitula ('chokes'). Young leaf stalks are sometimes blanched and eaten as a chard. The artichoke was domesticated and spread throughout the Old World very early and was grown by the Greeks and the Romans, later being introduced to the Americas by French and Spanish colonizers. Propagation is by suckers, since seedlings are highly variable and rarely of any value. The crop is most grown in central and southern Europe, and in California.

Niger seed
Guizotia abyssinica (Compositae)

Guizotia abyssinica ($2n = 2x = 20$) is a tall, annual herb bearing axillary yellow capitula. The black achenes are rich in oil which finds uses in cookery, in soap manufacture, as an illuminant and as cattle feed. The plant is native to Africa, from Ethiopia to Malawi. Ethiopia, where cultivation is best developed today, was probably the area of domestication. Early introduction of niger seed to India was followed by the development of sizeable commercial production. The crop does well on poor soils.

Hazel, cobnut, filbert
Corylus spp. (Corylaceae)

Corylus avellana ($2n = 2x = 28$; $2n = 22$ also recorded) is a deciduous shrub or small tree with unisexual flowers, bearing hard, ovoid nuts in a husk (involucre). It is wind pollinated. Native to Europe and southwestern Asia, the tree has been prized for its nuts and its property of warding off evil from early times. The nuts were once an important food in Ireland. Domestication probably first took place in seventeenth century Italy. Cultivation on a large scale began in the nineteenth century, numerous strains being selected and hybridizations performed with other species. This process continues. The nuts are eaten as a dessert and used extensively in confectionery.

Corylus maxima ($2n = 28$) the filbert, native to southeastern Europe and western Asia, is also grown for its nuts. It has been hybridized with *C. avellana* and some hazel cultivars are of hybrid origin. Filberts have longer involucres than hazel nuts, though the distinction is rather hard to maintain. *C. macquarri* has been suggested as the progenitor of *C. avellana*, migrating in late Miocene from Greenland to Europe, and there evolving into *C. avellana*. *Corylus americana* ($2n = 28$), a native American hazel, has been a source of North American cultivars. *Corylus colurna* is the source of Turkish cobnuts, which are widely exported.

Evreinof, V. A. (1963). Le passé géologique des noisetiers. *J. Agric. trop. Bot. appl.*, 10, 393–5.
Maurer, K. J. (1973). Versuch einer *Corylus* monograph. *Mitt. Rebe Wein, Ostbau Fruchteverwertung*, 23, 404–44.

Horseradish
Armoracia rusticana (Cruciferae)

Horseradish is a perennial with fleshy roots. The grated root, mixed with salt, oil and vinegar, is a pungent condiment used particularly with roast beef. Native to eastern Europe and Turkey, the plant is widely cultivated in northern temperate areas of the Old and New World and, to a limited extent, in the tropics at higher altitudes. Cultivation in the native area has been practised for at least 2,000 years and the plant was referred to by Dioscorides. Cordate and cuneate leaf-bases were illustrated by medieval herbalists and persist in modern cultivars. Clones, propagated by root cuttings, are old, perhaps ancient. Horseradish ($2n = 4x = 32$) is highly seed-sterile. Meiotic irregularities and variable seedling progeny have suggested a hybrid origin; there are two other species of *Armoracia*. Embryos commonly abort.

Courter, J. W. and Rhodes, A. M. (1969). Historical notes on horseradish. *Econ. Bot.*, **23**, 156–64.

Stokes, G. W. (1955). Seed development and failure in horseradish. *J. Hered.*, **46**, 15–21.

Garden cress
Lepidium sativum (Cruciferae)

Garden cress ($2n = 2x$, $4x = 16$, 32) is cultivated all over the world as a salad plant. The long succulent hypocotyls of young seedlings are eaten and often used as a garnish. The adult plant, which is sometimes seen as an escape, is an annual herb with white, fragrant flowers. *Lepidium sativum* seems to be a native therophyte in parts of western Asia, the Near East and Ethiopia. It must have spread very early to Europe and has been cultivated there for a very long time.

Cucurbits
Benincasa, Luffa, Momordica, Trichosanthes, Sechium (Cucurbitaceae)

Benincasa, Luffa, Momordica and *Trichosanthes* are indigenous to the Asian sub-tropics; *Sechium* has been spread by man from its native area, in southern Mexico and central America, throughout the Caribbean area. The nutritional value of their fruits is low, but they are widely used to enliven rice-based dishes consumed by Indian, Chinese, Malaysian and Indonesian peoples.

Benincasa hispida (wax gourd: $2n = 2x = 24$) is distributed from Japan to India. It is a monoecious vine bearing large (40–60 cm long) green fruits, globose-cylindrical in shape and covered with thick epicuticular wax. Fruits are used as a boiled vegetable or candied in small pieces. Varghese (1973) has observed secondary associations of chromosomes which led him to suggest that *B. hispida*, though apparently diploid, is in fact a stable tetraploid derived from an ancestor with a basic $x = 6$.

Luffa cylindrica and *L. acutangula* (both $2n = 2x = 26$) are monoecious vines of India bearing smooth and ridged fruits respectively, which contain many black seeds. Young fruits are used as a cooked vegetable. Dried, skeletonized fruits of *L. cylindrica* are used as sponges or filters (loofahs), mainly in Japan. Chromosomes of both species are morphologically identical but there is about 62 per cent of sterility in F_1 hybrids,

implying cryptic structural differences.

Momordica charantia (bitter gourd: $2n = 2x = 22$) is a monoecious vine of tropical Asia with 5-angled stems, which is variable in the form of its fruits. The variant most often grown has cylindrical green fruits about 10–25 cm long, with tuberculate skins. Mature fruits split, exposing a red aril. Fruits are used in the tropics as a vegetable, in pickles and curries and in salads. Meiosis is reported to be regular.

Sechium is monotypic. *Sechium edule* (chayoye: $2n = 2x = 28$) is an unusual cucurbit in that the fruits are one-seeded. It is a monoecious vine. The large, ribbed, pear-shaped fruits are boiled or eaten raw. The tuberous root is used as a starchy food. Varghese (1973) has shown *S. edule* to have $2n = 2x = 28$, in contradiction of earlier reports of $2n = 26$.

Trichosanthes cucumerina ($2n = 2x = 22$) is a monoecious vine of Indo-Malaysia with long, slender fruits. Young fruits are used as a boiled vegetable, especially popular in southern India. All reports suggest that meiosis is normal.

Little can be said of the early history of any of these minor cucurbits. All species of humid tropical lowlands, they do not seem to have left a fossil record. It is likely that most of them have been cultivated at least for several centuries, though evidence for this is virtually non-existent. Burkill reports *Luffa* as being in cultivation 400 years ago. Only *S. edule* seems to have an expanding role as a crop plant, increasingly found in markets of tropical America and as far north as southern California.

Choudhury, B. (1966). *Vegetables*. New Delhi, 156–62.

Herklots, G. A. C. (1972). *Vegetables in Southeast Asia.* London, 307–13, 326–42, 346–8.

Roy, R. P. (1972). *Cytogenetical investigations in the Cucurbitaceae*. PL 480 Research Project FG-IN-332 (Ad-CR-250). pp. 263.

Varghese, B. M. (1973). *Studies on the cytology and evolution of South Indian Cucurbitaceae*. Ph.D. Thesis, Kerala University, India.

Persimmon
Diospyros kaki (Ebenaceae)

Persimmons are berry fruits of certain species of *Diospyros*, *D. kaki* ($2n = 6x = 90$), the Japanese persimmon, being the most significant commercially. There is some evidence that cultivated *D. kaki* may have originated from *D. roxburghii*, a wild species in forests from Assam to Indochina. *Diospyros lotus*, another temperate Asiatic species, is cultivated to a small extent. *Diospyros kaki* is a large deciduous tree, widely grown in China and Japan and introduced to parts of America and southern Europe. The American persimmon, *D. virginiana* ($2n = 4x, 6x = 60, 90$), native to the hardwood forests of the eastern United States, yields a superior persimmon but is not much cultivated. It is hardier than *D. kaki*. The two species are cross-incompatible, hybrid seed development being arrested at an early stage.

Blueberry
Vaccinium spp. (Ericaceae)

Many species of this large, north temperate genus are of value for their attractively acid, red or blue berries. Fruits of many wild species have long been collected; such domestication as has occurred is recent (twentieth century). The plants are small bushes ('highbush' or 'lowbush' according to size) or creeping shrubs, often growing gregariously on acid soils. Almost pure stands on the blueberry barrens of New Brunswick and Maine have long been preserved and raked over for their berry harvest. On acid soils, cultivated blueberries outyield the wild populations but the labour of picking is still considerable, a factor in the delayed domestication of the group. In cultivation, seedlings of wild origin are sown, or plants are propagated by rhizomes or cuttings from selected strains.

Vaccinium corymbosum ($2n = 4x = 48$), the highbush blueberry, and *V. angustifolium* ($2n = 2x = 24$), lowbush blueberry, are cultivated North American species with large fruits, introduced into Europe. Cultivars of the former incorporate genes from *V. ashei* ($2n = 6x = 72$) and *V. australe* ($2n = 4x = 48$), both American species. *Vaccinium macrocarpon* ($2n = 2x = 24$, central and eastern North America) and *V. oxycoccus* ($2n = 4x = 48$, more or less circumboreal) are cranberry species (red fruits), both often being placed in the genus *Oxycoccus*. Their acid berries are used mainly in cranberry sauce. Both are cropped from the wild but also cultivated in North America (since 1840) and in Europe.

Lamb, J. G. D. (1973). High blueberries – a new fruit for Europe. *J. Roy. hort. Soc.*, 98, 401–4.

Coca
Erythroxylon coca (Erythroxylaceae)

Erythroxylon is a genus of shrubs native to tropical and sub-tropical South America. *Erythroxylon coca* ($2n = 24$) is the source of cocaine. It is grown at medium to higher altitudes in the Andes of Peru, Argentina, Bolivia, Colombia and Brazil. The leaves are the source of the narcotic, which is much used in modern medicine, but which have been long employed in South America as a masticatory affording stimulation and resistance to fatigue. Coca was distributed to the Old World tropics from Kew Gardens in the late nineteenth century and Java has become an important centre of production. Wild coca plants are unknown, but the presence of similar wild species points to the Montana region of Peru and Bolivia as the likely area of origin of the cultigen. Varieties or 'species' of cultivated coca (e.g. *E. novogranatense*) may reflect either the extreme polymorphism or the environmental plasticity of *E. coca*.

Martin, R. T. (1970). The role of coca in the history, religion and medicine of South American Indians. *Econ. Bot.*, 24, 422–38.

Chestnut
Castanea spp. (Fagaceae)

Both nuts and timber are produced by various species of *Castanea*, a genus widely native in the north temperate zone. *Castanea dentata* ($2n = 2x = 24$), the American chestnut, has been virtually exterminated by chestnut blight but the resistant Japanese chestnut, *C. crenata* ($2n = 2x = 24$), may offer hope of rescue by way of hybridization. The most important species is the Sweet or Spanish chestnut, *C. sativa*, native to southwestern Asia and widely introduced to southern Europe by the Greeks. It was introduced to Britain by the Romans, reaching the USA in the eighteenth century. Sweet chestnut has been grown for nuts and timber from ancient times, the nuts being eaten whole when cooked, or ground into flour for stews. The tree is monoecious but has unisexual flowers and most varieties are partly self-incompatible. Seedling establishment is simple but the progeny rarely yield good nuts. Cultivars are maintained by budding or grafting.

Jaynes, R. A. and Graves, A. H. (1963). Connecticut hybrid chestnuts and their culture. *Connecticut agric. Expt. Stn Bull.* 657.

Adlay, Job's tears
Coix lachryma-jobi (Gramineae)

Several species of *Coix*, a genus of monoecious grasses

with unisexual flowers, are used for human food or animal forage in parts of southeast Asia. Adlay is the Philippine name. *Coix lachryma-jobi* ($2n = 2x = 20$) is native to southeast Asia and was cultivated very early in India (1000–2000 B.C.) and, probably contemporaneously, in China and Japan. *Coix aquatica* and *C. gigantea* are both used in parts of southeast Asia. The spikelets are borne in threes (2 sterile, 1 fertile) in hard, beadlike 'involucres' (modified leaf sheaths). The beads are called Job's tears and are used for grains and ornament.

Little improvement work seems to have been done on the crop (though its grain is highly nutritious) and little is known about its origin. Greatest variation is found in Malaysia. Several varieties are known, varying in involucre size or shape, no doubt based on local selection. A dwarf form is known in Brazil. Adlay (var. *ma-yuen*) is a soft-shelled type which lacks the porcelain-like exterior of the wild-type bead. The grain is used as a source of ornamental beads, made into beer or eaten like rice.

Though known to Greek and Roman authors, the genus *Coix* became known widely in Europe only in the seventeenth century, after it had been introduced by Arab traders. The edible property of the seeds of *C. lachryma-jobi* was recorded in Miller's Gardener's Dictionary (1731). Though known in Asia since Vedic times, *Coix* was not stated to have medicinal properties until A.D. 1260. The advent of rice may have inhibited further development of adlay as a crop plant: it seems to have considerable potential.

Jain, S. K. and Banerjee, D. K. (1974). Preliminary observations on the ethnobotany of the genus *Coix*. *Econ. Bot.*, **28**, 38–42.

Schaffhausen, R. von (1952). Adlay or Job's Tears – a cereal of potentially greater economic importance. *Econ. Bot.*, **6**, 216–27.

Koul, A. K. (1974). Job's Tears. In J. B. Hutchinson (ed.), *Evolutionary studies in world crops. Diversity and change in the Indian subcontinent*. Cambridge, 63–6.

Japanese barnyard millet

Echinochloa frumentacea (Gramineae)

This minor cereal is the quickest growing of all the millets, producing a crop in as little as six weeks. It is grown in southeast Asia and eaten as porridge or sometimes mixed with rice. *Echinochloa frumentacea* ($2n = 4x = 36$; $2n = 56$ also recorded) is very similar to the widespread weedy grass *E. crus-galli*. ($2n = 4x$, $6x, = 36, 54$; 42 and 48 also recorded) which is itself occasionally cultivated (bharti). *Echinochloa crus-galli* is native over a large area of the Old World, in tropical, subtropical and temperate zones. *Echinochloa frumentacea* is awnless. Sino-Japanese and Indian hexaploid strains ($6x = 54$) have been found to have different genome constitutions and therefore must have at least partially different origins. The Sino-Japanese plants, now named *E. utilis*, are considered to have a Far Eastern hexaploid strain of *E. crus-galli* as their immediate ancestor. *Echinochloa frumentacea*, the Indian strain, is genomically identical to the wild species *E. colonum*, native to Java and Malaysia, and may well have been selected from that.

Echinochloa spp. have not been cultivated extensively in Europe for many years but *E. crus-galli* may once have had some importance as a crop plant. It was certainly cultivated in Iron age Skane, as evidenced by impressions on pottery.

Yabuno, T. (1962). Cytotaxonomic studies on the two cultivated species and the wild relatives in the genus *Echinochloa*. *Cytologia*, **27**, 296–305.

Teff

Eragrostis tef (Gramineae)

Teff is the most important crop in Ethiopia, which is the only country which grows it as a cereal grain. Elsewhere, as in Kenya, Australia and South Africa, it is occasionally cultivated for hay. Teff ($2n = 4x = 40$) is grown at altitudes of 1,700–2,500 m, where the annual rainfall is about 2,500 mm. The grain is ground into a brownish flour and made into pancakes called *ingera*. Cultivation in Ethiopia was established in prehistoric times but the details of its domestication are unknown. Legends involving *ingera* go back only as far as 100 B.C. but this use of the crop is certainly older. Teff straws are used in brick manufacture and there are reports of teff grains in pyramid bricks made in 3359 B.C., though some doubt exists as to the species of *Eragrostis* represented.

Common millet

Panicum miliaceum (Gramineae)

This millet, which is not known in the wild state, is of ancient cultivation. It was the *milium* of the Romans. Early cultivation in Europe (for instance by the early Lake Dwellers) has not survived on a large scale, and the crop is now important mainly in eastern and southern Asia. *Panicum miliaceum* ($2n = 4x = 36$) may have been domesticated first in central or eastern Asia. An Indian or eastern Mediterranean origin has also been suggested but little is known for certain. There are several diploid and tetraploid Asian species of *Panicum* among which an origin might be sought. The importance of the crop in arid areas is considerable. It has one of the lowest water requirements of any cereal and can be grown where the climate is too hot, the rainy season too short and the soil too poor for other cereals.

Foxtail millet
Setaria italica (Gramineae)

Foxtail millet is a significant grain crop in parts of south-eastern Europe, North Africa and Asia. It is widely cultivated in India and is the most important millet in Japan. *Setaria italica* ($2n = 2x = 18$) is unknown in the wild state though it is frequently naturalized as a weed in most warm parts of the world. The progenitor is thought to be the common Old World weed, *S. viridis* ($2n = 2x = 18$). Foxtail millet was one of five plants held sacred in China in 2700 B.C., so China is perhaps the likeliest place of domestication. Seeds appear in European archaeological sites. The grain was used extensively by the Lake Dwellers during a time when it was not known in Syria and Greece.

Gooseberry
Ribes spp. (Grossulariaceae)

Gooseberries are hardy, long-lived shrubs with stiff, often spiny branches. Propagated by cuttings, they are widely cultivated in cool, moist climates as a source of jams and preserves. A few species (e.g. *R. speciosum* from California) are grown for ornament. The crisp, often pubescent, fruits are true berries. Evidence of crossability indicates a close relationship between gooseberry species and the other members of *Ribes* (i.e. the currants), so the earlier generic distinction between *Ribes* and *Grossularia* is not supported.

Though the name gooseberry may have come from the use of the berries as a seasoning for goose, it is more likely (Grigson, 1974) that the 16th-century English names (grosiers, groser bush) derive from the French groseille à maquereau (mackerel currant, from its use with fish dishes). Groseille is a distant corrupt-

ion of the German kraus (= crisp). This may suggest support for the view that gooseberry is not native in Britain but there are older Welsh and English names, unrelated to groseille.

The European gooseberry is *Ribes uva-crispa* (syn. *R. grossularia*), now believed to be native from Britain eastwards throughout northern Eurasia, and also found in the Caucasus and N. Africa. Like all wild and cultivated gooseberries it has $2n = 2x = 16$. It is widely introduced and naturalized in Ireland and N. America.

The plant was not grown by the Greeks or Romans, who probably considered it a rare fruit of mountains, much inferior to the grape. Gooseberry bushes from France are mentioned in the fruiterer's bills (1276–92) of Edward I. By 1548, British varieties were in cultivation and, according to Turner's *New Herball*, were superior to the Continental article (Rake, 1958). By 1629, red, green, yellow, white and waxy-blue forms had been noted. Hedrick (1925) asserts that at least 23 cultivars were known by 1778. Domestication of the wild plant probably took place throughout its native range during the 16th century.

An unusual phase of crop evolution followed the development of 'gooseberry clubs' in the Midlands and North of England from the late eighteenth to the nineteenth centuries. Amateur gooseberry fanciers considerably increased fruit size and quality, under the goad of local competitions to find the finest gooseberry. By 1831, as many as 722 cultivars had been named (Lindley, quoted by Hedrick, 1925).

Not until the late nineteenth century did the tart gooseberry enter commercial cultivation, even in areas where grapes could not be grown. Commercial operations in Britain followed the abolition of the sugar tax in 1874. At first they were a great success, gooseberries being more resistant to gall mite than the established red currant. Introduction of American gooseberry mildew (*Sphaerotheca mors-uvae*) in 1905 proved disastrous for the crop, which was very susceptible. Gooseberry has never regained its earlier popularity in Britain.

Gooseberry breeding (comprehensively reviewed by Keep, 1975) in Britain started about 1911 (Crane and Lawrence, 1934) but all British strains proved susceptible to mildew and often to subsequent sulphur treatment as well. Prickliness, a wild character, was shown to be dominant in interstrain F_1 hybrids. Nineteenth century introduction of *R. uva-crispa* to N. America, the home of mildew, was unsuccessful. As early as 1833 the first hybrids of *R. uva-crispa* and *R. hirtellum* (a mildew resistant N. American gooseberry) were heralding the modern phase of gooseberry improvement, by interspecific hybridization. Commercial gooseberry cultivation has never been as important in N. America as in Europe except locally, in parts of Oregon.

Considerable interfertility exists between the gooseberry species of *Ribes* sect. Eugrossularia and even a few intersectional hybrids have been made. The large gene pool thus made accessible has enabled organized breeding for mildew and leaf spot resistance, spinelessness and other features since early in this century. Most gooseberry cultivars are self-fertile though they set fruit better with cross- or open-pollination. Intrasectional interspecific hybrids show more or less regular chromosome pairing. Most wild gooseberries are obligate outbreeders. The origin of self-compatibility in cultivars may well have been recent since they often still exhibit inbreeding depression.

Ribes divaricatum, a small-fruited N. American species, has proved to be a valuable donor for both mildew and leaf spot resistance. Russian breeders have successfully employed their native *R. aciculare* for the same purpose. Various N. American wild species are potentially excellent sources of mildew resistance which is normally a dominant character. Unarmed races of *R. oxyacanthoides* (N. America) have been used for some time as a source of spinelessness. Incorporation of several desirable features (e.g. spinelessness) into a new generation of cultivars should soon be completed, though pest resistances have proved slower to establish. Having a fairly robust fruit, the gooseberry is almost pre-adapted to mechanical harvesting, which is a further aim of recent improvement work.

Crane, M. B. and Lawrence, W. J. C. (1934). *The genetics of garden plants*. London.

Grigson, G. (1974). *A dictionary of English plant names*. London.

Hedrick, U. P. (1925). The small fruits of New York. *Rep. N.Y. State agric. Expt Sta.*, 33, 243–354.

Keep, E. (1975). Currants and gooseberries. In Moore, J. N. and Janick, J. (eds). *Advances in fruit breeding*, pp. 197–268. Purdue University Press.

Rake, B. A. (1958). The history of gooseberries in England. *Fruit Yearb.*, 10, 84–7.

Mangosteen

Garcinia mangostana (Guttiferae)

Mangosteen ($2n =$ ca 76; $x =$?) is a common village fruit tree of southeast Asia. It produces purple fruits with a thick rind, containing 4–8 segments of delicious whitish pulp. Wild forms occur in Malaysian forests.

The trees are dioecious and, in Malaysia and Java, only females are found, so seeds must be parthenogenetic. Vegetative propagation usually fails. Seeds have short and low viability. Two year-old seedlings are normally used for planting.

Pecan

Carya pecan (*C. illinoensis*) (Juglandaceae)

Better known as the source of hickory wood, the genus *Carya* also yields this nutritious nut, borne as the kernel of a drupaceous fruit. The pecan tree is large, growing to 65 m in the wild, but pruned to a height of about 15 m in orchards. It is native to Mexico and the southern-central United States. Selected clones with thin fruit shells have been isolated and are propagated by grafting on to seedlings. Wild trees, which are also used as a source of nuts, grow along streams in arid areas, forming sizeable stands where there is ample ground water. Cultivation began in the 1840s, most modern cultivars being based on the early clones selected from wild seedlings. The clone Centennial was first grafted in 1846 by a freed negro slave. Pecans are monoecious but the flowers are unisexual and variably dichogamous. Cross-pollination usually ensures a better harvest of mature nuts.

Walnut

Juglans spp. (Juglandaceae)

Several species are cultivated or exploited from the wild for their edible seeds ('nuts') which are enclosed within the hard drupaceous endocarp. *Juglans regia* ($2n = 2x = 32$), the most widely grown species, is native from southeastern Europe to China. It is a large, ornamental tree producing, as do many walnuts, a valuable timber. The species is cultivated throughout its native area and has been widely introduced, entering Britain in the fifteenth century. Introduced into California in the eighteenth century, it has there proved very successful, though the native American black walnuts have more strongly flavoured nuts. Eastern black walnut (*J. nigra*) has a thick endocarp and, even in thin-shelled selections, the kernel is much harder to withdraw intact than is the case with *J. regia*.

Propagation is by grafting or budding. Hybridization, between different *Juglans* species, and with *Carya pecan*, is being attempted in programmes to improve shell thickness, nut size and hardiness in cultivars. Many *Juglans* species are interfertile and the yield is often higher where different species grow together. Pollen from one species may successfully initiate nut production in another. Clonal mixtures are especially useful where self pollination is deficient owing to flowering dis-synchrony.

Wattle
Acacia spp. (Leguminosae)

Acacia is a large genus (800–900 spp.) of shrubs and trees, native over large areas of the tropics and subtropics. Arid areas of Africa and Australia are important centres of diversity. Numerous species are of value for many purposes (e.g. timber, gum, forage, tannin, perfume). Several species are cultivated.

Acacia senegal ($2n = 2x = 26$) is native from the Sudan to eastern India and is cultivated or semi-wild over much of Africa north of the River Senegal. Incisions in the outer bark lead to tear-like exudations of gum (gum arabic) which are collected when hard. Sudan has been a commercial source of the gum for thousands of years.

Acacia seyal ($2n = 4x = 52$) yields gum, forage and the timber (shittim wood) from which the Ark of the Covenant was fashioned. It is cultivated to a limited extent to supplement natural populations in parts of tropical Africa.

Acacia mearnsii ($2n = 2x = 26$) is the black wattle, the bark extract of which is a major source of tannin. Native to eastern Australia and Tasmania, the black wattle has been introduced to Natal, Kenya, Tanzania and Rhodesia and, less successfully, into the Asian and South American tropics. *Acacia mearnsii* is self fertile, though usually cross-pollinated by bees, and propagation is normally by seed. Vegetative propagation, by marcots, should, in future, permit selection of improved clones. Hybrids are possible with other Australian diploids. Those with green wattle (*A. decurrens*) a hardier, more rapidly maturing species, should prove useful.

Locust
Ceratonia siliqua (Leguminosae)

Known also as the Carob or St John's bread, *C. siliqua* ($2n = 2x = 24$) is a small, evergreen tree native to parts of the Mediterranean area and southwestern Asia. It is cultivated in the Old and New World (California) for its tough, edible legumes which are rich in sugar, protein and gum. The pods can be eaten as a sweetmeat, as a fodder or used as the basis of alcoholic beverages. The seeds, once used as jewellers' weights, are the original *carats*. The tree has been cultivated from ancient times in the Middle East and the fruit of wild populations has been historically significant as a source of food for needy travellers (e.g., John the Baptist, General Allenby's horses). Propagation by seed is usual, but good trees can be cloned from root cuttings.

Sunn hemp
Crotalaria juncea (Leguminosae)

Sunn hemp is a bast fibre grown on a large scale in India, for local use and export, as a raw material for sacking and cordage, though it is not so strong as hemp (*Cannabis*). *Crotalaria juncea* ($2n = 2x = 16$) is not found as a wild plant but its ancient cultivation in India may indicate that it originated there. *Crotalaria* is a large genus with many species native in the Asian tropics, but mostly concentrated in Africa. Sunn hemp is cultivated widely throughout the tropics as a green manure since it is a large, fast growing, rapidly rotting annual legume which can be ploughed in after 8–10 weeks. Propagation is by seed. The crop is almost completely cross-pollinated by bees, which are exploited by an intricate pollen release mechanism involving dimorphic stamens.

Guar, cluster bean
Cyamopsis tetragonolobus (Leguminosae)

Guar ($2n = 2x = 14$) is an annual monsoon crop and garden vegetable of India, Pakistan, Sri Lanka (Ceylon) and Burma, with numerous uses such as cattle forage, fuel, human food (as a legume or pulse), for soil improvement and as a medicine. The galactomannan gum of the endosperm is used in paper, textile, cosmetic and oil industries throughout the world, and is a useful absorbent for explosives. It was introduced to the USA originally for soil improvement and is now grown in Texas and Oklahoma as a cash crop. Guar was probably domesticated in a dry region of western Asia, where it thrives, but the species probably came originally from Africa. Arab traders are thought to have introduced it to Asia via southern India. The African *C. senegalensis* seems to be immediately ancestral to guar, which

311

is not anywhere found in the wild state.

Hymowitz, T. (1972). The transdomestication concept as applied to guar. *Econ. Bot.*, **26**, 49–60.

Derris, tuba root
Derris elliptica (Leguminosae)

Derris elliptica ($2n = 2x = 22, 24$) is the chief cultivated species of this genus of woody, evergreen climbers. *Derris ferruginea* and *D. malaccensis* ($2n = 2x = 22, 24$) are also used. The powdered root has been used as an insecticide and as a fish poison for many centuries. The active principle is rotenone. *Derris elliptica* is found wild from eastern India to New Guinea but is introduced in southern Malaysia, where it often does not flower. Derris cultivation based on introduced stocks is well established in the Philippines. Propagation is by stem cuttings. International trade in Derris powder began early in this century. Selection of strains for yield and high toxicity, followed by hybridisation, is the basis for improvement.

Lonchocarpus nicou ($2n = 4x = 44$), and one or two related species, are New World analogues of *Derris*, cultivated in Amazonian Peru and Brazil, where they are native. *Lonchocarpus* and *Derris* are closely related.

Lablab
Dolichos lablab (*Lablab niger*) (Leguminosae)

Also known as the hyacinth bean and by other names, the lablab is grown for its pods and seeds in India and other tropical areas. Numerous culinary preparations of the crop are known and the foliage provides hay, silage and green manure. Medicinal uses are also recorded.

Dolichos lablab ($2n = 2x = 22, 24$) is a herbaceous perennial but usually grown from seed as an annual from sea level to 2000 m in Asia. Uncertainty as to basic chromosome number ($x = 11, 12$) is, curiously, shared with the related genera *Phaseolus* and *Pueraria*. Numerous forms and cultivars are known including scramblers and (in Trinidad) a tall climbing type. The crop is cross-pollinated by insects. Thought to be of Asian origin, since it has long been cultivated in India, lablab has been introduced to Africa (Sudan and Egypt) and has since spread to many other tropical areas, becoming locally naturalized.

Schaffhausen, R. von (1963). *Dolichos lablab* or Hyacinth Bean. *Econ. Bot.*, **17**, 146–53.

Liquorice
Glycyrrhiza glabra (Leguminosae)

Glycyrrhiza glabra ($2n = 2x = 16$) is a perennial herb native to western and central Asia and southern Europe. It has been widely cultivated in its native area, at least from Greek times, for its sweet rhizomes and roots. The sweet principle, glycyrrhizin, is much sweeter than cane sugar. Extract of liquorice root is used as a flavouring in confectionery and medicinal preparations and the dried root is sometimes chewed. Liquorice was cultivated in Britain from the sixteenth century. The chief current sources of cultivated liquorice are the USSR, Spain and the Middle East.

Indigo
Indigofera spp. (Leguminosae)

Indigo (or anil) is a natural blue dye obtained from several of the 700-odd species of *Indigofera*; it has been used for over 4,000 years and is fast to water and light. The plants are tall herbs. They are cut before flowering and steeped in water, producing a yellowish solution. Aeration by stirring or beating causes the colouring agent (indican) to oxidize, forming a blue precipitate which is made into 'indigo cakes'.

Indigofera tinctoria ($2n = 2x = 16$), of India and Ceylon, is the principal source of commercial indigo. Other important species are *I. anil* (southern Asia, $2n = 2x = 12$), *I. arrecta* (Africa, $2n = 2x = 16$) and *I. sumatrana* (southeast Asia, $2n = 2x = 16$). *Indigofera arrecta* is still a local source of indigo in Africa and there is minor commercial production in India. Use of indigo was discouraged or prohibited in sixteenth century Europe in order to protect local woad-growing interests but indigo imports in the seventeenth century eventually obliterated the woad trade. A fleeting commerce based on introduced indigo in Carolina and Georgia existed in the eighteenth century. In its turn, indigo was supplanted by synthetic aniline dyes.

Lupin
Lupinus spp. (Leguminosae)

Lupinus is a large genus of 300–400 species, native mainly in North and South America and the Mediterranean area. Because of their vigour, rapid growth on poor land and their easily harvested seeds, they were widely cultivated at an early stage of agriculture. They have been used as grain foods, as animal forage, for soil improvement and for ornament. Lupins declined as agricultural technology improved and were displaced by other legumes, partly because lupin seeds generally contain alkaloids, which are difficult to remove. Some species may have reverted to being weeds. *Lupinus princei* ($2n =$?) and *L. atlanticus* ($2n =$?) have the large-seeded feature associated with domestication but are not known to be cultivated anywhere.

Lupinus albus ($2n = 50$; 30, 40, also reported), *L. pilosus* ($2n = 42$; 40 also reported) and *L. mutabilis* ($2n = 48$) have been cultivated for several thousand years. The first two are cultigens of the Old World (Europe and western Asia). *Lupinus mutabilis* is a very variable cultigen of the high Andes. The origins of these three domesticated species have not yet been worked out. Compared with wild lupins, they have larger seeds, lesser pigmentation (especially of the testa), faster growth and relatively indehiscent pods.

Lupinus augustifolius ($2n = 40, 48$) and *L. luteus* ($2n = 46, 48, 52$) have been domesticated more recently. *Lupinus luteus* has been cultivated longest in the Iberian peninsula and North Africa, where it seems to have been selected from the closely related (perhaps conspecific) *L. hispanicus* and *L. rothmaleri*. The probable progenitors have small, coloured seeds and dehiscent pods. *Lupinus luteus* enjoyed a vogue as an ornamental introduction to England (seventeenth–eighteenth centuries); it has aromatic yellow flowers. As a grain and forage it is grown particularly on the Baltic Coast. Kazimierski and Nowacki (1961) regarded *L. jugoslavicus* as a likely progenitor of some cultivated forms. However, it is now reduced under *L. augustifolius* subsp. *reticulatus*. It is native to maritime sands in southwestern Europe and is naturalized in Yugoslavia.

Lupinus pilosus was introduced to western Europe in the sixteenth century as a cultivated plant from the Mediterranean area. It is low in alkaloids and was used for flour in Canada as early as 1635.

Lupinus albus and *L. mutabilis* are the two most important cultivated lupins. *Lupinus albus* was almost certainly derived from wild forms native to the Balkan peninsula, now incorporated in its subsp. *graecus*. Selection for large seeds, permeable seed coat and non-shattering pods probably took place in different communities all round the Mediterranean. *Lupinus albus* has not lost its alkaloid content, which discourages pests and casual grazing.

Primitive New World lupins, possibly ancestral to *L. mutabilis*, include *L. douglasii* (eastern and central North America), the seeds of which were eaten by local Amerindians, and *L. ornatus* (western and central North America). These two species probably entered South America as semi-domesticated plants with Indian tribes migrating south in pre-Columbian times. *Lupinus mutabilis* may have originated in South America after a phase of hybridization. It is not known as a wild plant. American lupins certainly hybridize more readily than European species, which fall into four comparia. Central Mexico and California are the New World centres of lupin diversity.

Use of lupins for food has been somewhat frustrated by their alkaloid content. *Lupinus mutabilis* (chocho, tarwi, ullu) was once an important protein source in the Andes, following laborious leachings and steepings to remove alkaloids from the seeds. Since the introduction of *Vicia faba*, lupin cultivation has fallen rapidly. It is grown in Peru as a bitter hedge plant around pea and bean fields to discourage grazing (Hackbarth and Pakendorf, 1970).

Lupins nevertheless seem to have future potential as a protein source. The alkaloids would first have to be bred out and, since the first selections of alkaloid free lupins in 1928, lupin breeding has in fact expanded considerably. Variations in reported chromosome counts may indicate a need for careful cytotaxonomic research in *Lupinus*.

Brücher, H. (1970). Beitrag zur Domestikation protein-reicher und alkaloidarmer Lupinen in Südamerika. *Angew. Bot.*, **44**, 7–27.

Gladstones, J. S. (1970). Lupins as crop plants. *Field Crop Abstr.*, **23**, 123–48.

Hackbarth, J. and Pakendorf, K. W. (1970). *Lupinus mutabilis*, eine Kulturpflanze der Zukunft? *Z. Pflzücht.*, **63**, 237–45.

Kazimierski, T. and Nowacki, E. (1961). Indigenous species of lupins regarded as the initial forms of the cultivated species *Lupinus albus* and *Lupinus mutabilis*. *Flora*, **151**, 202–9.

Stepanova, S. I. (1969). [History of studies of the genus *Lupinus*]. *Rec. Wk. Postgrad. Jun. Sci. All Un. Sc.-res.*, **10**, 270–5 (in Russian).

Sainfoin

Onobrychis viciifolia (Leguminosae)

Sainfoin ($2n = 2x, 4x = 14, 28$) is a perennial herb native to Europe and western Asia. In its native area, sainfoin has been grown for centuries as a forage legume, doing particularly well on well drained soils. It has been introduced widely in temperate areas and is often naturalized. Though native in Britain, sainfoin as a crop was introduced to the country from Europe in the seventeenth century. *Onobrychis caput-galli* and

Onobrychis crista-galli, two Mediterranean diploids, are sometimes grown for fodder in their native areas, but have not been successful elsewhere.

Onobrychis viciifolia establishes less well than lucerne, but is nutritious and highly palatable, especially to sheep. Establishment is improved by milling the 'seeds' (one-seeded legumes). Single-cut sainfoin, a long lived strain, is slow growing and does not usually flower. Giant and double-cut varieties grow quickly, flowering and seeding even when two cuts of forage are taken each year. They were developed in continental Europe during the early nineteenth century. Sainfoin is less cultivated now than in the past. Improved strains of clover and lucerne, which are easier to grow on a wider variety of soils, have brought about its general decline.

Kul'tiasov (1967) has related the evolution of *Onobrychis* species to life forms and soil types. Landolt (1970) considers that *O. viciifolia* may have arisen following hybridization of *O. montana* (from mountains of southern Europe) with *O. arenaria* (with a sub-Mediterranean distribution) when the proposed parents were in contact during a post-glacial warm period. Hybridization and introgression would have been encouraged by the contemporaneous forest clearance.

Kul'tiasov, I. M. (1967). [Biomorphogenesis of species complexes in the genus *Onobrychis* on the basis of an historico-ecological analysis of their life forms]. *Bull. Moscow Soc. Nat. (Biol. Sect.)* (*Pl. Br. Abstr.*, **1968**, **38**, Abstr. 2769).

Landolt, E. (1970). Mitteleuropäschen Weisenpflanzen als hybridogene Abkömmlinge von mittel- und sude-uropäischen Gebirgssippen und sub-mediterranen Sippen. *Fedde Repert.*, **81**, 61–6.

Yam bean

Pachyrrhizus erosus (Leguminosae)

Pachyrrhizus is a tropical American genus of about six species of herbaceous vines. Pods may be eaten as string beans when young but the fibrous storage root is the chief product. *Pachyrrhizus erosus* ($2n = 2x = 22$) occurs wild in Mexico and northern Central America and has been cultivated in Central America from pre-Columbian times. Spaniards introduced it as a cultigen to the Philippines ('sinkamas') and it has since been grown over a large area of the Old World tropics. It is naturalized in southern China and Thailand. It is a market garden crop in Hawaii ('chopsui potato') and Singapore and is grown also in India, New Guinea and the Pacific. *Pachyrrhizus tuberosus*

($2n = 2x = 22$), apparently native around the headwaters of the Amazon, is grown in the highlands of Ecuador as a root crop. It is introduced and cultivated in the West Indies and China. *Pachyrrhizus ahipa* (jicama or aricoma) is a non-climbing Andean counterpart of *P. erosus* from Bolivia and northern Argentina. Like *P. erosus*, its seeds have insecticidal properties.

Winged pea

Psophocarpus tetragonolobus (Leguminosae)

Winged pea or Goa bean ($2n = $?) is a vine grown for its immature pods and tuberous roots. Mainly cultivated in tropical Asia, where it is probably native, it has been widely introduced (e.g. to Pacific Islands and West Indies). Bumble bees seem to be essential pollinators in Ghana. Burkill has suggested an insular origin (Mauritius or Madagascar), but the plant is not cultivated on the (nearer) mainland of Africa. *Psophocarpus palmettorum* ($2n = 20$), which is sometimes grown for its edible pods, is a wild species of tropical Africa.

Tamarind

Tamarindus indica (Leguminosae)

The tamarind ($2n = 2x = 24$) is a medium to large evergreen tree useful as an ornamental or shade plant, and with legumes yielding a sweet-acid pulp and edible seeds. Native to the tropical African savanna zone, the tamarind was introduced to India in prehistoric times. Indian cultivars have sweeter legume pulp.

Bambara groundnut

Voandzeia subterranea (Leguminosae)

Bambara groundnuts (njugo beans, kaffir pea) bear edible seeds formed in subterranean pods. It is an indigenous tropical African crop grown mainly as a supplementary vegetable in small plots and sometimes serving to mark field boundaries. Except on poorer sites it has been supplanted by *Arachis hypogaea*. The wild form (subsp. *spontanea*) is found from the Yos Plateau and Yola (northern Nigeria) to Cameroun and possibly to the Central African Republic. There are various differences from the cultivars (subsp. *sub-*

terranea) which have a clustered habit with short nodes, while subsp. *spontanea* has diffuse growth and more scattered pods. Both have $2n = 2x = 22$. First taken to Madagascar, perhaps by Arab traders, bambara groundnut reached Brazil in the seventeenth century and was subsequently introduced to the Far Eastern tropics and the Philippines. The crop is usually self pollinated (sometimes cleistogamous), though ant pollination of the more diffuse varieties is recorded in Ghana. Casual selection following hybridization seems to have been the basis of most cultivars and no improvement has been attempted. Perhaps it may follow the development of canning in Ghana and Rhodesia.

Hepper, F. N. (1970). Bambara groundnut (*Voandzeia subterranea*) *Field Crop Abstr.*, 23, 1–6.

Doku, E. V. and Karikari, S. K. (1971). Bambara groundnut. *Econ. Bot.*, 25, 255–62.

Asparagus
Asparagus officinalis (Liliaceae)

Asparagus is a dioecious perennial, native to Europe and eastern Asia (and perhaps also South Africa) though its long period of cultivation (since ancient Greek times) has obscured the boundary at which it ceases to be a wild plant. It has been widely spread by man and frequently becomes naturalized (as in Britain, where a native cliff ecotype, subsp. *prostratus*, has evolved). *Asparagus officinalis* ($2n = 2x = 20$) is grown from seed or rhizomes for its young, succulent shoots. There are some poorly defined cultivars. Shoots which are not cut bear scale leaves with numerous axillary photosynthetic cladodes and male or female (or rarely hermaphrodite) flowers. Female plants are supposed to produce fewer shoots than the males and are reported to have a shorter life. *Asparagus turkestanicus*, a central Asian taxon lacking cladodes, has been considered to be a relic species with affinities to South African species.

Work on breeding asparagus is a challenge since this dioecious crop is inevitably outpollinating. Useful variants can be maintained by vegetative propagation, but seed-grown crops are highly variable. It is possible to produce diploid homozygous hermaphrodite asparagus, by doubling up the chromosomes of haploid plants obtained from twin seedlings. Seed obtained from such diploidized hermaphrodites by selfing should be highly uniform, and it may be possible to stabilize improved variants of seed-propagated asparagus in this way.

Anon. (1972–4). *John Innes Inst. Ann. Rep.*, 1971, 1972, 1973.

Kenaf
Hibiscus cannabinus (Malvaceae)

Kenaf (Bimli jute, Deccan hemp) ($2n = 2x = 36$) is an important jute substitute. Cultivated forms are annual, erect herbs, largely self-pollinated. It is a common wild plant of tropical and subtropical Africa and is wild or naturalized in Asia. It has been widely introduced since the late war as a potential fibre crop. Cultivations have been established in China, the USSR, Thailand, South Africa, Egypt, Mexico and Cuba. The plant often escapes and persists as a weed.

An Indian origin for kenaf is now thought unlikely. Murdock (1959) argues that peoples of the western Sudan (the Nuclear Mande) domesticated *H. cannabinus* there before 4000 B.C. Wild forms occur in several parts of Africa, for example in the upper Niger and Bani valleys, near the putative centre of domestication. Wild collections have also been made in Angola and Tanzania. The Angolan collections appear primitive and may indicate Angola as the centre of origin, from which *H. cannabinus* migrated by way of eastern Africa to the Sudan. Direct Angola–Sudan migration would be blocked by rain forest. It is possible, on the other hand, that East Africa is the centre of origin from which kenaf migrated south-westwards to the Congo or Angola and northwestwards to the Sudan.

Wilson, F. D. and Menzel, M. Y. (1964). Kenaf (*Hibiscus cannabinus*), roselle (*Hibiscus sabdariffa*). *Econ. Bot.*, 18, 80–91.

Roselle
Hibiscus sabdariffa (Malvaceae)

Roselle (Jamaican sorrel) has two principal uses. The var. *sabdariffa* is a bushy shrub producing flowers with inflated calyces which are boiled with sugar to produce drinks and jellies. Var. *altissima* is a tall (5 m), scarcely branched fibre plant. There are intermediates.

Roselle is a self-pollinated, short-day plant grown for its fruit in various tropical countries but of chief importance as a substitute for jute. Some roselle varieties show higher resistance to root-knot nematodes than does kenaf (*Hibiscus cannabinus*, q.v.). *Hibiscus sabdariffa* ($2n = 4x = 72$; $2n = 36$ also recorded) is a tetraploid, probably domesticated by the Nuclear Mande of the western Sudan before 4000 B.C. (Murdock, 1959). African peoples probably used it only for its edible properties, for which it was introduced into the Asian tropics. It entered the New World with the slave trade, reaching Brazil in the seventeenth century and Jamaica in 1707. It has since been reintroduced to Africa as a fibre crop. Forms of *H. sabdariffa* combining characters of edible and fibre-plant roselle grow wild and in cultivation in Uganda and closely resemble *H. mechowii*, which seems to be the immediate wild progenitor of roselle. *Hibiscus mechowii* (or weedy *H. sabdariffa*) may have been domesticated for edible seeds and have later been selected for edible fruits. The fibre forms have probably developed from edible roselle on several occasions and may be introgressed by *H. meeusei* and *H. asper* (wild or cultivated for fibre in central and southern Africa). Crosses of roselle with kenaf have always failed; hybrids would, in any case, be sterile triploids. *Hibiscus mechowii* grows wild not only in the presumed centre of domestication but so widely in the rest of tropical Africa that its centre of origin cannot be assigned with conviction.

Murdock, G. P. (1959). *Africa, its peoples and their culture history*. New York.

Wilson, F. D. and Menzel, M. Y. (1974). Kenaf (*Hibiscus cannabinus*), roselle (*Hibiscus sabdariffa*). *Econ. Bot.*, 18, 80–91.

Topee tambu
Calathea allouia (Marantaceae)

The topee tambu, leren or sweet corn root ($2n = $?) is a perennial herb grown for its root tubers in tropical America and perhaps originating in the West Indies. The plant rarely flowers or seeds and propagation is by dividing the rootstock. Over a year of growth is necessary for the crop to mature but the tubers have a flavour like that of sweet corn, and keep well. The Spanish colonists in the New World found the topee tambu already casually cultivated by the Arawaks and Caribs.

Arrowroot
Maranta arundinacea (Marantaceae)

True arrowroot is a herbaceous perennial with thickened rhizomes, a starch producer rather than a root crop. The starch is supposedly easy to digest and is used in invalid foods and face powders.

Maranta arundinacea ($2n = 18, 48$; $x = $?) is found wild in Brazil, northern South America and parts of Central America. It is cultivated in the tropics of both Old and New Worlds, the major producer being St Vincent in the Lesser Antilles. Plants are propagated by rhizomes. It seems likely that arrowroot was first used (and gained its name) as an antidote to arrow poisons (especially the manchineel poison from *Hippomane manchinella*). Starch production from arrowroot is not referred to until the middle of the eighteenth century. It was not, therefore, domesticated as a starch plant by Amerindians, but first exploited as food by European colonists in the West Indies.

Sturtevant, W. C. (1969). History and ethnography of some West Indian starches. In P. J. Ucko and G. W. Dimbleby, *The domestication and exploitation of plants and animals*. London, 177–99.

Mulberry
Morus spp. (Moraceae)

Mulberries are multiple fruits, consisting of achenes surrounded by fleshy sepals mounted on a fleshy axis. The flowers are unisexual, most mulberry strains being dioecious. *Morus nigra*, the black mulberry ($2n = 22x = 308$), is native to Iran and Asia Minor, but has been cultivated in the Mediterranean area for many centuries. It is referred to in the Bible and by Greek and Roman writers. It is naturalized in parts of Europe and the Americas and is locally planted in the southern states of the USA. *Morus rubra* ($2n = 2x = 28$) is native to the eastern United States. It has larger fruits than the black mulberry and the wood is valuable. *Morus alba* ($2n = 2x = 28$), the white mulberry of China, is grown in tropical countries mainly for its leaves, which are eaten by silkworms. The fruits are inferior to those of black mulberry, which can be grown in the tropics only at high altitudes. *Morus alba* has been suggested as a progenitor of *M. nigra*.

Mulberries can be grown from seeds or propagated by cuttings or layering.

Nutmeg
Myristica fragrans (Myristicaceae)

Nutmegs are the seeds of *M. fragrans* ($2n = 6x = 42$), an evergreen, dioecious tree native to the eastern Moluccas. The seed aril is the source of mace. Nutmeg was not well known in Europe until the twelfth century and was a monopoly of Arab traders until the Portuguese and then the Dutch discovered and guarded the native nutmegs of Banda and Amboina islands. French and British botanists eventually established nutmeg plantations in Mauritius, French Guiana, Malaysia, Sumatra and the West Indies in the eighteenth and early nineteenth centuries. Indonesia and Grenada are the largest producers.

Nutmegs are normally grown from seed, the sexes being produced in about equal numbers. Male trees are felled to give a proportion of about one male to ten females and pollination is by insects. Approach-grafting of females on to seedlings is possible and should improve the efficiency of land use. Sex of seedlings can be determined using a colour reaction with ammonium molybdate.

Flach, M. and Cruickshank, A. M. (1969). Nutmeg. In Ferwerda and Wit (eds), *Outlines of perennial crop breeding in the tropics.* Wageningen, The Netherlands.

Pimento

Pimenta dioica (Myrtaceae)

Pimento (or allspice) is derived from the dried, unripe fruits of a small evergreen tree, native to the West Indies and Central America. Introductions of pimento to Ceylon and Singapore were unsuccessful and the Jamaican production is the most important, both from planted and from semi-wild trees. Pimento ($2n = 2x = 22$) thrives in Grenada and Jamaica, in which latter place it is native in wet limestone forest. The flowers are functionally dioecious, though apparently hermaphrodite. 'Male' and 'female' trees occur in about equal numbers, the pollen of 'females' being inviable. Insects and wind are both thought to be significant in pollen transfer.

Chapman, G. P. (1964). Some aspects of dioecism in Pimento (Allspice). *Ann. Bot.*, **28**, 451–8.
Rodrigues, D. W. (1969). Pimento. A short economic history. *Min. Agric. Fish. Kingston, Jamaica, Commodity Bull.*, **3**, pp. 52.

Guava

Psidium guajava (Myrtaceae)

Psidium is a large genus, indigenous to tropical America and the West Indies, of which many species have edible fruits. *Psidium guajava* ($2n = 2x = 22$; also $3x = 33$) is the best known of them.

Guava, a shrub or small tree, bears large yellow or green berries with numerous seeds embedded in sweet pulp. Trees are usually raised from seed and plantings are variable since cross-pollination (by bees and other insects) is very common. Vegetative propagation is possible and increasingly practised.

Guava is widely grown in the American tropics for local consumption and it spread into the Old World with Spanish and Portuguese explorations. It was established very early in Africa, western India and the Philippines. It has become locally naturalised in parts of the tropics and, in Fiji and New Caledonia, is a troublesome weed of pasture. There are commercial cultivations in India, Guyana, Brazil and Florida. The oldest known remains of guava (800 B.C.), in association with a human society, are from Peru. It is likely that domestication took place there originally, later spreading north. Oldest Mexican remains are 200 B.C.

Vanilla

Vanilla planifolia (Orchidaceae)

Vanilla planifolia (syn. *V. fragrans*) ($2n = 2x = 32$) is a climbing orchid native to Central America and widely introduced into tropical areas subject to high rainfall. In nature a rain-forest plant, trellis or slender trees are provided for support in cultivation. Vanilla spice is tincture of the cured, unripe capsules ('pods'). Exporting countries include Madagascar, the Seychelles and Réunion. In the West Indies, *V. pompona* ($2n = 2x = 32$), with short, fat pods, is the source of Pompona vanilla. Vanilla was used in pre-Columbian times by the Aztecs, as a flavouring for chocolate, and was known in Europe from about 1510. A difficulty with the cultivation of the crop on a large scale, especially in the absence of the natural pollinating insects, is the requirement for hand pollination to ensure a good fruit set. The crop is vegetatively propagated by stem cuttings and there seem to be rather few cultivars.

Oca

Oxalis tuberosa (Oxalidaceae)

Oca (also called iribia, cuiba, etc.) is a minor tuber crop of the high Andes (compare ullucu and isañu). It ranges from Venezuela to northern Argentina and presumably evolved from an unknown ancestor in that area. It is clonally propagated and quite variable. It is reported (G. E. Marks, pers. comm.) as having $2n$ $(=?6x)=66$. It is tristylous but reported to set seed on selfing; seed setting in the field, however, is rare. Apart from a certain Andean origin, its history is unknown and it would seem to have little future.

Leon, J. (1967). Andean root and tuber crops; origin and variability. *Proc. intnl Symp. trop. Root Crops, Trinidad*, 1(1), 118–23.

Leon, J. *et al.* (1958). Estudios sobre tubérculos alimenticios de los Andes. *Commun. Turrialba*, 63, pp. 49.

Betel nut

Areca catechu (Palmae)

The hard endosperm of the seeds of *A. catechu* releases a stimulating principle when chewed. Betel chewing is a habit of great antiquity and the tree has been taken into cultivation very widely (India, Burma, Thailand and the Malay Archipelago). It seems not to be found as a wild plant in primary forest. The palm flowers perpetually in Malaysia, but only in the dry months in India (November–February). Other species in the same section as *A. catechu* occur wild in Central Malaysia (Celebes), so it may be that the crop originated there. Corner states that there is historical evidence for such a view. *Areca catechu* ($2n = 4x = 32$) may be an allotetraploid (Barappa *et al.*, 1964).

Barappa, K. V. A., Ponnaiya, B. W. X. and Raman, V. S. (1964). Cytological studies in *Areca catechu* and *Areca triandra*. *Madras agric. J.*, 51, 363.

Ishwar Bhat, P. S. and Rao, K. S. N. (1963). On the antiquity of the arecanut. *Arecanut J.*, 13, 13–21.

Sugar palms

Arenga saccharifera and *Borassus flabellifer* (Palmae)

Arenga saccharifera, the Gomuti palm ($2n = 32$; 26 also recorded), is cultivated for sugar (jaggery). The young male inflorescence is wounded and the exudate collected and evaporated. *Borassus flabellifer*, the Palmyra palm ($2n = 36$) is reputed to have at least 801 uses, among which is its value as a sugar palm. The distribution of *A. saccharifera* is from India to Queensland. *Borassus flabellifer* grows in India and Malaysia. Little is known of the origins of the species or genera. Corner notes a tendency towards dioecy in *A. saccharifera*.

Sago palm

Metroxylon spp. (Palmae)

Metroxylon is a genus of suckering palms with about 15 species native from Thailand to Polynesia, from the Moluccas to Fiji. *Metroxylon rumphii* and *M. sagus* are often cultivated in villages as sago palms. Trees are felled before or when the inflorescence appears (after *ca.* 15 years), the pith is crushed and then washed to remove the starch. Both species are wild in the lowland swamps of New Guinea and New Britain, and may be native also in the Moluccas. They are introduced in western Malaysia. A continental Asian origin seems likely for the genus, from the restricted insular distribution of the different species.

Poppy

Papaver somniferum (Papaveraceae)

Papaver somniferum ($2n = 2x, 4x = 22, 44$) is an annual herb native to western Asia, widely introduced as a cultigen and naturalized as a weed. It is the source of opium (obtained from dried latex released from cuts made in the unripe capsules) and of poppy oil. The immediate wild progenitor of *P. somniferum* seems to be *P. setigerum* ($2n = 2x, 4x = 22, 44$) distributed from Persia to the Balkans. Opium poppy probably originated as a cultigen in Turkey. Its cultivation and use in medicine and as a narcotic are of great antiquity. Arab traders spread the opium habit and the plant westwards and it may have been used by the Swiss Lake Dwellers. The plant reached India and China by the eighteenth century A.D. and eventually became almost cosmopolitan. The socially damaging effects of opium consumption have encouraged restrictions on production but India, China, Turkey and the Balkans still raise large crops of the opium cultivars, which collectively form subsp. *somniferum*. Oil poppy varieties, the seeds of which are often eaten, belong to subsp. *hortense* and are grown in India and parts of Europe. Breeding for yield or for increased alkaloid or oil content involves hybridization of cultivars and also interspecific crosses (e.g. with *P. pilosum* ($2n = 2x = 28$)).

Duke, J. A. (1973). Utilization of Papaver. *Econ. Bot.*, 27, 390–400.

Passion fruit

Passiflora edulis (Passifloraceae)

Passion fruits are the yellow or purple berries of *P. edulis* ($2n = 2x = 18$), a woody tendril-climber. The berries are mostly used for their juice. The plant grows well in the subtropics or in tropical highlands with moderate rainfall. Many yellow passion fruit clones are self-incompatible and are cross-pollinated by bees and wasps (perhaps sometimes by humming birds). Yellow passion fruit (*f. flavicarpa*) originated as a mutant or hybrid of the purple form *edulis*. Hand pollination produces larger, juicier fruits than those which set naturally. Form *flavicarpa* will succeed at lower altitudes than *f. edulis*.

Passiflora edulis is native to southern Brazil; widely introduced into warm countries in the nineteenth century, it is now grown commercially in Australia, South Africa, New Zealand, Hawaii and Kenya. *Passiflora quadrangularis* ($2n = 2x = 18$), the Granadilla, is also native to South America but is a more tropical species and has been widely introduced into moist tropical lowlands. Outside its native area, hand pollination is necessary owing to the absence of the normal pollinating insects. The variability of native South American populations of passion fruits has been inadequately exploited.

Akamine, E. K. and Girolami, G. (1959). Pollination and fruit set in the yellow passion fruit. *Hawaii Agr. Expt. Sta. tech. Bull.*, 39, pp. 44.

Martin, F. W. and Nakasone, H. Y. (1970). The edible species of *Passiflora*. *Econ. Bot.*, 24, 333–43.

Rhubarb

Rheum rhaponticum (Polygonaceae)

Several of the 50 species of the Asian genus *Rheum* have been cultivated but only *R. rhaponticum* ($2n = 4x = 44$) is widely grown in temperate Eurasia and North America. Its precise origins as a cultigen are uncertain; *R. rhaponticum* occurs as a wild plant in the southeastern USSR but garden rhubarb does not breed true from seed and is clonally propagated. It is probably of hybrid origin, with *R. rhaponticum* as the most important progenitor.

Dried rhizomes of medicinal rhubarbs (*R. palmatum*, *R. officinale*), which have powerful laxative effects, were described in a Chinese herbal of 2700 B.C. They are natives of China and Tibet. Dried rhizomes ('rhei rhizoma') were imported to Europe via Smyrna (as 'Turkey rhubarb'). Living rhubarb was introduced to Britain for medicinal purposes before 1573 but, owing to confusion as to the geographical origin of drug rhubarb, it was *R. rhaponticum*, of scant medicinal value, which was grown. Only from the eighteenth century was the garden rhubarb recognized in Britain for its culinary properties. *Rheum palmatum* ($2n = 2x = 22$) was eventually introduced from Russia to Edinburgh in 1763, but failed as a commercial crop. This species may be implicated in the parentage of garden rhubarb clones.

Macadamia nut

Macadamia spp. (Proteaceae)

Australian or Macadamia nuts are the seeds of the ornamental, evergreen trees, *Macadamia integrifolia* and *M. tetraphylla*. They are native to a small, extratropical area of coastal Queensland and northern New South Wales. The seeds are borne in follicles on long, pendant racemes. Within is the fatty nutmeat consisting of the globular embryo bearing large, hemispherical cotyledons. Nutmeats are extracted from air-dried nuts and eaten raw, roasted or fried.

Confusion about the taxonomic status and nomenclature of Australian *Macadamia* species was resolved by Johnson (1954) and Smith (1956). Three distinct species have been recognized, forming a '*M. ternifolia* group' There are three other species not involved in the complex. The three nut species are *M. ternifolia*, *M. integrifolia* and *M. tetraphylla*. All have $2n = 2x = 28$. Storey (1965) has reviewed the history and typification of Australian *Macadamia* species. *Macadamia ternifolia* (Maroochy nut) bears small nuts with inedible, bitter kernels. It was not grown outside Australia until introduced into California in 1960. Both *M. integrifolia* ('smooth shell macadamia') and *M. tetraphylla* ('rough shell macadamia') bear edible nuts. Natural interspecific hybrids occur where their ranges overlap in south eastern Queensland (Smith, 1956) and where they have been grown together in California. Hybrids are fertile, the parental chromosomes seeming to be completely homologous. Presumably the parents have a closely related, common ancestor. This free genetic exchange between partially sympatric species raises the questions of how their multiple differences have arisen and of what is (or was) the isolating barrier separating them. Each species can be successfully cultivated in the endemic region of the other. Aboriginal peoples used the wild trees as a source of nuts (kindal-kindal) from prehistoric times but first plantings in Australia did not take place until about 1864. Wild trees are still used as a source of nuts. The planted trees were originally known as *M. ternifolia* but are now recognized as *M. tetraphylla*. Macadamia nuts were introduced to Hawaii in 1878, and again in 1893. The first introductions were of *M. integrifolia*, the second of *M. tetraphylla*. Both species were introduced to California in the late nineteenth and early twentieth centuries. Three naturally occurring hybrids were selected in Australia and have since been widely planted there and in California and Hawaii under the names Beaumont, Rankine and Teddington. In some places one to three crops can be produced each year and everbearing varieties of *Macadamia* have also been found.

Hawaiian plantations have been very successful. There, and in Australia, breeding experiments and selection have culminated in the establishment of named *Macadamia* cultivars, characterized by high yield, large kernels and thin shells. They are clonally propagated by grafting on to seedling roots. *Macadamia integrifolia* seems best suited to tropical island climates, while *M. tetraphylla* is better adapted to cultivation in Australia and California. F_2 hybrid progeny are an obvious source for further selections. *Macadamia* nut cultivation is now under trial in parts of Africa and Central and South America.

M. integrifolia nuts are sometimes called Queensland nut, a term applied also to the nuts of *Macrozamia*, with which Macadamia nuts should not be confused.

It is a pleasure to acknowledge the very lucid and comprehensive notes on *Macadamia* provided by W. B. Storey. This entry has been based almost entirely on them.

Johnson, L. A. S. (1954). *Macadamia ternifolia* and a related new species. *Proc. Linn. Soc. N.S.W.*, **79**, 15–18.

Ohler, J. G. (1969). Macadamia nuts. *Trop. Abstr.*, **24**, 781–91.

Smith, L. S. (1956). New species and notes on Queensland plants. *Proc. Roy. Soc. Qld.*, **67**, 29–40.

Storey, W. B. (1965). The ternifolia group of *Macadamia* species. *Pacific Sci.*, **19**, 507–14.

Pomegranate

Punica granatum (Punicaceae)

Punica granatum ($2n = 2x = 16, 18$), an ancient Old World crop plant, was associated by the Romans with the city of Carthage, but is native to Iran. It is widely grown and naturalized in the Mediterranean area, and has been introduced to California and most parts of the tropics and subtropics. It is a small deciduous tree or bush bearing large berries with tough rinds. The juicy flesh and pink seed arils can be eaten but are more conveniently expressed as a juice, which can be fermented into wine. There is a sterile double-flowered ornamental variety and several named cultivars with variously coloured flowers.

Gambier

Uncaria gambir (Rubiaceae)

Gambier is a resin extracted from the tanniniferous leaves of *U. gambir*, a shrubby plant formerly much cultivated in Malaysia. The extract is used in tanning, dye manufacture and various masticatories. Doubt surrounds the native status of the plant in Malaysia. In the wild, it reverts to being a slender climber, but it is not certain whether the plants found apparently wild in forests are truly wild or are relics of abandoned cultivation.

Akee

Blighia sapida (Sapindaceae)

A large tree occurring wild in West Africa, akee produces three-celled capsules which split to expose shiny, black, arillate seeds. The whitish arils from naturally matured fruits may be eaten raw or fried; unripe and overripe material is poisonous. The pink raphe connecting aril and seed is highly poisonous. Akee was probably domesticated first in West Africa, presumably near its native habitat, in or on the edge of forest. It was introduced to the West Indies in the eighteenth century and became well established as a local food crop in Jamaica. Nothing is known of changes under domestication and the chromosome number is not recorded.

Rambutan

Nephelium lappaceum (Sapindaceae)

A common village tree of Malaysia, rambutan is native to the central lowland regions. The seeds have a whitish aril which is sweet and juicy in good cultivars, scant and acid in wild forms. Though widely grown in Malaysia, rambutan does not thrive elsewhere. Two crops per years are usual, though it is probably not the same trees which bear fruit each time. Some trees are male, others bisexual. It is not known if pollen from bisexual plants is viable; if so, the male trees may be redundant. The origin of cultivated rambutan ($2n = 2x = 22$) is either from wild rambutan or from very closely related wild relatives in Western Malaysia, probably on the Sunda Shelf when Malaysia, Borneo, Sumatra and Java were covered with a single tract of rain forest. Village trees are not selected, but arise from chance seedlings which are very unlikely to breed true. Vegetative propagation is therefore indicated for types with good eating qualities.

Ramirez, D. A. (1961). Cytology of Philippine plants. VII. *Nephelium lappaceum* L. *Philippine Agric.*, **45**, 340–2.

Litchi
Nephelium litchi (*Litchi chinensis*) (Sapindaceae)

Litchi is native in southern China but has been widely introduced elsewhere for its seeds which have a sweet, white, fleshy aril easily separated from the smooth 'stone'. *Nephelium litchi* ($2n = 2x = 30$) is an evergreen, polygamous tree that fruits well only in cool, dry seasons. Lowland tropical introductions have not been successful. Propagation is usually by air layering.

Shea butter tree
Butyrospermum paradoxum subsp. *parkii* (Sapotaceae)

The shea butter tree is native to the West African savannas. *Butyrospermum paradoxum*, the only species, is a small tree bearing fleshy fruits each containing a few large brown seeds from which fat is extracted. Shea nuts are produced in commercial quantities from plantations, raised from seedlings, in several African countries, notably Nigeria and Ghana. The fat finds many local uses and is also exported.

Sapodilla
Manilkara zapota (*Achras zapota*) (Sapotaceae)

Sapodilla is a medium sized evergreen tree, native to Mexico and Central America, widely introduced in the tropics. The fruits are brown with black seeds in yellow pulp. Trees are often raised from seed but named varieties are maintained by grafting and other vegetative methods. The trunk of *M. zapota* ($2n = 2x = 26$) yields a durable wood and the bark can be tapped every 2–3 years for latex. When boiled, this yields chicle gum, used in the manufacture of chewing gum. It was an Aztec masticatory. Most chicle has been produced from wild trees in Mexico and British Honduras, where large stands occur naturally. Plantations of sapodilla have not been very successful in relieving pressure on wild populations and artificial chicle substitutes are being introduced to keep pace with the demand for chewing gum.

Quassia
Quassia amara (Simaroubaceae)

Quassia amara L. ($2n = ?4x = 36$) is the source of Surinam quassia, a drug extracted from chips or shavings of the wood. It has insecticidal properties and has been used for the treatment of threadworms and malaria. It is also a valuable timber tree. *Quassia amara* is native to tropical America and the West Indies.

Picrasma excelsa ($2n = ca.$ 60; also Simaroubaceae), native to the West Indies, is the source of Jamaican quassia.

Tree tomato
Cyphomandra betacea (Solanaceae)

Tree tomato ($2n = 2x = 24$) is a short-lived tree native to Peru but cultivated throughout the Andes at moderate to high altitudes for its oval berries which taste somewhat like tomatoes. The plant has been introduced to the Old World tropics and is grown to some extent in Pacific islands. Cultivation is most extensive in Ecuador but the history of domestication is unknown. No ancient names are recorded so it may have been established as a crop plant comparatively recently. A similar wild species, *C. bolivarensis*, has been found in Venezuela but its relationship to *C. betacea* has not been elucidated.

Heiser, C. B. (1969). *Nightshades, the paradoxical plants.* San Francisco.

Solanum fruits
Solanum spp. (Solanaceae)

Solanum is a large, cosmopolitan genus with a basic

$x = 12$. It is best known for its tuber-bearing species but many others are widely cultivated for their fruits, most of which have only very local significance. *Solanum diversifolium* (West Indies), *S. ellipticum* (Australia), *S. ferox* (Malaysia) and *S. triflorum*, a food plant of the Zuni Indians of Arizona, illustrate the widespread use which has been found for these fruits. A few other *Solanum* species have considerable actual or potential importance as berry crops.

Sunberry, *S. nigrum* ($3x$, $4x$, $6x$), the black nightshade, is a cosmopolitan weedy species which has probably been domesticated in several places. The taxonomy is very complex, crop taxa being referred to both by specific and varietal names and having rather uncertain geographical origins. The sunberry or garden huckleberry, used mainly in North America, is sometimes called var. *guineense*. Some improvement work has been devoted to garden huckleberry in terms of increasing the size, number and sweetness of its fruits. Wild or naturalized *S. nigrum* and its close ally in North America, *S. americanum*, may have been confused with garden huckleberry, so giving rise to reports that the fruits are poisonous.

Heiser (1969) provides an account of the supposed and the probable origins of the wonderberry, thought by its originator, Luther Burbank, to be a hybrid of *S. nigrum* var. *guineense* with *S. villosum*. It seems more likely that it arose from an introduced African species (perhaps now to be called *S. burbankii*) confused by Burbank with some of his hybrids.

Solanum muricatum ($2x$), the pepino, is a popular fruit grown from sea level in Peru to about 3,000 m in the Andes of Peru, Bolivia and Colombia. Many varieties of this highly variable, self-incompatible plant are in cultivation, most of them of very low fertility, producing 'seedless' fruits. Propagation is by cuttings; probably, several clones are planted, thus ensuring fruit set. Pepino was domesticated in pre-Columbian times and was widely grown at the time of the Spanish Conquest, and is still virtually confined to tropical America. Unknown in the wild state, pepino is thought by some to have originated in Colombia, where many-seeded (? primitive) varieties are common. Heiser (1964) considered that these types may have arisen from recent hybridization and that the great variability of the crop in Ecuador is a surer guide to its geographical origin. *Solanum muricatum* resembles the wild species *S. caripense* ($2x$, high Andes) and *S. tabanoense* ($2x$, Ecuador and Colombia). *Solanum caripense* readily forms hybrids with pepino.

Lulos are the berries of *S. quitoense* ($2x$) which is cultivated from 1,000–2,500 m, mainly in Colombia and Ecuador. The plant is very sensitive to frost. Lulos have been introduced to Costa Rica and Panama, occasionally escaping and becoming naturalized. Lulo is not found in the wild state though spiny (? primitive) varieties occur in northern Colombia. Two wild relatives (*S. tumo*, *S. pseudolulo*), also spiny, occur in Ecuador and Colombia. These facts may indicate a Colombian origin. The date of domestication is not known. Though the absence of *S. quitoense* as a wild plant might indicate that lulo is an ancient cultigen, Heiser (1969) believes it to be a comparatively recent one. The earliest written reference to the crop is as late as 1652. If the plant is indeed of Colombian origin, the Quechua name lulo cannot have been assigned to it earlier than the fifteenth century, when Colombia was conquered by the Incas.

Heiser, C. B. (1964). Origin and variability of the pepino (*Solanum muricatum*): A Preliminary Report. *Baileya*, 12 151–8.

Heiser, C. B. (1969). *Nightshades, the paradoxical plants.* San Francisco.

Isañu
Tropaeolum tuberosum (Tropaeolaceae)

Isañu (also called añu, mashua, etc.) is a minor tuber crop of the high Andes (compare oca and ullucu). It ranges from Colombia to Bolivia and is quite variable. It is clonally propagated and has $2n$ ($=? 8x$) $= 48$. (G. E. Marks, pers. comm., correcting an earlier, erroneous, record of $2n = 42$). Natural seed setting is abundant. An Andean origin (probably in the altiplano) is certain but the history of isañu is unknown.

Leon, J. (1967). Andean tuber and root crops, origin and variability. *Proc. intnl Symp. trop. Root Crops, Trinidad*, 1(1), 118–23.

Leon, J. *et al.* (1958). Estudios sobre tubérculos alimenticios de los Andes. *Commun. Turrialba*, 63, pp. 49.

Celery
Apium graveolens (Umbelliferae)

Apium graveolens ($2n = 2x = 22$) occurs as a wild plant of wet places in temperate Eurasia. Cultivated celery has been derived from it by selection for size and succulence of the petioles, which are used raw in salads or as a cooked vegetable. Celery may have been used first by the Greeks, first as a medicinal plant and later as a flavouring, before domestication, which probably took place in the Mediterranean area. By the seventeenth century, garden celery had become distinct from the wild plant. Leafy types (for garnishing) are referred to var. *secalinum*; blanched celery is var. *dulce* and celeriac (with swollen, edible roots) is var. *rapaceum*. Hybrids with *Petroselinum crispum* (parsley, *q.v.*) have been reported.

Helm, J. (1972). *Apium graveolens*, Geschichte der Kultur und Taxonomie. *Kulturpflanze*, **19**, 73–100.

Parsnip
Pastinaca sativa (Umbelliferae)

Pastinaca sativa ($2n = 2x = 22$). a native biennial of Europe and western Asia, has been cultivated as a root vegetable at least since Greek times. The tap-root is large and fleshy in good varieties, which developed from about the Middle Ages and are illustrated in most herbals. Escapes from cultivation reputedly revert to the wild condition, with tough, dry roots. Parsnip was introduced to the West Indies in 1564 but does well in the tropics only at high altitudes. Following introduction to Virginia in 1609, it was adopted by Indian tribes and subsequently spread widely among them.

Parsley
Petroselinum crispum (Umbelliferae)

Parsley ($2n = 2x = 22$) is a biennial or short-lived perennial native to rocky shores in southern Europe. The aromatic leaves are used as a flavouring and are rich in vitamin C. Plants are raised from seed, which is very slow to germinate. In the var. *tuberosum* (turnip-rooted parsley), the swollen roots are eaten boiled; this variety seems to be of recent origin and is used chiefly in Germany.

Parsley was used for flavouring by the Greeks and Romans. Following its spread as a cultigen to western Europe in the fifteenth–sixteenth centuries, it was introduced into most temperate countries (often becoming naturalized) and even to the tropics. Early cultivars had finely divided, flat leaves (like those of var. *tuberosum*) and this character sometimes segregates out in modern crisp-leaved parsley varieties. Intergeneric hybridization of parsley and celery (*Apium graveolens q.v.*) has recently yielded a range of new types of both crops.

Benbasat, E., Madzharova, D., Georgiev, G., Bubarova, M. and Chavdarov, L. (1972). *Otdal. Khibridatskiya na rasteniyata. Sofia, Bulgaria, BAN* (1972), 169–91 (*Pl. Br. Abstr.* (1973), **43**, Abstr. 4090).

Ramie
Boehmeria nivea (Urticaceae)

Ramie (or China grass) is a monoecious, wind-pollinated perennial with small unisexual flowers; the stems yield a very strong, lustrous bast-fibre. The species is diploid, with $2n = 2x = 14$. *B. nivea* subsp. *nivea* is native to eastern China and Japan, and has the undersides of the leaves covered with a felt of dense white hairs. Subsp. *tenacissima*, lacking the hairs, is native to Malaysia and is the type most grown in the tropics. Ramie has been cultivated longest in China, which is the major exporter. Introductions have usually failed to generate a viable industry, but the crop is produced commercially in the Philippines and Brazil. Propagation is by rhizome cuttings. Stems can be cut from the plants for several years before replanting becomes necessary.

Willimot, S. G. (1954). Ramie fibre, its cultivation and development. *World Crops*, **6**, 405–8.

Turmeric
Curcuma longa (Zingiberaceae)

Curcuma longa ($2n = 32, 62, 64; x = ?$) is the source of turmeric spice, used for colouring and flavouring curries and also imparting a yellow dye to cloth. It was known to Dioscorides (A.D. 77) and Marco Polo (1280). The plant is a rhizomatous perennial native to southeast Asia but widely introduced in cultivation in the tropics, especially in India. In India and Malaysia related wild species have sometimes been used for food since the rhizomes contain considerable quantities of starch. Species closely related to *C. longa* have been described from Java but Holttum states that they may be based on natural hybrids.

Holttum, R. E. (1951). Zingiberaceae cultivated in southern Asia. *Ind. J. Genet. Pl. Breeding*, **11**, 105–7.

Cardamoms

Elettaria cardamomum (Zingiberaceae)

Cardamoms are the dried or pickled ripe fruits of *E. cardamomum* ($2n = 4x = 48$), a large-leaved perennial herb, native of India. Cardamoms have long been an important spice in the Orient, most cultivation being in the mountains of southern India and Ceylon. The crop is first mentioned in European literature in the twelfth century. Plants are sometimes raised between tea and rubber on Indian plantations, doing well in partial shade. Wild plants are also a source of cardamoms for commerce. Cardamoms are grown in several tropical countries of the Old World, and have been very successfully introduced to Guatemala.

Samarawira, I. (1972). Cardamom. *World Crops*, 24, 76–8.

Ginger

Zingiber officinale (Zingiberaceae)

Ginger, a perennial herb, is an important spice and essential oil source. Both products are obtained from the finger-like rhizomes. Preserved (green) ginger, mainly of Chinese origin, is the peeled rhizome, boiled and packed in syrup. Sun-dried, peeled ginger is the black ginger of commerce.

Zingiber officinalis ($2n = 2x = 22$) is a small shade plant, probably native to southeast Asia, but widely introduced in the tropics and cultivated, often in home gardens, on a significant scale in China, Japan, Sierra Leone and the West Indies. Several varieties are grown in Malaysia. It rarely sets seed. The sites of original domestication and wild distribution have been obscured by long cultivation. Ginger was used from ancient times in India and China, perhaps first as a medicinal, and reached Asia Minor with trading caravans well before the hegemony of Rome was established. Ginger was an important spice in Europe in the early Middle Ages, first described in 1292. Because of its obvious physiological effects (dilation of blood vessels, perspiration) it was disseminated widely as a remedy for plague. It was introduced to the New World very early, and was exported from San Domingo in 1585. Jamaican ginger was established as an important item of commerce even earlier, large quantities being exported to Spain in 1547.

Holttum, R. E. (1950). The Zingiberaceae of the Malay peninsula. *Gardens Bulletin, Singapore 13*, (1), pp. 249.

Index of Authors

Names of authors contributing chapters to this book and authors cited in references are given here. The former are shown in bold face.

Index of scientific names

Family names in CAPITALS; generic names in **bold face**; specific and infra-specific names in *italics*. Page numbers of leading references in **bold face**. References to infra-specific taxa are limited and no infra-familial or infra-generic groupings are given. Chapters are short so, within chapters, first-page references alone are given.

Index
of common names

Page numbers of leading entries are shown in **bold face**.

Index of common names